Springer Proceedings in Complexity

Springer Proceedings in Complexity publishes proceedings from scholarly meetings on all topics relating to the interdisciplinary studies of complex systems science. Springer welcomes book ideas from authors. The series is indexed in Scopus.

Proposals must include the following:

- name, place and date of the scientific meeting
- a link to the committees (local organization, international advisors etc.)
- scientific description of the meeting
- list of invited/plenary speakers
- an estimate of the planned proceedings book parameters (number of pages/articles, requested number of bulk copies, submission deadline)

Submit your proposals to: christoph.baumann@springer.com

More information about this series at http://www.springer.com/series/11637

Anna Visvizi · Miltiadis D. Lytras
Editors

Research & Innovation Forum 2019

Technology, Innovation, Education, and their Social Impact

Springer

Editors
Anna Visvizi
Research & Innovation Institute (Rii)
Warsaw, Poland

Miltiadis D. Lytras
Deree College
The American College of Greece
Athens, Greece

ISSN 2213-8684 ISSN 2213-8692 (electronic)
Springer Proceedings in Complexity
ISBN 978-3-030-30808-7 ISBN 978-3-030-30809-4 (eBook)
https://doi.org/10.1007/978-3-030-30809-4

This Springer imprint is published by the registered company Springer Nature Switzerland AG
The registered company address is: Gewerbestrasse 11, 6330 Cham, Switzerland

Organization

Chairs

Visvizi, Anna, Ph.D., Research & Innovation Institute, Poland
Lytras, Miltiadis D., Ph.D., Deree College—The American College of Greece, Greece

Steering Committee Members

Visvizi, Anna, Ph.D., Research & Innovation Institute, Poland
Lytras, Miltiadis D., Ph.D., Deree College—The American College of Greece, Greece
Sarirete, Akila, Ph.D., Effat University, Saudi Arabia
Rohlíková, Lucie, Ph.D., University of West Bohemia, Czechia
Serradell Lopez, Enric, Ph.D., Open University of Catalonia, Spain

Program Committee

Abel, Marie-Hélène, Ph.D., Sorbonne Universités, Université de Technologie de Compiègne, France
Alowibdi, Jalal, Ph.D., University of Jeddah, Saudi Arabia
Ammetller-Montes, Gisela, Open University of Catalonia, Spain
Azorín-López, Jorge, Ph.D., University of Alicante, Spain
Burriel Coll, Verónica, Ph.D., Utrecht University, the Netherlands
Carbonaro, Antonella, Ph.D., University of Bologna, Italy
Chen, Mu-Yen, Ph.D., National Taichung University of Science and Technology, Taichung, Taiwan

rii forum ©

Rome, April 24-26, 2019

Research & Innovation Forum

Advancing research on technology, innovation, education and their social impact

www.rii-forum.org

Preface

Advances in sophisticated technology have an ever-growing impact on our societies, including communication, education, policy making, the modes of production, logistics, health care, entertainment, design, architecture, cities, etc. As the pace of technological advances accelerates, it is imperative that the dynamic nexus between technology and society is queried from a variety of perspectives. Only in this way we can identify not only nascent risks and emerging threats, but also the multiple opportunities that emerge. **Research and Innovation Forum (Rii Forum)** is driven precisely by this logic.

Research and Innovation Forum (Rii Forum) is an annual conference that brings together researchers, academics, and practitioners to engage in a conceptually sound, inter- and multi-disciplinary, empirically driven debate on key issues influencing the dynamics of social interaction today. The role of advances in sophisticated technology stands at the heart of discussions held during **Rii Forum**. Held annually, **Rii Forum** features in-depth cutting-edge research on both the most current and the emerging issues that unfold at the intersection of technology and society. The format of **Rii Forum**—consistent with traditional and flipped presentations, interactive workshops and featured roundtables—renders it a perfect venue to build bridges between the worlds of academia and policymaking to promote research-driven policy recommendations.

The **Rii Forum 2019** was held in Rome, April 24–26, 2019, and facilitated discussion, exchange of ideas, and networking. The **Rii Forum 2019** was attended by delegates from literally all over the world, including North and South America, Asia, the Arab Peninsula, and Europe. The conference opening speech was delivered by Dr. Akila Sarirete, Assistant Professor at the Computer Science Department and Dean for Graduate Studies and Research at Effat University, Jeddah, Saudi Arabia. Dr. Sarirete delivered a speech titled 'Fostering research, innovation, and education: the case of Effat University.' The kick-off speech during the second day of the conference was delivered by Dr. Saeed Ul Hassan, Director of Scientometrics Lab and Faculty Member at Information Technology University (ITU), Lahore, Pakistan. In his speech, Dr. Hassan addressed the question of 'How your research really matters?—Qualitative Assessment of Scholarly Impact using Citations from full-text Scientific Literature.' The conference closing speech was delivered by Dr. Enric Serradell Lopez, Director of Executive Education, Open University of Catalonia, Barcelona, Spain. Dr. Serradell offered a captivating speech on 'Being an academic: re-thinking critical thinking.'

With regard to the research workshop, its goal was to build functional connections among the conference participants. The workshop was coordinated by Dr. Akila Sarirete, Dean of Graduate Studies and Research at Effat University. During the workshop, early career and established researchers had the opportunity to meet administrators, program and projects' administrators and explore the prospect of joint projects. Several research and collaboration schemes were discussed in detail during the workshop.

The cutting-edge quality of scholarly discussion during the **Rii Forum 2019** was possible due to an arduous review, selection, and double-blind peer-review process. Specifically, nearly 200 extended paper proposals from all over the world had been originally submitted to be presented during the conference. These proposals were reviewed by the **Rii Forum Program Committee** that accepted 90 paper proposals and notified respective authors. Eventually, 70 papers were presented during the conference, while 52—following a rigorous double-blind review process—are featured in this volume.

The structure of the volume mirrors the key topics around which **Rii Forum 2019** discussions oscillated, i.e.,

- Technology-enhanced learning,
- Cognitive computing and social networking,
- Smart cities and smart villages,
- Information systems,
- Medical informatics, and
- Emerging issues at the cross section of technology, politics, society, and economy.

In recognition of the amount of work invested and the resultant quality of research presented during the **Rii Forum 2019**, to highlight outstanding contributions and performance, following lengthy deliberations, **Rii Forum Program Committee** decided on granting of the following awards.

The **Rii Forum award for best Ph.D. student paper** was granted to Mr. Rafael Mollá Sirvent, supervised by Dr. Higinio Mora (University of Alicante, Alicante, Spain).

The **Rii Forum award for social impact** was awarded Petr Hořejší, Jiri Vysata, Lucie Rohlikova, Jiri Polcar, Michal Gregor (University of West Bohemia, Pilsen, Czechia) for their paper titled 'Serious Games in Mechanical Engineering Education.'

The **Rii Forum award for best reviewer** was granted to Dr. Antoni Meseguer-Artola (Open University of Catalonia, Barcelona, Spain).

The **Rii Forum award for best ICT-enhanced presentation** was granted to Professor Placido Pinheiro (University of Fortaleza, Fortaleza, Brazil).

The **Rii Forum award for outstanding research** was granted to Wei Wang, Yenchun Jim Wu, and Ling He (National Taiwan Normal University, Taipei, Taiwan) for their paper titled 'Impact of Linguistic Feature Related to Fraud on Pledge Results of the Crowdfunding Campaigns.'

Finally, the **Rii Forum award for the best paper** was awarded to Adil E. Rajput, Akila Sarirete, Tamer Dessouky (Effat University, Jeddah, Saudi Arabia) for their paper titled 'Using Crowdsourcing to Identify a Proxy of Socio-Economic status.'

This collection of research papers presented during the **Rii Forum 2019** features cutting-edge research centered on technology-induced processes and developments that shape sociopolitical and economic processes today. We remain grateful to the **Rii Forum Steering Committee** and the **Rii Forum Program Committee** for their commitment, sound judgment, and hard work in the process of organizing the **Rii Forum 2019**. We are equally appreciative of the work of the panel chairs who ensured that timely progression of presentations and Q&A sessions. We would like to say 'thank you' to all contributing authors for their hard work and their patience in subsequent rounds of revise and resubmit. This would not be possible, of course, without the reviewers who devoted countless hours to evaluate papers submitted to this volume. Finally, we would like to express our gratitude to the entire Springer team and the Editors of Complexity for enthusiastically embracing our idea and for guiding us through the process.

Considering the quality of research presented during the conference and the amount of networking that was triggered during the conference, we would like to take this opportunity to invite you to join the **Rii Forum 2020** which will take place in Athens, Greece, in April 15–17, 2020. Please check the **Rii Forum** Web site (https://rii-forum.org) for updates.

Warsaw, Poland
Athens, Greece

Sincerely,
Anna Visvizi
Miltiadis D. Lytras
Chairs
Rii Forum 2019

Contents

Technology-Enhanced Learning

Simulating Peer Assessment in Massive Open On-line Courses 3
Filippo Sciarrone and Marco Temperini

Development Strategies and Trends in Educational Solutions for Deaf Students 15
Herleson Paiva Pontes, João Batista Furlan Duarte and Placido Pinheiro

Unpacking the Privacy Paradox for Education 27
Lorayne Robertson and Bill Muirhead

21st Century Skills. An Analysis of Theoretical Frameworks to Guide Educational Innovation Processes in Chilean Context 37
Ignacio Salamanca Garay and María Graciela Badilla Quintana

Cloud Platform for Enabling a Student Collaborative Assessment Process 47
Higinio Mora, María Teresa Signes-Pont, Rafael Mollá Sirvent and María L. Pertegal-Felices

Serious Games in Mechanical Engineering Education 55
Petr Hořejší, Jiří Vyšata, Lucie Rohlíková, Jiří Polcar and Michal Gregor

Integrating Makerspaces in Higher Education: Constructionism Approach to Learning 65
Tayeb Brahimi, Sajid Khalifa and Bensaid Benaouda

Definition of a Feature Vector to Characterise Learners in Adaptive Learning Systems 75
Alberto Real-Fernández, Rafael Molina-Carmona, María L. Pertegal-Felices and Faraón Llorens-Largo

Gender and Learning Outcomes in Entrepreneurship Education 91
Raquel Ferreras-Garcia, Ana Beatriz Hernández-Lara and Enric Serradell-López

Clickbait in Education—Positive or Negative? Machine Learning Answers 101
Adil E. Rajput

Improve Student Participation in Peer Assessment to Influence Learning Outcomes: A Case Study 109
Antonella Carbonaro and Roberto Reda

Learning with Wikipedia in Higher Education: Academic Performance and Students' Quality Perception 117
Antoni Meseguer-Artola, Inma Rodríguez-Ardura, Gisela Ammetller and Eva Rimbau-Gilabert

A Hybrid Machine Learning Approach for the Prediction of Grades in Computer Engineering Students 125
Diego Buenaño-Fernandez, Sergio Luján-Mora and David Gil

Computational Thinking in Basic Education in a Developing Country Perspective 135
Daniel Chagas and Elizabeth Furtado

Cognitive Computing and Social Networking

Cognitive Computing Approaches for Human Activity Recognition from Tweets—A Case Study of Twitter Marketing Campaign 153
Jari Jussila and Prashanth Madhala

Assessing Strategies for Sampling Dynamic Social Networks 171
Paolo Ceravolo, Francesco Ciclosi, Emanuele Bellini and Ernesto Damiani

Facebook Engagement—Motivational Drivers and the Moderating Effect of Flow Episodes and Age Differences 181
Inma Rodríguez-Ardura and Antoni Meseguer-Artola

Temporal Trend Analysis on Virtual Reality Using Social Media Mining 189
Chen-wen Shen, Jung-tsung Ho and Hung-wen Ma

Using User Contextual Profile for Recommendation in Collaborations 199
Siying Li, Marie-Hélène Abel and Elsa Negre

A Step Further in Sentiment Analysis Application in Marketing Decision-Making 211
Erick Kauffmann, David Gil, Jesús Peral, Antonio Ferrández and Ricardo Sellers

Knowledge Integration in Personalised Dietary Suggestion System Using Semantic Web Technologies 223
Antonella Carbonaro and Roberto Reda

Using Physical Activity Monitors in Smart Environments and Social Networks: Applications and Challenges 231
Jose-Luis Sanchez-Romero, Antonio Jimeno-Morenilla, Higinio Mora and Francisco Pujol-Lopez

Smart Cities and Smart Villages

Emotion Recognition to Improve e-Healthcare Systems in Smart Cities 245
Francisco A. Pujol, Higinio Mora and Ana Martínez

Smart City and Technology Transfer: Towards Sustainable Infrastructure in Bahrain 255
Rustom Mamlook, Odeh Al-Jayyousi, Samar Alabyooki and Suha Abdullah Ali Attallah

Smart Cities to Create Opportunities for Young People 269
Xose Picatoste, Isabel Novo-Corti and Diana Mihaela Ţîrcă

Automatic Speaker Verification, ZigBee and LoRaWAN: Potential Threats and Vulnerabilities in Smart Cities 277
Adil E. Rajput, Tayeb Brahimi and Akila Sarirete

Mexico City Traffic Analysis Based on Social Computing and Machine Learning 287
Magdalena Saldaña Pérez, Miguel Torres Ruiz and Marco Moreno Ibarra

Management City Model Based on Blockchain and Smart Contracts Technology 305
Higinio Mora, Raquel Pérez-delHoyo, Rafael Mollá Sirvent and Virgilio Gilart-Iglesias

Emerging Computer Vision Based Machine Learning Issues for Smart Cities 315
Muhammad Murtaza Khan, Muhammad U. Ilyas, Sajid Saleem, Jalal S. Alowibdi and Mohammed Saeed Alkatheiri

Information Systems

The CRISP-DCW Method for Distributed Computing Workflows 325
Marco Spruit and Stijn Meijers

Towards a Software Engineering Framework for the Design, Construction and Deployment of Machine Learning-Based Solutions in Digitalization Processes 343
Ricardo Colomo-Palacios

Applied Data Science in Financial Industry 351
Marco Spruit and Drilon Ferati

Handing a Hybrid Multicriteria Model for Choosing Specialists to Analyze Application Management Service Tickets 369
Karla Krupskaia Ferreira Lima, Placido Pinheiro and Marum Simão Filho

Data Strategy Framework in Servitization: Case Study of Service Development for a Vehicle Fleet 377
Jukka Pulkkinen, Jari Jussila, Atte Partanen and Igor Trotskii

Big Data Dimensionality Reduction for Wireless Sensor Networks Using Stacked Autoencoders 391
Muneeba Sirshar, Sajid Saleem, Muhammad U. Ilyas, Muhammad Murtaza Khan, Mohammed Saeed Alkatheiri and Jalal S. Alowibdi

Medical Informatics

BYOD, Personal Area Networks (PANs) and IOT: Threats to Patients Privacy 403
Samara Ahmed

Towards the Early Diagnosis of Dementia in People with HIV/AIDS Using Bayesian Networks 411
Placido Pinheiro, Luana Ibiapina Cordeiro, Maria Lúcia Duarte Pereira, Wilson Correa de Abreu, Mirian Caliope Dantas Pinheiro, Pedro Gabriel Caliope Dantas Pinheiro and Egidio Gomes Filho

A Hybrid Model to Guide the Consultation of Children with Autism Spectrum Disorder 419
Luciano Comin Nunes, Placido Pinheiro, Mirian Caliope Dantas Pinheiro, Monica Pompeu, Marum Simão Filho, Rafael Comin-Nunes and Pedro Gabriel Caliope Dantas Pinheiro

Use of EEG Technology with Based Brain-Computer Interface to Address Amyotrophic Lateral Sclerosis—ALS 433
Nighat Mir, Akila Sarirete, Jehad Hejres and Manar Al Omairi

Big Data-Assisted Word Sense Disambiguation for Sign Language 441
Luis Naranjo-Zeledón, Antonio Ferrández, Jesús Peral and Mario Chacón-Rivas

Emerging Issues: Technology and Society

Surfing the Waves of Digital Automation in Spanish Labor Market 451
Josep Lladós-Masllorens

Impact of Linguistic Feature Related to Fraud on Pledge Results of the Crowdfunding Campaigns 459
Wei Wang, Yenchun Jim Wu and Ling He

Sustainable Economic Development: Some Reflections on Access to Technology as a Matter of Social Engagement 469
Isabel Novo-Corti, Xose Picatoste and Diana Mihaela Țîrcă

Using Crowdsourcing to Identify a Proxy of Socio-economic Status 479
Adil E. Rajput, Akila Sarirete and Tamer F. Desouky

IoT-IaaS: A New Public Cloud Service to Unleash Innovation in the IoT Space 487
Muhammad U. Ilyas, Muhammad Murtaza Khan, Sajid Saleem and Jalal S. Alowibdi

The Climb to Success: A Big Data Analysis to Find Out Why Huawei Has Conquered the Market 495
Orlando Troisi, Mara Grimaldi, Francesca Loia and Gennaro Maione

Identification of Factors of Indigenous Ethnic Identity: A Case Study on the Waorani Amazonian Ethnicity 515
Aldrin Espín-León, Antonio Jimeno-Morenilla, María L. Pertegal-Felices and Jorge Azorín-López

Analysis of Virtual Currencies as Driver of Business Marketing 525
Higinio Mora and Rafael Mollá Sirvent

Conceptual Modeling as a Tool for Corporate Governance Support: State of the Art and Research Agenda 535
Yves Wautelet and Manuel Kolp

Price Determinants of Tourist Accommodation Rental: Airbnb in Barcelona and Madrid 549
Josep Lladós-Masllorens and Antoni Meseguer-Artola

Towards Personality Classification Through Arabic Handwriting Analysis 557
Mohamed A. Mostafa, Muhammad Al-Qurishi and Hassan I. Mathkour

Author Index 567

Contributors

Marie-Hélène Abel Sorbonne Universités, Université de technologie de Compiègne CNRS UMR 7253, HEUDIASYC, Compiègne, France

Samara Ahmed College of Medicine, King Abdulaziz University, Jeddah, Saudi Arabia

Manar Al Omairi Effat University, Jeddah, Saudi Arabia

Samar Alabyooki Innovation and Technology Management Department, Arabian Gulf University, Manama, Bahrain

Odeh Al-Jayyousi Innovation and Technology Management Department, Arabian Gulf University, Manama, Bahrain

Mohammed Saeed Alkatheiri College of Computer Science and Engineering, University of Jeddah, Jeddah, Saudi Arabia

Jalal S. Alowibdi Department of Computer Science and Artificial Intelligence, College of Computer Science and Engineering, University of Jeddah, Jeddah, Saudi Arabia

Muhammad Al-Qurishi Elm Company—Research & Innovation, Riyadh, Kingdom of Saudi Arabia

Gisela Ammetller Open University of Catalonia, (Universitat Oberta de Catalunya), Barcelona, Spain

Suha Abdullah Ali Attallah Innovation and Technology Management Department, Arabian Gulf University, Manama, Bahrain

Jorge Azorín-López Department of Computer Technology, Universidad de Alicante, Alicante, Spain

Emanuele Bellini Center of Cyber-Physical System, Khalifa University, Abu Dhabi, UAE

Bensaid Benaouda Faculty of Islamic Sciences, Istanbul Sabahattin Zaim University, Istanbul, Turkey

Tayeb Brahimi College of Engineering, Effat University, Jeddah, Kingdom of Saudi Arabia

Diego Buenaño-Fernandez Universidad de Las Américas, Quito, Ecuador

Antonella Carbonaro Department of Computer Science and Engineering, University of Bologna, Bologna, Italy

Paolo Ceravolo Sesar Lab, Università degli Studi di Milano, Milan, Italy

Mario Chacón-Rivas Inclutec, Instituto Tecnológico de Costa Rica, Cartago, Costa Rica

Daniel Chagas LUQS, Universidade de Fortaleza, Fortaleza, Brazil

Francesco Ciclosi Università di Macerata, Macerata, Italy

Ricardo Colomo-Palacios Østfold University College, Halden, Norway

Rafael Comin-Nunes Federal University of Ceara, Fortaleza, CE, Brazil

Luana Ibiapina Cordeiro State University of Ceará, Fortaleza, CE, Brazil

Ernesto Damiani Center of Cyber-Physical System, Khalifa University, Abu Dhabi, UAE

Wilson Correa de Abreu School of Nursing of Porto, Porto, Portugal

Tamer F. Desouky College of Engineering, Effat University, Jeddah, Kingdom of Saudi Arabia

João Batista Furlan Duarte Graduate Program in Applied Informatics, University of Fortaleza (UNIFOR), Fortaleza, CE, Brazil

Aldrin Espín-León Faculty of Sociology and Social Work, Central University of Ecuador, Quito, Ecuador

Drilon Ferati Department of Information and Computing Sciences, Utrecht University, Utrecht, The Netherlands

Antonio Ferrández University of Alicante, San Vicente del Raspeig, Alicante, Spain

Raquel Ferreras-Garcia Universitat Oberta de Catalunya, Barcelona, Spain

Marum Simão Filho Graduate Program in Applied Informatics, University of Fortaleza, Fortaleza, CE, Brazil

Egidio Gomes Filho University of Fortaleza, Fortaleza, CE, Brazil

Elizabeth Furtado LUQS, Universidade de Fortaleza, Fortaleza, Brazil

Ignacio Salamanca Garay Facultad de Educación, Centro de Investigación en Educación y Desarrollo (CIEDE), Universidad Católica de la Santísima Concepción, Concepción, Chile

David Gil University of Alicante, San Vicente del Raspeig, Alicante, Spain

Virgilio Gilart-Iglesias Department of Computer Technology and Computation, University of Alicante, Alicante, Spain

Michal Gregor Department of Industrial Engineering and Management, University of West Bohemia, Plzen, Czech Republic

Mara Grimaldi University of Salerno, Fisciano, ITA, Italy

Ling He College of Business Administration, Huaqiao University, Quanzhou, China

Jehad Hejres Effat University, Jeddah, Saudi Arabia

Ana Beatriz Hernández-Lara Universitat Rovira i Virgili, Reus, Spain

Jung-tsung Ho Department of Business Administration, National Central University, Jhongli District, Taoyuan City, Taiwan, Republic of China

Petr Hořejší Department of Industrial Engineering and Management, University of West Bohemia, Plzen, Czech Republic

Marco Moreno Ibarra Instituto Politécnico Nacional, Centro de Investigación en Computación, Nueva Industrial Vallejo, Gustavo a. Madero, Mexico

Muhammad U. Ilyas School of Electrical Engineering and Computer Science (SEECS), National University of Sciences and Technology (NUST), Islamabad, Pakistan;
Department of Computer and Network Engineering, College of Computer Science and Engineering, University of Jeddah, Jeddah, Saudi Arabia

Antonio Jimeno-Morenilla Department of Computer Technology, Universidad de Alicante, Alicante, Spain;
University of Alicante, San Vicente del Raspeig, Spain

Jari Jussila Häme University of Applied Sciences, Hämeenlinna, Finland

Erick Kauffmann University of Alicante, San Vicente del Raspeig, Alicante, Spain;
University of Costa Rica, San José, Costa Rica

Sajid Khalifa Design Department, Effat University, Jeddah, Saudi Arabia

Muhammad Murtaza Khan Department of Computer Science and Artificial Intelligence, College of Computer Science and Engineering, University of Jeddah, Jeddah, Saudi Arabia;
School of Electrical Engineering and Computer Science (SEECS), National University of Sciences and Technology (NUST), Islamabad, Pakistan

Manuel Kolp UCLouvain, Louvian-la-Neuve, Belgium

Siying Li Sorbonne Universités, Université de technologie de Compiègne CNRS UMR 7253, HEUDIASYC, Compiègne, France

Karla Krupskaia Ferreira Lima University of Fortaleza, Fortaleza, Brazil

Josep Lladós-Masllorens Open University of Catalonia (Universitat Oberta de Catalunya), Barcelona, Spain

Faraón Llorens-Largo Cátedra Santander-UA de Transformación Digital, University of Alicante, San Vicente del Raspeig, Spain

Francesca Loia University of Rome "La Sapienza", Rome, ITA, Italy

Sergio Luján-Mora Universidad de Alicante, Alicante, Spain

Hung-wen Ma Department of Business Administration, National Central University, Jhongli District, Taoyuan City, Taiwan, Republic of China

Prashanth Madhala Häme University of Applied Sciences, Hämeenlinna, Finland

Gennaro Maione University of Salerno, Fisciano, ITA, Italy

Rustom Mamlook Innovation and Technology Management Department, Arabian Gulf University, Manama, Bahrain

Ana Martínez Department of Computer Technology, University of Alicante, San Vicente del Raspeig, Alicante, Spain

Hassan I. Mathkour Computer Science Department, King Saud University, Riyadh, Kingdom of Saudi Arabia

Stijn Meijers Department of Information and Computing Sciences, Utrecht University, Utrecht, The Netherlands

Antoni Meseguer-Artola Open University of Catalonia, (Universitat Oberta de Catalunya), Barcelona, Spain

Nighat Mir Effat University, Jeddah, Saudi Arabia

Rafael Molina-Carmona Unidad Científica de Innovación Empresarial "Ars Innovatio", University of Alicante, San Vicente del Raspeig, Spain

Higinio Mora Department of Computer Technology and Computation, University of Alicante, Alicante, Spain;
Department of Computer Technology, University of Alicante, San Vicente del Raspeig, Alicante, Spain

Mohamed A. Mostafa Computer Science Department, King Saud University, Riyadh, Kingdom of Saudi Arabia

Bill Muirhead Ontario Tech University, Oshawa, ON, Canada

Luis Naranjo-Zeledón University of Alicante, Alicante, Spain;
Inclutec, Instituto Tecnológico de Costa Rica, Cartago, Costa Rica

Elsa Negre Paris-Dauphine University, PSL Research University, CNRS UMR 7243, LAMSADE, Paris, France

Isabel Novo-Corti Universidade da Coruña, A Coruña, Spain;
Faculty of Economy & Business, A Coruna, Spain

Luciano Comin Nunes Graduate Program in Applied Informatics, University of Fortaleza, Fortaleza, CE, Brazil

Atte Partanen Häme University of Applied Sciences, Hämeenlinna, Finland

Jesús Peral University of Alicante, Alicante, Spain;
University of Alicante, San Vicente del Raspeig, Alicante, Spain

Maria Lúcia Duarte Pereira State University of Ceará, Fortaleza, CE, Brazil

Magdalena Saldaña Pérez Instituto Politécnico Nacional, Centro de Investigación en Computación, Nueva Industrial Vallejo, Gustavo a. Madero, Mexico

Raquel Pérez-delHoyo Department of Building Sciences and Urbanism, University of Alicante, Alicante, Spain

María L. Pertegal-Felices Developmental and Educational Psychology Department, University of Alicante, Alicante, Spain

Xose Picatoste Department of Economics, University of A Coruna, Coruna, Spain

Placido Pinheiro Graduate Program in Applied Informatics, University of Fortaleza (UNIFOR), Fortaleza, CE, Brazil

Mirian Caliope Dantas Pinheiro Graduate Program in Applied Informatics, University of Fortaleza, Fortaleza, CE, Brazil

Pedro Gabriel Caliope Dantas Pinheiro University of Fortaleza, Fortaleza, CE, Brazil

Placido Pinheiro Graduate Program in Applied Informatics, University of Fortaleza, Fortaleza, CE, Brazil

Jiří Polcar Department of Industrial Engineering and Management, University of West Bohemia, Plzen, Czech Republic

Monica Pompeu Graduate Program in Applied Informatics, University of Fortaleza, Fortaleza, CE, Brazil

Herleson Paiva Pontes Graduate Program in Applied Informatics, University of Fortaleza (UNIFOR), Fortaleza, CE, Brazil;
Federal Institute of Education, Science and Technology of Ceara (IFCE), Jaguaribe, CE, Brazil

Francisco A. Pujol Department of Computer Technology, University of Alicante, San Vicente del Raspeig, Alicante, Spain

Francisco Pujol-Lopez University of Alicante, San Vicente del Raspeig, Spain

Jukka Pulkkinen Häme University of Applied Sciences, Hämeenlinna, Finland

María Graciela Badilla Quintana Facultad de Educación, Centro de Investigación en Educación y Desarrollo (CIEDE), Universidad Católica de la Santísima Concepción, Concepción, Chile

Adil E. Rajput College of Engineering, Effat University, Jeddah, Kingdom of Saudi Arabia

Alberto Real-Fernández Unidad Científica de Innovación Empresarial "Ars Innovatio", University of Alicante, San Vicente del Raspeig, Spain

Roberto Reda Department of Computer Science and Engineering, University of Bologna, Bologna, Italy

Eva Rimbau-Gilabert Open University of Catalonia, (Universitat Oberta de Catalunya), Barcelona, Spain

Lorayne Robertson Ontario Tech University, Oshawa, ON, Canada

Inma Rodríguez-Ardura Open University of Catalonia, (Universitat Oberta de Catalunya), Barcelona, Spain;
University of Miami, Coral Gables, FL, USA

Lucie Rohlíková Department of Computer Science and Educational Technology, University of West Bohemia, Plzen, Czech Republic

Miguel Torres Ruiz Instituto Politécnico Nacional, Centro de Investigación en Computación, Nueva Industrial Vallejo, Gustavo a. Madero, Mexico

Sajid Saleem School of Electrical Engineering and Computer Science (SEECS), National University of Sciences and Technology (NUST), Islamabad, Pakistan;
Department of Computer and Network Engineering, College of Computer Science and Engineering, University of Jeddah, Jeddah, Saudi Arabia

Jose-Luis Sanchez-Romero University of Alicante, San Vicente del Raspeig, Spain

Akila Sarirete College of Engineering, Effat University, Jeddah, Kingdom of Saudi Arabia

Filippo Sciarrone Department of Engineering, Roma TRE University, Rome, Italy

Ricardo Sellers University of Alicante, San Vicente del Raspeig, Alicante, Spain

Enric Serradell-López Universitat Oberta de Catalunya, Barcelona, Spain

Chen-wen Shen Department of Business Administration, National Central University, Jhongli District, Taoyuan City, Taiwan, Republic of China

María Teresa Signes-Pont Department of Computer Technology and Computation, University of Alicante, Alicante, Spain

Muneeba Sirshar School of Electrical Engineering and Computer Science (SEECS), National University of Sciences and Technology (NUST), Islamabad, Pakistan

Rafael Mollá Sirvent Department of Computer Technology and Computation, University of Alicante, Alicante, Spain

Marco Spruit Department of Information and Computing Sciences, Utrecht University, Utrecht, The Netherlands

Marco Temperini DIAG-Dipartimento di Ingegneria Informatica, Automatica e Gestionale, Sapienza University of Rome, Rome, Italy

Diana Mihaela Țîrcă Constantin Brancusi University, Târgu Jiu, Romania; Faculty of Economics Sciences, Târgu Jiu, Romania

Orlando Troisi University of Salerno, Fisciano, ITA, Italy

Igor Trotskii Häme University of Applied Sciences, Hämeenlinna, Finland

Jiří Vyšata Department of Machining Technology, University of West Bohemia, Plzen, Czech Republic

Wei Wang College of Business Administration, Huaqiao University, Quanzhou, China

Yves Wautelet KU Leuven, Brussels, Belgium

Yenchun Jim Wu Graduate Institute of Global Business and Strategy, National Taiwan Normal University, Taipei City, Taiwan

Technology-Enhanced Learning

Simulating Peer Assessment in Massive Open On-line Courses

Filippo Sciarrone and Marco Temperini

Abstract Peer Assessment is a powerful tool to enhance students high level meta-cognitive skills. In this paper we deal with a simulation framework (K-OpenAnswer) allowing to support peer assessment sessions, in which peers answer a question and assess some of their peers' answers, with the enrichment of "teacher mediation". Teacher mediation consists in the possibility for the teacher to add information into the network of data built by the peer assessment, by grading some answers. This can be useful to enhance the automated grading functionality of an educational system supporting peer assessment. We present a software system allowing to apply the K-OpenAnswer simulation framework on simulated Massive Open On-line Courses (MOOCs). The system allows to guide the dynamic of the student models and grades evolution, according to the teacher's intervention. It also allows to appreciate such dynamic and make observations about it. The aim of this paper is to show the functionalities that the teacher can use, and their usefulness on simulated MOOCs, planning the use of the same functionalities in the case of real MOOCs.

1 Introduction

In a student's activity, having to answer questions, having such answers assessed, and assessing others' answers, are all crucial tasks and opportunities. They allow to verify and enhance different (and differently high level) meta-cognitive skills, in connection with getting to master the studied concepts [1, 3, 9, 11]. Peer Assessment (PA) is a way to achieve such enhancements [10, 14]. In a PA session the student can be exposed to open-ended questions, a tool widely usable when one wants not to go

F. Sciarrone (✉)
Department of Engineering, Roma TRE University, Via della Vasca Navale, 79, Rome, Italy
e-mail: sciarro@ing.uniroma3.it

M. Temperini
DIAG-Dipartimento di Ingegneria Informatica, Automatica e Gestionale, Sapienza University of Rome, Via Ariosto 25, Rome, Italy
e-mail: marte@diag.uniroma1.it

A. Visvizi and M. D. Lytras (eds.), *Research & Innovation Forum 2019*, Springer Proceedings in Complexity,
https://doi.org/10.1007/978-3-030-30809-4_1

to the extremes of either multiple-choice quizzes or full fledged essays. Open-ended questions can implement a wide variety of tasks: examples are short essays presenting a solution and its justifications, programming code and its technical description, free text answers proposing solutions to other kinds of exercises, possibly using complex formulae. Open-ended questions result in a more challenging and informative source of analysis for the student and the teacher, than multiple-choice tests [13]. One aspect of PA is that, in the educational practice, it can be complemented and possibly decisively enriched by the teacher's intervention, i.e. grading. Asking the teacher to grade all the answers in a PA session entails, of course, the reverse side of the coin, with respect to the PA advantages mentioned above: especially with large classes, the task of grading the answers in several PA sessions during the semester can be too taxing, making the whole PA operation less likely to occur and provide its benefits. In Massive Open On-line Courses (MOOCs), such an approach is altogether impractical. In this paper we present an approach to PA, that we call K-OpenAnswer, and a standalone computer system supporting its application in simulated MOOCs. In K-OpenAnswer approach, we assume that, in a PA session, each peer has answered a question, and peer-assessed some of her/his peers' answers. Then, the final grading of the answers is based on the peers' assessments, and is mediated by the teacher's grading work operated on a subset, as minimal as possible, of the answers. Based on the PA data, and on the information added into the system by the teacher's grading of such answers, the remaining answers are given automated grading. Basically, in K-OpenAnswer the PA provides an initial student modeling, which is enhanced through the use of the teacher's grades, and then used to infer the remaining grades. The student modeling is a representation of the individual learner competence on the topic of the question (K) and of her/his capability to assess correctly the peers' answers (J). In previous work we have presented the OpenAnswer framework operating along the lines above described, grounded on a Bayesian Network (BN) based student modeling process [4, 15, 16]. The use of BN in student modeling can be effective when the number of peers is not too high. In the case of a MOOC such an approach is made expensive by the computational complexity of the algorithms involved in its management (and by the sheer amount of data to manage). To overcome such difficulties we have previously proposed an approach (K-OpenAnswer, [6]) based on the use of a K-NN machine learning method [8, 12] to support the representation of the data coming from a PA session, the computation of the student modeling, and the grades inferences. In this paper we elaborate on the K-NN approach, presenting a software system that allows to simulate PA sessions in MOOCs of parameterized dimension (in terms of number of students). The system allows to simulate the use of K-OpenAnswer in the class and to appreciate the behavior of the approach, visualizing the students network dynamics, and supporting observations such as revealing students at risk. Looking at such simulations we can understand how useful could the same software be, when applied on a real MOOC. The structure of the paper is as follows: in the next section we recall briefly the network learning process. In Sect. 3 we present the above mentioned statistical system and discuss some of its characteristics. The last section provides some conclusions and discussion of future work.

2 The Network Learning Process

In this section we briefly illustrate the algorithms on which the MOOC dynamics are based. For further information, the reader can refer to [5, 6]. The algorithms are based on the K-NN *learning* algorithm [7, 12]. Each student is represented by a Student Model (SM), $SM \equiv \{K, J, Dev, St\}$, where $K \equiv [1, 10]$ is the grade that the teacher has assigned to this student through the correction of one or more structured open-ended exercises. From a learning point of view, it represents the learner's competence (*Knowledge level*). The variable $J \equiv [0, 1]$ is a measure of the learner's assessing capability (*Judgement*) and depends on K. The standard Deviation Dev that represents the credibility of the value of K. The higher this value, the less the value of K of the student is credible. It is calculated as the standard deviation generated, for each i-th learner as follows:

$$Dev_i = \sqrt{\frac{\sum_{l=1}^{n}(K_i - K_l)^2}{n}} \tag{1}$$

where each K_l is one of the group of students that graded her in the PA step. Finally, $St \equiv \{CORE, NO\_CORE\}$ represents the student status. Each student can be in two different states: CORE and NO_CORE. Initially all the students are NO_CORE. If a student is voted by the teacher then she becomes a CORE student. Each NO_CORE student is represented as s^- while a CORE student is represented as s^+. Consequently, the community of students is, at any given moment, dynamically parted into two groups: the *Core Group* (CG), and its complement $\overline{CG}$. CG is composed by the students whose answers have been graded directly by the teacher: for them K is given (fixed). In the following we also call this set as S^+, and call its elements the s^+ students. On the contrary, S^- is the set of students whose grade is to be inferred (so, they have been graded only by peers). Finally, by this SM representation, each learner can be represented as a point in a 2-dimensional space (K, J). Each SM is initialized by means of a PA where the tutor assigns an open-ended question to all the students and each student grades the answers of n different peers, while she receives n peer grades. So, each SM is initialized as follows:

$$K_l^- = \frac{\sum_{i=1}^{n} K_i^-}{n} \tag{2}$$

where K_i^- is the grade received by the i-th of the n peers who graded the s_l^- student.

For each s_l^- student, J_l^- is initialized as follows:

$$J_l^- = \frac{1}{1 + \sqrt{\sum_{i=1}^{n} \Delta_i^2}} \tag{3}$$

$\underline{\Delta_i^2} = (K_{lj} - \overline{K_j})^2$, being K_{lj} the grade assigned by the student s_l to the student s_j and $\overline{K_j}$ the arithmetic mean, i.e., the initial K^- of the student s_j, computed by (2). All students are initialized to $St = NO\_CORE$. At each learning step the module *learns* to (hopefully) better classifying the students in S^-, until a termination condition suggests to stop cycling, and the S^- students converge to the grades finally inferred by the module. After the SM initialization, all learners belong to the S^- set. Each learner evolves in the (K, J) space as follows:

- The teacher is suggested a ranked list of students/answers to grade, sorted by the Dev key;
- The teacher selects a group of students in the ranked list, and grades their answers. Such grades are the new, final, K^+ values for such students;
- The graded students become s^+ students, and their position in the (K, J) space changes;
- All peers who had voted for the student who became s^+ change their model. The model updating algorithm follows recursively a graph path starting from the voted students and so on backwards. For each learner, first K, and J are updated. Once all the students influenced by the teacher's vote have been updated, all their Dev updated.

In the following we will use *KMIN* and *KMAX* to denote the minimum and maximum values for K (i.e. here respectively 1 and 10). *IMAX* will denote the maximum difference between two values of K, i.e. here 9. Moreover *JMIN* and *JMAX* will denote the minimum and maximum values for J (i.e. here resp. 0 and 1). Finally, Dev_{min} and Dev_{max} represent the lowest and highest values for the variable Dev, i.e., $DevMIN = 0$ and $DevMAX = 9$. The graded learner model is updated. First the K value and after the other values:

$$K^+ = K_{teacher} \tag{4}$$

Secondly the J value:

$$\begin{aligned} J^+_{new} &= J_{old} + \alpha(JMAX - J_{old}) \;\; with \;\; (0 \leq \alpha \leq 1) \\ J^+_{new} &= J_{old} + \alpha J_{old} \;\; with \;\; (\alpha < 0) \\ \alpha &= \frac{K_{teacher} - K^-_{old}}{IMAX} \end{aligned} \tag{5}$$

Notice, in (5):

1. A convex function has been adopted for J update, providing the two cases according to the possible value of α. In particular J_{old} could stand for J^+_{old} or J^-_{old}, depending on the student being already in S^+ (case J^+_{old}), or being just entering in S^+ (case J^-_{old}) or remaining in S^- (case J^-_{old} again);
2. In general we assume that the assessment skill of a student depends on her *Knowledge Level K*, so the J value is a function of K. We used this type of evolutionary

form as it is the easiest to treat as a first approach and also because it is used very often in automatic learning as an update of statistical variables in a machine learning context (see for example [2]).

Subsequently, the value of *Dev* is modified recalculating it on the student graded by the teacher. All the students, s^-, who are influenced by the graded student are modified, according to the following rules (students s^+ are fixed because graded by the teacher):

$$\begin{aligned} K^-_{new} &= K^-_{grading} + \alpha(KMAX - K^-_{graded}) \quad (0 \le \alpha \le 1) \\ K^-_{new} &= K^-_{graded} + \alpha K^-_{graded} \quad (\alpha < 0) \\ \alpha &= \frac{1}{IMAX}(K^-_{grading} - K^-_{graded})\frac{Dev_{graded}}{IMAX} \end{aligned} \tag{6}$$

where: K_{new} is the new value of K of the intermediate student. The $\frac{Dev_{graded}}{IMAX}$ factor expresses a kind of inertia of the value of K to change. Each J value is changed as follows:

$$\begin{aligned} &J^-_{new} = J^-_{grading} + \beta(JMAX - J^-_{grading}) \quad (0 \le \beta \le 1) \\ &J^-_{new} = J^-_{grading} + \beta J^-_{grading} \quad (\beta < 0) \\ &J^-_{new} = J^-_{grading} + (K^-_{grading} - K^-_{graded}) \\ &\qquad (\beta = 0 \wedge J^-_{grading} = J^-_{graded}) \\ &\qquad with: \\ &\beta = \frac{1}{IMAX}(K^-_{new} - K^-_{grading})|J_{grading} - J_{graded}|\frac{Dev_{graded}}{IMAX} \end{aligned} \tag{7}$$

After, in order to complete the SMs, all the *Dev* variables are updated.

Finally, after that the teacher has graded some s^- students, become s^+ students, the modified K-NN algorithm can start. The learning process changes by means of the laws:

$$\begin{aligned} K^-_{new} &= K^-_{old} + \alpha(KMAX - K^-_{old}) \quad (0 \le \alpha \le 1) \\ K^-_{new} &= K^-_{old} + \alpha(1 - K^-_{old}) \quad (\alpha < 0) \\ \alpha &= \frac{1}{I_{max}} \frac{\sum_{i=1}^{k} \frac{1}{d_i}(K^+_i - K^-_{old})}{\sum_{i=1}^{k} \frac{1}{d_i}} \frac{Dev_i}{IMAX} \end{aligned} \tag{8}$$

where:

1. d_i is the Euclidean distance between the s^-_{old} student under update, and the i-th student in the Core Group (s^+_i);
2. The K^-_{new} value is given as a convex function, to keep K in [1, 10];
3. the acronym K-NN features a K, possibly misleading here, so we are using k for the number of nearest neighbors to be used in the learning algorithm.

4. The $\frac{Dev_i}{IMAX}$ factor has the same meaning as in (6).

$$
\begin{aligned}
J^-_{new} = J^-_{old} + \frac{(K^-_{new} - K^-_{old})}{IMAX} J^-_{old} \quad (\beta = 0 \wedge J^+_i = J^-_{old}, i = 1 \ldots k)\\
J^-_{new} = J^-_{old} + \beta(JMAX - J^-_{old}) \quad (0 \le \beta \le 1)\\
J^-_{new} = J^-_{old} + \beta J^-_{old} \quad (\beta < 0)\\
\text{with } \beta = \frac{(K^-_{new} - K_{old})}{IMAX} \frac{\sum_{i=1}^{k} \frac{1}{d_i} |J^+_i - J^-_{old}|}{\sum_{i=1}^{k} \frac{1}{d_i}} \frac{Dev_i}{IMAX}.
\end{aligned}
\tag{9}
$$

where:

1. As mentioned earlier, we assume J depending on K: this is expressed through the difference between the K^-_{new} value, obtained by (8), and the K^-_{old} value.
2. d_i is the Euclidean distance between the s^-_{old} student under update, and the $i-th$ student in the Core Group (s^+_i);
3. The J_{new} value is given as a convex function, to keep J in its normal range [0, 1];
4. k is as explained in the previous equation.
5. About the coefficient β, some notices are due, for the cases when $\beta = 0$. On the one hand, when the J^+ of the k nearest neighbors is equal to the J^-_{old} value of the s^-_i student under update, J^-_{new} is computed by the difference between K^-_{new} and K^-_{old} only. The rationale is that when the s^- student changes her K^- value, her assessment skill should change as well (by the assumption of dependence of J on K). On the other hand, when the K^- value for the student under update is not changed, the assessment skill stays unchanged as well.

3 The Statistical Environment

In order to prepare an experimentation, we have implemented the K-OpenAnswer framework in a standalone software system. Several functionalities are available, to provide the teacher with a statistical environment for the analysis of a simulated MOOC behavior. The implementation has been produced using the C programming language, with the idea to have best computational efficiency. We intend to expose it, in future, on the cloud, so to allow supporting the general architecture shown in Fig. 1.

In Fig. 1 we see the main "modules" of the educational system supporting MOOC with K-OpenAnswer. The blocks of interest here are the *Learning Process* and the *Statistical Engine*. The former is the logical framework recalled in the previous section. It provides the student modeling based on the PA and mediated by the available teacher's grades. The latter is the software discussed in this section. It provides the functionalities to build simulated MOOC classes, add PA and teacher's data, and analyze the consequent dynamic of the student models network. Furthermore, the

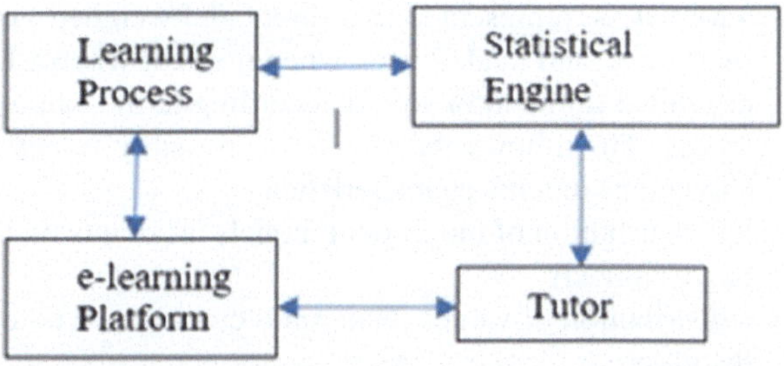

Fig. 1 The architecture of the overall system

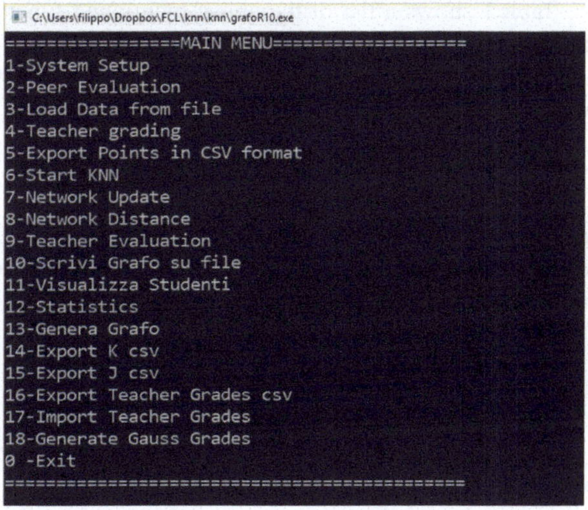

Fig. 2 The functions of the simulation module

teacher (*Tutor*) operating with the peers in the PA sessions, and the Learning Management System, in which the sessions are organized (*E-learning platform*), are the remaining necessary blocks in the overall educational environment.

Figure 2 shows the software system running. In this phase of the implementation we did not care to add a graphical user interface; on the other hand, as will be seen later, a graphical part of the system is implemented, coupling the software with the statistical system R, in order to support data analysis.

The functionalities offered by the system are basically implementing the following workflow

- a MOOC simulation (MOOC-s henceforth) can be defined by specifying the number of students in the class and the number of peer assessments that each peer is expected to provide;
- the students in a MOOC-s can be given an initial distribution of models;
- a number of teacher's grades can be added into the system, determining a phase of network update (student models and grades); When n students are graded in a turn, they become members of S^+ (referring to the previous section) and the value of the Dev model variable is recalculated;

– a MOOC-s, represented as a direct and weighed graph, can be saved in a file for future use, and loaded back when needed; when a MOOC is loaded, students are presented to the tutor sorted according to the value of the *Dev* parameter of their model. This allows the teacher to possibly select answers to grade, basing the choice on students characteristics.
– the distribution of the student models, as points on the (K, J) plan, and grades can be visualized;
– a distribution of values (Gaussian) can be used to assign grades to the students in the class.

Table 1 shows a list of the above mentioned functionalities (each one is implemented in a module of the software system).

Table 2 presents a list of sample definitions of MOOCs simulations used in our experimental work.

Figure 3 illustrates the students distribution in the (K, J) space, produced by the visualization functionality.

When the K-NN algorithm is applied to reconfigure the network after the incident caused by the teacher's evaluation, for each point in space (K, J), the nearest Ks are computed, by Euclidean distance, and the model update follows the rules mentioned in the previous section. The termination condition for the update process is that the

Table 1 MOOC simulation environment

Module	Goals
Peer evaluation	Simulation of the peer evaluation
Load MOOC from file	The MOOC can be uploaded from a file
Tutor grading	The tutor can grades a set of students
Export students	All the students are exported into a file
Start K-NN	K-NN management
Network updating	Students models updating
Evaluation	All students grading
Statistics	Some statistical analysis
Build a new MOOC	A new MOOC is created
Generate Gaussian grades	A Gaussian grades distribution is created

Table 2 MOOC simulation trials

# Students	# Peer evaluations
100	3
500	4
1000	5
1500	3
1500	4
2000	3

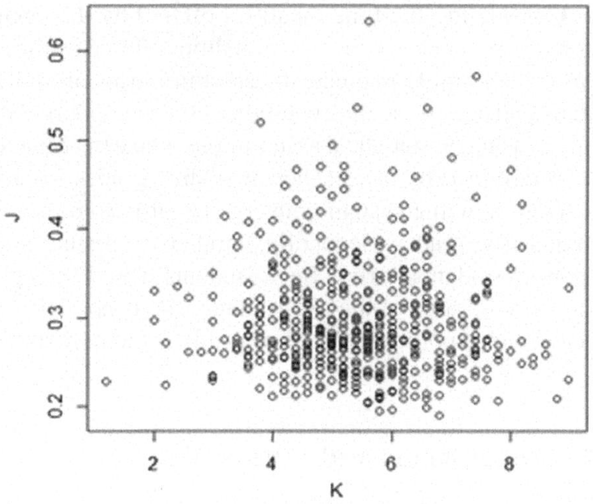

Fig. 3 The MOOC with 1000 students and 3 peers evaluations

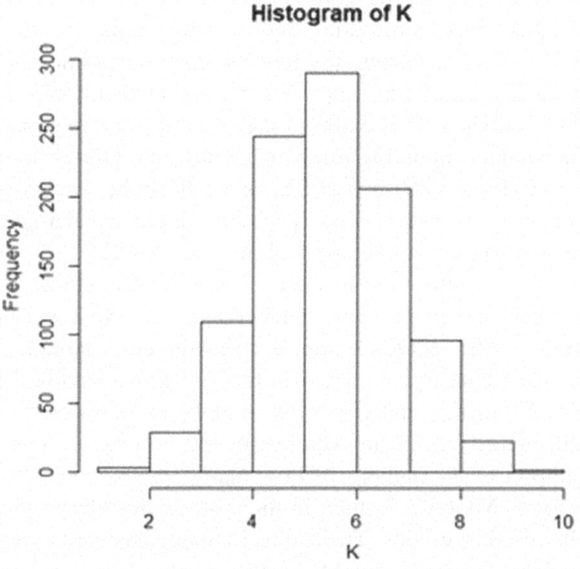

Fig. 4 The tutor grade Gaussian distribution

difference between two successive network configurations is less than a prefixed ϵ. The distance between two configurations of a network is calculated as the average distance between all points.

As mentioned earlier, a functionality allows to set up a distribution of grades on the students in the class. This is done according to a Gaussian distribution. An example is illustrated in Fig. 4.

This distribution can serve to analyze how the addition of grades in the system actually modifies the graphical layout of the models (cf. Fig. 3).

Considering the functionalities offered by the environment, interesting applications are possible, for the functionalities offered by the environment. First, the teacher has the possibility to create different instances of MOOCs, and for each one of them study and determine the evolution of the network of data coming from the PA, in real time, basing on statistical assumptions. Once the MOOC is defined, the evolution can be determined by the teacher by adding grades, so another interesting observation is about how the system converge towards some stable state, making unnecessary to add new grades. Secondlly, another interesting application is the possibility of spotting students at risk. Such students would be represented by isolated points in the "low-K" area of the (K, J) plane: she is not being reasonably related with other peers, and is not learning, so probably some remedial intervention is to plan.

4 Conclusions and Future Work

In this paper we have illustrated a simulation environment, deemed to allow the teacher (tutor) simulating and monitoring the evolution of the student models in a MOOC class, during sessions of Peer Assessment. The environment implements a student modeling approach based on the K-NN method of machine learning, K-OpenAnswer. The aim of the framework is to support teacher-mediated PA, with the student modeling allowing to infer the grades not provided by the teacher. This allows for a wider use of PA, as it relieve the teacher from the burden of grading all the answers in a session. The computational characteristics of the framework also provide a reasonable applicability in a MOOC. The aim of the simulation environment is to allow the teacher to study the dynamic of the network of data built by the peer evaluations and teacher's grades. We have shown that this analysis is possible, on the MOOCs simulated in the environment. The same functionalities can be used also in the case of a real MOOC, providing for interesting analysis tools. Several further analysis tools are to be implemented, such as the possibility of using different distribution policies for the initialization of the MOOC-s PA data, or the support of an analysis of how many grades would be needed by the teacher in the various MOOCs defined in the system, in order to reach a satisfactory precision of the inferred grades. These developments, as well as a pilot application to a real case of MOOC, are planned for future work.

References

1. Anderson, L.W., Krathwohl, D.R.e.: A Taxonomy for Learning, Teaching, and Assessing: A Revision of Bloom's Taxonomy of Educational Objectives. Allyn and Bacon (2000)
2. Bishop, C.M.: Pattern Recognition and Machine Learning (Information Science and Statistics). Springer, Berlin (2006)

3. Bloom, B., Engelhart, M., Furst, E., Hill, W., Krathwohl, D.: Taxonomy of Educational Objectives: The Classification of Educational Goals. Handbook I: Cognitive Domain. McGraw-Hill Inc., New York (1956)
4. De Marsico, M., Sciarrone, F., Sterbini, A., Temperini, M.: Supporting mediated peer-evaluation to grade answers to open-ended questions. EURASIA J. Math. Sci. Technol. Educ. **13**(4), 1085–1106 (2017)
5. De Marsico, M., Sciarrone, F., Sterbini, A., Temperini, M.: Peer assessment and knowledge discovering in a community of learners. In: IC3K 2018—Proceedings of 10th International Joint Conference on Knowledge Discovery, Knowledge Engineering and Knowledge Management, pp. 119–126 (2018)
6. De Marsico, M., Sterbini, A., Sciarrone, F., Temperini, M.: Modeling a peer assessment framework by means of a lazy learning approach. In: Huang, T.C., Lau, R., Huang, Y.M., Spaniol, M., Yuen, C.H. (eds.) Emerging Technologies for Education, pp. 336–345. Springer International Publishing, Cham (2017)
7. Gasparetti, F., Limongelli, C., Sciarrone, F.: Exploiting wikipedia for discovering prerequisite relationships among learning objects. In: 2015 International Conference on Information Technology Based Higher Education and Training (ITHET), pp. 1–6 (2015). https://doi.org/10.1109/ITHET.2015.7218038
8. Gasparetti, F., Medio, C.D., Limongelli, C., Sciarrone, F., Temperini, M.: Prerequisites between learning objects: automatic extraction based on a machine learning approach. Telematics Inform. **35**(3), 595–610 (2018)
9. Li, L., Liu, X., Steckelberg, A.L.: Assessor or assessee: how student learning improves by giving and receiving peer feedback. Br. J. Educ. Technol. **41**, 525–536 (2010)
10. Limongelli, C., Sciarrone, F., Temperini, M.: A social network-based teacher model to support course construction. Comput. Hum. Behav. **51, Part B**, 1077–1085 (2015)
11. Limongelli, C., Mosiello, G., Panzieri, S., Sciarrone, F.: Virtual industrial training: joining innovative interfaces with plant modeling. In: The 11th International Conference on Information Technology Based Higher Education and Training—ITHET 2012, pp. 1–6. IEEE (2012)
12. Mitchell, T.M.: Machine Learning, 1st edn. David McKay, New York (1997)
13. Palmer, K., Richardson, P.: On-line assessment and free-response input-a pedagogic and technical model for squaring the circle. In: Proceedings of 7th CAA Conference, pp. 289–300 (2003)
14. Sadler, P., Good, E.: The impact of self- and peer-grading on student learning. Educ. Assess. **11**(1), 1–31 (2006)
15. Sterbini, A., Temperini, M.: Supporting assessment of open answers in a didactic setting. In: 2012 IEEE 12th International Conference on Advanced Learning Technologies (ICALT), pp. 678–679 (2012)
16. Sterbini, A., Temperini, M.: Analysis of open answers via mediated peer-assessment. In: 2013 17th International Conference on System Theory, Control and Computing (ICSTCC), pp. 663–668 (2013)

Development Strategies and Trends in Educational Solutions for Deaf Students

Herleson Paiva Pontes, João Batista Furlan Duarte and Placido Pinheiro

Abstract Over the last decade, there has been a notable movement towards the inclusion of deaf students in academic environments, most of them heavily based in Computer Science and its technologies. Despite the high number of researches in this area, few present or discuss guidelines and best practices that should be taken in account when developing solutions for the deaf audience. This work proposes strategies that researchers should follow when planning, developing and delivering learning systems aiming the inclusion of deaf students, considering their communicational, sensory and cultural particularities. Systems that follow these guidelines can enhance the learning process by the use of technology. The approaches presented in this study support innovative systems in the offer of a truly meaningful learning environment, and can be applied in researches areas like augmented and virtual reality, natural user interfaces, and gamification.

H. P. Pontes (✉) · J. B. F. Duarte (✉) · P. Pinheiro (✉)
Graduate Program in Applied Informatics, University of Fortaleza (UNIFOR),
Av Washington Soares 1321, Fortaleza, CE 60811-905, Brazil
e-mail: herleson@gmail.com
URL: https://unifor.br/

J. B. F. Duarte
e-mail: furlan@unifor.br
URL: https://unifor.br/

P. Pinheiro
e-mail: placido@unifor.br
URL: https://unifor.br/

H. P. Pontes
Federal Institute of Education, Science and Technology of Ceara (IFCE),
Rua Pedro Bezerra de Menezes 387, Jaguaribe, CE 63475-000, Brazil
URL: http://ifce.edu.br/jaguaribe

A. Visvizi and M. D. Lytras (eds.), *Research & Innovation Forum 2019*,
Springer Proceedings in Complexity,
https://doi.org/10.1007/978-3-030-30809-4_2

1 Introduction

Assistive technologies have, for well over a decade, been applied for the inclusion of students with disabilities in educational environments, as an attempt to overcome individual and social barriers, besides promote the same socialization process for all people [2, 3, 15]. When the first systems for inclusive education were developed, most of them was basically computer programs which fundamentally used I/O devices like keyboard, mouse, microphone and speakers in particular ways to enable message exchange. These systems were difficult to build and use by students with disabilities, besides user interfaces and interactions were not optimized. However, the proof-of-concept programs demonstrated that the development of assistive solutions could dramatically increase educational processes like knowledge acquisition and meaningful learning to this target audience [6].

Nowadays, assistive solutions have now matured to a point where there are countless applications and devices, which propose reduction of barriers that disrupt the participation of these people in daily activities in areas like Health, Economics, Engineering, Transportation and, mainly, Education [9]. In Education area, while adjustments in physical locations to solve spatial and motion problems for students with disabilities are notable in almost every school, faculty and university; inclusion technologies focus the empowerment of communication processes between all actors during the learning process: students with disabilities, their teachers and colleagues [18]. Initiatives in communicational area are still far from being noticed in academic organizations when compared to motion and spatial ones and, in this context, technology plays a vital role mitigating communicational barriers quickly and offering an effective scenario for learning activities.

The key to the success of these inclusive technologies has partly been their intuitive interactions, effective interfaces, complex language-driven algorithms and massive computing performance, resulting in improved accessible platforms for people with disabilities [13]. Essentially, solutions to students with disabilities reduce the gap between virtual and real worlds by enabling seamless interfaces, intelligent systems and inherent actions. Therefore, create assistive systems has never been easier.

Together with the development of assistive technologies, a huge revolution in interdisciplinarity and inclusive approaches for knowledge acquisition and meaningful learning was also seen. It is clear that the learning process requires a complex relationship between human senses and mental activities to arrange and structure a person's knowledge, featuring how this strategy is not linear nor uniform [12]. Moreover, sensation and perception processes overlap themselves so the mind can receive all external stimulus, interpret these data and build mental maps that structure and integrate the information generated [7].

In this context, whilst getting started with educational assistive technology can be simple, being able to fully understand its requirements and develop an effective solution is an art that can take months or years to master, and strongly relies in which type of disability is targeted, what learning goals must be met, and what properties of the academic environment should be considered [15].

Worldwide, a considerable number of students with disabilities in educational spaces are deaf. There are more than 466 million people have disabling hearing loss and 34 million of these are children [19]. Caused by the lack of the ability to communicate with others, deaf students commonly present negative effects on the academic performance, such as increased rates of grade failure and problems with socialization. For that reason, deaf students need education assistance for optimal learning experiences.

One of the main efforts in many countries to help deaf students is the universalization of sign languages, commonly recognized as an official alternative language and used as the primary mean of communication among deaf. Unfortunately, use sign languages is still a challenge in most academic environments and, despite the existence of systems and devices for sign recognition, as far as the authors are aware, there is practically non-existent proposals that takes in account sign languages' phonological and morphological aspects, deaf culture particularities, meaningful learning requirements and assistive technologies.

Much of the information that supports this article can be found in a variety of different sources, including books, journals, manuals, conference presentations, and on Internet. However, consolidate all this information and build strategies for inclusive systems development from it are arduous exercises that require a substantial effort. The aim of this work is, therefore, to give an overview of state-of-the-art strategies and best practices to develop effective assistive solutions, considering and combining the learning process requirements when involving deaf students, linguistic and cultural specificities of these students, principles behind human senses and innovative devices available nowadays.

The rest of this paper is sectioned as follows. First, Sect. 2 gives a short overview of sensory processing and the role of human senses in this process, followed by deaf's linguistic and cultural aspects in Sect. 3. Then, an overview of meaningful learning for deaf students is discussed in Sect. 4. Finally, strategies and best practices for the build of educational inclusive solutions are offered in Sect. 5, and concluded with a short summary in Sect. 6.

2 Sensation and Perception

Sensory analysis is a tool used by an individual to interpret information and trigger reactions based on events and activities that are performed through human senses [7]. This technique combines sensation and perception processes so a person can perceive the physical world and interact with it. Figure 1 illustrates the overall architecture of the sensory analysis and its components.

The sensation process is basically a response of a sensory receptor to external stimuli, in other words, it is a physiological response of the organism [11]. It is a process in which the human senses convert the energy of a stimulus into neural messages and provoke reactions. Although human senses are constantly receiving stimuli and are specialized in record specific kinds of energy, their abilities to sense

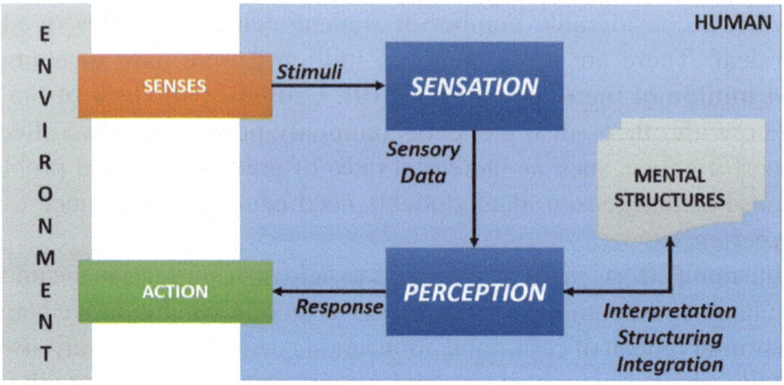

Fig. 1 Human's sensory analysis architecture

the environment are limited. According to Weber's Law, changes in stimuli may be noticeable according to a person's psychological experience during the sensation process and, therefore, it is a relative activity [7]. In other words, there is not a simple and unidirectional relationship between sensations and mental structures that represent an individual's knowledge.

The perception process, on the other hand, is the outcome generated by a person based on the sensory data from the sensation process [7]. It is the interpretation, arrangement and integration of what was captured by the senses, and it can be influenced by physiological and psychological factors, as well as by external issues such as cultural and social aspects [11]. A same sensation, for instance, can lead to distinct perceptions in different people, since each individual has his/her own experiences and build distinct perceptions in various aspects from these records. However, it is expected that people living in the same place and sharing similar traditions show almost identical patterns during the perception process, because these individuals share a similar background.

For people with disabilities, the sensory analysis activity suffers adjustments in its operation, particularly when involving communicational disabilities like deafness and blindness [7]. If an individual lack one of his/her senses, the organism compensates this disability by enhancing the other senses and building distinct mental structures when compared to those with all senses. Because the senses' thresholds of a person with disability is different and his/her perception arrangement is directly affected by the unusual sensory data collected, learning experiences may differ from person to person and from culture to culture, describing the perceptual context of an individual [11].

3 Deaf Language and Culture

To provide a way to send and receive messages using other senses than hearing, a set of visuospatial motions that use hand shapes, facial expression, gestures and body language were developed as languages that enables communication among deaf people. These languages are known as sign languages and, like the spoken ones, they have syntax and grammar of its own [1, 8].

The conception of visual and spatial languages based on signs for communication is possible thanks to the ability that human mind has to establish associations between meanings and patterns [16]. These patterns can be assimilated in various ways by the human consciousness: words, images and motions are some examples of these methods. This key aspect of the brain, supported by the sensory analysis and its sensation and perception processes [7, 11], enables the ability in humans to link different representations to a same meaning. As a person that rearranged his/her sensory analysis process because the lack of the hearing sense, deaf are benefited by their uncommon perceptual context when using sign languages during communication with others [5].

3.1 Linguistic Aspects

In Linguistics, Phonology is the study area of the phonemes in a language [4]. From a sign language's point of view, phonology tries to identify what minimal unit sets can be applied to develop signs for message exchange, to establish what possible combination patterns can be created between these minimal units, and to understand how the variations in the phonological environment promote the generation of larger linguistic units [16].

Regardless the sign language, phonology elements for visuospatial communications are basically made up from the combination of articulated motions performed by the hands and the location in which these actions are executed in a finite space, as described in Table 1.

Another important concept in Linguistics is Morphology, which represents the study area responsible for evaluating the rules applied during word structuring and the components used to create this term [5]. Considering sign languages' perspective, the join process of motions and spatial locations provides a huge set of signs and, consequently, generates visual representation for a wide number of words [8].

Table 1 General phonological components of a sign language

Component	Description
Hand configuration	Refers to the shapes a hand assumes during the execution of a sign
Hand orientation	Identifies to which direction the palm of the hand aims in a sign
Articulation point	Establishes the spatial location within the sign is being produced, using the body as the reference point
Motion	Describes the motion path that the hands travel during the execution of a given sign
Non-manual expressions	Gathers all properties complementary of the sign performed, such as facial expressions, body motions and looks

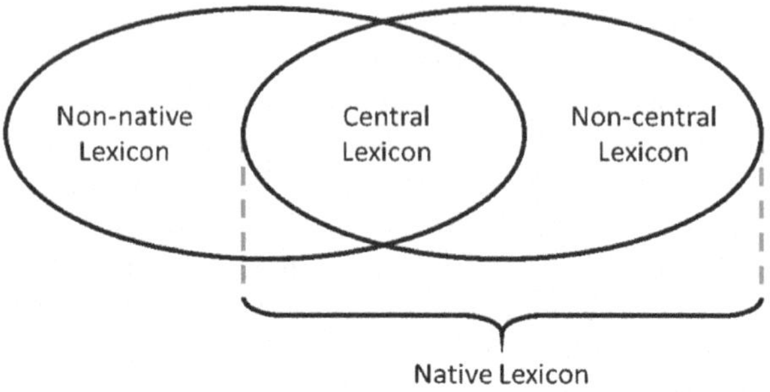

Fig. 2 Lexical classification for sign languages

The morphological complexity contained in sign languages encourages a lexical classification, similar to those in oral languages, which categorize sign into classes, such as noun, verb, adjective, and others. Generally speaking, the lexical structure of a sign language has three main elements: central lexicon, non-central lexicon and non-native lexicon. Figure 2 illustrates these components.

3.2 Cultural Aspects

A sign language evolves according the environment where it is used and the experiences of its users, so the vocabulary grows following cultural and technologies changes [14]. In this context, deaf describe themselves as individuals with their own and unique culture, who fight constantly to overcome their disabilities to live harmonically in society. Thus, deaf's living context is noticeable by social movements that describe deafness simply as a difference in human sensation and perception experiences, avoiding relationship with stereotypes like freak, incapable and powerless.

Understand the particularities that make deaf culture unique is essential to ensure that inclusive proposals respect deaf identity and address all needs required by deaf people [17]. In fact, a sign language is the most important symbol of the deaf culture because it shows, by means of its high number of signs, the identity of this group to society [14].

Concluding, integration efforts between deaf and hearing people must check the details of the deaf culture, must comprehend how deaf deal with their deafness, and must understand how communication happens through sign languages. The success of an inclusion initiative relies on the correct and effective combination of all these factors.

4 Meaningful Learning for Deaf

Currently, there is a series of questions about the learning styles and the methodological approaches to be used for deaf students in search of meaning in the process of knowledge acquisition [10]. Because there are many pedagogical, temporal, sensory, environmental and technological variables in learning process, find the right strategy to overcome negative outcomes on the academic performance of deaf students is still challenging.

Surely, every student must take part actively during the learning process to ensure it effectiveness [14]. However, to achieve positive results in academic scenarios, this student needs a language, a way to interact with others and exchange messages [17]. A language plays the main role during the development of superior psychological processes in a student, like attention, memory and consciousness. Therefore, meaningful learning relies essentially on the dynamic and interactive social relationships established between a student and his/her colleagues, teachers and parents during academic endeavors.

Since meaningful learning uses dynamic interactions in academic environments, sensory analysis is the biological process that provides the required interface between the student and the external world. Sensation and perception processes work together to translate messages received in mental models and to set responses for the stimuli captured by human senses.

In this context, deaf students are benefited from this process just like hearing students, since the differences in sensory process' architecture are concentrated in the enhancement of available senses and the generation of distinct mental models from these senses. In other words, a meaningful learning can be achieved by either deaf or hearing students, if the right senses receive the right stimuli at the right time.

Moreover, it is already known that sight and touch are highly stimulating senses in humans [7]. Touch is the first sense developed by a person and uses the skin to capture several stimuli and send them to the brain. Sight is the sense that perceive things through the eyes by capturing light rays reflected from those objects. Because deaf compensate their disabling hearing loss with enhanced sight, touch, taste and smell, these people create additional unusual mental structures to represent acquired knowledge, thereby developing better memory and language skills than those born with hearing. The brain of a deaf person rearrange itself in a way to use the information at its disposal so that it can interact with the environment in a more effective manner with the senses available [10].

5 Strategies and Best Practices

Develop an educational assistive technology is unlike a traditional educational system, because the academic scenario and requirements are dramatically different when involving deaf students: their sensory analysis activities use enhanced senses to process stimuli during sensation and combine exclusive mental structures to provide different insights during perception, their communication process is based on visuospatial languages with complex linguistic aspects, their culture affects directly the experiences and changes their perceptual context, and their mind models are structured in a way that a meaningful learning relies in other senses than hearing. In this context, achieve effectiveness in an educational inclusive solution is still a difficult task that can take months or years to master.

5.1 Communication Strategies

In educational contexts, bidirectional communication processes are the foundation for a meaningful learning. Therefore, one concern when developing educational systems for deaf students is the language used to establish the communication mean between sender and receiver.

A common approach present in many solutions is the use of captions, labels and other text components to show the desired content. In roaming environments, like distance learning scenarios, this strategy provides good performance when retrieving the content from servers. However, linguistic constructions in a sign language are totally different from those of spoken languages: the former is visual and spatial, based on body motions; the latter is oral and hearing, based on the sound waves. Moreover, a dialogue using signs cannot be directly translated to a spoken language, because many grammar structures and components used in spoken languages have different use or even don't exist in a sign language. If the use of texts is unavoidable, give preference for small sentences and reduced messages.

Another usual strategy to communicate with deaf students is the simultaneous offer of spoken and sign languages. In this case, the message spoken by the educational application is displayed, at the same time, in a video with a sign interpreter. Although this method communicates with deaf students in their native language, it has three drawbacks: first, the space used to display the video on the screen reduces the operational area of the application, especially in mobile devices; second, the network bandwidth required to remotely access this kind of content is very high; third, whilst sight is one of the enhanced senses in a deaf student, it is impossible for a person to look simultaneously to two different points in a single screen: the operational area of the application and the sign interpreter window. If the split-screen approach is the selected for an educational application, be sure to dimension operational and sign video areas to promote communication with deaf students without sacrifice the application educational tools.

A more complex technique is the use of images and videos to enable communication with deaf students. In this strategy, no text nor audio is used and all content is show in videos and images. While this approach can cover both deaf and hearing students in a single educational platform, the efforts required to develop this kind of media are very high. Network traffic is very high in this approach, and it may not fit to distance learning scenarios. Also, all material developed must share the content in a cognitive and intuitive way, avoiding texts and sounds at all cost, which may be hard to design according the topic that must be covered by these media. In addition, images and videos require massive computing resources (storage, processing and energy consumption), which can prevent the use of this strategy in mobile devices.

5.2 Learning Improvements

The learning process of deaf students is reasoned on the sensory analysis process and its particularities for those with disabilities. Therefore, improvements in the learning process for deaf students include heavy use of visual elements to share the educational content. The outcomes from the development skills and expertise to deaf are more expressive, thanks to the sight sense's role during the facts and events lived by these students and the lack of manifestations guided by the hearing sense. In other words, educational systems that uses images, videos and animations are most welcome as learning tools to improve knowledge acquisition. However, developers must be careful about the time deaf students will be exposed to these visual solutions, to avoid sight problems caused by long exposures in front of screens.

It is important that all computer solutions related to education of deaf students consider the domain in which the representations will be displayed. In this context, share educational contents using sign languages reinforces the bond between all actors of the learning process, because deaf students will see their identities and singularities being noticed and respected by teachers and other colleagues, reducing socialization barriers and motivating engagements. It is important to ensure that all linguistic aspects of the sign language are being followed during communication with sign languages, encouraging students to better understand their native language. Since the most important symbol of a deaf culture and its members is their sign language, all software that uses a sign language must present the identity of this group and its relation to the society.

In some cases, deaf students start their academic careers only in adulthood, because the family did not motivate the learning of sign languages when he/she is a child. Therefore, another strategy to improve learning process by deaf students is the planning and development of inclusive computer systems focused in children. In addition, educational solutions for children can improve learning process by stimulating sight sense with colorful screens and animations.

Another important sense that has potential to improve learning process is touch. Although this sense is a highly stimulating one in humans, few educational systems use it in their learning activities. Devices with feedback forces, like haptics, joysticks and wearables, can be used to improve student interaction with educational solutions during their operations. However, it is important to note that deaf students have higher sensibility caused by the adjustments in their sensory analysis and, therefore, educational applications must evaluate how comfortable the deaf students are when using these devices and when receiving feedback forces on their skin.

It is also possible to create solutions that combine sight and touch sensations to improve the learning process. In this case, developers must be aware to balance these stimuli during the execution of the educational system, so deaf students can interact with these solutions without being exposed to excessive number of different stimuli.

6 Conclusion

In this article, an overview of main variables involved in the creation of educational systems for deaf students was given. Furthermore, it was also given an overview of strategies and best practices to develop educational solutions for deaf. The focus is to simplify and support the build of innovative inclusion systems that offer a truly meaningful learning environment, and can be applied in researches areas like augmented and virtual reality, natural user interfaces, and gamification.

In this context, the guidelines proposed by this work will help developers and educators when planning and designing educational solutions involving deaf students as users. The strategies presented enable social and cultural alignments of these systems to the deaf culture, so the learning process can be enhanced by technology.

As future work, there is a plan to measure the impact of these strategies in educational games for children and teenagers, by evaluating the educational efficiency of these games when interacting with the target audience. Also, with the increase of new natural immersive interfaces, enhancements of the current practices and the proposal of new ones can be designed, so these educational applications would be adjusted according students' knowledge, resulting in the evolution of the learning process in academic environments.

References

1. Capovilla, F.C., Raphael, W.D.: Dicionário Enciclopédico Ilustrado Trilíngüe da Língua de Sinais Brasileira: Sinais de A a L, vol. 1. EdUSP (2001)
2. Carter, E.W., Asmus, J., Moss, C.K., Biggs, E.E., Bolt, D.M., Born, T.L., Brock, M.E., Cattey, G.N., Chen, R., Cooney, M., et al.: Randomized evaluation of peer support arrangements to support the inclusion of high school students with severe disabilities. Except. Child. **82**(2), 209–233 (2016)
3. Cole, M.: Education, Equality and Human Rights: Issues of Gender', Race', Sexuality. Routledge, Disability and Social Class (2011)
4. Dias, R.: Língua Brasileira de Sinais : Libras. Pearson Education do Brasil (2015)
5. Figueira, A.d.S.: Material de apoio para o aprendizado de Libras. Editora Phorte (2011)
6. Garzotto, F., Gonella, R.: Children's co-design and inclusive education. In: Proceedings of the 10th International Conference on Interaction Design and Children, pp. 260–263. IDC '11, ACM, New York (2011). http://doi.acm.org/10.1145/1999030.1999077
7. Hockenbury, D.H., Hockenbury, S.E.: Discovering Psychology. Macmillan (2010)
8. Honora, M., Frizanco, M.L.E.: Livro ilustrado de Língua Brasileira de Sinais: desvendando a comunicação usada pelas pessoas com surdez. Ciranda Cultural (2011)
9. Kelly, B., Lewthwaite, S., Sloan, D.: Developing countries; developing experiences: approaches to accessibility for the real world. In: Proceedings of the 2010 International Cross Disciplinary Conference on Web Accessibility (W4A), pp. 3:1–3:4. W4A '10, ACM, New York (2010). http://doi.acm.org/10.1145/1805986.1805992
10. Kelman, C.: Significação e aprendizagem do aluno surdo. MARTINEZ, AM TACCA, MCVR (Orgs.) Possibilidades de Aprendizagem: ações pedagógicas para alunos com dificuldade e deficiência. Alínea, Campinas, pp. 175–207 (2011)
11. MacDonald, M.: Your Brain: The Missing Manual: The Missing Manual. O'Reilly Media, Inc. (2008)

12. Madeira-Coelho, C.M.: Linguagem, fala e audição nos processos de aprendizagem: desafios e estratégias. Possibilidades de aprendizagem: ações pedagógicas para alunos com dificuldade e deficiência. Editora Alínea, Campinas (2011)
13. Newman, L., Browne-Yung, K., Raghavendra, P., Wood, D., Grace, E.: Applying a critical approach to investigate barriers to digital inclusion and online social networking among young people with disabilities. Inf. Syst. J. **27**(5), 559–588 (2017). https://onlinelibrary.wiley.com/doi/abs/10.1111/isj.12106
14. Perlin, G.: A cultura surda e os intérpretes de língua de sinais. Educação Temática Digital **7**(2), 136–147 (2006)
15. Pontes, H.P., Duarte, J.B.F., Pinheiro, P.R.: An educational game to teach numbers in brazilian sign language while having fun. Comput. Hum. Behav. (2018). http://www.sciencedirect.com/science/article/pii/S0747563218305892
16. Quadros, R.M., Cruz, C.R.: Língua de Sinais : Instrumentos de Avaliação. ArtMed (2011)
17. Rodrigues, C.H., de Quadros, R.M.: Diferenças e linguagens: a visibilidade dos ganhos surdos na atualidade. Rev. Teias **16**(40), 72–88 (2015)
18. Schmengler, A.R., Freitas, S.N., de Oliveira Pavão, S.M.: Acessibilidade no atendimento educacional de alunos público-alvo da educação especial em uma classe hospitalar do estado do rio grande do sul. Práxis Educativa (Brasil) **13**(1), 128–144 (2018)
19. World Health Organization.: Deafness and hearing loss site (2019). https://www.who.int/news-room/fact-sheets/detail/deafness-and-hearing-loss

Unpacking the Privacy Paradox for Education

Lorayne Robertson and Bill Muirhead

Abstract While people claim that they want to control access to their personal information, the ease and convenience of online applications put their personal data at risk with a keystroke. The sale of personal information and subsequent behavioral tracking happens in invisible ways not fully understood by most end users. Younger online participants, such as students, who are accustomed to sharing information online, may not be aware of digital footprints or digital permanence. Similarly, student users may be unaware of risks to participating online at school or how to manage these risks. Selected countries and international organizations have offered policy solutions that, when analyzed expose policy gaps and vulnerabilities. The authors unpack the issue of digital privacy for schools with a three level approach to the research. First, they report on what is known about the identified risks to students of vulnerable ages who access online programs in schools and reported levels of awareness of these risks. Next, the authors disaggregate privacy issues related to the paradox or tradeoff of trust versus convenience in click-through agreements with providers. Third, the authors analyze national and international policy approaches designed to protect the digital privacy of students and offer recommendations for moving forward.

1 Introduction

This paper looks at the paradox of privacy when security is weighed against convenience, and considers this paradox in light of the increasing use of technology by teachers and school districts. Today more than 4 billion people are *internet users* [1] representing 53% of a global population of 7.7 billion people, indicating that internet users are now the majority population. This global movement to online is reflected in schools, although implementation has been slower in some jurisdictions. The majority of students who attend schools in developed countries have access to

L. Robertson (✉) · B. Muirhead
Ontario Tech University, 11 Simcoe St. North, Oshawa, ON L1G4R7, Canada
e-mail: lorayne.robertson@uoit.ca

A. Visvizi and M. D. Lytras (eds.), *Research & Innovation Forum 2019*,
Springer Proceedings in Complexity,
https://doi.org/10.1007/978-3-030-30809-4_3

the Internet [2]. There has been a recent surge in technology uptake in US schools due to the affordability and flexibility of smaller laptop computers such as Chromebooks and iPads for education and cloud-based storage [3, 4]. Present estimates are that over 30 million American students use Google apps for education or the G-suite [5]. App development is exponential with more than 2 million apps available for both Android and Apple [6].

There is also growing awareness that Information and Communication Technologies (ICT's) are helpful to students in schools [4]. Research in Europe indicates that ICT's in schools provide benefits for student learning, positively impacting student motivation, promoting more student-centered learning approaches, and broadening the teaching and learning strategies in use in the schools [7]. The US government promotes technology use in schools, noting that it increases the tools for learning, expands the course offerings, accelerates students learning, and increases student motivation and engagement [8]. Some US research finds that a 1:1 student to computer ratio, properly implemented, also positively impacts student learning outcomes [9].

Historically, the cost of equipment has affected the growth of ICT's in schools [4]. As more affordable devices emerged, this was reflected in lower learner-to-computer (LCR) ratios. Gradually, schools began to introduce laptops, placing more portable devices into students' hands. With the availability of wireless technology, schools began to introduce one-to-one (1:1) computing, or Bring Your Own Device (BYOD) programs. As issues with bandwidth sufficiency resolve and ubiquitous internet access in schools happens, technology use in schools increases [3, 4, 8]. The introduction of cloud-based technology services allows schools more choice in applications that can be tailored to more closely match student and program needs in a cost-effective way [4].

The trend to connect more students to individual devices has taken hold in multiple countries. The LCR is lowest in Swedish schools at 2:2, while Denmark, Estonia, Norway and Spain have a reported LCR of 3:1 [10]. Canadian schools have had access to the internet for the past 15 years, and the LCR at that time was 5:1 [11]. With the advent of more affordable devices the LCR will predictably drop. Current global data on connectivity indicates that Canada and other countries are slowly but steadily closing the urban/rural gap for broadband access, increasing internet availability for more schools [10]. Internet usage continues to increase. More than 80% of the population is online in 38 countries globally [12]. Increasingly, users are school-aged children and adolescents who spend their out-of-school time online. This shift took place over the past decade; 99% of Canadian adolescents surveyed have internet access [13]. Against this backdrop of ever-increasing online participation, digital privacy concerns emerge. We document digital privacy risks for students and present a user framework capturing the paradoxical nature of digital disclosure. We also analyze some international policy responses and suggest ways forward.

2 Digital Privacy

Imagine the following scenario from an earlier era... *Children swim at a seaside resort and parents take photos with an analog camera. Physical copies of these photos are stored in boxes. Only family can access the photos and family easily controls the access. Even a lost photo cannot be connected to the child, providing a sense of security and safety. Later in life, nothing connects these childish images to an adult who wants his childhood to remain private.*

This is the backdrop for understanding the significance and speed of the changes that impact student digital privacy in an era of internet access ubiquity. Privacy in the past was about who had access and who had control over someone else's personal information. We define digital privacy in a broader sense, as *an expectation of privacy unless the user has given consent that includes an awareness of the risks associated with online services and individual control over the collection, distribution and retention of personal information*. The expectation of privacy means that, unless someone has granted another person or organization access to their information, they can reasonably expect that their private information will be held private. People want the right to decide who has access to their information through mechanisms such as *consent for access*, or *waiving the right to privacy*.

There are dominant players who profit from information sharing and selling associated with online participation. Stoddart, Canada's former Privacy Commissioner, likens online tracking to *someone following you around the shopping mall and tracking where you shop* [14]. Users online leave a *digital footprint*, which is a trail of data including websites visited, emails sent and information given to access online services [15]. Trackers are typically third parties who participate in *behavioral tracking* without the consent of the end user [14]. This information tracking is known by other names such as *list brokering* or *personalization* [16]. (See https://epic.org/privacy/profiling/ for a more extensive list of information collected.) Collecting information for list brokerage infringes on the privacy rights of populations who are unaware that their personal information is for sale [16]. Youth generally are not aware that their online activities create a *digital dossier* and that they also may be information brokers who provide data about friends [17]. Supervising adults may not realize that digital dossiers can be sold without alerting end users [18].

There is also the issue of the longevity of the data collected with permanency. Solove [19] explains that, even when someone is not at the home computer, digital dossiers become deeper and longer through the tracking of purchases, credit cards, bank cards, and loyalty cards. Again, there is no overt invasion of privacy, but simple, innocuous steps that add to personal digital dossiers of increasing length and permanence. Tracking companies construct massive data bases that connect to civic and government databases with information about your property taxes or the market assessment of your home, to personal information such as race, gender, political affiliation and income, and your online and shopping habits. Members of the public may not be aware of their dossier until they are targeted for specific marketing or refused a job or a loan; however, they contribute hourly to an industry that makes billions of dollars per year. Solove claims that three major companies hold dossiers on every American adult through hidden devices that collect clickstream data, set cookies to monitor web downloads, or using web bugs attached to emails; one of these three is Equifax [19].

In the foreseeable future, unknown numbers of smart objects will also collect data and communicate information with or without the awareness of end users. Closed circuit television (CCTV) cameras and Global Positioning systems (GPS) presently track movement in vehicles and through location tracking on devices. The Internet of Thing (IoT) is growing daily, and will track more information about our whereabouts and habits. This change has taken place over the course of a generation, creating what we suspect may be inter-generational differences in approaches with respect to expectations of digital privacy.

3 Risks to Students Through Third Party Apps

Recently, an international privacy collaborative [20] conducted a sweep of educational websites under the theme of *user control over personal information* in order to identify privacy risks for students. The goal was to document transparency practices of online educational services. The sweep examined whether or not educational services globally were informing educators and students about how they collected, used or disclosed their personal information and how much control the educators and students could exercise over that information. They found that, in almost all cases, teachers had to provide their name and their email address in order to create an account. Sometimes, they were also required to provide the name of their school. The sweep also found that, "Most but not all" of the online services *required* students to log in using their name and email address. For younger children, their parent's email address was required. Some applications allowed the creation of a user profile and image, but this type of personalization was discouraged [20]. In other words, in order to access online educational services, children and their teachers had to contribute to the online collection of their digital dossiers. There were other, important findings from the GPEN sweep [20]. Some applications and some teachers employed privacy-protective practices, however, only one in three of these online educational services permitted the teacher to create a virtual classroom and assign code names for the students (a privacy-protective practice). The creation of pseudonyms for students allows them to access online learning without providing personal information [20].

In general, the GPEN sweep found that many free educational applications on mobile devices require students to provide personal information in order to access them. When educational services ask students to provide their social media login identification, this allows their information to become linked across the web. Most of the online educational services set cookies to track the browsing histories and preferences of students [20]. These findings raise important issues about the risks to students' digital privacy requiring practical and policy responses, and they have implications for educational app providers as well as parents, teachers and schools. There is a need for clearer guidelines and best practices to guide online learning.

4 Theorizing the Privacy Paradox

Earlier attempts to theorize the paradoxical nature of privacy decisions have shown that there are contradictions in people's purchasing habits despite the risk that comes with online shopping, for example [21, 22]. Online users are more or less aware that there is a degree of digital tracking of their transactions but the convenience provided by the digital tools or access outweighs their concern or their intention to protect privacy. Despite wanting to protect privacy, they act in ways that do not protect privacy and hence, this *privacy paradox* [22]. Because of this finding, researchers are reluctant to see policy as the full solution to the deterioration of privacy control by the consumer [22]. Within education, some research has indicated that teachers care about privacy, but they are also not aware of how to protect student privacy or how to encourage students to protect their privacy [23, 24].

Privacy is often understood in terms of applications, infrastructure and risk associated with the use or unintended use of personal data. This orientation to risk as the loss of data omits decisions about what technologies, services and personal choices including beliefs, inform decisions made by individuals, groups or institutional services that collect data. While services are often seen as software and applications, increasingly, corporations are developing hybrid infrastructures that bundle services with specific hardware such as Google with Chrome Books and a G-Suite of applications for schools and students, or Apple with Ipads and iCloud or Amazon with their set of learning solutions including Amazon-Inspire, LMS and Amazon AWS (web services that run Cloud applications).

The framework below applies the concept of convenience, ease of use, and implementation to the a priori decisions that influence concepts of privacy. Recent data breaches and intrusions into corporate accounts have established some risks inherent in online participation and the erosion of trust concerning the external collection of personal data. Recent data breaches have alerted the general public to risks of disclosing their data to service providers.

As well, recent breaches regarding data have brought to light the data that are collected by large corporations such as Facebook when using its services. Disclosure of data by Facebook to Cambridge Analytica is now an infamous incident, as well as the individual profiling and targeting of users in the recent America election cycle. Google has also been highlighted for collecting personal data and data mining users' activities when using its search function, its Chrome Browser and its ubiquitous email service (Gmail) under the rationale of better serving its customers/users. Google's expansion into educational contexts with its G-Suite for students and its low cost Chromebook hardware has further placed Google at the center of the educational enterprise in many locations in the US and Canada. Even where Chromebooks are not provided to students, the availability of G-Suite Services to schools at no cost has

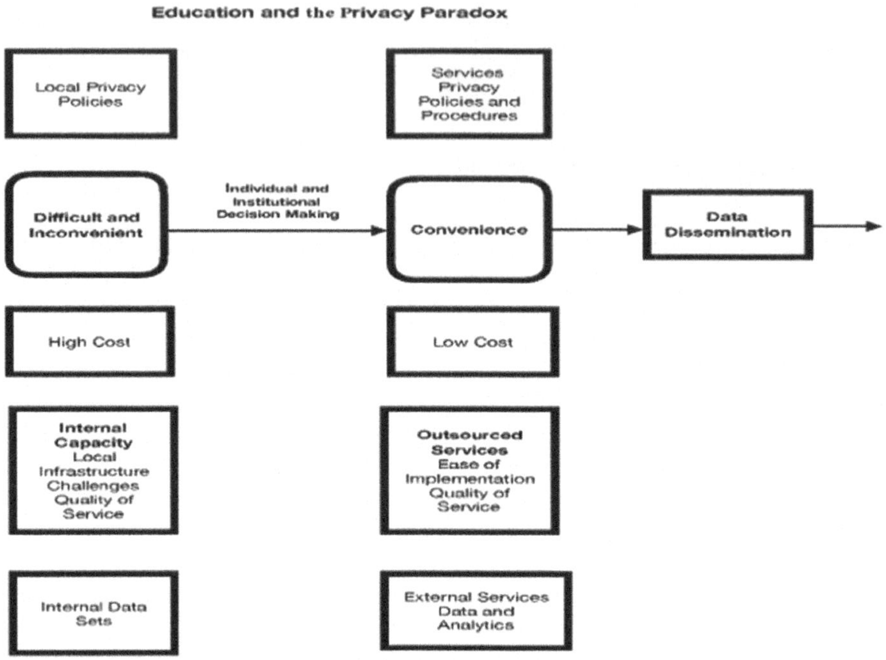

Fig. 1 Education and privacy paradox

made its use pervasive. While teachers have spoken and written about the benefits of such tools to enhance learning [3], less has been written about the educational paradox of convenience vs privacy in educational settings.

A complex picture emerges. Following an initial reluctance to implement technology based on costs, the desirability and rise of 1-1 programs has created a situation where more students go online in school. Teachers want to use technology but are not fully understanding the implications of internet-based and cloud-based technologies [23, 24]. While 1:1 programs and BYOD have grown in schools, anecdotal reports indicate a continuing reluctance to fully integrate technology because not all students have equitable access to digital hardware and software. In our view, the combination of sluggish implementation of technology in schools followed by rapid implementation creates a complexity that has constrained policy development on digital privacy for schools (see Fig. 1).

5 Policy Responses

Multiple policy responses across the globe are designed to protect privacy such as the General Data Protection Regulation [25] in Europe. It allows for the collection of personal data by providers only under strict conditions. In Canada, the *Personal Information Protection and Electronic Documents Act* (PIPEDA) [26] requires private sector organizations to respect privacy law. PIPEDA promotes eight key principles for fair information practices for service providers, which include notice (informing users when information is collected); choice (allowing users to opt in); protection (from unauthorized access); and limits of the scope and purposes for how information can be collected [26].

Unlike the approaches of Canada and Europe to address legislation to supervise commercial interests on the internet, the US approach has been directed toward child protection. The *Children's Online Privacy Protection Act* (COPPA) [27] requires that children must be 13 to have an online profile, although there is some debate that the approach instead should be to empower students to protect their own privacy [28]. The US also has enacted the *Children's Internet Protection Act* (CIPA) [29] and the *Family Educational Rights and Privacy Act* [30] intended to protect children from harmful internet content or data exposure.

Policy responses globally are scattered, likely due to the nature and pace of technological innovation [4] and the reality that policies are not the full solution [15, 23, 24]. Risk-abatement researchers question the efficacy of age restriction criteria for social media sites. While Facebook has set the age of 13 as a requirement for enrolment, more than 1/3 of Canadian students ages 10–12 report that they are already have a Facebook account [13]. US physicians report that young people's constant online presence increases their risk from oversharing information or posting incorrect information about others [31]. Multiple studies indicate that students lack understandings about *digital permanence* and cannot see the future implications of what they post online today [17, 32]. More co-ordination is needed to protect students' personally-identifiable information (PII). The issue of who is responsible for supervising student's online access and protecting their data is complex. While safeguards such as firewalls can serve a purpose in school districts, they are an incomplete solution. Despite education's potential through universal access to teach data protection competencies, a Canadian study recently reports that 41% of students in grades 4–11 learn about privacy settings from their parents and 15% report that they learn it from their schools [13]. The view of the Canadian Office of the Privacy Commissioner is that solutions to digital privacy issues will require both policies and education for the end users [33].

6 Risk Abatement Through Education

One global collective has developed principles for educating end users in schools. The International Working Group on Digital Education [34] has a framework that encourages schools globally to be more digitally responsible and civic-minded. Their framework is designed to work with all countries' data protection policies. They advocate common sense principles, beginning with the premise that privacy is a human right that needs to be protected. Students are encouraged to work with their parents and gain consent for online activities. Students should understand the nature of personal data, and how this can change from individual to individual. Students also need to be aware of their digital footprint and be respectful of their own privacy and the privacy of other people. One of the goals of this framework is to help students understand the key players who commercialize data through the offer of free services. Students should know who is collecting information and how long their data can be shared. Students should know how to monitor their own digital dossier online [34]. These principles can serve to guide schools and governments to educate all end users, and in particular, students in schools.

7 Conclusion

A report designed to restore Canadian's confidence in Canada's privacy regime [33] contains a warning that, without improving how privacy is protected online, Canadians could have insufficient trust to enable the digital economy to grow, and this could compromise the ability of Canadians to benefit from innovation. This national report considers multiple forms of risk abatement solutions for children, adolescents, and teachers who pursue online learning. These solutions fall into two categories: increasing the accountability of commercial enterprises through *policy compliance*, and a focus on *educating the end users of technology*. For example, this report recommends that students should learn about privacy at a young age, and learn how to "de-identify" or anonymize their personal information. These new directions, if taken seriously, have implications for curriculum policies in schools worldwide.

As more students learn online, countries need to consider universal principles of privacy protection that include rights and responsibilities for students as well as limits for corporations. Students need to know how their online presence is being traced and that they should exercise care selecting what is put online because it is permanent. They should learn that privacy has value and is worthy of consideration as a human right, and that their actions can put their own privacy at risk as well as the privacy of others. They should understand the tradeoff of privacy for convenience. Most importantly, students should be encouraged to work with adults for supervision of their online activities until they are old enough to appreciate the risks and consequences of online participation. In our analysis, privacy protection for education requires both end user empowerment and comprchensive policy solutions.

References

1. Internet usage and social media statistics (2019). Retrieved at http://www.internetlivestats.com
2. Trucano, M.: How many schools are connected to the Internet? EduTech. Retrieved at http://blogs.worldbank.org/edutech/how-many-schools-are-connected-Internet/ (2014)
3. Ahlfeld, K.: Device-driven research: the impact of chromebooks in American schools. Int. Inf. Libr. Rev. **49**(4), 285–289 (2017). https://doi.org/10.1080/10572317.2017.1383756

4. Pierce, G.L., Cleary, P.F.: The K-12 educational technology value chain: apps for kids, tools for teachers and levers for reform. Educ. Inf. Technol. **21**(4), 863–880. Retrieved at https://link.springer.com/article/10.1007/s10639-014-9357-1 (2016)
5. Singer, N.: How Google Took Over the Classroom. Retrieved from https://www.nytimes.com/2017/05/13/technology/google-education-chromebooks-schools.html?ref=dealbook&target=comments&_r=1 (2017)
6. Number of apps available in leading app stores as of 3rd quarter 2018. Retrieved at https://www.statista.com/statistics/276623/number-of-apps-available-in-leading-app-stores/
7. Balanskat, A., Blamire, R., Kefala, S.: The ICT impact report. Eur. Schoolnet **1**, 1–71. Retrieved at http://colccti.colfinder.org/sites/default/files/ict_impact_report_0.pdf (2006)
8. US Department of Education.: Use of Technology in Teaching and Learning. Retrieved from https://www.ed.gov/oii-news/use-technology-teaching-and-learning (n.d.)
9. Greaves, T.W., Hayes, J., Wilson, L., Gielniak, M., Peterson, E.L.: Revolutionizing Education Through Technology: The Project RED Roadmap for Transformation. International Society for Technology, Eugene (2012). in Education
10. Partnership on Measuring ICT for Development.: Final WSIS Targets Review: Achievements, Challenges and the Way Forward. International Telecommunication Union (2014): Geneva. Retrieved at https://www.itu.int/en/ITU-D/Statistics/Pages/publications/wsistargets2014.aspx (2014)
11. Ertl, H., Plante, J.: Connectivity and Learning in Canada's Schools: Science, Innovation and Electronic Information Division. Connectedness Series, 11. Statistics Canada. Retrieved at https://www150.statcan.gc.ca/n1/pub/56f0004m/56f0004m2004011-eng.pdf (2004)
12. Internet Users by Country.: Internetlivestats.com. Retrieved at http://www.internetlivestats.com/internet-users-by-country/ (2016)
13. Steeves, V.: Young Canadians in a Wired World. Phase III: Life Online. MediaSmarts, Ottawa, Canada. Retrieved at http://mediasmarts.ca/sites/mediasmarts/files/pdfs/publication-report/full/YCWWIII_Life_Online_FullReport.pdf (2014)
14. Stoddart, J.: Respecting Privacy Rights in the World of Online Behavioural Advertising. Retrieved at https://www.priv.gc.ca/en/opc-news/speeches/2011/sp-d_20111206/ (2011)
15. Robertson, L. P., Leatham, H., Robertson, J., Muirhead, B.: Digital privacy across borders: Canadian and American perspectives. In: Emerging Trends in Cyber Ethics and Education, pp. 234–258. IGI Global (2019)
16. Electronic Profiling Information Center (EPIC).: Retrieved from https://epic.org/privacy/profiling/ (n.d.)
17. Berson, I.R., Berson, M.J.: Children and their digital dossiers: lessons in privacy rights in the digital age. Int. J. Soc. Educ. **21**(1), 135–147 (2006)
18. Robertson, L., Muirhead, B., Leatham, H.: Protecting students online: international perspectives and policies on the protection of students' digital privacy in the networked classroom setting. In: 12th International Technology, Education and Development (INTED) Conference, pp. 3669–3678. Valencia, Spain, 5–7 Mar 2018. doi: https://doi.org/10.21125/inted.2018.0705
19. Solove, D.J.: The Digital Person: Technology and Privacy in the Information Age. NyU Press (2004)
20. Global Privacy Enforcement Network (GPEN).: GPEN Sweep 2017: User Controls Over Personal Information. Retrieved at https://www.privacyenforcement.net/sites/default/files/2017%20GPEN%20Sweep%20-%20International%20Report.pdf (2017)
21. Miyazaki, A.D., Fernandez, A.: Consumer perceptions of privacy and security risks for online shopping. J. Consum. Aff. **35**(1), 27–44 (2001)
22. Norberg, P.A., Horne, D.R. and Horne, D.A.: The privacy paradox: Personal information disclosure intentions versus behaviors. J. Consum. Aff. **41**(1), 100–126 (2007)
23. Leatham, H.: Digital Privacy in the Classroom: An Analysis of the Intent and Realization of Ontario Policy in Context (Master's Thesis). Retrieved from https://ir.library.dc-uoit.ca/xmlui/handle/10155/816 (2017)
24. Leatham, H., Robertson, L.: Student digital privacy in classrooms: teachers in the cross-currents of technology imperatives. Int. J. Dig. Soc. (IJDS) **8**(3). (2017). ISSN: 2040-2570 (Online) http://infonomics-society.org/ijds/

25. European Commission: Justice.: Protections of Personal Data. Retrieved from https://ec.europa.eu/info/law/law-topic/data-protection_en (n.d.)
26. Personal Information Protection and Electronic Documents Act (PIPEDA).: Retrieved from https://www.priv.gc.ca/en/privacy-topics/privacy-laws-in-canada/the-personal-information-protection-and-electronic-documents-act-pipeda/ (2016)
27. Federal Trade Commission.: Complying with COPPA: Frequently Asked Questions. Retrieved from https://www.ftc.gov/tips-advice/business-center/guidance/complying-coppa-frequently-asked-questions (2015)
28. Boyd, D., Hargittai, E., Schultz, J., Palfrey, J.: Why parents help their children lie to Facebook about age: unintended consequences of the 'children's online privacy protection Act'. First Monday, **16**(11). Retrieved from http://journals.uic.edu/ojs/index.php/fm/article/view/3850/3075 (2011)
29. Federal Communications Commission. Children's Internet Protection Act (CIPA).: Retrieved at https://www.fcc.gov/consumers/guides/childrens-internet-protection-act (2000)
30. FERPA.: Family Educational Rights and Privacy Act of 1974, Pub. L. 93-380, title V, Sec 513, Aug. 21, 1974,88 Stat. 571. Retrieved from https://www2.ed.gov/policy/gen/guid/fpco/ferpa/index.html (1974)
31. O'Keeffe, G. S., Clarke-Pearson, K.: The impact of social media on children, adolescents, and families. Pediatrics **127**(4), 800–804. Retrieved from http://research3.fit.edu/sealevelriselibrary/documents/doc_mgr/1006/O'Keeffe_and_Pearson._2011._The_Impact_of_Social_Media_on_Children,_Adolescents,_and_Families.pdf (2011)
32. Palfrey, J. G., Gasser, U., Boyd, D.: Response to FCC notice of inquiry 09-94: Empowering Parents and Protecting Children in an Evolving Media Landscape. Retrieved from https://papers.ssrn.com/sol3/papers.cfm?abstract_id=1559208 (2010)
33. Office of the Privacy Commissioner of Canada (OPC).: Real fears, real solutions: a plan for restoring confidence in Canada's privacy regime. 2016–17 Annual Report to Parliament on the Personal Information Protection and Electronic Documents Act and the Privacy Act. Retrieved from https://www.priv.gc.ca/media/4586/opc-ar-2016-2017_eng-final.pdf (2017a)
34. International Working Group on Digital Education (IWG).: Personal data protection competency framework for school students. In: International Conference of Privacy and Data Protection Commissioners. Retrieved from https://icdppc.org/wp-content/uploads/2015/02/International-Competency-Framework-for-school-students-on-data-protection-and-privacy.pdf (2016)

21st Century Skills. An Analysis of Theoretical Frameworks to Guide Educational Innovation Processes in Chilean Context

Ignacio Salamanca Garay and María Graciela Badilla Quintana

Abstract Society faces multiple challenges in this millennium and education should promote skills that allow students and teachers to face them. Students will have to be prepared for jobs that have not yet been created, for technologies that have not yet been invented, to solve problems that have no yet been anticipated. As educators will be a shared responsibility to take advantage of the opportunities and find solutions. In this context, international organizations linked to educational change are rethinking the role of the current education. One initiative is the creation of frameworks to promote and develop skills for the 21st century which are defined as a set of skills necessary to interact in today's world, mainly in study and employment, where the management of technology is crucial. In Chile, initiatives in this area are being foreshadowed and it is necessary to define a reference framework that complement this process. The aims of this paper is to propose a frame of reference of skills for the 21st century that synthesizes the diversity of referents to have a useful referent to guide educational innovation processes in our Latin American context. Based on a comparative methodology, it was applied a descriptive-comparative analysis of frameworks formulated by international organizations that promote these skills. Theoretical analysis results show concordance in five categories: Cognitive, Social, Emotional, Information and Communication Technologies (ICT), and Curricular Contents for the 21st century. With this background, we are able to present a referential framework maintaining the five categories, and adding 38 essential skills.

I. S. Garay (✉) · M. G. B. Quintana
Facultad de Educación, Centro de Investigación en Educación y Desarrollo (CIEDE), Universidad Católica de la Santísima Concepción, 4090541 Concepción, Chile
e-mail: isalamanca@doctoradoedu.ucsc.cl

M. G. B. Quintana
e-mail: mgbadilla@ucsc.cl

A. Visvizi and M. D. Lytras (eds.), *Research & Innovation Forum 2019*, Springer Proceedings in Complexity,
https://doi.org/10.1007/978-3-030-30809-4_4

1 Introduction

One of the challenges of the knowledge society lies in the promotion of an educational system that can solve the needs and emerging problems of S21. This has become a subject of debate for some authors [2, 8, 9], arguing that the central knot of this problem arises from the decontextualization between the constitutive elements of the educational system (curriculum, methodologies and attitudes) conceived under the modern educational paradigm and the new subject that needs to be educated so that it develops effectively and efficiently in the knowledge society [1–3, 5, 7, 8, 19]. Today it is suggested to rethink the curriculum, putting the accent on the development of other skills that will be fundamental to develop in a globalized world dominated by information and its transformation into knowledge. What are these skills? The so-called skills for the 21st century (hereinafter HS21) [13], These are defined as "a set of skills necessary to function in today's world, mainly in study and employment, where the management of technology is central" [6, p. 3]. That is why various international organizations linked to the improvement of education are promoting them through referential frameworks. In the Latin American and Chilean context, the work in this line is at a starting point, and it is imperative to define and systematize a new referential framework that promotes this type of skills and helps in future educational innovation initiatives.

The first part of the document describes the main problems and challenges that the student will face in 21st century society (hereinafter S21), laying the foundations for the promotion and development of the HS21 in the world; in addition, the most influential international organizations linked to the educational change and innovation that work in the promotion of this skills are described.

In the second part a descriptive-comparative analysis of the 4 most influential referential frameworks of HS21 were conducted.

The third part synthesizes the main conclusions of the analysis carried out and proposes a new referential framework for the promotion of the HS21 in a Chilean context.

2 Background

2.1 The Main Problems and Challenges of S21

Today's students will be the citizens of tomorrow and will have to face various challenges and problems. They will have to develop in a complex and dynamic world [4]. Within the challenges to come we can find: (i) the disproportionate growth of the world population that brings with it the challenge of housing, work and food; (ii) the environmental pollution caused by plastic and global warming that will be seen in the melting of the poles and the increase in the level of the oceans, the extinction of various terrestrial and marine flora and fauna; (iii) the increase of automation in the industry that will leave millions of people without work and will push the labor force to specialize in knowledge production [12, 15, 16]; (iv) the external and internal warlike confrontations of nations that will lead to uncontrolled immigration of large population volumes; (v) the globalization and expansion of the internet into all areas of our lives; (vi) cybercrime among others.

2.2 Organizations that Promote HS21

The most influential international organizations that promote the incorporation of HS21 in education systems are: the Organization for Economic Cooperation and Development (OECD) is generating a project called Education 2030: The Future of Education and Skills Project, which aims to help countries to know the skills that will be needed in 2030 and how to develop them [11]. Likewise, the United Nations Organization for Education, Science and Culture (UNESCO) in its last assembly E2030: Education and skills for the 21st century, held in Argentina in January 2017, the ministers of education from several nations said they have a great challenge in this issue, and one of the first actions will be to modify the educational structures to promote in students the tools to perform satisfactorily in S21 [17, 20]. In the United States there is the Partnership for 21st Century Skills (P21), which since 2009 has proposed a set of initiatives to align traditional learning with the reality of the world of work; creates and proposes a systemic and holistic conceptual framework to place the HS21 in the US educational system and its public policies [14] and finally, Assessment and Teaching of 21st Century Skills (ATC21S), a public-private project led by the University of Melbourne in Australia in partnership with Cisco, Microsoft and Intel and participation of other countries such as: Australia, Finland, Singapore, the United States, Sweden and Costa Rica where the goal is to incorporate the HS21 in classrooms and promote collaborative problem solving and ICT literacy. In Chile, the Ministry of Education (MINEDUC) is currently implementing an educational innovation center that is incipiently promoting the incorporation of the HS21, seeking to develop and scale innovative solutions for the learning of all students [10].

3 Methodology

3.1 Aim and Investigation Questions

The aim is to propose a new referential framework of HS21 that synthesizes the diversity of existing frameworks by establishing new categories and skills to help complement the processes of educational innovation that are taking place in Chile. The research questions are:

i. What are the most influential frameworks of HS21 internationally?

Table 1 Frequency of categories and skills referential frameworks (*Source* Own preparation)

Framework	Categories	Skills
OECD	4	25
UNESCO	4	27
P21	4	17
ATC21S	4	14

ii. What are the main categories in which the referential frames of HS21 agree?
iii. What are the main skills in which the referential frames of HS21 agree?
iv. Is it possible to establish points of union between the categories and skills and generate a new frame of reference for HS21?

3.2 Descriptive—Comparative Analysis of 4 Referential Frameworks HS21

As explained above, there are 4 influential organizations that encourage HS21: (i) Education 2030: The Future of Education and Skills Project (OECD), (ii) The four pillars for the education of the future (UNESCO) [18], (iii) Partnership for 21st Century Skills (P21) and (iv) Assessment and Teaching of 21st Century Skills (ATC21S) establishing the frequencies of the categories and skills that make them, each one of them were compared, and agreements and disagreements one to another were established (Table 1).

Table 2 presents the description of the main categories and skills found in the four reference frameworks studied.

From the different referential frames exposed, we can infer the following concordances:

(1) The organizations that promote HS21 consider that S21 faces today, and will face in the future (with greater force) various global problems such as: pollution, migration, use and abuse of Information and Communication Technologies and global citizenship, this is why the students of today and future citizens of tomorrow will have to solve them. That is why the various educational systems must update and focus their curriculums and methodologies to meet these demands in the training of their students.
(2) The organizations examined can demonstrate a consensus in determining categories to group the different HS21 within which we can identify the types: cognitive, social, personal or emotional, informational, digital and curricular contents for S21.
(3) In the cognitive category, the most recurrent skills are: creative thinking, critical thinking, metacognition and systemic or holistic forms of thinking.

Table 2 Description of the referential frameworks HS21 (*Source* Own preparation)

Organization	Categories	21st century skills
P21	For personal and professional life	Flexibility and adaptability, initiative and self-direction, social and transcultural skills, productivity and responsibility, leadership and responsibility
	Learning and innovation	(4C) Creativity and innovation, critical thinking and problem solving, communication and collaboration
	Information and communication technologies	Information literacy, media literacy and ICT literacy
	Basic topics and themes of 21st century	English, world languages, letters, mathematics, economic science, science, geography, history, government and civic education, global awareness, financial, economic, commercial and business literacy, civic literacy, healthy living and environmental literacy
ATC21S	Ways of thinking	Creative thinking, critical thinking and meta cognition
	Ways to work	Communication and collaboration
	Tools to work	Information literacy and ICT literacy
	Ways to live in the world	Local and global citizenship, life and career and social and personal responsibility
UNESCO	Learning to know in 21st century	Global awareness, literacy (finance, economy, business and entrepreneurship), civics, health and well-being
	Learning to do in 21st century	Critical thinking, problem solving, communication, collaboration, creativity, innovation and ICT literacy
	Learn to be in 21st century	Socialization, interculturality, initiative, autonomy, personal responsibility, production of meaning, meta cognition, entrepreneur mentality and learning to learn
	Learning to live together in 21st century	Search and value diversity, teamwork, civic and digital citizenship, global competence and intercultural competence
OECD	Types of knowledges	Disciplinary knowledge, interdisciplinary knowledge, epistemic knowledge and procedural knowledge

(continued)

Table 2 (continued)

Organization	Categories	21st century skills
	Cognitive and metacognitive skills, social and emotional and physical-practical	Critical thinking, creative thinking, learning to learn and self-regulation, empathy, self-efficacy and collaboration, use of technological devices (hardware and software)
	Personal, local, social and global values and attitudes	Confidence, respect for oneself, trust and virtue, Respect for the environment, respect for diversity, empathy, respect for the environment, for human and animal dignity
	Transformative competences	Creation of new value, reconciliation of tensions and dilemmas, taking responsibility

(4) In the social category we find two subcategories: interpersonal and intrapersonal. In the first, communication skills, collaborative skills, leadership, citizenship, responsibility and tolerance for diversity are highlighted.
(5) In the category of information and communication technologies and digital we find two subcategories: information literacy with search skills, identification, analysis, evaluation, synthesis and creation of information in digital or physical format; computer literacy subcategory expresses software and hardware utilization skills.
(6) The curricular contents for the S21 that stand out are intercultural, focus on law and gender, financial literacy, environmental literacy, democracy and citizenship.

From the different referential frameworks exposed, we can infer the following discrepancies:

(1) The organizations that promote the HS21 define in different ways the categories that make up the referential frames, however they refer to the same things.
(2) In relation to the different skills that make up the referential frames are grouped into different categories without clarity to what specific category correspond, grouping for example: cognitive skills with social skills or personal skills with emotional skills.

As we can see there are great points of agreement between the various referential frameworks and some discrepancies, however, all these serve as a starting point to build a new referential framework that comes to organize the categories, skills and their respective definitions.

The following presents the new reference framework of HS21; to determine the categories and skills, the main concordances of the analysis carried out were used, concluding that the most important ones were 5: cognitive, social, emotional, ICT and XXI century contents and 38 HS21 more frequently found (Table 3).

Table 3 Referential framework of HS21: a proposal for Chilean education (*Source* Own preparation)

Categories	Skills	Definition
Cognitive	Critical thinking	Cognitive process of a rational, reflective and analytical nature, oriented to the systematic questioning of reality
	Creative thinking	Cognitive process that allows generating a large number of different ideas that allow solving a problem of reality
	Meta cognition	Cognitive process that allows to reflect on the way used by oneself to learn something
	Learn to learn	Cognitive process that allows to generate different personal strategies to learn something
	Logical thinking	Cognitive process that allows to generate abstract concepts, establish relationships between them and draw coherent conclusions
	Systemic thinking	Cognitive process that allows to relate different levels of a phenomenon and consider it in its entirety
	Complex thinking	Cognitive process that allows to relate different levels and aspects of a phenomenon and consider it in its entirety
Social	Collaboration	Interactive process in an activity with the mission to specify shared objectives
	Cooperation	Work done in harmony with other people to achieve a goal
	Communication	Social interaction that allows to transmit ideas, feelings or emotions through verbal or non-verbal language
	Empathy	Ability to perceive, share and understand what another person feels, mainly feelings and emotions
	Flexible character	Adaptation to different circumstances, capable of yielding, as opposed to a rigid person
	Respect	Consideration and special assessment that is held by someone or something
	Responsibility	Compliance with the agreements or obligations taken personally or in groups
	Leadership	A person who distinguishes himself naturally from the rest, organizes them and makes favorable decisions for the group
	Tolerance	Acceptance of opinions, ideas or attitudes of other people even if they do not coincide with their own
	Citizenship	Correct way of behaving in society knowing the rights and obligations of it
Emotional	Autonomy	A person's ability to make decisions independently of the opinion or desire of others
	Proactivity	Ability to contribute their own ideas, take the initiative and generate changes and/or achievement of objectives
	Self-control	Ability to master behavior and emotions, being able to remain calm, in unfavorable situations

(continued)

Table 3 (continued)

Categories	Skills	Definition
	Self-efficacy	Confidence in one's own ability to achieve personal or professional goals predisposed
	Self-confidence	Ability to feel confident about the talents and attributes that one has
	Self-care	Ability to carry out favorable activities in favor of oneself
	Resilience	Ability to overcome critical or difficult moments and continue with the life project
	Stress management	Ability to manage and reduce physical and emotional stress in problematic situations
ICT	Literacy in ICT	Mastery of concepts and tools to generate new knowledge through information and communication technologies
	Literacy in media and internet	Mastery of concepts and tools to generate new knowledge through means of communication and internet
	Interpretation and data analysis	Ability to generate relationships, draw conclusions and generate new information
	Programming	Ability to use the principles of computer language and create programs
	Creation of mobile applications	Ability to use the principles of computer language and create applications for mobile telephony
	Robotics and electronics	Ability to use the principles of computer language and create programs that interact with physical elements
XXI century contents	Financial literacy	Mastery of concepts and tools to generate new knowledge in relation to the good use of money
	Interculturality	Building equitable relationships between people, communities or cultures of different countries
	Environmental literacy	Mastery of concepts and tools to generate new knowledge through the care of the environment
	Formulation of the life project	Conceptual, procedural and attitudinal orientations to guide personal actions to people's objectives
	Gender approach	Study of the cultural and social constructions for men and women, what identifies the feminine and the masculine
	Rights approach	Study of international and national standards and promote and defend human rights
	Literacy in ethics and social justice	Mastery of concepts and tools to generate new knowledge about personal and social ethical behavior and appreciation of justice

4 Conclusion and Discussion

Due to the existence of several international reference frameworks on HS21 that undoubtedly are a starting point. The need arose to analyze and compare them in order to order the skills that are confused among each other, moving from one category to another, as well as determining clear categories that would prefigure a frame of reference for our context and thus collaborate with the various initiatives that are being carried out at the national in this area. That is why a model is prefigured that at its center incorporates, categorizes and defines various HS21 that are expected to serve to complement educational initiatives aimed at innovation and educational change.

Going back to our research questions:

What are the most influential frameworks of HS21 internationally?
From the diverse literature reviewed it can be concluded that the most influential frameworks at the international level are those that have an impact on the public policies of the countries that share the principles of educational innovation within which it is determined that the Partnership for 21st Century Skills. (P21) of the USA, the Assessment and Teaching of 21st Century Skills. (ATC21S), Australia, The four pillars for the education of the future. (UNESCO) and Education 2030: The Future of Education and Skills Project (OECD).

What are the main categories in which the reference frames of HS21 agree?
From the different reference frames studied, five main categories have been identified: Cognitive, Social, Emotional, ICT and Curricular Contents for the 21st century.

What are the main skills in which the reference frames of HS21 agree?
We identified a group of 38 skills that are detached from the frames of reference examined.

Is it possible to establish points of union between the categories and skills and generate a new frame of reference for HS21?
As we have seen, there were points of union between the various frameworks examined, thanks to this, a new framework of HS21 could be formulated.

The prefiguration of a framework of skills for the 21st century in education is a reality, however, it can and should be improved. This is why it is necessary not only to identify the categories and define the skills that compose them, but also to determine hierarchically the level of importance between them and thus delimit more the area of action for future implementation in the processes of educational innovation in our country. That is why we are working on a second stage materialized in a survey that aims to measure the degree of importance attributed by the educational communities on the different skills exposed in this referential framework and thus hierarchically order and exclude the various skills that are not they are relevant in the future.

Acknowledgements Thanks to the support of the National Commission for Scientific and Technological Research, CONICYT, Ministry of Education, Chile, through the Doctorate Scholarship Abroad (Beca de Doctorado Nacional, 2017) Granted to Ignacio Salamanca Garay (File number: 21170028) and CIEDE-UCSC.

References

1. Aedo, C., Walker, I.: Skills for the 21st Century in Latin America and the Caribbean. The World Bank, Washington, DC (2012)
2. Bellei, C., Morawietz, L.: La Difícil Introducción de Competencias para el siglo XXI en la educación Chilena. Santiago, Metropolitana, Chile (2016)
3. Brunner, J.: Habilidades del siglo XXI en Chile. Santiago, Metropolitana, Chile (2017)
4. Carneiro, R.: The big picture: understanding learning and meta-learning challenges. Eur. J. Educ. **42**(2), 151–172 (2007)
5. Cobo, C.: Educar Chile. https://www.escuelaenmovimiento.cl/temas-educativos/cristobal-cobo-todos-necesitamos-hoy-las-habilidades-para-el-siglo-xxi. Last accessed 14 Mar 2019
6. Jenson, J., Taylor, N., Fisher, S.: Critical Review and Analysis of the Issue of "Skills, Technology and Learning". Final report, York University (2010)
7. Meller, P.: Una introducción a las habilidades escolares para el siglo 21. Cieplan, Santiago (2016)
8. Meller, P.: Pulso. http://www.pulso.cl/economia-dinero/patricio-meller-educacion-la-reforma-no-apunto-ensenar-las-habilidades-se-necesitan-siglo-xxi/. Last accessed 14 Mar 2019
9. Meller, P.: Fundación Chile.https://fch.cl/patricio-meller-la-metodologia-ensenanza-esta-fallando-todas-las-escuelas-chile/. Last accessed 14 Mar 2019
10. Ministerio de Educación.: http://innovacion.mineduc.cl/acerca-del-centro-de-innovacion/. Last accessed 14 Mar 2019
11. OECD.: Education 2030: The Future of Education and Skills Project. OECD (2018)
12. OIT.: Global Employment Trends 2012: Preventing a Deeper Jobs Crisis. Organización Internacional del Trabajo, Ginebra. www.ilo.org/wcmsp5/groups/public/—dgreports/—dcomm/—publ/documents/publication/wcms_171571.pdf. Last accessed 14 Mar 2019
13. Partnership for 21st Century Skills.: 21st Century Skills, Education & Competitiveness. Tucson: Partnership for 21st Century Skills (2008)
14. Partnership for 21st Century Skills. Framework for 21st century learning. www.21stcenturyskills.org/documents/framework_flyer_updated_april_2009.pdf. Last accessed 14 Mar 2019
15. Redecker, C.: The Future of Learning: Preparing for Change. Oficina de Publicaciones de la Unión Europea, Luxemburgo (2011)
16. Reynolds, R., Tavares, N., Notari, M.: 21st century skills and global education roadmaps. In: Chu, S., Reynolds, R., Notari, M., Taveres, N., Lee, C. (eds.) 21st Century Skills Development through Inquiry Based Learning From Theory to Practice, pp. 17–32. Springer Science, Singapore (2017)
17. Scott, C.L.: El futuro del aprendizaje 2 ¿Qué tipo de aprendizaje se necesita en el siglo XXI? Investigación y Prospectiva en Educación UNESCO, París. Documentos de Trabajo ERF, vol. 14 (2015)
18. UNESCO.: Los cuatro pilares de la educación en La educación encierra un tesoro. Informe de la Comisión internacional sobre la educación para el siglo XXI, pp. 91–103. Santillana/UNESCO, Madrid, España (1996)
19. UNESCO.: Tic en Educación en América Latina y Caribe. UNESCO- Acción Digital, Santiago (2014)
20. UNESCO.: Declaración de Buenos Aires: E2030: Educación y habilidades para el siglo 21. UNESCO, Buenos Aires (2017)

Cloud Platform for Enabling a Student Collaborative Assessment Process

Higinio Mora, María Teresa Signes-Pont, Rafael Mollá Sirvent and María L. Pertegal-Felices

Abstract In this paper, an innovative cloud platform for implementing a collaborative assessment process in higher education is introduced. This platform provides a useful tool to support a peer review method for student assessment. The students are involved in the process of assessment by means of crossed reviews which aim to identify the lacks or mistakes in their works. The suggested improvements will be performed and assessed by the students themselves in successive deliveries through the web platform created. This method provides several advantages in the teaching-learning process for engineering students. The works' review process cultivates the critical thinking and allows to achieve better results. In addition, students are engaged more actively in their educational process which increases their motivation in learning. The system is based on web 2.0 features and enables the collaboration among students and instructors to build learning along this process.

1 Introduction

One of the major challenges in the educational world is the generation of collective intelligence instead of multiple individual intelligences. The development of complex solutions in today's global world requires the collaboration of many people.

H. Mora (✉) · M. T. Signes-Pont · R. M. Sirvent
Department of Computer Technology and Computation,
University of Alicante, Campus de San Vicente del Raspeig, 03690 Alicante, Spain
e-mail: hmora@dtic.ua.es; hmora@ua.es

M. T. Signes-Pont
e-mail: Teresa@dtic.ua.es

R. M. Sirvent
e-mail: rmolla@dtic.ua.es

M. L. Pertegal-Felices
Developmental and Educational Psychology Department,
University of Alicante, Campus de San Vicente del Raspeig, 03690 Alicante, Spain
e-mail: ml.pertegal@ua.es

A. Visvizi and M. D. Lytras (eds.), *Research & Innovation Forum 2019*,
Springer Proceedings in Complexity,
https://doi.org/10.1007/978-3-030-30809-4_5

Thus, learn to work in teams and cooperate to reach further is a very important skill for future professionals. Experimental studies have pointed out the existence of some collective intelligence when people work together [1, 2, 3]. Thus, collective intelligence arises when groups of individuals acting collectively in ways that seem intelligent [4].

The peer review process of scientific works goes towards achieving these goals. Reviewing the work of other students helps them find the strengths and weaknesses of the proposals, make a comparative work and get better understanding of the problems.

To properly implement this method in an educational environment, an appropriate tool is required. The integration of the Information and Communication Technologies (ICT) has enabled a paradigm shift in education which provides new opportunities to reach better learning outcomes [5, 6].

The evolution of the web as a collaborative and semantic platform [7, 8] where users can share information, collaborate and take advantage of the interactions of other users to enhance their experience and offer great possibilities for going further. There are many success cases of these platforms for social interactions, building media repositories and other collaborative ventures in the construction of information and knowledge [9].

The aims of the preliminary work presented in this paper consists precisely in exploiting the features and capabilities of these platforms for creating collective intelligence and improving the teaching-learning process. The hypothesis of the work considers that the combination of the peer review methodology and the collaborative platforms could lead to the generation of collective intelligence among other well-known benefits such as critical thinking growth.

This paper is organized as follows: in Sect. 2, we review the representative tools for peer review in education; in Sect. 3, the methodology and the procedure are described; next in Sect. 4, discusses the main results of the works; and finally, Sect. 5 draws some conclusions of the research.

2 Collaborative Tools for Education

The assessment of students works made by themselves is negatively perceived by participants due to conflict of interests when assessing friends' works. The computer-aided peer assessment can easily overcome this conflict and make anonymity possible [8]. Therefore, most of the peer review methods for education use some electronic platform to perform the assessment activities. There are a lot of generic online platforms for performing the peer review process. For example, some of the most frequent used for scientific conference management are EasyChair (http://easychair.org/), EDAS (https://edas.info/doc/), Open Conference Systems (https://pkp.sfu.ca/?q=ocs), Acamedics (http://acamedics.com/index.html), etc. These tools are designed for providing support to the management of the papers, and they provide several resources for all the process including the publication of final works, the proceeding and the generation of brochures with the program [10]. Some of them have also support for educational purposes.

In academic context, there are studies that illustrate research on virtual peer review using highly interactive, web-based software for educational purposes. Among these platforms, the *Calibrated Peer Review* (CPR) (http://cpr.molsci.ucla.edu) is one of the most widely used. This is a tool for formative and summative assessment of student works assignments in targeted courses [11, 12].

In addition, there are works that design specific eLearning platforms for facilitating online peer review and making a student-friendly collaborative social learning environment. In this line, there are examples based on Moodle [13] and web technology [14].

Other researches develop their own online platforms for specific purposes, for example, for Master students [15], for mobile devices [16, 17]; and for variable level of knowledge of each student [18]. In all cases, the results are positive and these platforms increase the learning achievements.

3 Methods

3.1 Research Questions

The research questions of this work are the following. These are in line with the aims of this work.

1. What is the effect of using the new communication and technology tools in the continuous learning from each other by means of the peer review methodology?
2. Is collective intelligence generated from the collaboration in the learning by means of the peer review methodology?

3.2 Participants

This work has been developed for a first course group of Computer Science Degree at University of Alicante. A total of 143 students participated in the study along the course. The first course has been chosen to test the proposed method to involve students in this methodology from the beginning of their university studies.

The continuous learning is very useful along the degree career. The students have generally an advanced technological skill from a user point of view [19], but they have significant shortcomings in methodologies for professional and scientific fields.

3.3 Methodology

The achievement of the teaching objectives by the authors of this work led to the design of an assessment model that could exploit the collaborative work of students. This method is based on a previous work where the implementation of a participative assessment method for student peer review was conducted [20]. There, the students must review and assess their classmates' works in a similar way than the peer review methodology for scientific publications. Previous works in this matter shows that this methodology has improvements in learning process. Through peer review, the students can build their knowledge by themselves with little intervention from instructors [13].

Now, in this research, further objectives are stated to make the most of the student peer review method and the new ICT capabilities for achieving collective intelligence and improving the teaching-learning process.

The collaborative evaluation process is an iterative process that comprises the steps described in Fig. 1. It is important to highlight that the good functioning of the model requires that students commit to do their job well. Experience to date shows no significant problems in this regard since students are aware of their responsibility in the process of evaluation of the subject. However, although the delivery of works and reviews are compulsory for all students, experience shows that not all end up delivering them on time due to problems with some of the exercises or lack of time due to conflicts with other subjects. In those cases, if the teaching staff notes any difficulty to finish on time, often, the deadline is extended a few more days.

To assist in this process, a cloud platform has been developed. This tool enables the proposed collaborative assessment methodology and provides a simple means to measure achievement results.

After completing the above iterative process, the supervision of the teaching staff must verify whether the works are compliant with the initial expectancies or not.

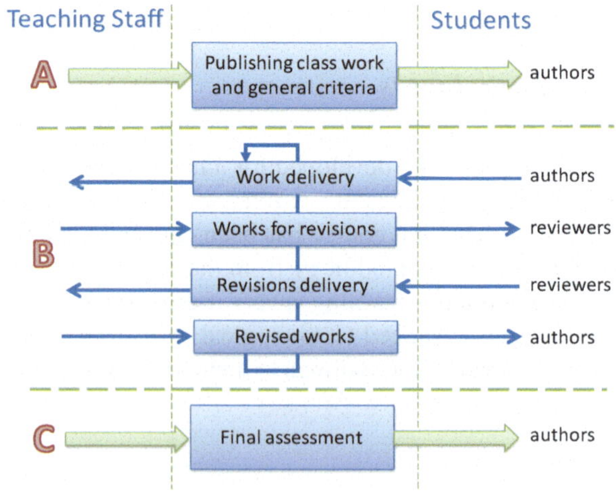

Fig. 1 Assessment process

Then, the works are individually assessed. The teacher calculates the grade of each student based on several aspects: the initial quality of work, the progression that has manifested in the process, the final quality of the work and recommendations of the reviewers. The work made as reviewers and rigor used to do it are taken into account in the assessment of the students. The works that have not passed after the revision process are carefully examined to rule out deficiencies in the work of the reviewers. If it is the case, the corresponding works are considered as particular cases and the students are more directly helped by teachers.

The suitable methodology is facilitated in order to allow them to succeed in solving the difficulties. A new delivery is then allowed. The acquired knowledge is transmitted because everyone can learn through the other's reviews received as well as through the effort made reviewing the other works. In all cases, the acquired knowledge is highlighted. Each student has really improved the initial knowledge on the subject because he/she has reviewed three works and has been reviewed and helped three times for his/her own work. It is important to outline that the capabilities of the reviews are also improved by the effort they have made when they improve their own work as authors. So, the benefits provided by the double condition author-reviewer are optimized.

4 Results and Discussion

The results of the peer review assessment as learning methodology are quite satisfactory and many pedagogical advantages have been observed. In few weeks, the students can understand the new organization of the teaching-learning process: as they must review each other their class works, they have to update their knowledge every day and so they are more able to understand the new concepts explained by the lecturer. In general, an increase of the student's capabilities has been observed in the theoretical classes as well as in the practical classes. In addition, the feedback obtained by the proposed model has allowed the teaching staff to have a more precise knowledge of the difficulty of the contents and learning paces. Thus, the duration and intensity of the lecturer's explanations has been adapted to the real needs of students.

The following general advantageous characteristics have also been observed: The students must study regularly, perhaps daily, so that their effort can be distributed and they don't forget so easily the acquired knowledge. They are also able to follow the explanations of the lecturer and to assimilate the new concepts that are needed to understand the next lesson. This learning method aids to overcome bad habits such as copy-paste or copy each other, as well as to improve the quality of the bibliographic references.

The experience conducted in this work has enabled us to observe the advantages of the ubiquitous tools in this process and their effect on the student motivations for learning. Experience shows us that participating in this evaluation process is perceived by engineering students as something interesting that catches their attention, especially when they discover that this is the system followed for the review of scientific work. Collaborate in the assessment of their peers is exciting.

Regarding to the collective intelligence emergence, this experience shows that the collaboration between different reviewers and authors produce better final works than those made by students in an individual way.

5 Conclusions

The integration of the web 2.0 technologies in the academic activity facilitated innovative explorations and new learning methods. It would not have been possible to carry out this method without a modern web platform. The interaction of many students making and evaluating works creates interesting data in order to study their evolution and needs.

In addition, with collaborative tools, it has been demonstrated that ability to work in teams is increased and the quality of the final classworks is improved. We can conclude that learning of the students grow when comparing with traditional methodologies.

However, this methodology may consider the substantially increase of the workload of the students, as they have to make their own works and the assigned reviews. Therefore, the instructors must take it into account and use this method only in more appropriate cases.

Future work can focus on developing motivation features to further involve the students in their own learning. In addition, we will extend this methodology to other advanced subjects of Computer Science Degree where students have more knowledge and experience in the university.

References

1. Hansen, M.J., Vaagen, H.: Collective intelligence in project groups: reflections from the field. Procedia Comput. Sci. **100**, 840–847 (2016). https://doi.org/10.1016/j.procs.2016.09.233
2. Schut, M.C.: On model design for simulation of collective intelligence. Inf. Sci. **180**(1), 132–155 (2010). https://doi.org/10.1016/j.ins.2009.08.006
3. Basso-Aranguiz, M.S., Badilla-Quintana, M.G.: ICT resources to improve learning in higher education. Int. J. Knowl. Soc. Res. **7**(4) (2016). https://doi.org/10.4018/ijksr.2016100101
4. Malone, T.W., Bernstein, M.S. (eds.). Handbook of Collective Intelligence. MIT Press (2015). ISBN: 9780262029810
5. Bahçeci, F., Gürol, M.: The effect of individualized instruction system on the academic achievement scores of students. Educ. Res. Int. **2016**, Article ID 7392125 (2016). https://doi.org/10.1155/2016/7392125
6. Anderson, P.: Web 2.0 and Beyond. Principles and Technologies. CRC Press, Taylor and Francis Group (2012). ISBN 9781439828670

7. Mahmoud, C.B., Azaiez, I., Bettahar, F., Gargouri, F.: Discovery mechanism for learning semantic web service. Int. J. Seman. Web Inf. Syst. **12**(1) (2016). https://doi.org/10.4018/ijswis.2016010102
8. Wen, M.L., Tsai, C.C.: Online peer assessment in an inservice science and mathematics teacher education course. Teach. High. Educ. **13**, 55–67 (2008). https://doi.org/10.1080/13562510701794050
9. Mora, H., Ferrández, A., Gil, D., Peral, J.: A computational method for enabling teaching-learning process in huge online courses and communities. Int. Rev. Res. Open Distrib. Learn. **18**(1) (2017)
10. Parra, L., Sendra, S., Ficarelli, S., Lloret, J.: Comparison of Online Platforms for the Review Process of Conference Papers, International Conference on Creative Content Technologies (2013)
11. Likkel, L.: Calibrated peer review™ essays increase student confidence in assessing their own writing. J. Coll. Sci. Teach. **41**(3), 42–47 (2012)
12. Carlson, P.A., et al.: Improving engineering education with enhanced calibrated peer review: assessment of a collaborative research project. In: Annual Conference of the American Society of Engineering Education (2012)
13. Zenha-Rela, M., Carvalho, R.: Work in progress: self evaluation through monitored peer review using the moodle platform. In: Proceedings. Frontiers in Education. 36th Annual Conference, pp. 26–27. IEEE (2006). https://doi.org/10.1109/fie.2006.322458
14. Li, C., Dong, Z., Untch, R. H., Chasteen, M.: Facilitating peer review in an online collaborative learning environment for computer science students. In: Proceedings of the International Conference on Frontiers in Education: Computer Science and Computer Engineering (FECS), p. 1. The Steering Committee of the World Congress in Computer Science, Computer Engineering and Applied Computing (WorldComp) (2012)
15. Costa, D., Mullan, B., Kothe, E., Butow, P.: A web-based formative assessment tool for masters students: a pilot study. Comput. Educ. **54**, 1248–1253 (2010). https://doi.org/10.1016/j.compedu.2009.11.011
16. Chen, C.H.: The implementation and evaluation of a mobile self- and peer-assessment system. Comput. Educ. **55**(1), 229–236 (2010). https://doi.org/10.1016/j.compedu.2010.01.008
17. Hwang, G.J., Chang, H.F.: A formative assessment-based mobile learning approach to improving the learning attitudes and achievements of students. Comput. Educ. **56**(4), 1023–1031 (2011). https://doi.org/10.1016/j.compedu.2010.12.002
18. Xanthou, M.: An intelligent personalized e-assessment tool developed and implemented for a greek lyric poetry undergraduate course. Electron. J. e-Learn. **11**(2), 101–114 (2013)
19. Mora, H., Signes Pont, M.T., De Miguel Casado, G., Gilart Iglesias, V.: Management of social networks in the educational process. Comput. Hum. Behav. **51**(B), 890–895 (2015). https://doi.org/10.1016/j.chb.2014.11.010
20. Mora, H., Signes Pont, M.T., Camps Jordá, R., García Chamizo, J.M.: Learning method based on collaborative assessment performed by the students: an application to computer science. ACM SIGCSE Bull.—ITiCSE '09 **41**(3) (2009). https://doi.org/10.1145/1562877.1563014

Serious Games in Mechanical Engineering Education

Petr Hořejší, Jiří Vyšata, Lucie Rohlíková, Jiří Polcar and Michal Gregor

Abstract The aim of this paper is to explore the possibilities and limits of using serious computer games in mechanical engineering education. Mechanical engineering has long struggled around the world due to a shortage of students, so it is necessary to constantly look for opportunities to make learning more attractive. Contemporary students are equipped with sufficient competences and technical equipment for effective learning in the virtual environment of computer games, so a whole range of sub-projects in the field of gaming education are being carried out throughout the world. A serious game or applied game is a game designed for a primary purpose other than pure entertainment, most likely for education. A reference serious game called 'Manager Simulator' was developed over three years. The technical means of development were verified through this. In addition, another serious game, 'Workshop', was developed, that guides the student through a virtual production process. The student has to produce machine parts according to the specified production processes and blueprints. Machines and tools are available for the purpose. Some of the machine equipment have commentary within the interactive tutorial. The student then tries their own production. The game responds to their activities and actions (for example, if the product is forgotten when turning or milling, then the part may be rendered unusable, etc.).

P. Hořejší · J. Polcar · M. Gregor
Department of Industrial Engineering and Management,
University of West Bohemia, Univerzitni 8, 306 14 Plzen, Czech Republic
e-mail: tucnak@kpv.zcu.cz

J. Vyšata
Department of Machining Technology,
University of West Bohemia, Univerzitni 8, 306 14 Plzen, Czech Republic

L. Rohlíková (✉)
Department of Computer Science and Educational Technology,
University of West Bohemia, Univerzitni 8, 306 14 Plzen, Czech Republic
e-mail: lrohlik@kvd.zcu.cz

A. Visvizi and M. D. Lytras (eds.), *Research & Innovation Forum 2019*,
Springer Proceedings in Complexity,
https://doi.org/10.1007/978-3-030-30809-4_6

1 Introduction

Engineering production is currently experiencing a worldwide boom, and this is reflected in the increased demand for skilled labour at all levels. This also calls for the training and education of such workers. It is desirable not only to teach them theoretical knowledge, but also practical skills and experience. Graduates from non-technical schools who have no practical experience in the field apply less often to technical universities. Equipment and safety regulations in some institutions, however, limit the possibility of practical training. In this situation, a partial solution is to use a suitable software product to best understand the experience of manufacturing work. It can, however, also be an appropriate aid even if practical instruction is an option.

It is extremely time consuming and expensive to teach students a complete understanding of manufacturing processes, so that they meet real job requirements both theoretically and practically, as was found by Poor et al. [1]. Because of the problem in understanding, it is usual in our Central European secondary schools and universities to demonstrate various cutting processes on practical examples for each cutting machine and each procedure. For better understanding, it is suitable to let students work by themselves. This means having a number of machines to work on, and several components for each student are also required. During the very first trials we have to expect mistakes caused by not following instructions correctly. In such situations, students start to understand the main issues and problems of machining. This paper deals with using the serious games concept in order to improve the didactic level of machine engineering education. 'Serious games' is a term belonging to the so-called 'edutainment' concept. These are games (board, computer, etc.) whose goal is not only to entertain, but also to educate.

2 Literature Review

The traditional method of studying the machining process by 'learning from your mistakes' which was used in the 70s and 80s in polytechnics for training workshop workers was economically inefficient, requiring many tools and components. Virtual training is presented [2], where a training environment for high precision manufacturing was created with the support of online lecturers and students. The machining cutting simulator idea is also one of the virtual training possibilities mentioned [3, 4].

Technical measurements presented [5] shows a method for designing an environment with specific tasks, depending on the study requirements. Similar model training manuals are presented [6, 7]. These training environments should be enhanced with technical knowledge, such as [8].

Virtual training is very common and highly advanced in medical environments, *for example discussed in*: [9–11]. The level of sophistication of virtual training is also visible in research [12]. Another example of a serious mechanical engineering, but non-computer, game is evaluated by Pourabdollahian et al. [13]. Our training simulator for mechanical engineering students does not require higher than average computer skills, and it aims to teach machining and machine tool dynamics, like [14], based on intuitive proband behaviour.

Computer games as learning tools can be observed from the very beginning of computer games, in the form of language learning aids, numerical computations, and natural science concepts. Today's didactic games combine the sophisticated graphics of games, a didactic contribution, and engagement by inserting the player into the action. Some interesting research on this subject was dealt with, e.g. [15].

Markopoulos et al. [16] describes the benefits of computer games for teaching in engineering, making a difficult subject more manageable, increasing intrinsic motivation, scientific knowledge, collaboration, interest, and reduction or better management of workloads. Likewise, they focused on demonstrating the benefits of serious games, using the elements of virtual reality in engineering [17, 18], which list the additional benefits of serious games in the form of a gaming experience comparable to the real process, and thus active knowledge acquisition. Arnab et al. [19] deals with the use of elements of collaboration and competition.

3 Methodology

In view of the above, two didactic games (serious games) have been developed in the framework of the innovation of teaching in the Faculty of Mechanical Engineering at the University of West Bohemia in Pilsen to more closely acquaint the student—the player—with some practical aspects of engineering operations. In addition to introducing games and bringing their content closer to a didactic function, it is worth mentioning how they were created. However, a significant part of the contribution is a test of one of the games, the aim of which was to find out the intersubjective degree of attractiveness of the game, as well as its teaching potential, primarily for the purpose of creating a higher or more advanced version of the software.

This is, therefore, an evaluation of research, but the information may provide a picture of a more general nature for at least a partial solution to the question of how computer games can impact teaching. However, a more reliable answer will be needed in the future to carry out further research, preferably with upgraded software, in order to evaluate and compare the results of student groups, one using didactic play, and the other not. Research will no longer only be based on intersubjective questionnaire data, and will have a greater informative value.

3.1 The Concept of the Game

The Manager Simulator game from [20] was created first. It aims to combine theoretical knowledge about lean production methods, such as Just in Time and Kanban, and its practical deployment in a simulated business, as well as to clarify its meaning. Its development was the source of the first experience, and became the motivation for the development of the second game. The second game is called 'The Workshop', [21], and simulates the production of a particular component. The player must first get the production documentation, and then he/she goes through various workplaces in the workshop and manufactures the component according to the relevant technological process. The genre of both didactic games falls into the adventure field—games where the player gradually solves quests in a typically tangled story. Tasks are fun, in order to achieve the main goal. The didactic effect is actually a by-product in this structure, but it is very important too.

3.2 The Workshop Game

The Didactic Adventure Workshop was created using the experience we gained from developing and deploying the Manager Simulator game. The game is solved this time in the first person in 'pseudo 3D view', and 3D models were used. In this game, the student can produce a component of a certain type for a trial, using virtual machine equipment in a virtual workshop. Figure 1 shows a view of the entire workshop on the Home screen. All three machine workstations with workpiece boxes are visible here. The student has one lathe, one pillar drill, and one cantilever milling machine. On the left there is a cupboard and a table, with a notice board above the table. This is the foreman's workplace. The equipment also includes a machine vice on the table

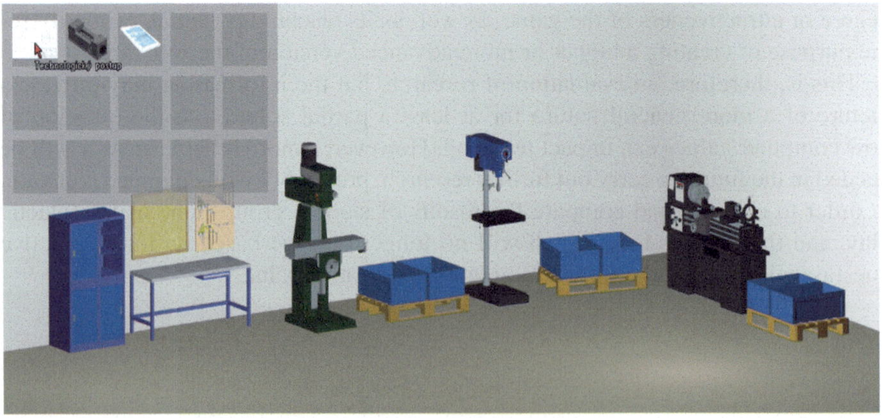

Fig. 1 View of the entire workshop

Fig. 2 Lathe workstation

under the notice board, which the player must take to his inventory (see Fig. 2 top left). It is also possible to take the production blueprints and manufacturing plan from the wall. The work material can be found ready at the first workplace. Apart from placing the machine vice on the drill and the milling machine, the player does not have to prepare any machines. Even the tools are already in the spindles of the milling machine and the drill, as well as in the toolpost of the lathe.

The student has to use the machines to produce a machine component according to the drawing and the manufacturing process s/he has in the inventory.

By clicking on a machine, you can move it to the workplace (Fig. 2). When you click on the image of the gear in the top right corner, a detailed view appears of the machine's workspace. Here you can control the machine by clicking on the individual controls. The game is accompanied by spoken comments that usually refer to general facts and lessons from the field of machining, or commentary on the course of the game. Game control is intuitive, and runs through principles known from adventure games. The player also has either a pre-fabricated product or a finished product in the inventory, which interacts with individual parts of the machine during the production process. Like [22], s/he also uses various machine controls. The production triggers manufacturing animations.

Let's look at an example of one situation that can occur in virtual production: for example, if the process requires a tailstock for supporting the workpiece, and the player attempts to work without a support, it is likely that the dimensions will be out of tolerance. The player is informed by an audio/text comment about the successful completion of production of the components.

The implementation of this training software is based on the experience of modelling production systems and virtualization of real production areas, and several comments from machining technology were also taken into consideration in the current version. However, many improvements are still to be made in future developments. The player/student, learns that there is a relationship between production documentation and the production process itself. The player also obtains their first experience of working with certain types of machines, and can learn, for example, the importance of firmly and rigidly clamping the workpiece. They will also be aware of the organization of the workplace. For later versions, however, it will be useful to deepen some didactic effects (see below).

3.3 Development of the Game

The MS1 management simulator and Workshop game were developed using the Wintermute Engine Development Kit. This is a computer game development tool, specialized for the creation of point-and-click style adventure games with 2D backgrounds and with player and non-player characters, either in 2D or in 3D. The engine consists of a player—the Wintermute Engine (WME) and the Wintermute Engine Development Kit content creator. It is similar to [23].

The content is made in various software tools, such as 2D image editors, 3D modellers, sound editing software, etc. Such assets are brought to life using scripts written in our own, C-like language. The development kit includes scripts for managing commonly used functions, such as controls and menus.

The resulting product is developed using various tools. Each sub-scene is separated, and can be edited separately: the developer can insert a 2D image within the Scene Editor. In our case, these were 3D scenes rendered in 3DS Max. Within this scene, 2D regions—'click maps'—are inserted. These maps represent named interactive entities. Figure 3 shows a detail of the lathe workspace, where all the buttons, the carriage with a toolpost, and the tailstock are interactive—these can be moved, and on the left side you can see the rotary chuck for attaching the workpiece. Each script creates a local script that controls complex interactivity (for example, interacting with a sub-object to start the animation, play the sound, put the subject in a different state). In the above scenario, the script controls the interaction of the workpiece with the clamping elements. Global scripts then control common functionality (and communicate with local ones), such as inventory management.

A major advantage of this approach for schools is that the builds do not require recent high-end gaming computers, but can even be set up to run smoothly on older and mobile hardware.

The Management Simulator serious game was developed over the course of more than a year by two programmers and one graphic designer. The Workshop program was, for the same set of functions, developed over the same period of time. The validation by pedagogues was performed prior to the actual implementation. Responses to comments were implemented into the application over the next six months.

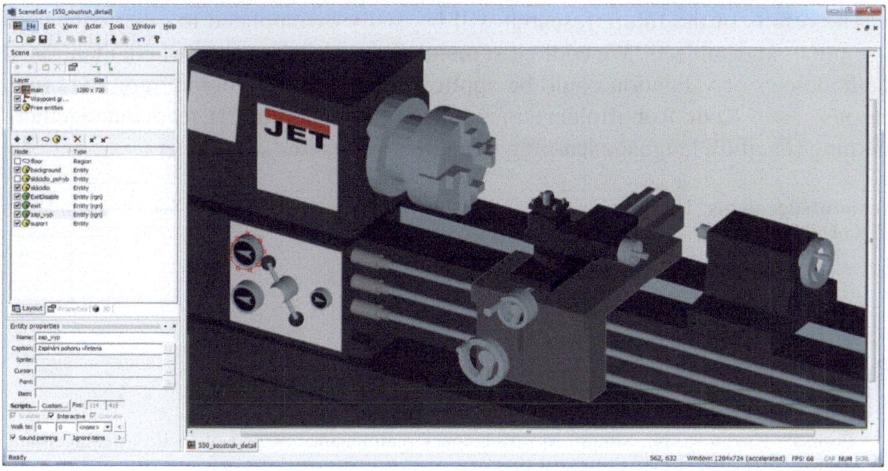

Fig. 3 Detail of lathe support in WME Scene Editor

4 Conclusion and Future Work

Similarly as [24, 25], this work deals with the creation of a virtual environment for use as an aid in teaching mechanical engineering. Based on the first pilot serious game, the second one serious game: "Workshop" was developed.

For the following work, we will conduct an exploratory validation research. For the research we expect to prepare questionnaires. These questionnaires will be developed on the basis of experience gained from a previous European project CZ.1.07/2.2.00/15.0397—Product life cycle in digital factory environment. We expect to include a statistically significant group of students of the first year of mechanical engineering. There is an expectation that such a group will contain experienced students (e.g. coming from mechanical engineering high school) and students will no previous machine operating experience. In the other words we will tag each student by his/her historical education background, by using manufacturing machines experience and previous experience with computer games.

We expect to track down all those factors dependency on a fun factor and benefit of didactic game. Questionnaire will contain also open question. We will collect mainly the ideas for improvement. The we will summarize all the findings and compare them to other similar researches like: the research [26] which mainly focuses on Thai students and finds virtual environment benefits not dependent on time, place or device, which is also visible in [27] where the study is conducted with students who had never performed a winter sport. The memorization benefit of learning in a virtual environment is demonstrated in [28]. The problematics of using a virtual environment as a research platform is discussed in [29] and safety obtained by using a virtual world is discussed in [30], where simulation of clinical practice is demonstrated.

The results of the future research will be easily extrapolated to other education disciplines and subjects. We could just mention the serious games where mentioned methodology of validation could be applied, for example, programming [31] (e.g. serious games Kodi, RoboBuider), in medicine [26] (e.g. Pulse!), team management learning (Pacific), language learning (e.g. Duolingo), etc.

Acknowledgements This paper was prepared with the support of the Internal Science Foundation of the University of West Bohemia SGS-2018-031.

References

1. Poor, P., Kuchtova, N., Simon, M.: Machinery maintenance as part of facility management. Procedia Eng. **69**, 1276–1280 (2014)
2. Lee, W.B., Li, J.G., Cheung, C.F.: Development of a virtual training workshop in ultra-precision machining. Int. J. Eng. Educ. **18**(5), 584–596 (2002)
3. Fillatreau, P., Fourquet, J.-Y., Le Bolloc'H, R., Cailhol, S., Datas, A., Puel, B.: Using virtual reality and 3D industrial numerical models for immersive interactive checklists. Comput. Ind. **64**(9), 1253–1262 (2013)
4. Lu, S.C.-Y., Shpitalni, M., Gadh, R.: Virtual and augmented reality technologies for product realization. CIRP Ann.—Manufact. Technol. **48**(2), 471–495 (1999)
5. Neamţu, C., Hurgoiu, D., Popescu, S., Dragomir, M., Osanna, H.: Training in coordinate measurement using 3D virtual instruments. J. Int. Measur. Confideration **45**(10), 2346–2358 (2012)
6. Mukhanov, B., Omirbekova, Z., Alimanova, M., Jumadilova, S., Kozhamzharova, D., Baimuratov, O.: A model of virtual training application for simulation of technological processes. Procedia Comput. Sci. **56**(1), 177–182 (2015)
7. Torres, F.G.: Model engineering of ancient and historical machines as a training tool for mechanical engineering students. Int. J. Eng. Educ. **27**(3 Part 1), 591–596 (2011)
8. Cruz-Neira, C.: Making virtual reality useful: a report on immersive applications at Iowa State University. Future Gener. Comput. Syst. **14**(3–4), 147–155 (1998)
9. Foroudi, F., Wong, J., Kron, T., Roxby, P., Haworth, A., Bailey, A., Rolfo, A., Paneghel, A., Styles, C., Laferlita, M., Tai, K.H., Williams, S., Duchesne, G.: Development and evaluation of a training program for therapeutic radiographers as a basis for online adaptive radiation therapy for bladder carcinoma. Radiography **16**(1), 14–20 (2010)
10. Rodríguez-Andrés, D., Juan, M.C., Mollá, R., Méndez-López, M.: A 3D serious game for dental learning in higher education. In: IEEE 17th International Conference on Advanced Learning Technologies (ICALT), Timisoara, 2017, pp. 111–115 (2017). https://doi.org/10.1109/icalt.2017.29
11. Addy, T.M., Dube, D., Croft, C., Nardolilli, J.O., Paynter, O.C., Hutchings, M.L., Reeves, P.M.: Integrating a serious game into case-based learning. Simul. Gaming **49**(4), 378–400 (2018). https://doi.org/10.1177/1046878118779416
12. Verner, I.M., Gamer, S.: Robotics laboratory classes for spatial training of novice engineering students. Int. J. Eng. Educ. **31**(5), 1376–1388 (2015)
13. Pourabdollahian, B., Taisch, M., Kerga, E.: Serious games in manufacturing education: evaluation of learners' engagement. Procedia Comput. Sci. **15**, 256–265 (2012). ISSN 1877-0509
14. Jönsson, A., Wall, J., Broman, G.: A virtual machine concept for real-time simulation of machine tool dynamics. Int. J. Mach. Tools Manuf. **45**(7–8), 795–801 (2005)
15. De Gloria, A., Bellotti, F., Berta, R.: Serious games for education and training. Int. J. Serious Games, **1** (2014). https://doi.org/10.17083/ijsg.v1i1.11

16. Markopoulos, A.P., Fragkou, A., Kasidiaris, P.D., Davim, J.P.: Gamification in engineering education and professional training. Int. J. Mech. Eng. Educ. **43**(2), 118–131 (2015)
17. De Freitas, S., Jarvis, S.: Serious games—engaging training solutions: a research and development project for supporting training needs: Colloquium. Br. J. Educ. Technol. **38**(3), 523–525 (2007)
18. Kopecek, P.: Job shop scheduling in mechanical engineering. In: Annals of DAAAM and Proceedings of the International DAAAM Symposium, pp. 1293–1294 (2011)
19. Arnab, S., Bhakta, R., Merry, S.K., Smith, M., Star, K., Duncan, M.J.: Competition and collaboration using a social and gamified online learning platform. In: Boyle, L., Connolly, T.M. (ed.). Proceedings of the European Conference on Games-based Learning, pp. 19–26. Academic Conferences and Publishing International Limited, Reading (2016)
20. Suchý, M.: Diploma thesis "Tvorba vážné hry v konceptu digitální továrny" University of West Bohemia (2013)
21. Andrle, T.: Diploma thesis "Tvorba výukového programu ve Wintermute Engine" University of West Bohemia (2014)
22. Marayong, P., Li, M., Okamura, A.M., Hager, G.D.: Spatial motion constraints: Theory and demonstrations for robot guidance using virtual fixtures. In: Proceedings—IEEE International Conference on Robotics and Automation, pp. 1954 (2003)
23. Boeker, M., Andel, P., Vach, W., Frankenschmidt, A.: Game-based e-learning is more effective than a conventional instructional method: A randomized controlled trial with third-year medical students. PLoS ONE **8**(12) (2013)
24. Liu, D.: Information technology teaching methods in NC teaching. In: 2017 3rd International Conference on Information Management, ICIM 2017, pp. 526 (2017)
25. Wilson, C., Bennett, W., Jr., Guarino, S., Bove, K., Cain, T.L.: Applying a user-centered design approach to developing game-based training environments for aircraft maintainers. In: End-User Considerations in Educational Technology Design, pp. 217–236 (2017)
26. Khlaisang, J., Songkram, N.: Designing a virtual learning environment system for teaching twenty-first century skills to higher education students in ASEAN. Technol. Knowl. Learn. **24** (2019). https://doi.org/10.1007/s10758-017-9310-7
27. Topu, FB., Goktas, Y.: The effects of guided-unguided learning in 3d virtual environment on students' engagement and achievement. Comput. Hum. Behav. **92** (2019). https://doi.org/10.1016/j.chb.2018.10.022
28. Voinescu, A., David, D.: The effect of learning in a virtual environment on explicit and implicit memory by applying a process dissociation procedure. Int. J. Hum. Comput. Interact. **35** (2019). https://doi.org/10.1080/10447318.2018.1424102
29. Grech, A., Cassar, M.: Can virtual learning environments be used as research platforms? Nurse Res. **26** (2018). https://doi.org/10.7748/nr.2018.e1568
30. King, D., Tee, S., Falconer, L., Angell, C., Holley, D., Mills, A.: Virtual health education: scaling practice to transform student learning using virtual reality learning environments in healthcare education to bridge the theory/practice gap and improve patient safety. Nurse Educ. Today **71** (2018). https://doi.org/10.1016/j.nedt.2018.08.002
31. Gorbanev, I., Agudelo-Londoño, S., González, R.A., Cortes, A., Pomares, A., Delgadillo, V., Muñoz, Ó.: A systematic review of serious games in medical education: quality of evidence and pedagogical strategy. Med. Educ. Online **23**(1) (2018). https://doi.org/10.1080/10872981.2018.1438718
32. Bowman, D.A., Hodges, L.F.: Evaluation of techniques for grabbing and manipulating remote objects in immersive virtual environments. In: Proceedings of the Symposium on Interactive 3D Graphics, pp. 35 (1997)

Integrating Makerspaces in Higher Education: Constructionism Approach to Learning

Tayeb Brahimi, Sajid Khalifa and Bensaid Benaouda

Abstract This study explored the impact of using makerspaces in higher education. The paper sought to investigate the effects of constructionism approach on students learning outcomes in the setting of makerspaces which allows community members to design, prototype and manufacture items using tools that would otherwise be inaccessible or unaffordable such as 3-D printers, laser cutters, CNC machines, and CAD/CAM software. The case study involves students in the Design Program at Effat University, Jeddah, Saudi Arabia. In such a makerspaces environment, results based on course learning outcomes in product design showed that students perform better learn creative ways to problem-solving, and engage effectively through creative experimentation. Further empirical research into the effectual relations between design and 3D constructions may further demonstrate the vital importance of makerspaces on students' learning performance and mastery of skills in the context of higher learning.

1 Introduction

The development and easy access to new technology continue to have an impact on teaching and learning methodologies [1–4]. Makerspaces known as hackerspaces, hack labs, or FabLabs, is a novel approach of product development carried out in an atmosphere of shared interests, thinking, knowledge, resources, and equipment,

T. Brahimi (✉)
College of Engineering, Effat University, Jeddah, Kingdom of Saudi Arabia
e-mail: tbrahimi@effatuniversity.edu.sa

S. Khalifa
Design Department, Effat University, Jeddah, Saudi Arabia
e-mail: sakhalifa@effatuniversity.edu.sa

B. Benaouda
Faculty of Islamic Sciences, Istanbul Sabahattin Zaim University, Istanbul, Turkey
e-mail: bbensaid@gmail.com

A. Visvizi and M. D. Lytras (eds.), *Research & Innovation Forum 2019*,
Springer Proceedings in Complexity,
https://doi.org/10.1007/978-3-030-30809-4_7

especially in computing or technology, which encourage learners to design, experiment, and build, thus facilitates interactive participation in innovative development while leading to new business and economic ventures [5–7]. Makerspaces can be defined as a physical location where learners or community members gather and share knowledge and tools to develop, innovate, and create solutions [8, 9]. Makerspaces also represent one of the integral parts of a modern education system that brings together and facilitates the community of interdisciplinary individuals.

Makerspaces improve students' learning when makerspace used. They provide places for people to work on hands-on projects building in a setting of community. The maker movement attracts more individuals into product design, generates dense but diverse networks, creates new ideas and innovative thinking, and lowers the costs for prototyping [8]. According to Saorín et al. [9], the emergence of education with digital fabrication techniques offers an opportunity for the development of creativity. For them, research results show that activities with digital editing tools and three-dimensional printing are valid for the development of creative competence [9].

Makerspaces also prove to be a powerful tool to teach design and manufacturing concepts in parallel, which is especially important for product designers, and is also believed to help students develop their personal soft skills, such as teamwork, communication and learning to work with students from other fields while also building physical prototypes for research projects and their theses [10]. Fisher [11] argued that Maker spaces and the integration of 3D printing provide ample opportunities for students to use critical thinking and problem solving, and further engage students in hands-on learning and experimentation [11].

The key point is that learning through making new constructions, as Papert [12] explained in his theory of constructionism. This learning-by-making is in line with Piaget [13] experiential learning theory where people learn effectively through making things. In the makerspaces sphere, this movement has been recognized to have its root in the learning theory of Papert's constructionism as reported by Martinez and Stager in their book [14] "Invent to Learn: Making, Tinkering, and Engineering in the Classroom". Martinez and Stager argued that makers construct knowledge as they build physical artifacts with real-world value. The culture or rather mindset of maker space can be described as a medium for cross-curricular, higher-order thinking in the classroom. Makerspaces are not only about a space being used by participants to engage in making, but also seen as a mindset [15].

In our context of learning, students are encouraged to acquire increased exposure to the environment of scientific engagement, innovative thinking, and creativity in order to promote the development of maker experiences. This started in a learning space FabLab, as model fabrication unit at Effat University. With emphasis on constructionism theory [12] where learning is positioned in the context of social participation and individual knowledge, the FabLab attracts students from different engineering and technology academic fields, and echoes the objective set by the Accreditation Board for Engineering and Technology (ABET) which states as follows: "provides students with technology, manufacturing equipment and the ability to design a system, component, or process to meet desired needs within realistic constraints such as economic, environmental, social, political, ethical, health and safety, manufacturability, and sustainability".

The basic objective of the FabLab is to enhance students' ability to produce professional and accurate models and hosts different design machines such as laser cutting machine which helps students accelerate the process of model-making, materials such as wood, acrylic, cardboard, and MDF which can be engraved quickly and precisely by means of AutoCAD drawings, 3D printers producing complicated prototypes. Students can also use software like as Sketch Up, 3D AutoCAD, Solid works and 3D Max to prepare their own 3D files and be able to export data as STL files to available printers. Because the 3D printers can produce more complicated shapes, students are able to combine larger scale models with the laser cutters and then add more detailed and complex components to them. 3D printers and laser cutter are considered the most common equipment used in makerspaces labs as reported by the American Society for Engineering Education [16].

This study explored the advantages of using makerspaces for students in the design program and sought to investigate some of the best practices of constructionism approach to learning and its impact on students' learning (Fig. 1).

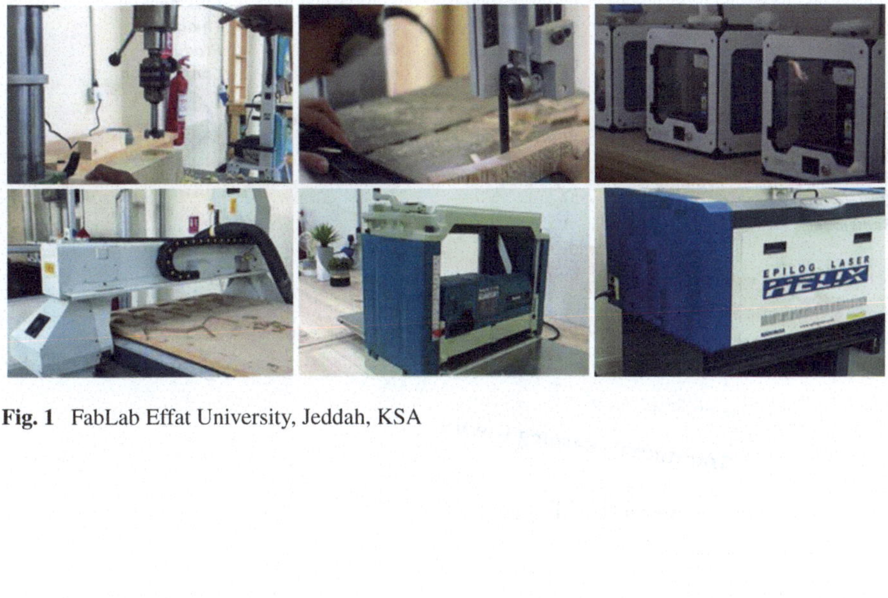

Fig. 1 FabLab Effat University, Jeddah, KSA

2 Domain of Learning and Makerspaces

Traditionally, learning is seen as a vertical process where the instructor serves as both the repository and transmitter of knowledge while the practice is viewed as an application of the theory being learned. However, with the rapid development and easy access to information and communication technology, new methods of teaching and learning have been developed, particularly in engineering education where educational institutions are changing the way faculty and students learn, work, and establish collaborations [17]. The domain of learning [18–23] is based upon five approaches (Fig. 2): (i) behaviorist learning which focuses on the "what" through positive and negative reinforcement leading the learner to react to external stimulus; (ii) cognitive learning which teaches the "how" including procedures and principles, and views the learner as an organized processor of knowledge and information; (iii) constructivist learning which teaches the "why" and considers learning as a process of knowledge construction, the learner is considered active and meaning is created by the learner from the experience gained; (iv) connectivism learning emerged as a result of learners' ability to acquire and share knowledge, not only inside as per the traditional learning theories, but also outside using technology through communications, nodes and connections, and finally, (v) constructionism learning as developed by Papert et al. [12] where students learn creative ways to problem-solving and engage through creative experimentation and more hands-on projects. The constructionism learning approach is derived from the maker movement or makerspaces where learners learn effectively through making things while the role of the instructor would be a facilitator who coaches and help students attaining their own goals, produce a construction that other learners can see, improve, or criticize [14].

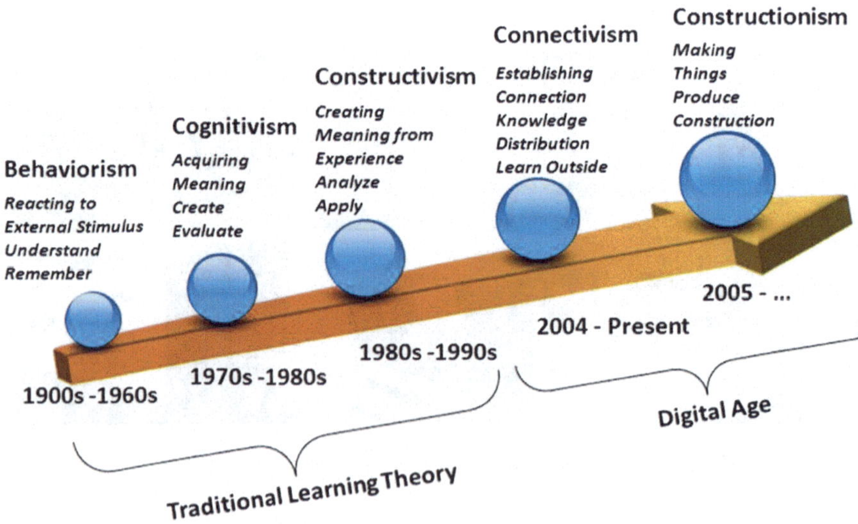

Fig. 2 Domain of learning and learning timeline

3 Knowledge Delivery for the Design Courses

Effat University is a non-profit higher education institution for women in Saudi Arabia, established in 1999, operating under the umbrella of King Faisal's Charitable Foundation. It is the first private institution of higher education for women in the Kingdom, with steady academic and aspiration for research leadership in the Kingdom. The design program of Effat University provides three fundamental academic tracks: Environmental Design, Interior Design, and Product Design. In this paper, we discuss the cases and curriculum related to the product design as they are directly related to the makers. The studio courses are structured to develop learners' knowledge and understanding from the fundamental to the advanced level where in all stages the makerspaces learning method is used and learners test ideas and develop prototypes.

The first course of product design studio 3 focuses on the development of systematic design processes through an array of different design projects, done individually and in small teams, providing discussion and illustrations of concepts, theories, terminology and methodologies of industrial design. Basic competence in shop techniques is established through the creation of simple objects and artifacts that communicate a visual narrative of value and/or function through their form. The course of product design studio 4 is based on the results of observational research, whereby students design simple, single or limited-use objects for various client groups including special populations such as children, elderly and the disabled. By limiting the complexity of the projects, the focus of the course is on the generation of products forms while developing meticulous attention to every aspect of the design process. Complete, clear, compelling design documentation and construction of multiple studies and presentation models at different scales are required.

During the journey of four academic semesters, effective semester 3, students get exposed to all of the basic skills required for product design while subsequent study studios cover disciplines like Ethnography, Human factors, Digital modeling, and Bonding. All of the design studios are delivered with a specific theme for new product development where students are constantly challenged to come up with innovative solutions. This unique environment of makerspaces integrated learning creates "experiential learning environment" [24]. When assigning the design question or problem, students use various design research methods including but not limited to ethnography.

4 Method

The objective of this research was to examine the use of makerspaces in the discipline of product design through qualitative analysis of students' experience in studio 3 and 4. The criteria of assessment focused on the concept of makerspaces. Projects were developed as a sample demonstrating the effects of makerspaces on student's comprehension and skills development. The collected data in this research allowed for observing a makerspaces' experience, demonstrating the effects of using makerspaces as teaching, educational and experiential methods. Two studies have been selected to explore students' learning and their experience while designing their projects in the FabLab. The first case involves fabrication of a bicycle, a product applicable for two users' types or generations while in the second one students were assigned to design and produce a torch applicable for two users' types or generations.

In the Design programs, studio 4 takes places in semester 4. Learners were challenged to re-design a wish bane bike. Using a number of criteria, students came up with different alternatives and solutions. This product could be developed for children under the age of 5 years on plastic scale modes as a toy. A next age group for the same concept design was proposed for children above five years of age where the product would be manufactured from sustainable materials such as wood and plastic. Studio 3 is the first level product design studio where students are taught basic methods as an introduction and then get challenged within simple products and work in makerspaces environment where they use 3D printing and digitals modeling as primary tools. Students were given the fundamentals mechanics to proceed with their design involving some mechanical engineering and technology knowledge. Viewed according to the entrepreneurship perspective of the product, a scaled size product could be produced as a toy for children to play with some modification in product features and options.

5 Results and Discussion

In order to investigate the effects of makerspaces on design studio, one group has been identified at two levels. A group of students at Effat University studied Studio 3 in spring 2017 and then Studio 4 in fall 2017, in subsequent semesters where their psychomotor skills were measured. Since the product design program is new at the university the enrolment number is below 10 students. At the time of conducting this research, there were 7 students in the group, whose results are discussed herewith. Effat University sets 60% as the passing grade for students in all programs. Figure 3 shows the grade percentage of the final project with and without makerspace. Upon analysis and comparison of the final project data, Figs. 3 and 4, in both studios, significant improvement was recorded, which can be credited to use of the Makerspaces facility in the later studio. In Fig. 3, the overall distribution shows improvement for each student, with the best student increasing score by more than 5% as a result of makerspaces.

Research and experience point to the existence of a correlation between the makerspaces environment and design education. While analyzing various factors, it emerged that in design education and other domains of education the practical skills are measured under the category of "Psychomotor Skills" which is represented by fifth domain in the National Qualifications Framework (NQF) for Higher Education in the Kingdom of Saudi Arabia [25]. According to NQF, psychomotor skills

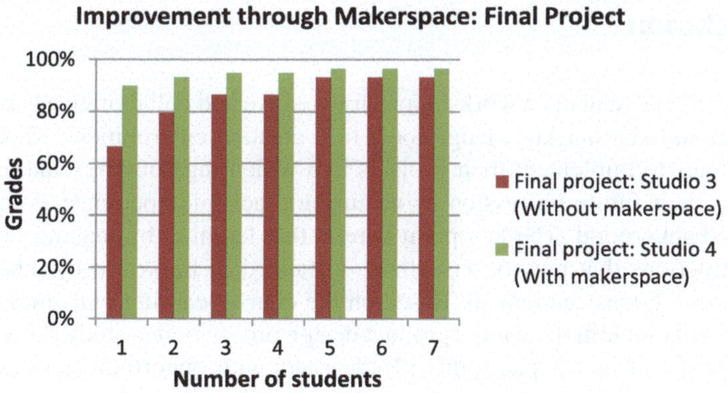

Fig. 3 Final project results for spring 2017 (studio 3) and fall 2017 (studio 4)

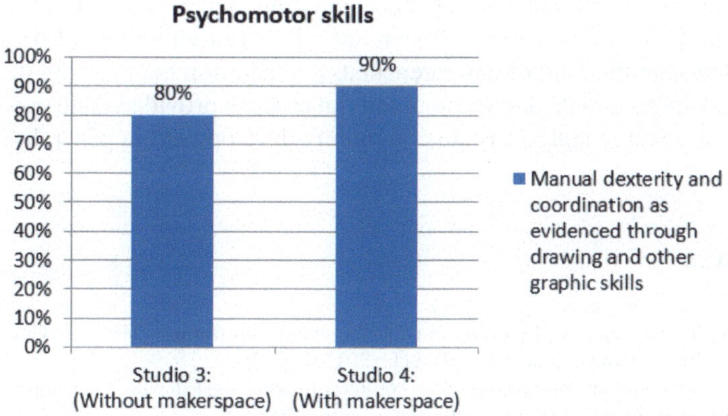

Fig. 4 Psychomotor skills in spring 2017 (studio 3) and fall 2017 (studio 4)

involve manual dexterity that is extremely important in some fields of study. Manual dexterity and coordination are evidenced through drawing and other graphic skills and also through craftsmanship which is demonstrated in the construction of study models and presentation models.

In studio 3 level, when students were assigned the task, they would use basic 3D modeling software such as 3DMax which is limited in functionality and analysis. Students, would go through the development of basic skills where sketching, manual rendering, 2D drawing, and 3D basic modeling are expected to be excelled that can be seen in the "Torch" example, contrary to Studio 4, where students develop higher level skills with Makerspaces facility, working together with other students as well as the makers community. At this level, clear evidence is seen in the form of learning and using Solidworks software, working on a complex product with a high number of variables in addition to higher basic skills, such as manual sketching and hand rendering.

6 Conclusion

Makerspaces represent open workstations for creative and collaborative learning and innovation, and sharing knowledge tools in a creative environment. Makerspaces allow students to implement their projects in a wide range of ways and help them prepare for their future profession by acquiring practical experience alongside the theoretical background. The key point here is that learning by making to produce new constructions that may be visualized, criticized or improved by others based on the theory of constructionism. Based on the course learning outcomes and psychomotor skills set initially for the product design program, this study shows that the implementation of makerspaces provides learners with opportunities to use digital product techniques and tools with three-dimensional printing, and hence improves their creativity and innovation in the design of products. Makerspace helped students turn their innovative ideas into new designs and as result further supporting the contributions to entrepreneurship and community development. Further research on the possible relations between design and 3D construction would be useful for students' learning the field of entrepreneurship in addition to exploring the accessibility of makerspace to the large community in order to provide viable opportunities of leaning by making and to sustain community development in general.

References

1. Brahimi, T., Sarirete, A.: Learning outside the classroom through MOOCs. Comput. Hum. Behav. (2015). https://doi.org/10.1016/j.chb.2015.03.013
2. National Academy of Engineering of the National Academies: Educating the engineer of 2020: adapting engineering education to the new century (2005)
3. Bates, T.: Teaching in a Digital Age, Guidelines for Designing Teaching and Learning for a Digital Age. Tony Bates Associates Ltd. (2015)
4. Duderstadt, J.J.: Engineering for a Changing World a Roadmap to the Future of Engineering Practice, Research, and Education. Millennium Project, University of Michigan (2008)
5. Anderson, V.: Makers: the New Industrial Revolution. Crown Business, New York (2012)
6. Hatch, H.: The Maker Movement Manifesto," Rules for Innovation in the New World of Crafters, Hackers, and Tinkerers, 1 ed. McGraw-Hill Education (2013)
7. Khalifa, S., Brahimi, T.: Makerspace: a novel approach to creative learning. In: Learning and Technology Conference (L&T), 2017, The MakerSpace: from Imagining to Making (2017). https://doi.org/10.1109/lt.2017.8088125
8. Van Holma, E.J.: Makerspaces and contributions to entrepreneurship. In: ScienceDirect. Elsevier Ltd. (2015)
9. Saorín, J.L., Melian-Díaz, D., Bonnet, A., Carrera, C.C., Meier, C., De La Torre-Cantero, J.: Makerspace teaching-learning environment to enhance creative competence in engineering students. In: Thinking Skills and Creativity, vol. 23, pp. 188–198. Elsevier Ltd. (2017)
10. Weinmann, J.: Makerspaces in the University Community. Master's thesis. Institute of Product Development. Technische Universität München, Germany (2014). https://web.stanford.edu/group/design_education/wikiupload/0/0a/Weinmann_Masters_Thesis.pdf
11. Fisher, E.: Makerspaces move into academic libraries. ACRL TechConnect, available at: http://acrl.ala.org/techconnect/?p¼2340 (2012). Accessed 2 Apr 2014

12. Papert, S., Harel, I.: Constructionism: Research Reports and Essays 1985–1990 by the Epistemology and Learning Research Group, the Media Lab. Massachusetts Institute of Technology, Ablex Pub. Corp, Norwood (1991)
13. Piaget, J., Inhelder, B.: The Child's Conception of Space. See especially "Systems of Reference and Horizontal-Vertical Coordinates", pp. 375–418. W. W. Norton & Co., New York (1967)
14. Martinez, S.L., Stager, G.: Invent to Learn: Making, Tinkering and Engineering in the Classroom. Kbh., Nota (2019)
15. Hlubinka, M., Dougherty, D., Thomas, P., Chang, S, Hoefer, S., Alexander, I., McGuire, D.: Makerspace Playbook. Maker Media. Available from: http://makerspace.com/wp-content/uploads/2013/02/MakerspacePlaybook-Feb2013.pdf (2013)
16. Barrett, T., Pizzico, M.C., Levy, B., Nagel, R.: A review of university maker spaces. In: 122 ASEE Annual Conference & Exposition, June 14–17, 2015, Seattle (2015)
17. Education 2030.: Towards Inclusive and Equitable Quality Education and Lifelong Learning for All. ED-2016/WS/28, Incheon Declaration and SDG4—Education 2030 Framework for Action (2015)
18. Cohen, L.L., Manion, L., Morrison, K.: Research Methods in Education. Routledge, London (2007)
19. Will, H.: Makerspace: Towards a New Civic Infrastructure". Retrieved Jan. 2017, from https://placesjournal.org/article/makerspace-towards-anew-civic-infrastructure/
20. Sarirete, A.: Knowledge and learning issues related to engineering education: a constructivist approach. Int. J. Teach. Case Stud. **2**(1), 17–28 (2009)
21. Wenger, E.: Communities of Practice: Learning, Meaning, and Identity. Cambridge University Press. Willey, Cambridge (2013)
22. Siemens, G.: A Learning Theory for the Digital Age. Retrieved January 2017 from http://www.elearnspace.org/Articles/connectivism.htm (2005)
23. Peggy, A.E., Timothy, J.N.: Behaviorism, cognitivism, constructivism: comparing critical features from an instructional design. Perspective **6**(4), 50–72 (1993)
24. Kolb, D.A.: Experiential Learning: Experience as the Source of Learning and Development. Pearson Education, Upper Saddle River (2015)
25. https://www.ncaaa.org.sa/en/Releases/NationalQualifFrameworkDocuments/National%20Qualification%20Framework.pdf

Definition of a Feature Vector to Characterise Learners in Adaptive Learning Systems

Alberto Real-Fernández, Rafael Molina-Carmona, María L. Pertegal-Felices and Faraón Llorens-Largo

Abstract Adaptive learning can be defined as a learning model based on technology that can detect the students individual situation, context, learning needs and style, and the state of their learning process dynamically, and act according to them. So, it is necessary to define a student or learner model, that is, the set of information obtained and retained by the learning system about the learner so that the learner is characterised, and the learning process is adapted. In this work, we propose a learner model made of three main types of information: behavioural features, performance features and personal features. For this model to be useful in automatic learning systems, a formal feature vector must be then obtained. The features in the vector must be meaningful, discriminating and independent so that effective machine learning algorithms can be applied.

A. Real-Fernández · R. Molina-Carmona (✉)
Unidad Científica de Innovación Empresarial "Ars Innovatio", University of Alicante, San Vicente del Raspeig, Spain
e-mail: rmolina@ua.es

A. Real-Fernández
e-mail: albertorealfdez@gmail.com

M. L. Pertegal-Felices
Developmental and Educational Psychology Department, University of Alicante, San Vicente del Raspeig, Spain
e-mail: ml.pertegal@ua.es

F. Llorens-Largo
Cátedra Santander-UA de Transformación Digital, University of Alicante, San Vicente del Raspeig, Spain
e-mail: faraon.llorens@ua.es

A. Visvizi and M. D. Lytras (eds.), *Research & Innovation Forum 2019*, Springer Proceedings in Complexity,
https://doi.org/10.1007/978-3-030-30809-4_8

1 Introduction

In a digital society as ours, where knowledge is no longer static, concepts are continuously changing, and students must be prepared for an uncertain environment. We need to change the way learning is, so that our current learners are prepared for a continuous learning and formation cycle. Besides, each student presents different learning needs, learning styles and learning rhythms. For that reason, we should assume new personalised and adaptive learning paradigms.

Adaptive Learning (AL) can be described as a learning model based on technology that can detect the students individual situation, context, learning needs and style, and the state of their learning process dynamically, and act according to them [1].

In this context, artificial intelligence can play a crucial role. We are already used to intelligent systems recommending a book or a song, to applications that adapt to our needs and our way of interacting. Simply speaking to a machine, it is capable of finding a solution to the problem we are posing. The challenge we face now is to make artificial intelligence algorithms also capable of knowing what learning needs our students have, what skills they have acquired and how they behave in order to provide them with a truly personalised and adapted learning experience. Then we can talk about real Adaptive Learning.

The core of an AL system is usually a machine learning algorithm that is able, from a feature vector that characterises the learner, to provide each individual with the most suitable activity for his or her characteristics, while maintaining the maximum possible motivation. The aim of this paper is defining a useful model or profile of the learner to be used in the future in such AL systems.

The model of a learner is a compact definition of the features that characterise the individual as learner. A useful learner profile must include variables describing characteristics of the individuals about their traits (learning styles, intelligence type, natural skills, circadian cycles), their behaviour during the learning tasks (use of the learning platform) and the results of their learning process (learning history). The objective of this research is proposing a learner model that represents the characteristic profile of a learner and it is made of variables that are meaningful, measurable, discriminating and independent.

Section 2 presents the background of this research, focused on the definition of Adaptive Learning, the features of learners that are able to characterise them and the concept of feature vector. The definition of the learner model is presented in Sect. 3, including features about behaviour, performance and personal traits. Section 4 is devoted to discussing how to obtain a feature vector from the learner model. Finally, conclusions and future works are included in Sect. 5.

2 Background

2.1 Adaptive Learning

As described previously, our current educational environment needs to assume new learning paradigms, a digital transformation to fulfil the learning needs of our current learners. A transformation that should reside on the core concepts of adaptive and personalised learning.

The concept of adaptive learning is based on measuring the learners progress, their learning styles and adapting the learning process to each of them. It consists on a continuous flow of lessons and activities presented to the learners according to their individual features and measuring their responses. This concept also implies providing different paths by giving learners the chance of choosing their own learning way. Including the personalised learning concept means to focus the learning process in the each individual learners, in their needs and context, and providing the suitable content and features for each of them [2]. Combining these both concepts, a learning system could process course analytics and learning assessments data, and use them to feed suitable algorithms that will allow it to cover the learners need and improve their skills [3].

When the learning model that adapts the students learning process to their individual situations, contexts, learning needs and styles, uses technology, we usually refer to it as an Adaptive Learning System or Smart Learning System [4–6].

2.2 Learner Features

Recent meta-analytic investigations made by several researchers [7–10] have identified the individual features of the students themselves as the most important factor in academic performance. Hattie [9] states that the individual features of students, such as their intellectual capacity and motivation, suppose a 50% of their performance, and the teachers, teaching methods, the school and family suppose the other 50%. Within these individual features, many researches state that each person has different skills and learns in a different way [11, 12].

So, when designing a learning process, it is important to assume that in the same group there could be students with a different learning level and different types of intelligence. Regarding this last aspect, some studies state that there are different types or styles of intelligence [13, 14]. According to Gardner and his theory of multiple intelligences, it is divided in eight styles: linguistic, logical-mathematical, visual-spatial, musical, bodily-kinaesthetic, interpersonal, intrapersonal and naturalistic [13, 15]. On the other hand, the named tri-archic theory of intelligence categorised intelligence into three subtypes: componential that refers to the individuals mental mechanisms-, experimental define intelligence as the relations between the individual and their gained experience- and contextual relates the intelligence of the individual with their context and the way they put their componential processes into practice [14]. Furthermore, there is another indicator that reflects the difference between the individual features of learners, concretely the personality. Myers-Briggs states four indicators to categorise personality: extraversion-introversion, sensing-intuition, thinking-feeling and judging-perceiving [16].

On the other hand, regarding other of the features mentioned, motivation, according to Webb, Sheeran and Wilms it is stated as one of the most determining factor in learning results [17, 18]. Concretely, between the different motivation types that are involved in a learning process, is the achievement motivation the one that presents a greatest influence is performance, and implies dedication, persistence and self-commitment. To increase this motivation, it is important that the learner feels capable enough for completing a concrete task. The learners have to success in academic process to motivate themselves towards a goal, and to do so they need to face tasks both challenging and with some success guaranteed. As Csikszentmihalyi states, when a student is doing a task suitable according to their skills, they are in a state of flow, in which motivation is at its highest level [19]. However, if the task is too difficult the learners enter in a state of anxiety, while if it is too easy, they get bored. Therefore, an adaptive learning system must keep the students in a flow state and increase progressively the difficulty of the task given to them, while improving their skills.

Great part of our intelligence is being learned and defined both by the amount and the quality of the strategy we follow in learning, solve problems and deal along our lives. Intelligence is no longer a static or inherited feature, but a skill or set of skills that uses different types of strategies. This is where the interest in learning strategies reside on [20]. The trend to use one strategy or another is related with the learning style of each person. These styles determine the organisation and management of leaning strategies, and the willing to use a strategy independently from the learning task [21]. Besides, the learning strategy is the set of activities the person completes to process all the information when facing a concrete learning task [22]. Regarding learning styles, Kolb identifies four types of them: diverging, assimilating, converging and accommodating, registered in the known as Learning Style Inventory [23].

Furthermore, physiological aspects must also be taken into account, since it has proved that circadian rhythms of each person present a set of cyclical daily patterns that take an important part in many processes, concretely in learning. This is known as chronotypes, divided in two types: morning, also called lark, and evening, also called owl. Indeed, some studies conclude that chronotype plays an important part in the learning strategies learners use beyond their personality [24].

Besides these mentioned studies, there are also evidences observed from neuronal activity which reveal that not everyone feels the learning process in the same way, and learners prefer different types of strategies to compete this process. Strategies that can be visual, technical, linguistic, reformulation, logical organisation, and more. It therefore seems that teaching needs to be adapted to learners individual feature. And to do so, learning processes should embrace different strategies, according to the variety of skills, interests or motivations the learners present, aiming to satisfy their personal preferences and that show them the utility of what they are learning [25].

2.3 Feature Vectors

In a learning system, a learner model is a knowledge resource that contains all the learners features considered significant for the system performance [26]. That is, the learner model stores all the details about learners preferences, their behaviour while using the system and the context the learning process is developed, for helping the system to adapt to each of them. It is formally defined as a feature vector that contains all the variables that defines each learner, in which each variable represents a type of information stored and the concrete value of it.

A feature is an explanatory variable used to characterise an aspect of a phenomenon. Therefore, a feature vector is a set of measurable variables of a phenomenon being observed [27]. Feature vectors are used in machine learning algorithms to characterise the individuals and to perform tasks such as classification, regression, clustering and so on. Features must be meaningful, discriminating and independent so that effective machine learning algorithms can be applied.

One discipline that has received much attention in recent times within the research community is Learning Analytics [28]. Within this field, the students behaviour within a Learning Management System (LMS) is used to evaluate them [29]. Other studies, in addition to behavioural data, also use data corresponding to learning outcomes [30]. In other cases, it is proposed to go further and try to predict the performance of students at the end of the course from their performance in the LMS after the first steps in that course [31] or even redesign the course from an instructional perspective [32]. All these systems propose, in short, the definition of a feature vector, although in most cases only external factors are taken into account (behaviour and performance) but not internal factors of the individual (preferences, learning styles, personality …). Some of these works are an inspiration for this research, although we propose an extension of the feature vector to consider individuals internal factors.

3 Learner Model

The student or learner model, in our context, is the set of information obtained and retained by a learning system about the learner. It may include information about achievement, learning level, knowledge, skills, preferences, interaction, behaviour, progress, and so forth [33, 34]. This set represents the core of the adaptation process in an adaptive learning system, since the system uses each corresponding model to adapt to a learner individually.

Three main types of information are included in our learner model: behavioural features, performance features and personal features. In the following sections these three types of information or features are explained in detail, defining the corresponding concepts and proposing measurement instruments.

3.1 Behavioural Features

Behavioural features are made up of all that information that indicates how the student behaves during his or her learning tasks. It includes, therefore, information about the time spent on tasks, the start and end times of these tasks, if there are frequent interruptions, changes of activity, and so on.

In an automatic learning system, collecting these data is very simple: it is enough to take a log of the student's activities in the system. The events that occur during the interaction between students and the system are logged in an event database, with their appropriate timestamp and related information. They made up the behavioural data. This is what most Learning Analytics systems do: they get all this data from the LMS and try to obtain the students' performance from their behaviour while using the platforms [30, 35].

3.2 Performance Features

The performance features are made of all the domain specific data, that is, information obtained from the learners status and the level of their knowledge and skills in a concrete learning process [36], what we have also called external or contextual factors. It involves all the skills achieved by the learner, their concrete status of the existing learning element (such as a topic, concept, knowledge, etc.) at any time, and the possible mistakes or fails made by them while progressing in the learning process. Besides, it is necessary to include specific data about the learner performance, that means data such as time taken to complete a learning activity, the number of successfully and unsuccessfully completed ones, or prior knowledge of the learner.

All in all, performance features make up the student's learning history. Learning Analytics systems have not paid as much attention to this type of data as to behavioural data. However, it is not complicated to obtain performance data from most LMS, since some events logged in these automatic learning systems can be used to make up performance features: grades, activity results, percentage of correct results when performing the tasks, time to solve each problem, and so on. In general, all these learning outcomes are qualified are good characterisers of the competences and skills of the learners.

3.3 Personal Features

Personal features include domain independent information, that is information of a learner that includes their learning goals, background, motivational state, and their own preferences [36]. This type of information also embraces learners personal features, which involves their learning style, types of intelligence, personality or even chronotypes. In few words, what we have called individuals internal factors.

For defining the domain independent information or internal factors of the learners, four aspects appear to be especially important in cognitive functioning [37]:

- Spatial qualities, which refer to the concrete space (with which we connect through the senses) and the abstract one (with which we connect through intelligence, imagination, emotions and intuition).
- Time, controlled by the order and structure of reality: sequential (linear, serialised) or random (non-linear, multidimensional).
- The mental processes of deduction and induction.
- Relationships, which reinforce individuality or collaboration and sharing.

Gallego [38] makes an interesting overview of different points of view on the definition of learning styles that can be found in literature, to end up proposing his own definition: Learning styles are the cognitive, affective and physiological traits that serve as relatively stable indicators of how learners perceive, interact and respond in their learning environments.

Several authors [38–41] consider that learning styles, in the broadest sense, are fundamental in deciding which teaching strategies to choose and, therefore, they are the main element in defining the student profile. In the following paragraphs a selection of the main theories and their instruments for measuring learning styles is presented. These instruments will be the starting point for the definition of our student model.

- **Canfield's Learning Styles Inventory** [42]
  - Learners styles:
    Social learners.
    Independent learners.
    Conceptual learners.
    Applied learners.
  - Self-report questionnaire, to evaluate the reactions or feelings about some learning situations.
  - 30 questions, to be ranked 1 to 4 (less to more preference).

- **Dunn and Dunn Learning Style Model Productivity Environmental Preferences Survey** [43]
  - Environmental preferences:
    Immediate environment (sound, light, temperature and design).
    Emotionality (motivation, persistence and structure).
    Sociological needs (self-oriented, peer-oriented or authority-oriented).
    Physical needs (perceptual preference(s), food intake and mobility).
  - Self-report questionnaire, to evaluate productivity environmental preferences.
  - 100 statements to be ranked in a Likert-type scale (agree/disagree).

- **The Grasha-Riechmann Student Learning Styles Scale** [44]

– Learners styles:

Avoidant.
Collaborative.
Competitive.
Dependent.
Independent.
Participant.

– Self-report questionnaire, to evaluate the student attitudes toward learning, classroom activities, teachers, and peers.
– 60 items that measure the 6 styles. For any style, there are 10 statements to be ranked in a 5-degrees Likert scale (agree/disagree). Each style is ranked as through the ten questions each that are averaged together to measure dominance in one or more of the six learning styles.

- **Rezler and Rezmovick learning preference inventory** [45]

– 3 bipolar pairs (6 dimensions) to register the preferences for the following kinds of learning:

Individual/Interpersonal.
Student-structured/Teacher- structured.
Abstract/Concrete.

– Self-report questionnaire, to evaluate the preference about the different kinds of learning. Two parts:

Part 1: 6 groups of 6 words, each word to be ranked 6 to 1 (promotes learning most, promotes learning least).
Part 2: 9 questions, 6 possible answers each, each answer to be ranked 6 to 1 (promotes learning most, promotes learning least).

- **The Revised Two Factor Study Process Questionnaire (R-SPQ-2F)** [46]

– 2 dimensions to study:

Motives.
Strategies.

– Each dimension is approached with 2 possible depths:

Deep approach.
Surface approach.

– Self-report questionnaire, to evaluate attitudes towards studies and usual way of studying.
– 20 statements, to give one of five possible answers for each statement:

A: this item is never or only rarely true of me.
B: this item is sometimes true of me.
C: this item is true of me about half the time.
D: this item is frequently true of me.
E: this item is always or almost always true of me.

- **Approaches and Study Skills Inventory for Students** [47]

– 3 possible students approaches to learning and studying:

Deep approach.
Strategic approach.
Surface approach.

– Self-report questionnaire, to evaluate the attitudes of the students about learning and studying.
– 18 statements (6 statements about each approach) to be ranked in a 5-degree Likert scale (agree/disagree).

- **The Kolb Learning Style Inventory 4.0 (KLSI 4.0)** [48]

– 9 learning styles:

Initiating.
Experiencing.
Creating.
Reflecting.
Analysing.
Thinking.
Deciding.
Acting.
Balancing.

– Self-report questionnaire, to evaluate the attitudes of the students about learning and studying.
– 20 items:

12 short questions to be finished with one up of four possible sentence endings (corresponding to the four learning modes: experiencing, reflecting, thinking and doing).
8 additional items about learning in different contexts, used to assess learning flexibility.

- **Cuestionario Honey-Alonso de Estilos de Aprendizaje (CHAEA)** [37]

– 4 learning styles:

Active.
Reflexive.

Theoretical.
Pragmatic.

- Self-report questionnaire, to evaluate the identification with the four learning styles.
- 80 statements to be ranked + (agree) or (disagree)

• **Myers Briggs Type Indicator (MBTI)** [16]

- 4 dichotomic dimensions (not a polar opposite, but a gradual continuum):

  Extraversion/Introversion.
  Sensing/Intuition.
  Thinking/Feeling.
  Judging/Perceiving.

- Sixteen personality types result from the cross-products of these four dimensions.

• Self-report questionnaire, to evaluate the psychological preferences in how people perceive the world around them and make decisions.
• 93 forced-choice format questions (the respondent is required to choose between two answers in order to identify which is naturally preferred). The options are pairs of words or short sentences.
• The MBTI is scored using psychometric techniques.

• **The Matching Familiar Figures Test (MFFT)** [49]

- 4 possible personalities (related to reflexivity or impulsivity) represented in a matrix with four quadrants:

  Errors in the horizontal axis.
  Latency in the vertical axis.

- Possible personalities:

  Efficient: few errors, low latency.
  Reflexive: few errors, high latency.
  Impulsive: many errors, low latency.
  Inefficient: many errors, high latency.

- Skill questionnaire to identify the degree of reflexivity or impulsivity.
- 20 items, each one made of one model figure, and 6 different figures to compare to (only one is identical to the model). Correctness and time are measured.

Most automatic learning systems do not take these personal characteristics into account when characterising the student. We will take these tests described in the literature as a starting point to try to obtain these characteristics automatically in the learning platform. Table 1 summarises the mentioned categories of features, showing their type, what they represent and some examples of the information included.

Table 1 Comparative feature overview

Type	Description	Examples
Behavioural	How the student behaves during the learning tasks	Time spent, starting time, changes of activity
Performance	The learners status and level of their skills in the learning process	Achieved skills, time taken to complete activities, completed activities
Personal	Information that depends only on the learner, independently of the system	Learning goals, preferences, learning style

4 Towards a Feature Vector

Once all the necessary learners features to be analysed have been identified and defined and so the learner model, it is time to bunch them all in some way that a computing algorithm can interpret and manage, and this can be done with a so-called feature vector. In this vector, each contained feature is defined as a property of a concrete event to be observed, and it is used in machine learning algorithms to characterise the individuals, or in this case, the learners [27]. For being suitable to be automatically processed, features in the vector must be meaningful, measurable, discriminating and independent.

In our context, a feature is meaningful if it makes sense to a human observer, that is, if the mathematical representation it assumes is interpretable by a person and if it corresponds to what it actually represents. In this case, the features presented are based on models proposed by researchers to explain the complex learning process. Although a mathematical representation is necessary, the basis on which this representation is developed is solid from the point of view of educational research.

All these features contained in the feature vector must also be measurable, so that they will have an interpretable value for the algorithms, which can be quantitative, such as integer or real; or qualitative, also named categorical, that contains a value inside a pre-defined range, e.g. a variable that represents if an activity have been completed successfully or not could have two possible values like failed or passed [50].

And to be able to correctly classify the individuals, the features must also be discriminating to allow differentiate them into categories. A deep analysis of the features must be done to select the more discriminating features.

Finally, features in the vector must be independent, that is, the value of every feature is not related with the value of the other variables. Eliminating the dependent variables simplifies the problem, makes it easier from a computational point of view and avoids redundant information.

In order for these data to be really useful, it is necessary to compile a big volume of quality data because the larger amount of information the system has to analyse and interpret, the better the results will be. But there is another critical aspect these collected data require, their relevance. The data need to provide the best possible information about the features being represented and that are wanted to be studied. So, if a really significant learner model is pretended to be defined, it is necessary to identify the features to be extracted, represent each of them with a suitable type of data and collect as much information as possible.

The problem comes when the type of data that is pretended to be extracted is not available and then any information left is used, turning out to an amount of less relevant data, so the system is not able to really characterise the individuals. In the educational environment, for example, it is the case of Learning Management Systems or LMS, which store a large amount of data about the learners activity while using the platform, but it is not complete enough to analyse. This is due to that the data provided only reflect the use of the platform, the behaviour of the learner, but not the results of their learning process. Behavioural data captured from LMSs are well structured and easy to be obtained, but learning is a very complex human activity, which cannot be simplified. As it has been mentioned previously, to correctly describe each learner with a suitable set of characteristics, the created learner model must contain data extracted from different sources, both from platform specific data and from the learner itself. Long and Siemens [51] consider that there is a risk to return to behaviourism as a learning theory if we confine analytics to behavioural data.

Once all the corresponding data have been correctly collected and the desirable features identified, there is a required work in transforming all those features into measurable values, one of the mentioned requisites of them. Since the feature vector must be able to correctly represent these values, they must be tangible. So, after detecting the features that will compose the vector, there is a conversion process. It would consist in detecting the type of value that each feature represents, that is, if it can be measured as a number or as a value inside a defined range. For example, the number of successfully completed activities could be expressed as an integer value, but the learning style would be inside a range of predefined values. Then, the system must be able to detect the indicators that define the feature and compute them to create the vector.

5 Conclusions and Further Work

Adaptive learning system require the definition of the learner model in order to characterise the several facets of an individual when learning. A first contribution of this paper is the identification of the main types of information to be included in the learner model: behavioural features, performance features and personal features. The behavioural features include information on how the learner performs during the learning process: time spent on tasks, start and end times, changes of activity, and so on. In an automatic learning system, these data are obtained from the events that occur during the interaction between students and the system. The performance features represent the learners status and the level of their knowledge and skills, that is, their learning history. The automatic learning systems can store information about grades, activity results, percentage of correct results and so on to make up the performance features. Finally, the personal features refer to the learners goals, background, motivational state, preferences and learning styles, for instance. Although most automatic learning systems do not take these personal characteristics into account, the wide research on measurement instruments can be a starting point to obtain them automatically from the learning platforms.

A second important contribution is the definition of some guidelines for converting the learner model into a feature vector that can be used in the future in an automatic adaptive learning system. For a features vector being suitable to be automatically processed, features in the vector must be meaningful, measurable, discriminating and independent.

This research is the starting point for designing an automatic adaptive learning system. In the future, we intend to design the features vector and validate it to meet the properties mentioned. It will probably be necessary to revise the features vector to reduce the dimensions and obtain a more efficient system. Finally, the optimised feature vector should allow us to develop a learning system based on artificial intelligence that becomes a true adaptive learning system.

Acknowledgements This research is partially supported by *Unidad Científica de Innovación Empresarial "Ars Innovatio"*, *Agència Valenciana d'Innovació* and University of Alicante, Spain.

References

1. Hwang, G.J.: Definition, framework and research issues of smart learning environments–a context-aware ubiquitous learning perspective. Smart Learn. Environ. **1**(1), 4 (2014)
2. Greany, K.: Adaptive learning: how to personalize your learning strategy (2018)
3. Shelle, G., Earnesty, D., Pilkenton, A., Powell, E.: Adaptive learning: an innovative method for online teaching and learning. J. Extension **56** (2018)
4. Molina-Carmona, R., Villagrá-Arnedo, C.: Smart learning. In: Proceedings of the Sixth International Conference on Technological Ecosystems for Enhancing Multiculturality—TEEM'18, pp. 645–647. ACM Press, Salamanca, Spain (2018)
5. Real-Fernández, A., Molina-Carmona, R., Llorens-Largo, F.: Aprendizaje adaptativo basado en competencias y actividades–(Adaptive learning based on competences and activities). La innovación docente como misión del profesorado : Congreso Internacional Sobre Aprendizaje. Innovación y Competitividad, pp. 1–6. Servicio de Publicaciones Universidad, Zaragoza, Spain (2017)
6. Real-Fernández, A., Molina-Carmona, R., Llorens-Largo, F.: Smart system based on adaptive learning itineraries. In: Poster Presentation in the Sixth International Conference on Technological Ecosystems for Enhancing Multiculturality—TEEM'18, pp. 654–659. ACM Press, Salamanca, Spain (2018)
7. Ahn, S., Ames, A.J., Myers, N.D.: A review of meta-analyses in education: methodological strengths and weaknesses. Rev. Educ. Res. **82**(4), 436–476 (2012)
8. Hattie, J.: Visible Learning: A Synthesis of Over 800 Meta-Analyses Relating to Achievement. Edición, 1 edn. Routledge, London; New York (2008)
9. Hattie, J.: Visible Learning for Teachers: Maximizing Impact on Learning. Routledge (2013)
10. Hattie, J., Anderman, E.M. (eds.).: International Guide to Student Achievement, 1 edn. Routledge, New York (2012)

11. Castejon, J.L., Perez, A.M., Gilar, R.: Confirmatory factor analysis of project spectrum activities. A second-order g factor or multiple intelligences? Intelligence **38**(5), 481–496 (2010)
12. Sternberg, R.J., Castejón, J.L., Prieto, M.D., Hautamäki, J., Grigorenko, E.L.: Confirmatory factor analysis of the Sternberg Triarchic Abilities Test in three international samples: an empirical test of the triarchic theory of intelligence. Eur. J. Psychol. Assess. **17**(1), 1–16 (2001)
13. Gardner, H.: Intelligence Reframed: Multiple Intelligences for the 21st century. Basic Books, New York (2000) OCLC: 247819868
14. Sternberg.: Beyond IQ Paperback: A Triarchic Theory of Human Intelligence. Cambridge University Press, Cambridge (2009)
15. Gardner, H.: Multiple Intelligences: Reflections After Thirty Years. National Association of Gifted Children Parent and Community Network Newsletter (2011)
16. Quenk, N.L.: Essentials of Myers-Briggs Type Indicator Assessment, 2nd edn. Essentials of Psychological Assessment Series. Wiley, Hoboken (2009) OCLC: ocn320494042
17. Webb, T.L., Sheeran, P.: Does changing behavioral intentions engender behavior change? A meta-analysis of the experimental evidence. Psychol. Bull. **132**(2), 249–268 (2006)
18. OECD.: Student Engagement at School: A Sense of Belonging and Participation: Results from PISA 2000. PISA. OECD (2003)
19. Csikszentmihalyi, M.: Flow: The Psychology of Optimal Experience. Perennial Modern Classics. Harper & Row (1990)
20. Castejón Costa, J.L.: Introducción a la psicología de la instrucción. Editorial Club Universitario (1997)
21. Messick, S.J.: Structural relationships across cognition, personality, and style. In: Snow, R.E., Farr, M.J., Farr, M.J. (eds.) Aptitude, Learning, and Instruction: Volume 3: Cognitive and Affective Process Analyses, vol. 3. Routledge, Hillsdale (1987)
22. Schmeck, R.R. (ed.).: Learning Strategies and Learning Styles. Perspectives on Individual Differences. Springer, US (1988)
23. Kolb, D.A.: Facilitator's Guide to Learning. Hay Group Transforming Learning (2000)
24. Ruffing, S., Hahn, E., Spinath, F.M., Brünken, R., Karbach, J.: Predicting students' learning strategies: the contribution of chronotype over personality. Pers. Individ. Differ. **85**, 199–204 (2015)
25. Jensen, A.R.: The g Factor: The Science of Mental Ability. The Science of Mental Ability. Praeger Publishers/Greenwood Publishing Group, Westport, The g Factor (1998)
26. Carberry, S., Carbonell, J.G., Chin, D.N., Cohen, R., Lehman, J.F., Finin, T.W., Jameson, A., Jones, M., Kass, R., Kobsa, A., McCoy, K.F., Morik, K., Paris, C.L., Quilici, A.E., Rich, E., Jones, K.S., Wahlster, W.: User Models in Dialog Systems. Softcover reprint of the original, 1st edn. (1989 edn.) Springer (2011)
27. Bishop, C.M.: Pattern Recognition and Machine Learning. Information Science and Statistics. Springer, New York (2006)
28. Siemens, G.: Learning analytics: the emergence of a discipline. Am. Behav. Sci. **57**(10), 1380–1400 (2013)
29. Niles-Hofmann, L.: Data-Driven Learning Design
30. Villagrá-Arnedo, C., Gallego-Duraìn, F.J., Compañ-Rosique, P., Llorens-Largo, F., Molina-Carmona, R.: Predicting academic performance from behavioural and learning data. Int. J. Des. Nat. Ecodyn. **11**(3), 239–249 (2016)
31. Villagrá-Arnedo, C.J., Gallego-Durán, F.J., Llorens-Largo, F., Compañ-Rosique, P., Satorre-Cuerda, R., Molina-Carmona, R.: Improving the expressiveness of black-box models for predicting student performance. Comput. Hum. Behav. **72**, 621–631 (2017)
32. Molina-Carmona, R., Villagrá-Arnedo, C., Gallego-Durán, F., Llorens-Largo, F.: Analytics-driven redesign of an instructional course. In: Proceedings of the 5th International Conference on Technological Ecosystems for Enhancing Multiculturality—TEEM 2017, pp. 1–7. ACM Press, Cádiz, Spain (2017)
33. Fröschl, C., Nguyen, L., Do, P.: Learner Model in Adaptive Learning, vol. 35. Paris (2008)
34. Bull, S., Kay, J.: Open learner models. In: Kacprzyk, J., Nkambou, R., Bourdeau, J., Mizoguchi, R. (eds.) Advances in Intelligent Tutoring Systems, vol. 308, pp. 301–322. Springer, Berlin Heidelberg (2010)

35. Macfadyen, L.P., Dawson, S.: Mining LMS data to develop an "early warning system" for educators: a proof of concept. Comput. Educ. **54**(2), 588–599 (2010)
36. Fröschl, C.: User Modeling and User Profiling in Adaptive E-learning Systems: An Approach for a Service-Based Personalization Solution for the Research Project AdeLE. VDM Verlag Dr, Müller (2008)
37. Alonso, C.M., Gallego, D., Honey, P.: Los Estilos de Aprendizaje: Procedimientos de diagnóstico y mejora, 7th edn. Ediciones Mensajero, S.A., Bilbao (2007)
38. Gallego, D.: Diagnosticar los estilos de aprendizaje (2019)
39. Gallego Rodríguez, A., Martínez Caro, E.: Estilos de aprendizaje y e-learning. Hacia un mayor rendimiento académico. Rev. de Educación a Distancia (7) (2003)
40. García Cué, J.L., Santizo Rincón, J.A., Alonso García, C.M.A.: IInstrumentos de medición de estilos de aprendizaje. J. Learn. Styles **2**(4) (2009)
41. Palomino Hawasly, M.A., Strefezza, M., Contreras Bravo, L.E.: Sistema Difuso Para la Detección Automática de Estilos de Aprendizaje en Ambientes de Formación Web. Ciencia, Docencia y Tecnología **27**(52), 9 (2016)
42. Canfield, A.A.: Western Psychological Services (Firm): Canfield Learning Styles Inventory (LSI) Manual. Western Psychological Services, Los Angeles (1988)
43. Price, G.E., Dunn, R., Dunn, K.J.: Productivity Environmental Preference Survey: An Inventory for the Identification of Individual Adult Preferences in a Working Or Learning Environment. Peps Manual, Price Systems (1991)
44. Grasha, A.F.: Teaching with Style : A Practical Guide to Enhancing Learning by Understanding Teaching and Learning Styles. Alliance Publishers (1996)
45. Rezler, A.G., Rezmovic, V.: The learning preference inventory. J. Allied Health **10**(1), 28–34 (1981)
46. Biggs, J., Kember, D., Leung, D.Y.: The revised two-factor study process questionnaire: R-SPQ-2f. Br. J. Educ. Psychol. **71**(1), 133–149 (2001)
47. Entwistle, N., Tait, H.: Approaches and Study Skills Inventory for Students (ASSIST) (Incorporating the Revised Approaches to Studying Inventory—RASI) (2013)
48. Kolb, D., Kolb, A.: The Kolb learning style inventory 4.0: guide to theory. Psychometr. Res. Appl. (2013)
49. Kagan, J., Rosman, B.L., Day, D., Albert, J., Phillips, W.: Information processing in the child: significance of analytic and reflective attitudes. Psychol. Monogr. Gen. App. **78**(1), 1–37 (1964)
50. Aggarwal, C.C.: Machine Learning for Text, 1st edn. Springer (2018)
51. Long, P., Siemens, G.: Penetrating the fog: analytics in learning and education. Educase Rev. (2011)

Gender and Learning Outcomes in Entrepreneurship Education

Raquel Ferreras-Garcia, Ana Beatriz Hernández-Lara and Enric Serradell-López

Abstract This paper aims to analyze whether the competences worked in a business plan have any kind of influence on students' learning outcomes by means of gender. In order to analyze these effects we present an empirical study to know the relationship between gender, competences and learning results in a business plan course. To do so, we use data obtained from questionnaires distributed among 425 students of the Bachelor in Business Administration at the Universitat Oberta de Catalunya. The questionnaires measure the students' perception of competences achievement and their learning results. Our main aim is to analyze whether or not the competences developed using the business plan methodology influence the students' learning results, as well as the influence of gender, comparing male and female students. The regression analyses show that acquiring competences related to "time management" have a positive influence on the assessment of the business plan as a good learning tool. By the other hand, acquiring "specific competences of the business plan" have a positive influence on students' level of satisfaction, moreover gender affects positively over the "time management" competences. However, in both cases, women achieve lower learning results than men and older students get higher learning outcomes. The findings show a negative effect of being female on learning results, which reveals that the learning results of women were poorer than those of men using the business plan methodology. Another key finding is the great positive influence of age and competences related to business plans and time management interacted with gender on learning outcomes.

R. Ferreras-Garcia · E. Serradell-López (✉)
Universitat Oberta de Catalunya, Barcelona, Spain
e-mail: eserradell@uoc.edu

R. Ferreras-Garcia
e-mail: rferreras@uoc.edu

A. B. Hernández-Lara
Universitat Rovira i Virgili, Reus, Spain
e-mail: anabeatriz.hernandez@urv.cat

A. Visvizi and M. D. Lytras (eds.), *Research & Innovation Forum 2019*,
Springer Proceedings in Complexity,
https://doi.org/10.1007/978-3-030-30809-4_9

1 Introduction

The competence-based educational approach implemented by the European Higher Education Area (EHEA) [8] implies a shift of focus from contents to competences, which become the crucial element of the learning process [11, 17]. This new approach assesses competences as a result of learning and has fostered sensible changes in the university model, affecting its organization, objectives and activities, relocating the interest from the acquisition of knowledge to the individual who learns [14].

All these reforms have called the attention of researchers who have been studying considerably the role of competences within the teaching/learning process [11, 12, 18].

Within the field of business studies, entrepreneurial competences, emerge as key skills to promote employability and exert a relevant impact on the emergence of new firms and on companies' growth and success [2, 31, 32]. Previous research has defined and identified entrepreneurial competences [4, 7, 27], have justified their relevance [6, 23, 31, 35], as well as the pedagogy and learning methodologies used to develop them [25], highlighting the prevalence of business plans [6, 10, 19].

What has remained mostly unexplored is the effectiveness of business plans in terms of their learning value, translating the competences fostered by them into real learning outcomes.

Another unexplored aspect of previous research is related to the gender dimension in the learning value of business plans in entrepreneurship education. Gender differences have attracted the attention of research in the fields of education, management and entrepreneurship.

The objective of this work is to analyze whether or not the competences developed using the business plan methodology influence the students' learning results, as well as the influence of gender, comparing male and female students.

2 Literature Review

2.1 Entrepreneurs' Competences and Learning Outcomes

Like in the general concept of competence, entrepreneurs' ones refer to a set of knowledge, abilities, attitudes and features for effective and successful work performance [3, 27], in this case referring to start or transform enterprises [4]. Therefore, entrepreneurs or individuals who start or transform a business, presumably have entrepreneurs' competences [27].

Despite the considerable amount of research carried out on entrepreneurs' skills, it is still difficult to find a precise identification of them, existing an open debate on the equivalence between entrepreneurial competences and the skills that entrepreneurs' need.

Chandler and Jansen [7] described the competences needed by entrepreneurs, distinguishing three main families: managerial, technical/functional and entrepreneurial. Thus, entrepreneurial competences are only one of several types that the entrepreneur needs, although there is agreement in the literature [2, 31] on their relevance to start and transform enterprises.

More recently Mitchelmore and Rowley [27], proposed a list of key competences that entrepreneurs should have, after considering different frameworks proposed by previous research [7, 26, 34], and divided them in four groups that entail entrepreneurial skills, business and management skills, human relations skills and conceptual and relationship skills. Penchev and Salopaju [28] added a fifth group of competence in the bunch of entrepreneurs' skills, which included attitudes and features.

These competences demand a learning foundation to be properly acquired [20, 23, 29]. Therefore, entrepreneurship teaching need dynamic methodologies that help future entrepreneurs to develop the competences that they will need to start and transform businesses successfully [31, 33]. Business plans are one of these methodologies typically used in entrepreneurship education.

Previous research has emphasized the value of business plans, considered essential for completing projects successfully and for increasing the chances of success in new business, especially in their initial start-up phase [5, 13]. However, despite the previous studies defending the usefulness and effectiveness of business plans, little attention has been paid to their learning value on the basis of how the competence fostered by them are translated into learning outcomes.

Other studies have explored previously the connection between competences and learning outcomes using other learning methodologies. For example [12], explored the link between the generic competences acquired by students using business games and learning outcomes [16, 18], analyzed the influence of students' interactivity on learning results.

To sum up, there is scarce knowledge on which are the most efficient competences in terms of learning.

2.2 Gender in Entrepreneurship Education

The first step to become a successful entrepreneur is to train and acquired entrepreneurial competences. In this sense, some differences can be also observed with regards to how men and women learn, as well as their specific abilities and features to learn.

In particular, in higher education, gender differences influence students' behaviors and their performance in terms of learning [15].

In the specific case of entrepreneurial education previous findings also reveal gender differences [21, 37, 38]. Other findings of previous research that has considered gender as a segmentation variable, centered in the comparison between men and women, have stated that there are few differences in the perception of entrepreneurial skills between men and women [20]. However [24], assert that women perceive that they have higher entrepreneurial skills than men, while [30, 36] show that men are the ones most prone to create new businesses.

These inconclusive results of previous studies show the necessity to go further in the understanding of the role of gender on entrepreneurial education.

This paper aims to analyze whether the competences worked in a business plan have any kind of influence on students' learning outcomes by means of gender.

3 Methodology

3.1 Data Collection

The data used to carry out this study was collected from students on different courses at the degree in Business Administration and Management at the Universitat Oberta de Catalunya (UOC), specifically students enrolled on the final bachelor's degree project course in the entrepreneurship specialization during the 2014/15 second semester, 2015/16 and 2016/17 academic years.

The questionnaire was designed to obtain the students' perception of competences achievement and the learning results obtained from the development of a business plan. In total the questionnaire was completed by 425 students.

The first part of the questionnaire includes two questions to gather information on gender and age. The second, third, fourth and fifth parts correspond to the different typologies of entrepreneurs' competences adapted from previous research, which refer to generic competences, specific managerial competences, cross-disciplinary competences and specific competences of the business plan [1, 9–11]. The final part collects the learning results of the students who took part on the business plan. The competences and learning results are described in Table 1.

3.2 Measurement of Variables

The competences acquired by students were measured using an exploratory factor analysis with varimax rotation in order to reduce the large number of competences into a few interpretable underlying factors.

With regards to competences, the factorial analysis was applied over their different typologies. The generic competences' factor analysis generated three factors, related to decision making, attitude and ICT and time management with a total variance explained of 63.9%. The cross-disciplinary competences' factor analysis generated two factors, related to individual and teamwork respectively with a total variance explained of 63.919%. The factor analysis for the specific managerial competences'

Table 1 Competences and learning results

Generic competences
[C1] Process and analyze a body of general information referring to a company
[C2] Process and analyze partial information referring to parts of a company
[C3] Make decisions
[C4] Draw conclusions from the information obtained or provided
[C5] Relate information or data
[C6] Apply theoretical decision-making concepts
[C7] Manage time
[C8] Solve problems related with deadlines
[C9] Use new technologies
[C10] Creativity
[C11] Capacity for innovation
[C12] Ability to work with uncertainty
Specific managerial competences
[C13] Improve a company's competitive position
[C14] Develop strategies
[C15] Manage risk
[C16] Process and analyze financial information
[C17] Identify and work with sources of relevant financial information
[C18] Integrate ethics in organizational decisions
Cross-disciplinary competences
[C19] Show attitudes and behaviors that are consistent with ethical, responsible professional practice
[C20] Search, identify, organize and make adequate use of information
[C21] Optimally organize and plan the professional activity
[C22] Interpret and assess the information critically and synthetically
[C23] Work as a team, in on-site or online environments, in multidisciplinary environments
[C24] Negotiate in a professional environment
[C25] Communicate correctly, verbally and/or in writing, both in the mother tongue and in a foreign language, in the academic and professional spheres
[C26] Use and apply information and communication technologies in the academic and professional spheres
[C27] Undertake entrepreneurial ventures and innovate
Specific competences of the business plan
[C28] Understand the workings of the economy, its agents and institutions, with particular emphasis on corporate behavior
[C29] Generate relevant economic knowledge from data, applying the appropriate technical tools

(continued)

Table 1 (continued)

[C30] Manage efficiently a company or organization, understanding its competitive and institutional position and identifying its strengths and weaknesses
[C31] Perform efficiently administrative and management tasks in any key company or organizational area
[C32] Evaluate critically specific business situations and establish possible business and market evolutions
[C33] Plan, manage and evaluate business projects
[C34] Focus on results, meeting internal and external customer requirements
Learning results
[R1] The business plan is a good learning tool
[R2] I am satisfied with the experience

and the specific competences of the business plan generated only one factor each one with a total variance explained of 56.185 and 58.63%.

This study controls Gender, measured by dichotomous variable encoded with '0' for men and coded with '1' for women, and Age, defined as a categorical variable with 2 levels in terms of the median, where 0 includes students who are aged between the minimum 22 and 32 and 1 between 33 and the maximum 61.

All the competences and learning results are evaluated using a 5-point Likert scale (from 1 'Strongly disagree' to 5 'Strongly agree').

4 Results

Two regressions analysis were made. The dependent variables of this study are: the assessment of the business plan as learning tool, and the level of satisfaction with the learning experience.

The independent variables of the regressions are the seven factors obtained during the factorial analysis exploration: decision making, attitude and ICT, time management, individual work, teamwork, specific managerial competences and specific competences of the business plan.

The regressions carried out include the control variables, the competences' factors and the interaction terms between competences' factors and Gender.

The aforementioned regressions allow us to compare and contrast the impact of gender and competences developed in the business plan on students' level of learning results.

Regarding the first regression, the variance explained 34.6% of the dependent variable (assessment of the business plan as learning tool). Gender has a negative impact on the results while Age has a positive impact. Both variables are significant. Moreover, the competences' factor related to "time management" has a positive influence on the assessment of the business plan as a good learning tool.

Regarding the second regression, the variance explained 35.9% of the dependent variable (level of satisfaction with the learning experience). As in the first regression, gender has a negative impact, while Age has also a positive impact. Additionally, specific competences of the business plan" have a positive influence on students' level of satisfaction. The regression over the level of satisfaction establishes the moderating effect of gender on the relationship between competences related to "time management" and learning outcomes. The coefficient of this interaction term is statistically significant.

Therefore, from these results, we can conclude that the learning outcomes of women are poorer than those of men and that the learning outcomes of younger students are poorer than those of older students when using the business plan methodology, although significant differences in the impact of the competences on learning outcomes have not been contrasted/observed when we compare men and women except for the competences related to "time management".

5 Discussion and Conclusions

In the introduction of this paper, we highlighted the lack of studies on this subject. This study has enabled us to broaden knowledge on the development of entrepreneurial competencies and the quantification of the achievement of learning results in an entrepreneurship education environment taking into account the impact of gender.

With the reforms involved in the Bologna Process, studies on skills and their relationship with the teaching/learning process have risen considerably. Thus, we have seen that there are many experiences in the scientific literature addressed to the study of competencies and their relationship to education. Until now the literature has not paid special attention to the relationship between competencies and learning outcomes. Although similar studies have been conducting using other methodologies, like business simulation games [12, 16], to our knowledge there is not any prior work that analyses the link between competencies developed in a business plan and learning outcomes.

With this work we aim at solving these limitations analyzing the impact of competencies acquired through business plans on students' learning outcomes.

With regards to the impact of entreprencurs' competencies acquired through a business plan on learning outcomes, we identified two competencies that exert the greatest impact, which are those related to time management and the specific competencies related to the business plan. Our results are partially consistent with those of previous research centered in other learning methodologies, like business simulation games. Another key finding is the great positive influence of age on learning outcomes.

With regards to the gender variables, our results confirm that women obtained poorer learning outcomes than men when using business plans. This result is in line with the poor entrepreneurial activity developed by women in comparison to men [22, 30, 36], which could mean to use training and learning methodologies more adapted to the needs of women. However, our findings in terms of how is the impact of competencies on learning results when men and women are compared do not confirm differences between them, except for the competences related to "time management".

References

1. Agencia Nacional de Evaluación de la Calidad y Acreditación (ANECA).: Libro Blanco. Título de grado en Economía y empresa. Agencia Nacional de Evaluación de la Calidad y Acreditación, Madrid (2005)
2. Bakkali, C., Messeghem, K., Sammut, S.: Les structures d'accompagnement à la création d'entreprise à l'heure de la gestion des compétences. Manage. et Avenir **9**, 149–162 (2010)
3. Bartlett, C.A., Ghoshal, S.: The myth of the generic manager: new personal competencies for new management roles. Calif. Manag. Rev. **40**(1), 92–116 (1997)
4. Bird, B.: Towards a theory of entrepreneurial competency. Adv. Entrepreneurship Firm Emergence Growth **2**(1), 51–72 (1995)
5. Brinckmann, J., Grichnik, D., Kapsa, D.: Should entrepreneurs plan or just storm the castle? A meta-analysis on contextual factors impacting the business planning-performance relationship in small firms. J. Bus. Ventur. **25**, 24–40 (2010)
6. Carrier, C.: L'enseignement de l'entrepreneuriat: au-delà des cours magistraux, des études de cas et du plan d'affaires. Rev. de l'Entrepreneuriat **8**(2), 17–34 (2009)
7. Chandler, G.N., Jansen, E.: The founder's self-assessed competence and venture performance. J. Bus. Ventur. **7**, 223–236 (1992)
8. European Higher Education Area (EHEA).: Bologna Declaration, http://www.ehea.info/Uploads/Declarations/BOLOGNA_DECLARATION1.pdf (1999). Last accessed 10 Oct 2017 (2016)
9. Ferreras-Garcia, R., Serradell-López, E.: Anàlisi de les percepcions de les competències assolides en entorns virtuals d'aprenentatge: El cas dels plans d'empresa de la UOC. Rev. d'Innovació Docent Universitària **8**, 142–154 (2016)
10. Ferreras-Garcia, R., Hernández-Lara, A.B., Serradell-López, E.: Entrepreneurial competences in a higher education business plan course. Education + Training (2019) (Forthcoming)
11. Fitó-Bertran, À., Hernández-Lara, A.B., Serradell-López, E.: Comparing student competences in a face-to-face and online business game. Comput. Hum. Behav. **30**, 452–459 (2014)
12. Fitó-Bertran, À., Hernández-Lara, A.B., Serradell-López, E.: The effect of competences on learning results: an educational experience with a business simulator. Comput. Hum. Behav. **51**, 910–914 (2015)
13. Giunipero, L.C., Denslow, D., Melton, H.L.: Risk propensity, risk perception and business plan formalisation: a conceptual investigation. Int. J. Entrepreneurship Innov. Manage. **8**(4), 397–416 (2008)
14. González, J., Wagenaar, R.: Tuning Educational Structures in Europe: Universities' Contribution to the Bologna Process: An Introduction. University of Deusto (2008)
15. Harris, F., Harper, S.R.: Masculinities go to community college: understanding male identity socialization and gender role conflict. New Direct. Community Coll. **142**, 25–35 (2008)
16. Hernández-Lara, A.B., Serradell-López, E.: Student interactions in online discussion forums: their perception on learning with business simulation games. Behav. Inf. Technol. **37**(4), 419–429 (2018)
17. Hernández-Lara, A.B., Perera-Lluna, A., Serradell-López, E.: Applying learning analytics to students' interaction in business simulation games. The usefulness of learning analytics to know what students really learn. Comput. Hum. Behav. (2018) https://doi.org/10.1016/j.chb.2018.03.001

18. Hernández-Lara, A.B., Serradell-López, E., Fitó-Bertran, À.: Do business games foster skills? A cross-cultural study from learners' views. Intangible Capital **14**(2), 315–331 (2018)
19. Honig, B.: Entrepreneurship education: toward a model of contingency-based business planning. Acad. Manage. Learn. Educ. **3**(3), 258–273 (2004)
20. Kakkonen, M.L.: Students' perceptions of their business competences and entrepreneurial intention. Management **6**(3), 25–243 (2011)
21. Komulainen, K., Korhonen, Räty, M.: Risk-taking abilities for everyone? Finnish entrepreneurship education and the enterprising selves imagined by pupils. Gender Educ. **21**(6), 631–649 (2009)
22. Langowitz, S., Nan, S., Minniti, M., Arenius, P.: Global Entrepreneurship Monitor: 2004 Report on Women and Entrepreneurship. University of Illinois at Urbana-Champaign's Academy for Entrepreneurial Leadership Historical Research Reference in Entrepreneurship (2005)
23. Lans, T., Hulsink, W., Baert, H.: Entrepreneurship education and training in a small business context: insights from the competence-based approach. J. Enterprising Culture **16**(4), 363–383 (2008)
24. Ljunggren, E., Kolvereid, L.: New business formation: does gender make a difference? Women Manage. Rev. **11**(4), 3–12 (1996)
25. Loué, C., Laviolette, E.-M., Bonnafous-Boucher, M.: L'entrepreneur à l'épreuve de ses compétences: éléments de construction d'un référentiel en situation d'incubation. Rev. de l'Entrepreneuriat **7**(1), 63–83 (2008)
26. Man, T.W., Lau, T., Chan, K.F.: The competitiveness of small and medium enterprises: a conceptualization with focus on entrepreneurial competencies. J. Bus. Ventur. **17**(2), 123–142 (2002)
27. Mitchelmore, S., Rowley, J.: Entrepreneurial competencies: a literature review and development agenda. Int. J. Entrepreneurial Behav. Res. **16**(2), 92–111 (2010)
28. Penchev, P., Salopaju, A.: Entrepreneurial Competencies Needed by Managers in their Work. Master thesis, Jönköping International Business School, Jönköping University, Sweden (2011)
29. Pfeifer, S., Borozan, D.: Fitting Kolb's learning style theory to entrepreneurship learning aims and contents. Int. J. Bus. Res. **11**(2), 216–222 (2011)
30. Rodrigues, R., Raposo, M., Ferreira, J., do Paço, A.: Entrepreneurship education and the propensity for business creation: testing a structural model. Int. J. Entrepreneurship Small Bus. **9**(1), 58–73 (2010)
31. Sánchez, J.C.: University training for entrepreneurial competencies: its impact on intention of venture creation. Int. Entrepreneurship Manage. J. **7**(2), 239–254 (2011)
32. Servicio Público de Empleo Estatal: Key competences in Vocational Education and Training—Spain. Cedefop ReferNet Thematic Perspectives Series, http://libserver.cedefop.europa.eu/vetelib/2016/ReferNet_ES_KC.pdf. Last accessed 10 Oct 2017 (2016)
33. Sitzmann, T., Ely, K., Brown, K.G., Bauer, K.N.: Self-assessment of knowledge: a cognitive learning or affective measure? Acad. Manage. Learn. Educ. **9**(2), 169–191 (2010)
34. Smith, B., Morse, E.: Entrepreneurial Competencies: Literature Review and Best Practices. Industry Canada: Small Business Policy Branch (2005)
35. Taatila, V.P.: Learning entrepreneurship in higher education. Educ. + Train. **52**(1), 48–61 (2010)
36. Urbano, D.: New Business Creation in Catalonia: Support Measures and Attitudes Towards Entrepreneurship. Generalitat de Catalunya, Ministry of Employment and Industry, Centre for Business Innovation and Development (2006)
37. Villasana, M., Alcaraz-Rodríguez, R., Álvarez, M.M.: Examining entrepreneurial attributes of Latin American female university students. Gend. Educ. **28**(1), 148–166 (2016)
38. Vivakaran, M.V., Maraimalai, N.: Feminist pedagogy and social media: a study on their integration and effectiveness in training budding women entrepreneurs. Gend. Educ. **29**(7), 869–889 (2017)

Clickbait in Education—Positive or Negative? Machine Learning Answers

Adil E. Rajput

Abstract The topic of clickbait has garnered lot of attention since the advent of social media. Meriam-Webster defines Clickbait as something designed to make readers want to click on a hyperlink especially when the link leads to content of dubious value or interest. Clickbait is used synonymously with terms with negative connotations such as yellow journalism [1], tabloid news etc. Majority of the work in this area has focused on detecting clickbait to stop being presented to the reader. In this work, we look at clickbait in the field of education with emphasis on educational videos that are authored by individual authors without any institutional backing. Such videos can become quite popular with different audiences and are not verified by any expert. We present findings that despite the negative connotation associated with clickbait, the audience value content regardless of the clickbait techniques and have an overall favorable impression. We also establish initial metrics that can be used to gauge the likeness factor for such educational videos/MOOCs.

1 Introduction

Clickbait has gained attention of researchers for a long time and advent of social media has intensified the debate further. In simple terms, the goal of click bait is to attract attention of the users by attractive text or picture or both. Many researchers consider the term clickbait synonymous to yellow journalism where the content is sensationalized without providing any well-researched facts. Yellow Journalism has had a negative connotation as the mainstream population welcomes well-researched news in a society. The attitude towards clickbait is akin to that of yellow journalism as many authors have proposed ways to avoid clickbait on the net [2].

While traditional media such as print media, television etc. would attract mainly corporations who could be able to afford such advertising media, the Internet in general and search engines allowed regular users to advertise for their services as

A. E. Rajput (✉)
College of Engineering, Effat University, Jeddah, Saudi Arabia
e-mail: aiilahi@effatuniversity.edu.sa

A. Visvizi and M. D. Lytras (eds.), *Research & Innovation Forum 2019*, Springer Proceedings in Complexity,
https://doi.org/10.1007/978-3-030-30809-4_10

well. The social media experience amplified the model even further resulting in a network that would allow a wave of global advertisements. Machine Learning, Big Data and Natural Language Processing took the model a step further where users would be targeted per their choices on the Internet (both from a commercial and entertainment perspective). Furthermore, advent of Personal Area Networks (PANs) [3, 4] provided plenty of opportuities to gather massive data pertaining to society including health, education etc. that can provide meaningful insights into users' behavior [5].

Authors in [6] discuss the impact of clickbait in education and mention the case of an academic who garnered massive citations due to the controversial nature of colonialism and why the author in [7] believed that colonialism got a bad reputation. Critics claimed that the author used a controversial topic to gain extra attention in a quick fashion and was akin to the techniques used in clickbait/yellow journalism. This begs the following question: Does clickbait exist in education? Academic institutions globally have always made claims to attract students such as "Quality Education", "Better Job prospects" etc. However, we could not come across any study that would denounce such behavior, as is the case for other industries.

The open courseware project announced by MIT about 15 years ago [8] proved to massively popular with both active and lifelong learning students. However, in this case the reputation of the academic institution backed up the material placed online and gained instant credibility. This ushered an era of Massive Online Open Courses (MOOCs) and educational videos made popular through various social media channels especially YouTube. Various training classes that were traditionally expensive and required physical presence and at times required expensive setup were posted on YouTube—mostly in the hopes of monetizing the knowledge that various people possessed. Given the study in [9], authors of such videos would divide the training material into shorter segments to ensure that users maximize the benefit from such videos. The authors advertise for their educational content by (1) either embedding the link to other videos or their channel in the Description field that can serve content on various topics or (2) at times embedding it in the comments section of their own videos. In both the cases, the author writes text (and images in some cases) that would be enticing to end users. Examples include "Learn to hack into a network", "Learn Python in an hour" etc. The techniques used fulfill all the criteria of clickbait as the majority of the authors make claims that were neither well researched nor verifiable in an objective fashion (at least not easily).

In this paper, we aim to evaluate whether the clickbait model in social media is necessarily perceived as negative. Specifically, we present the following:

1. Briefly establish the case that clickbait has a negative connotation (Sect. 2—Literature Review)
2. Establish the metrics that we use to gauge the interest of viewership (Sect. 3—Experimental Setup)
3. Results of our evaluation (Sect. 4—Results and Discussion)

2 Literature Review

Lowenstein in [10] argues about the psychological factor that underlies the theory of clickbait. Specifically, the author argues about the information-gap of the reader/viewer that would propel the person to click on the underlying link as the title would prove to be enticing to the reader/viewer and they would click on it to satisfy the curiosity. Authors in [11] study listicles from Buzzfeed and notice that 85% use a cardinal number and employ the use of strong nouns. Authors in [12] establish that the use of discourse deixis and cataphora is widely common in ad-centric clickbait links. Specifically, cataphora in linguistic refers to pointing to something later in the text. As an example, "When he arrived, Smith smiled". In this sentence the pronoun, "he" defers the mentioning of Smith to later part of the sentence engaging the reader. Discourse Deixis means using part of a text/speech to refer to part of the utterance that came or has yet to come. As an example, consider the statement "This story will leave you mesmerized". In this sentence, the speaker/writer is referring to a part of speech/text that has not come yet. Authors in [13] develop a corpus by gathering about 3000 popular tweets and present a model on detecting clickbait links. Specifically, the study gathered data on 215 features and divide the data into three categories namely (1) the teaser message (2) the linked webpage and (3) meta information. The authors use machine-learning algorithms and present a novel technique to detect click baits. Authors in [14] refer to clickbait as tabloid journalism and deem it as a form of deception. This reiterates the work of this paper where majority of the literature (if not all) consider click-bait as a negative tactic. The authors review both textual and non-textual techniques to detect clickbait and establish that a hybrid approach is the best way to detect such links. In [15], Silverman contends that tabloid journalism is a detriment to professional journalism. Authors in [16] tackle the topic of verification of news item and discuss the methodological limitations of such techniques. The work done in [17] describes an initiative launched in 2017 where a challenge was launched to see how clickbait can be detected in social media. Based on this initiative, [18] take the efforts against clickbait to another level and establish the need of a corpus that can be utilized to detect clickbait. The topic of corpus has garnered lot of attention in various field as authors in [19] also use twitter to build a corpus to detect mental health symptoms. In [20], the authors build a browser extension that can help detect and block clickbait links. The work done in [21] describes the solution of Zingel Clickbait detector, which evaluates each tweet for the presence of clickbait. Please note that majority of the techniques rely upon Natural Language Processing techniques (NLP) and such techniques' application goes beyond into various areas such as sentiment analysis etc. [22]. As opposed to the stance by the aforementioned work, authors in [23] do talk about the positive aspect of clickbait in academic articles. The authors limit the work in Frontiers in Psychology and the authors own articles and contend that positive framing of an academic article and avoiding playing with words result in a positive dissemination of the work. The authors confirmed the findings with their own articles. However, the result of authors is specific to text articles and does not consider academic/educational videos. Furthermore, the articles used by the authors are mostly refereed and hence the reader has some level of confidence when reading such articles. Our work looks into the topic of educational videos specifically and none of the videos we looked at have been evaluated by experts in the subject field. Learning Analytics, as a field, is not new as researchers have always been concerned about gathering data about learners to help optimize the delivery of course material. Authors in [24] discuss the importance of establishing the teacher as a mentor and reevaluate the passive mode in which classes are being taught. They look at ways where technology can help in this vein. The work done in [25] proposes a conceptual framework similar to a maturity model focused on integrating social network research with importance of technology and explores how the concept of smart education has evolved. Furthermore, the work done in [26, 27] takes these concepts and focuses on establishing metrics and key performance indicators that would provide feedback to gauge the effectiveness of smart education.

3 Experimental Setup and Metrics

We conducted our experiments by writing crawlers that would scavenge YouTube for a particular topic. Specifically, given the attention cybersecurity has garnered in the lately, we chose the following two terms "Replay Attacks" and "MITM attacks" (Man in the Middle Attacks).

3.1 Metrics Used

The problem faced during the experiments was to establish metrics that can serve a proxy to the feedback provided by the user. Many studies have used the "Like" and "Dislike" feature available on YouTube to reflect the sentiments of the user. However, not all the users who view a video would necessarily take the initiative to click on the like/dislike button. Another factor that we considered was the comment section where viewers can describe their opinion about the content being provided. Based on this, we came up with the following metrics to serve a proxy to user feedback.

- Number of Views
- Number of Likes
- Number of Dislikes
- Likes to Dislikes ratio (Higher ratio reflects a well-liked content)
- Number of comments (Higher number indicates a well-liked or not-liked all video at all as user will only go out of their way generally to write a comment in such cases).

3.2 *Experimental Setup*

For this paper, we used the Python language along with the API provided by YouTube to collect the requisite data and perform a depth-first search approach to collect the metrics. Specifically, we do the following:

1. Search for videos under the titles "Replay Attacks" and "MITM attacks". For each of the video, we collect the aforementioned metrics
2. We also check the description field and see if the author has embedded a link to another YouTube video
3. If the video is also authored by the same author, we repeat steps 1 and 2.

Once we collected the data, we wanted to see if the clickbait technique (embedding links in the description field) turns away the users, as is the perception of clickbait or the viewers become more interested in the topic at hand.

4 Results and Discussion

For this test, searched for ten videos initially and performed a depth-first search algorithm as described above. One of the underlying assumptions of the search is that the popular videos will come up first as YouTube rewards viewership. However, what is not obvious right away from such a search is whether the author of a particular video would continue to provide high quality videos or otherwise. The following tables summarizes the results for two videos along with their embedded links during our search.

Based on the data from performing a depth first search on 10 videos for two topics, we obtained the following average. The first row of the table indicates the initial video that came up in our search. Subsequent rows show the number of embedded videos discovered by our crawler under the clickbait model described in Table 1.

Please note the following:

1. The high number of views remain consistent for the author of popular videos irrespective of the topic of the embedded videos
2. The Likes to Dislike Ratio remains high consistent with point number 1

Table 1 Popular video statistics

Views	Likes	Dislikes	Comments	Likes/Dislikes	Comments/Views
25,034	325	18	53	18.05	472.4
17,345	165	21	33	23.21	525.6
19,012	83	8	13	19	1569
5467	27	0	0	10	11
320	16	0	0	10	11

3. The Comments/Views ratio indicate that despite a video is generally liked by the viewers, very few will actually go ahead and write a comment concerning the video
4. For average of all the results,

   a. The number of embedded videos was 5 videos
   b. The number of comments go down as the depth of the embedded link increases
   c. The number of comments overall stay quite low regardless of the likes/dislikes ratio
   d. The likes/dislikes ratio remained favorable over about 47 videos.

From the above we infer that viewers do not appear to be affected by the click-bait model if they view the material positively. Given the strength of few videos, an author can attract lots of visitor to his/her channel using the clickbait model in the educational videos environment—an exact opposite behavior of the yellow journalism connotation. The results presented in [23] are consistent with what we have found and we believe that the educational videos do not seem to conform to the traditional negative connotation associated with yellow journalism.

5 Conclusion and Future Work

In this paper, we have addressed the existence of clickbait model in the educational videos environment. Furthermore, we believe that we have established enough evidence to Form a basis that the viewers in such an environment display opposite behavior to the negative connotation attached to the clickbait/yellow journalism. The videos we evaluated had both a high viewership and high like count even though experts in the field did not independently verify the scientific content contained by the videos. We would like to explore the following in the future work:

1. Perform the above experiments on a larger dataset
2. Confirm the high like/dislike ratio by performing sentiment analysis using NLP techniques
3. Repeat the experiment with videos that do not have a high viewership or like/dislike ratio and see if the clickbait model exists in such videos
4. Compare the results of English videos to those in another language to see whether the behavior is consistent across cultures.

References

1. https://www.britannica.com/topic/yellow-journalism
2. Thiel, K.: Avoiding Clickbait. Cavendish Square Publishing, LLC (2018)
3. Rajput, A., Brahimi, T.: Characterizing IOMT/personal area networks landscape. In: Lytras, M., et al. (eds.) Innovation in Health Informatics: A Smart Healthcare Primer. Amsterdam, Elsevier, Netherlands (2019). (earlier version available as arXiv preprint arXiv:1902.00675)
4. Ahmed, S.: BYOD, personal area networks (PANs) and IOT: threats to Patients Privacy. arXiv preprint arXiv:1902.06462 (2019)
5. Rajput, A.E., Ahmed, S.M.: Big data and social/medical sciences: state of the art and future trends. arXiv preprint arXiv:1902.00705 (2019)
6. Roelofs, P., Gallien, M.: Clickbait and impact: how academia has been hacked. Impact Soc. Sci. Blog (2017)
7. Gilley, B.: The case for colonialism. Acad. Quest. **31**(2), 167–185 (2018)
8. https://ocw.mit.edu
9. Harrison, D.J.: Assessing experiences with online educational videos: converting multiple constructed responses to quantifiable data. Int. Rev. Res. Open Distrib. Learn. **16**(1) (2015)
10. Loewenstein, G.: The psychology of curiosity: a review and reinterpretation. Psychol. Bull. **116**(1), 75 (1994)
11. Vijgen, B.: The listicle: an exploring research on an interesting shareable new media phenomenon. Stud. Univ. Babes-Bolyai, Ephemer. **59**(1) (2014)
12. Blom, J.N., Hansen, K.R.: Click bait: forward-reference as lure in online news headlines. J. Pragmat. **76**, 87–100 (2015)
13. Potthast, M., Köpsel, S., Stein, B., Hagen, M.: Clickbait detection. In: European Conference on Information Retrieval, pp. 810–817. Springer, Cham (2016)
14. Chen, Y., Conroy, N.J., Rubin, V.L.: Misleading online content: recognizing clickbait as false news. In: Proceedings of the 2015 ACM on Workshop on Multimodal Deception Detection, pp. 15–19. ACM (2015)
15. Silverman, C.: Lies, damn lies, and viral content. how news websites spread (and debunk) online rumors, unverified claims, and misinformation. Tow Cent. Digit. J. **168**(4), 134–140 (2015)
16. Rubin, V.L., Conroy, N.J., Chen, Y.: Towards news verification: deception detection methods for news discourse. In: Hawaii International Conference on System Sciences (2015)
17. https://clickbait-challenge.org
18. Potthast, M., Gollub, T., Komlossy, K., Schuster, S., Wiegmann, M., Fernandez, E.P.G., Hagen, M., Stein, B.: Crowdsourcing a large corpus of clickbait on twitter. In: Proceedings of the 27th International Conference on Computational Linguistics, pp. 1498–1507 (2018)
19. Rajput, A., Ahmed, S.: Making a case for social media corpus for detecting depression. arXiv preprint arXiv:1902.00702 (2019)
20. Chakraborty, A., Paranjape, B., Kakarla, S., Ganguly, N.: Stop clickbait: detecting and preventing clickbaits in online news media. In: 2016 IEEE/ACM International Conference on Advances in Social Networks Analysis and Mining (ASONAM), pp. 9–16. IEEE (2016)
21. Zhou, Y.: Clickbait detection in tweets using self-attentive network. arXiv preprint arXiv:1710.05364 (2017)
22. Rajput, A.: Natural language processing, sentiment analysis and clinical analytics. In: Lytras, M., et al. (eds.) Innovation in Health Informatics: A Smart Healthcare Primer. Elsevier, Amsterdam, Netherlands (2019). (available as arXiv preprint arXiv:1902.00679)
23. Lockwood, G.: Academic clickbait: articles with positively-framed titles, interesting phrasing, and no wordplay get more attention online. Winnower **3** (2016)
24. Visvizi, A., Lytras, M.D., Daniela, L.: The future of innovation and technology in education: a case for restoring the role of the teacher as a mentor. In: The Future of Innovation and Technology in Education: Policies and Practices for Teaching and Learning Excellence, pp. 1–8. Emerald Publishing Limited (2018)

25. Lytras, M., Visvizi, A., Daniela, L., Sarirete, A., Ordonez De Pablos, P.: Social networks research for sustainable smart education. Sustainability **10**(9), 2974 (2018)
26. Lytras, M.D., Aljohani, N.R., Visvizi, A., Ordonez De Pablos, P., Gasevic, D.: Advanced decision-making in higher education: learning analytics research and key performance indicators (2018)
27. Lytras, M., Visvizi, A., Damiani, E., Mthkour, H.: The cognitive computing turn in education: prospects and application (2018)

Improve Student Participation in Peer Assessment to Influence Learning Outcomes: A Case Study

Antonella Carbonaro and Roberto Reda

Abstract A variety of technology enhanced teaching strategies and learning activities have been applied in education, including assessment mechanisms. In this paper, we aim to examine the extent to which peer assessment promotes deep learning and favours the incremental learning of the concepts presented throughout a traditional introductory programming course. In particular, we want to examine how the enactment of the peer-assessor and peer-assessee roles is associated with students' learning improvements, after enacting reciprocal peer assessment. We illustrate a case study based on a novel web-based peer assessment tool to improve engagement and learning outcomes in a programming course. We want to assess whether programming skills are growing through peer review and whether exposure to different programming techniques is helpful to students. One of the most remarkable findings of our experience was that students reported that assessing others' work was an extremely valuable learning activity.

1 Introduction

A variety of technology enhanced teaching strategies and learning activities have been applied in education, including assessment mechanisms to guarantee that students gain sufficient practice, as well as providing feedback on the quality of students' solutions, for example on a traditional introductory programming course. One major approach being used to evaluate and mark students' programming exercises is the automatic assessment of programming assignments. This approach is mainly used to manage a large number of students or a large number of programs to be evaluated. Research into automatic programming assessment has a long history. It has been an area of research of considerable interest since the 1960, and it continues

A. Carbonaro (✉) · R. Reda
Department of Computer Science and Engineering, University of Bologna, Bologna, Italy
e-mail: antonella.carbonaro@unibo.it

R. Reda
e-mail: roberto.reda@unibo.it

A. Visvizi and M. D. Lytras (eds.), *Research & Innovation Forum 2019*,
Springer Proceedings in Complexity,
https://doi.org/10.1007/978-3-030-30809-4_11

to be an important aspect and to attract the attention of researchers. Review code, write and read comments and see how other students tackle the same problems are important skills in order to provide students with opportunities to learn from one another, improve their learning experience and reach efficient learning outcomes. In the current study, we want to focus on the impact of our peer assessment web-based programming-assisted system [1–3] on students' learning outcomes, time management capabilities and student engagement. The system supports students' programming skills through peer code review and the delivery of feedback to peers. We examine the extent to which peer assessment promotes deep learning and favours the incremental learning of the concepts presented throughout the course. Peer assessments require a good faith effort on the part of each student not only to submit their own original work, but also to then anonymously evaluate the work of others responsively and constructively. Therefore, for each assignment that they submit, students are generally then asked to evaluate the work of their peers. This is not a negligible amount of work or time, especially in a course that requires weekly peer-assessed assignments and when the students' prior programming experience varies significantly.

The paper is organized as follows. The next Section introduces the automatic assessment of programming assignments and explores research efforts related to the student engagement mechanism in this field. Section 3 proposes the adopted peer approach. Section 4 describes the questionnaire survey conducted to evaluate the system, focusing on its impact on students' programming competence and student engagement. Finally, Sect. 5 provides some thoughts on the case study and the experimental results.

2 Related Works

In the context of programming assignments, several approaches to automatic programming assessment can be found in related literature; they are typically based on either static analysis or dynamic testing. This refers to whether a program needs to be executed while it is being assessed and focus on which features of programming assignments are automatically assessed. Dynamic analysis (assessment based on executing the program) is often used to assess functionality, efficiency, and testing skills, while static checks that analyze the program without executing it are used to provide feedback on style, programming errors, software metrics, and even design. Tools that cover both static and dynamic testing are also well presented in the survey [4]; they already reported several variations of output comparison including running the model solution and student's code side by side and the use of regular expressions to match the output. On top of that, the assessment process can be done by looking into a code structure (white-box) or simply based on a functional behavior of a program (black-box) [5]. Output comparison is the traditional approach used by many of the systems we found [6]. In the peer assessment process, students are involved both in the learning and assessment processes. Peer assessment plays an extremely important role in helping students see work from an assessor's perspective, with potential additional benefits [7]. For example, it exposes students to solutions, strategies, and insights that they otherwise would likely not see. Evaluating peers' work also helps students reflect on gaps in their understanding, making them more resourceful, confident, and higher achievers. Peer assessment has been used for many different kinds of assignments, including design, programming [8, 9] and essays.

We want to use peer assessment to promote learning in programming courses rather than for summative assessment [10]. Sitthiworachart and Joy [11] remarks that "peer assessment is not only a tool to provide a peer with constructive feedback which is understood by the peer. Above all, peer assessment is a tool for the learner himself." Recent technological advances, such as wireless transmission and mobile devices, allow learners to access the learning management systems anytime and anywhere, thus facilitating learner mobility and the interaction and collaboration between learners, and learners and instructors [12].

3 Weekly Peer Assessment Mechanism

The author of this paper has many years of experience in the modelling and implementation of teaching support systems [13–17]. We recently proposed a web-based programming-assisted system, which improved students' programming skills through peer code review and the delivery of feedback to peers. Its purpose was to investigate the extent to which peer assessment in programming courses promotes deep learning in order to assess the accuracy of students' judgements during a peer assessment exercise and provide evidence that peer assessment in computer programming has a positive instructive effect [18]. Now we want to focus on the use that the students of the academic year 2018/2019 made of this tool by describing the data from a survey conducted at the end of the course. The considerations that emerge from the data analysis can be used to further improve student involvement and learning performance. In addition, the present use case could be a starting point for other programming courses or for courses on different topics.

The coursework consists of weekly incremental programming assignments of increasing difficulty. The author of this paper maintains a website dedicated to the course, available at www.programmazione.info. Each student who participates in a weekly peer assessment mechanism acts as an author on account of having to write the weekly assigned program, as a reviewer when she/he reviews a program written by another student or as a reviser when she/he revises her/his program as suggested by reviewers' comments. During the course, authors must submit their codes before a given weekly deadline, as well as submit their revision of other students' programs within a given deadline, which is set one week later. We decided to not use the peer assessment process to evaluate students in their final exam. First, the aim of the peer review process is to increase students' engagement over the duration of the course. This is a crucial aspect of a 12-week strength program dedicated to introducing C language tools with lots of programming examples and to fostering algorithmic thought processes. It should be considered that student engagement during the first year of academic study is particularly important on academic persistence and achievement [19]. In addition, skills acquired during the course will be better assimilated during the suspension of lessons before exams when students have time to assimilate the more complex concepts. In addition, students do not have to respect the weekly deadlines set for the delivery of their programs and for the evaluation of the programs of their peers. However, it was explained to them that the system as a whole will work efficiently if they decide to participate in each activity with constancy and commitment, and in the same way their skills will progress as the course topics unfold. The only constraint is the delivery of all the weekly exercises before being able to take the written exam test.

In the following section, we will evaluate all the choices made by analyzing the data resulting from the survey.

4 Questionnaire Survey

A web-based survey with direct contact during last two lessons of the course was chosen for time- and cost-efficiency purposes. The questionnaire runs to around 20 questions in length and takes around 30 min to complete The questionnaire administered included questions such as: "On average, how many hours do you use to complete delivery (of the exercise of the week)?", "On average, how many hours do you use to complete a review (of one exercise of the week)?", "Evaluate your experience after considering all the sources related to an exercise (express your degree of agreement from 1, the weakest, to 5, the strongest) [Boring because of the repetitions] [I learned from another] [Happy to have helped the class], "As a reviewer, how many sources do you need to learn problem-solving skills from others?", "As a student, evaluate the degree of satisfaction in having participated in the peer review process along with the weekly administration of the program assignments", "As a student, evaluate the degree of satisfaction in having periodically obtained the evaluations of your exercises [not satisfied] [satisfied] [very satisfied]", "As a student, evaluate the improvement you think you have achieved in terms of programming ability through the constant exercise imposed by weekly deliveries". 80 students answered the questionnaire.

Impact on students' programming competence. 97.5% of the students agreed that being involved in the assessment process helped them to improve their programming skills and to learn different programming techniques, thanks to reviewing programs written by other students. In particular, 50% of the students stated that their programming abilities has increased considerably (over 75%) and only two students stated that their programming skills has not increased. Additionally, 97.5% of the students were very happy or happy to support the rest of the class with the review process; again, only two students stated they were not satisfied with the involvement in the automatic peer review process. These are positive results in terms of influencing cooperation among students, improving their thinking skills and deepening their understanding.

Impact on students' programming skills awareness. This is a very important aspect in a course where the difference between the students' initial programming ability is very high. In fact, about 40% of the students come from a technical course at high school where several of the topics of the programming course have already been covered, although generally in a less in-depth way. Another 40% of the students come from scientific high schools; some of them have attended the applied sciences course, which includes the study of programming. In this case, however, their competence is much more basic. The remaining 20% of the students have never approached concepts related to computer programming in their previous course of study. It is therefore important to manage the course in such a way as to stimulate the most prepared students by providing them with new insights without ever ignoring the needs and learning times of other students. It is part of this policy of attention to make students aware of the progressive improvement that they can achieve with the exercise of programming without inferiority to those students who initially had more opportunities for programming. As shown by the questionnaire responses, 97.5% of the students agreed that their awareness of their own programming skills has increased. In particular, more than 38% of the students stated that their awareness has increased considerably and only two students stated that it has not increased. One of these two students is one of those who said he had not made any improvements in his programming skills. It would be interesting to know the reasons behind these answers, but the questionnaires were collected anonymously.

Impact on students' engagement. A very interesting moment to draw conclusions about the effectiveness of the proposed method in involving students in the activities of the course is the first session of exams. The semester lessons end with the Christmas holidays, after which the first written verification is fixed. Generally, about 25% of the students do not enroll in this written test because they have not been able to prepare in time, having left out the practice of programming during the course of the lessons. After the administration of the weekly peer assessment mechanism, 97 students out of 100 participated in the first session of examinations. These numbers are very important for the entire curriculum where programming is the first course and a delay on this course would affect the entire university career.

The students also left some useful considerations and suggestions in the free space provided for this purpose. In particular, one student found that the workload related to the evaluation of the exercises, which he quantified in 4 h per week, was excessive. The average number of hours used per week was two. Another student particularly appreciated the freedom left to participate in the delivery of the programs and assessments, a factor that cancels both the stress of the students and the phenomenon of plagiarism more common in those who have to meet strict deadlines in order to take the exam.

5 Conclusions

Research has made many efforts to positively influence students' learning outcomes in computer programming. In fact, different approaches have been tested to support the traditional teaching methodology, including those involving peer evaluation mechanisms. This paper reports on the good results achieved in the programming competence of students when called upon to revise their peers' code, as they are encouraged to learn different programming techniques. This approach also provides them with the opportunity to evaluate programs, helping students to improve their programming style. As evidence of this, students reported that assessing others' work was an extremely valuable learning activity.

References

1. Carbonaro, A.: Good practices to influence engagement and learning outcomes on a traditional introductory programming course. Interact. Learn. Environ. (2018)
2. Carbonaro, A.: Collaborative and semantic information retrieval for technology-enhanced learning. In: Social Information Retrieval for Technology-Enhanced Learning, p. 535 (2009)
3. Carbonaro, A., Ravaioli, M.: Peer assessment to promote deep learning and to reduce a gender gap in the traditional introductory programming course. J. E-Learn. Knowl. Soc. **13**(3) (2017)
4. Ala-Mutka, K.M.: A survey of automated assessment approaches for programming assignments. Comput. Sci. Educ. **15**(2), 83–102 (2005)
5. Surendra, G., Dubey, S.K.: Automatic assessment of programming assignment. Comput. Sci. Eng. 2(1), 67 (2012)
6. Ihantola, P., et al.: Review of recent systems for automatic assessment of programming assignments. In: Proceedings of the 10th Koli Calling International Conference on Computing Education Research. ACM, (2010)
7. Kulkarni, C., et al.: Peer and self assessment in massive online classes. In: Design thinking research, pp. 131-168. Springer International Publishing (2015)
8. Chinn, Donald: Peer assessment in the algorithms course. ACM SIGCSE Bulletin **37**(3), 69–73 (2005)
9. Wang, Y., et al.: A multi-peer assessment platform for programming language learning: considering group non-consensus and personal radicalness. Interact. Learn. Environ. 1–20 (2015)
10. Carbonaro, A., Ferrini, R.: Personalized information retrieval in a semantic-based learning environment. In: Social Information Retrieval Systems: Emerging Technologies and Applications for Searching the Web Effectively, pp. 270–288
11. Sitthiworachart, J., Joy, M.: Deepening computer programming skills by using web-based peer assessment. In: Proceedings of the 4th Annual Conference of the LTSN Centre for Information and Computer Sciences. LTSN Centre for Information and Computer Sciences (2003)
12. Andronico, A., Carbonaro, A., Colazzo, L., Molinari, A., Ronchetti, M., Trifonova, A.: Designing models and services for learning management systems in mobile settings. Mobile and Ubiquitous Information Access, pp. 90–106. Springer, Berlin Heidelberg (2004)
13. Carbonaro, A., Ferrini, R.: Ontology-based video annotation in multimedia entertainment. In: Consumer Communications and Networking Conference, pp. 1087–1091 (2007)
14. Riccucci, S., Carbonaro, A., Casadei, G.: Knowledge acquisition in intelligent tutoring system: a data mining approach. In: Mexican International Conference on Artificial Intelligence, pp. 1195–1205. Springer
15. Carbonaro, A.: WordNet-based summarization to enhance learning interaction tutoring. J. E-Learn. Knowl. Soc. **6**(2), 67–74 (2010)

16. Carbonaro, A.: Interlinking e-learning resources and the web of data for improving student experience. J. E-Learn. Knowl. Soc. **8**(2), 33–44 (2012)
17. Reda R., Piccinini, F., Carbonaro, A.: Towards consistent data representation in the IoT healthcare landscape. In: Proceedings of the 2018 International Conference on Digital Health, pp. 5–10. ACM (2018)
18. Carbonaro, A.: Improving web search and navigation using summarization process. Commun. Comput. Inf. Sci. **111** CCIS (PART 1), 131–138 (2010)
19. Kuh, D.G., Cruce, T.M., Shoup, R., Kinzie, J., Gonyea, R.M.: Unmasking the effects of student engagement on first-year college grades and persistence. J. High. Educ. **79**(5) (2008)

Learning with Wikipedia in Higher Education: Academic Performance and Students' Quality Perception

Antoni Meseguer-Artola, Inma Rodríguez-Ardura, Gisela Ammetller and Eva Rimbau-Gilabert

Abstract Despite some quality concerns, the success of Wikipedia as a resource of information is influencing education. New technology-enhanced learning strategies are recently developed to include this open educational resource in courses' design. Although research about the use of Wikipedia in higher education and its quality perception is scarce, there are some empirical evidences showing its positive effect on the students' academic performance and its positive quality perception. In line with this, the main aim of this paper is to prove statistically that Wikipedia's usage improves the final marks of the students enrolled in topics from different knowledge areas. Additionally, we study the Wikipedia's quality perception among students. Based on an experimental research design with 2330 students, and using data from a questionnaire and from their course marks, we prove through different statistical tests that (1) the academic active use of Wikipedia has a positive influence on the student's academic performance, which is moderated by knowledge areas, and that (2) Wikipedia's quality perception is positive and it does not depend on the student's performance.

A. Meseguer-Artola (✉) · I. Rodríguez-Ardura · G. Ammetller · E. Rimbau-Gilabert
Open University of Catalonia, (Universitat Oberta de Catalunya), Av. Tibidabo 39-43, 08035 Barcelona, Spain
e-mail: ameseguer@uoc.edu

I. Rodríguez-Ardura
e-mail: irodriguez@uoc.edu

G. Ammetller
e-mail: gammetller@uoc.edu

E. Rimbau-Gilabert
e-mail: erimbau@uoc.edu

I. Rodríguez-Ardura
University of Miami, 5250 University Dr, Coral Gables, FL 33146, USA

A. Visvizi and M. D. Lytras (eds.), *Research & Innovation Forum 2019*, Springer Proceedings in Complexity,
https://doi.org/10.1007/978-3-030-30809-4_12

1 Introduction and State of the Art

There is a recent and intense debate about how advances in social networks research and the corresponding advances in information and communication technologies can be integrated in order to employ them in the domain of a better education [14]. Within this debate, Wikipedia can become a driver and an example to other educational initiatives. Wikipedia provides learning and research resources through a public online platform, and it is widely used to find information and to collaborate in knowledge creation. It is considered as a central element of the Open Educational Resources (OER) movement and one of the biggest OER in the world [4]. Although Wikipedia is frequently used by higher education students as a source of information [24], there are a few courses in which it plays an actual active role within the learning process [1]. Nevertheless, the number of cases where Wikipedia is implemented in education is rapidly increasing, and it is starting to become a common practice within different academic levels [26]. There are several initiatives, such as the Wikipedia Global Education Program from the Wikimedia foundation (http://education.wikimedia.org), with the objective of introducing faculty to Wikipedia and therefore expand its academic use. The fact that Wikipedia is founded on the idea of wikis, one of the most important Web 2.0 components, plays an important role in enhancing the students' learning process [18].

Many recent teaching experiences have begun to use Wikipedia as an active learning tool. Most of them show very satisfactory academic results and substantial improvement in various students' skills and motivations [10, 11, 25]. Taking into account that student performance is strongly influenced by the lecturers' teaching strategies [15], there is a clear interest in exploring the possible influence of Wikipedia's usage on the students' academic performance.

Although faculty in higher education still have many concerns regarding the use of Wikipedia in the student's learning process [10, 12], there is evidence of its significant benefits. We can find several works in the literature [6, 7, 19, 23] showing that the academic use of Wikipedia improves important basic skills of the students: interaction, communication, collaboration, writing and comprehension, research and innovation.

A pilot developed in 2013 offered the first evidence about the effects of the use of Wikipedia in a course in Statistics [16]. It showed that Wikipedia had a weak positive effect on the student's academic performance. Following this line of research, in this paper we have performed a more comprehensive study, and analyzed four undergraduate introductory courses taken by a total of 2330 students. Each of these courses situated in a specific knowledge area: human resources, statistics, marketing, or consumer behavior.

Apart from the academic performance, the active use of Wikipedia as an educational resource has a positive impact on its quality perception. Information and communication technologies applied to a concrete framework have a direct impact on its quality perception [13]. Previous research shows that the inclusion of Wikipedia in the learning process has a positive and significant influence on the students' quality perception of Wikipedia [21]. In this paper we will also be interested in analyzing this relationship between Wikipedia usage and quality perception; and to find out if it varies across knowledge areas.

2 Research Hypotheses

Through our study, we sought: to explore the student's perceptions about the quality of Wikipedia, to show the potential positive impact of the active use of Wikipedia on the student's academic performance, and to explore whether its influence depends on the knowledge area or not.

2.1 Impact of Wikipedia's Usage on the Student's Academic Performance

Although there are few empirical studies about Wikipedia's usage in higher education, a pilot study developed in 2013 offered statistical results about the effects of this open resource on the students' final marks [16]. Based on this result, we expect in this paper to confirm the positive effect of Wikipedia on the academic performance of the students and to show how this influence vary across knowledge areas. According to these objectives, we will be interested in checking the following two initial hypotheses:

H1: The academic use of Wikipedia has a positive impact on the academic performance.

H2: The influence of Wikipedia usage onto the academic performance varies across knowledge areas.

2.2 Perceptions of Wikipedia's Quality

As a source of information, Wikipedia's quality, in terms of its completeness, reliability, up-to-datedness and usefulness, is an important element to take into account to explain attitudes and practices towards Wikipedia. The perceived quality of many of its articles, jointly with the accessibility of its content, its hypertextual structure and the abundance of references and sources of information are key factors in Wikipedia's usage [12]. The perception that Wikipedia is inaccurate and possibly lacks of credibility discourage students and faculty members from actively using it as a reliable source of information [2, 5, 8]. In fact, one of the main factors that explain the teaching uses of Wikipedia among university faculty is the perceived quality of the information in this free encyclopedia [17]. A positive quality perception has a direct impact on the perceived enjoyment and usefulness of Wikipedia, which in turn play a key role in the decision to use it in the teaching process.

The evidence of an intensive use of Wikipedia by students [9], especially for their academic work [20], leads us to ask ourselves about the quality perception of the students and its relationship with their academic performance, and to hypothesize that there exist a positive effect between them. Since previous results [20] show that the use and the perceived usefulness of Wikipedia differs by subject studied, we will be also interested in knowing whether this effect depends on the knowledge area of the course in which students are enrolled.

Accordingly, we hypothesized that:

H3: Students' academic performance has a positive effect on Wikipedia's quality perception.

H4: The impact of academic performance onto Wikipedia's quality perception varies across knowledge areas.

3 Methodology

In this study we analyzed four undergraduate introductory courses taken by a total of 2330 students. Each of these courses situated in a specific knowledge area: human resources, statistics, marketing, or consumer behavior. In order to adequately assess the influence of Wikipedia on the students' final marks, we randomly divided the students enrolled in each course into two groups. In one of them, students actively used Wikipedia in their learning process; in the other not. As a result of this split, a total of 1232 students were involved in courses using Wikipedia.

In order to collect information about quality perception, students using Wikipedia as a primary learning were asked to compare the standard learning materials of the course with Wikipedia, and to provide their perceptions on the basis of four quality facets: completeness, reliability, up-to-datedness and usefulness.

The validity of the hypotheses of this research has been analyzed using different quantitative methodologies. We tested hypotheses H1 and H2 using t-tests on data gathered from all students in the 4×2 groups. This statistical analysis allowed us to check whether there are significant differences between groups (the ones defined by Wikipedia usage and the ones determined by the four knowledge areas) or not.

To test H3 and H4 we used information from a subsample. We analyzed information obtained from those students included in the groups where there has been an active use of Wikipedia. Hence, 1232 registers have been considered in this part of the analysis. Perceived quality is the objective variable in the two hypotheses. It is an unobserved (latent) variable that has been measured through 4 observed variables: completeness, reliability, up-to-datedness and usefulness. Once we checked the internal reliability of the latent variable, we assessed the validity of H3 and H4 using ANOVA analyses.

4 Results

The difference between the overall mean grade of the students using Wikipedia and the mean value for those students not using Wikipedia was +0.72. The t-test confirmed that this difference is statistically different from zero (p-value = 0.00) and permitted to assess the validity of H1.

Across knowledge areas, we found the greatest differences in the case of the students following the consumer behavior course (+0.90). At the other end we got the students enrolled in marketing, with a difference between the means of their final marks of +0.41. The differences in the case of human resources and statistics were +0.71 and +0.63, respectively. In all cases the p-value of the corresponding t-test was lower than 0.05, showing that all differences were significant. Consequently, we found out that there were relevant differences between knowledge areas, which supported the validity of H2.

In order to analyze H3 and H4, students were divided into three different groups: low performance (final marks lower that 6), medium performance (final marks between 6 and 8) and high performance (final marks higher than 8). For each group we computed the mean of the quality perception. To test differences between groups statistically in the general case and also for each knowledge area, we used the ANOVA test. In the general case, which includes all students in the subsample, the p-value of the ANOVA test was higher than 0.05. Consequently, there are no significant differences between the three groups, and this implies that H3 is not fulfilled: Wikipedia's quality perception does not depend on the performance of the students.

In contrast, we found significant differences between groups in some knowledge areas. The ANOVA test in the case of marketing and statistics resulted with a p-value lower than 0.05. According to this, we can conclude that H4 is confirmed.

5 Conclusions and Managerial Implications

This study explored the academic use of Wikipedia in higher education and analyzed their influence on the course design and the students' learning process. It also studied students' perception on Wikipedia's quality and its relationship with academic performance. Through an experiment which involved 2330 undergraduate students enrolled in four different topics, data on students' academic performance and Wikipedia's quality perception was gathered.

Statistical results showed that the academic use of Wikipedia has a significant positive influence on the student's academic performance. Students enrolled in courses whose design contained an active use of Wikipedia received higher final marks than those not using Wikipedia. This effect on the academic performance was moderated by the knowledge area. Independently of the topic, all students got better marks when Wikipedia is included in the syllabus. Nevertheless, results were different between groups. Students in consumer behavior and human resources showed the biggest differences. These results extended the preliminary results obtained in Meseguer-Artola [16], showing that the positive influence on the academic performance is also valid across different knowledge areas.

Concerning Wikipedia's quality perception, results in this paper showed that students agree with the idea that Wikipedia has an acceptable quality level. This finding coincides with the results in Meseguer et al. [17] about faculty's quality perception. This perception does not depend on the student's academic profile (defined through its academic performance). Students reported that the most important facet of Wikipedia is that it is updated, containing recent information and recent references. The least valued quality component was the completeness of the articles in Wikipedia. Students reported that Wikipedia articles provided less information than other learning materials from the course.

Although in general there was not a significant correlation between quality perception and academic performance, we got some significant differences in two out of the four knowledge areas. High performance students in marketing and statistics gave the lowest values in the four quality dimensions: completeness, reliability, up-to-datedness and utility. To better understand these results in the statistics case, we have to consider that it is traditionally a hard topic. Probably the big differences between the academic performance of the best students, where information meets the requirements of their particular activity [22], and the rest of students can explain the moderating effect we have obtained.

These conclusions suggest that the academic use of Wikipedia in Higher Education can significantly improve students' performance in different areas of knowledge. Hence, faculty have to pay attention to the new opportunities that its usage during the learning process can offer. Although in this paper it is shown a specific very simple way of working different subjects through Wikipedia, there are lots of alternatives that can be also considered during the course design. Professors can find in Aibar and Lerga [1] lots of examples with best practices which can suit the specific requirements of each course.

Taking into account the positive effect on the academic results and the positive quality perception of Wikipedia, we may suggest to academic institutions to encourage its usage among their members. Apart from being an educational resource, it can also be used as a collaborative platform for knowledge creation and management [3]. Institutions can also benefit from considering Wikipedia as a key factor in their organization.

For further research, we are considering two new lines of work which can complement and deepen the main findings of this paper. Since our research presents a static view of the Wikipedia's academic use, it would be interesting to get new data for different periods of time and compare then the evolution of the results along time. In this first line of work we will be interested in knowing whether the students who get used of having Wikipedia recursively in their learning process obtain even better academic results.

Another future line of work could be oriented to obtaining new information about what happens with the Wikipedia's use in new topics and in different academic levels.

This paper focuses in one specific higher education institution and analyses just four knowledge areas. Hence, in order to infer more robust results, it would be important to compare results between different academic institutions and among new knowledge areas.

References

1. Aibar, E., Lerga, M.: Best practice guide to use Wikipedia in university education. Universitat Oberta de Catalunya. http://hdl.handle.net/10609/41662 (2015)
2. Azer, S.: Is Wikipedia a reliable learning resource for medical students? Evaluating respiratory topics. Adv. Physiol. Educ. **39**(1), 5–14 (2015). https://doi.org/10.1152/advan.00110.2014
3. Biasutti, M., EL-Deghaidy, H.: Using Wiki in teacher education: impact on knowledge management processes and student satisfaction. Comput. Educ. **59**(3), 861–872 (2012). https://doi.org/10.1016/j.compedu.2012.04.009
4. Caswell, T., Henson, S., Jensen, M., Wiley, D.: Open content and open educational resources: enabling universal education. Int. Rev. Res. Open Distrib. Learn. **9**(1) (2008)
5. Chen, H.: The perspectives of higher education faculty on Wikipedia. Electron. Libr. **28**(3), 361–373 (2010). https://doi.org/10.1108/02640471011051954
6. Dawe, L., Robinson, A.: Wikipedia editing and information literacy: a case study. Inf. Learn. Sci. **118**(1/2), 5–16 (2017). https://doi.org/10.1108/ILS-09-2016-0067
7. Di-Lauro, F., Johinke, R.: Employing Wikipedia for good not evil: innovative approaches to collaborative writing assessment. Assess. Eval. High. Educ. **42**(3), 478–491 (2017). https://doi.org/10.1080/02602938.2015.1127322
8. Dooley, P.: Wikipedia and the two-faced professoriate. In: Wikisym'10 Proceedings of the 16 th International Symposium on Wikis and Open Collaboration, Article No. 24. ACM, New York (2010). https://doi.org/10.1145/1832772.1832803
9. Knight, C., Pryke, S.: Wikipedia and the university, a case study. Teach. High. Educ. **17**(6), 649–659 (2012). https://doi.org/10.1080/13562517.2012.666734
10. Konieczny, P.: Rethinking Wikipedia for the Classroom. Contexts **13**(1), 80–83 (2014). https://doi.org/10.1177/1536504214522017
11. Lim, S.: How and why do college students use Wikipedia? J. Am. Soc. Inform. Sci. Technol. **60**(11), 2189–2203 (2009). https://doi.org/10.1002/asi.21142
12. Lladós-Masllorens, J., Aibar, E., Meseguer-Artola, A., Minguillón, J., Lerga, M.: Explaining teaching Uses of Wikipedia through faculty personal and contextual features. Online Inf. Rev. **41**(5), 728–743 (2017). https://doi.org/10.1108/OIR-10-2016-0298
13. Lytras, M.D., Visvizi, A.: Who uses smart city services and what to make of it: toward interdisciplinary smart cities research. Sustainability **10**(6), 1998 (2018). https://doi.org/10.3390/su10061998
14. Lytras, M.D., Visvizi, A., Daniela, L., Sarirete, A., Ordonez De Pablos, P.: Social networks research for sustainable smart education. Sustainability **10**(9), 2974 (2018). https://doi.org/10.3390/su10092974
15. Margaryan, A., Littlejohn, A., Vojt, G.: Are digital natives a myth or reality? University students' use of digital technologies. Comput. Educ. **56**(2), 429–440 (2011). https://doi.org/10.1016/j.compedu.2010.09.004
16. Meseguer-Artola, A.: Learning by comparing with Wikipedia: the value to students' learning. RUSC. Univ. Knowl. Soc. J. **11**(2), 57–69 (2014). https://doi.org/10.7238/rusc.v11i2.2042
17. Meseguer-Artola, A., Aibar, E., Lladós, J., Minguillón, J., Lerga, M.: Factors that influence the teaching use of Wikipedia in higher education. J. Assoc. Inf. Sci. Technol. **67**(5), 1224–1232 (2016). https://doi.org/10.1002/asi.23488
18. Parker, K.R., Chao, J.T.: Wiki as a teaching tool. Interdiscip. J. Knowl. Learn. Objects **3**(1), 57–72 (2007)

19. Ricaurte-Quijano, P., Álvarez, A.C.: The Wiki learning project: Wikipedia as an open learning environment. Comunicar **49**(24), 61–69 (2016). https://doi.org/10.3916/C49-2016-06
20. Selwyn, N., Gorard, S.: Students' use of Wikipedia as an academic resource—Patterns of use and perceptions of usefulness. Internet High. Educ. **28**, 28–34 (2016). https://doi.org/10.1016/j.iheduc.2015.08.004
21. Soler-Adillon, J., Pavlovic, D., Freixa, P.: Wikipedia in higher education: changes in perceived value through content contribution. Comunicar **26**(54), 39–48 (2018). https://doi.org/10.3916/C54-2018-04
22. Taylor, R.S.: Value-Added Processes in Information Systems. Ablex, Norwood (1986)
23. Walker, M.A., Li, Y.: Improving information literacy skills through learning to use and edit Wikipedia: A chemistry perspective. J. Chem. Educ. **93**(3), 509–515 (2016). https://doi.org/10.1021/acs.jchemed.5b00525
24. Wannemacher, K., Schulenburg, F.: Wikipedia in academic studies: corrupting or improving the quality of teaching and learning? In: Ebner, M., Schiefner, M. (eds.) Looking Toward the Future of Technology Enhanced Education: Ubiquitous Learning and the Digital Native, pp. 295–310. IGI Global, Hershey (2010)
25. Whalley, B.: Wikipedia: reflections on use and acceptance in academic environments. Ariadne **69** (2012). http://www.ariadne.ac.uk/issue69/whalley. Last Accessed 14 May 2018
26. Wikimedia education program. https://outreach.wikimedia.org/wiki/Education. Last Accessed 14 May 2018

A Hybrid Machine Learning Approach for the Prediction of Grades in Computer Engineering Students

Diego Buenaño-Fernandez, Sergio Luján-Mora and David Gil

Abstract The growing application of information and communication technologies (ICTs) in teaching and learning processes has generated an overload of valuable information for all those involved in education field. Historical information from students' academic records has become a valuable source of data that has been used for different purposes. Unfortunately, a high percentage of research has been developed from the perspective and the need of teachers and educational administrators. This perspective has left the student in the background. This paper proposes the application of a hybrid machine learning approach, with the aim of laying the groundwork for a future implementation of a recommendation system that allows students to make decisions related to their learning process. The work has been executed on the historical academic information of students of computer engineering degree. The results obtained in this article show the effectiveness of applying a hybrid machine learning approach. This architecture is composed of, on the one hand, techniques of supervised learning applied with the objective of classifying the data in clusters, and on the other hand, having this initial classification, unsupervised learning techniques applied with the objective of carrying out a predictive analysis of students' historical grade records.

1 Introduction

Online educational platforms and academic management systems can capture, with different levels of granularity, all types of data generated from the interaction of

D. Buenaño-Fernandez (✉)
Universidad de Las Américas, Quito 08544, Ecuador
e-mail: diego.buenano@udla.edu.ec

S. Luján-Mora · D. Gil
Universidad de Alicante, 03690 Alicante, Spain
e-mail: sergio.lujan@ua.es

D. Gil
e-mail: david.gil@ua.es

A. Visvizi and M. D. Lytras (eds.), *Research & Innovation Forum 2019*,
Springer Proceedings in Complexity,
https://doi.org/10.1007/978-3-030-30809-4_13

the student with these environments [1]. The literature review, in the next section, indicates that academic information about students has been processed using different methods and algorithms related to educational data mining (EDM). However, these findings tend to benefit teachers and administrators, underestimating the usefulness that a real-time analysis of academic grades could have for students going forward. For example, student academic data have been used to develop predictive systems mainly aimed at reducing student dropout rates or improving the academic content of a course, without considering a more detailed analysis of the student's learning process. The application of data mining techniques in the educational field should generate applications that support the student's learning process.

The main objective of this work is to demonstrate the viability of using data mining and machine learning techniques to provide an accurate prediction method for historical datasets of student grades. By doing this we obtain new rules to track the students, individually, by subject, by area, etc. The main aim of this tracking is to reduce the dropout rate. Furthermore, we provide a real-time student follow-up to improve the education system. The early identification of students at risk to drop their courses is essential information for successfully applying student maintenance strategies.

This paper proposes the application of a hybrid machine learning approach, where supervised and unsupervised learning techniques are combined. This type of learning is usually used when there are more unlabeled than labeled datasets [2]. In the present work, historical datasets of student grades in the degree of Computer Systems Engineering were clustered by techniques of unsupervised learning. Following this initial classification, supervised learning techniques were applied with the aim of determining a predictive model that would lay the foundations for the future development of a recommendation system for students. Predicting the academic performance of students is considered one of the most common problems and, at the same time, it represents a complex educational data mining task.

Recommendation systems have been developed to deal with the information overload generated from information and communication technologies (ICT) mediated environments. In education, these systems focus on generating recommendations for activities aimed at students and teachers. The recommendation systems are within a student-centered educational model where activities oriented to the development of self-learning are prioritized [3]. In this sense, recommendation systems have been used for different purposes in the field of education, for example, they can be used to help students choose their specializations and to adequately plan courses in the semester calendar [4, 5]. Based on predictions of a student's previous grades, a teacher can help focus student efforts on areas of potential problems [6]. In addition, curriculum committees can use the prediction results to guide curriculum changes and evaluate the effects of those changes for the benefit of students [7].

2 Related Work

In the literature reviewed, several methods and algorithms have been identified for predicting historical datasets of student grades. The following is a description of works that evidence the use of different machine learning methods for the analysis of educational data.

There is a wide variety of work in which various supervised learning algorithms are applied in order to predict student performance based on previous grades. For example, Mashiloane and Mchunu [8] evaluate the performance of three known classification algorithms—namely, J48 Decision Tree, Naïve Bayes and Decision Table—, to predict the dropout rates of students at the University School of Computing of Witwatersrand. This study found that the algorithm of Decision Trees is a powerful and precise tool to predict the performance of students, given that 92% of the cases were correctly predicted.

In [5, 9, 10] the use of collaborative filtering methods is proposed. This method assumes that similar users share similar interests. In the educational field, collaborative filtering methods are based on the hypothesis that a student performance can be predicted from grade history of all courses successfully completed. The results show that this approach is as effective as commonly used machine learning methods, such as support vector machine (SVM).

Kostopoulos [11] researches the application of semi-supervised machine learning techniques for predicting the performance of distance higher education students. The study compares supervised learning algorithms such as C4.5 and Decision Tree with semi-supervised Tri-Training algorithms. The results show the advantage of semi-supervised methods and especially the performance of the Tri-Training algorithm.

In [12] the authors propose the development of methods that use historical datasets of student grades by courses, with the objective of estimating student performance. Their proposal was based on the use of dispersed linear models and low-range matrix factorizations. This work showed that focusing on course-specific data improves the accuracy of grade prediction. In the paper presented by Sweeney et al. [4] historical datasets of student grades were used as a reference, with the objective of predicting the performance of each student in their upcoming courses for the next academic semester/term. The research applied the factoring machine (FM) model that combines the advantages of SVM with factoring models. Unlike the SVM algorithm, FM models all interactions between variables using factored parameters [13].

3 Methods and Materials

The proposed method evaluates the extent to which partial grades (PG) impact the pass or fail score in a given subject. In addition, it is planned to analyze the importance of the grades of the previous semester in the current grades. For the experimentation,

Table 1 Sample of the dataset

Period	Teacher	Subject name	PG1	PG2	PG3	FG	Area	Situation
2016–1	Teacher 1	General physics	8.0	4.4	6.3	6.2	Maths and physics	Pass
2017–1	Teacher 2	Communications theory	6.0	5.6	5.3	5.7	Infrastructure	Fail
2017–1	Teacher 3	Digital electronics	4.4	8.1	6.9	6.4	Electronics	Pass

we use weka,[1] which has proven to be an excellent tool that includes many data mining methods as well as machine learning algorithms. The methodology proposed for this paper is described in the following five steps:

1. The data collection and data cleaning of the historical datasets of student grades is carried out.
2. Selection of subset to test the method. In this step, a subset of the students that meet the requirements established for the subsequent analysis is selected.
3. The methods of machine learning and data mining are selected.
4. The model for predicting student grades is generated from previously processed data.
5. The results obtained are analyzed and visualized.

3.1 Description and Debugging of the Data Set

The dataset used for the present work is composed of 6358 historical datasets of student grades corresponding to the periods from 2016-1st semester to 2018-2nd semester in the Computer Systems Engineering degree. The dataset comprises the academic records of 335 students and a total of 79 subjects organized into 7 knowledge areas (Programming and Development, Mathematics and Physics, Information Network Infrastructure, Databases, Economy and administration, Electronics, Languages and General Education). The data were extracted from the institution's database and stored in an Excel file. In order to pass a course, the student must obtain a grade equal to or higher than 6. Table 1 shows a sample of the dataset. The final grade (FG) is made up of three differently weighted components: PG1 35%; PG2 35%; and PG3 30%.

Once the information was collected, a process of anonymizing the data was carried out to comply with international data protection standards. This process consisted of eliminating or substituting the personal data fields (citizenship card, names and surnames) of both students and teachers.

[1] https://www.cs.waikato.ac.nz/ml/weka/.

PG1			PG2			PG3			Final grade		
Min.	:	0.000	Min.	:	0.000	Min.	:	1.000	Min.	:	1.00
1st Qu.	:	5.900	1st Qu.	:	5.600	1st Qu.	:	5.200	1st Qu.	:	6.10
Median	:	7.100	Median	:	7.000	Median	:	6.700	Median	:	6.90
Mean	:	6.902	Mean	:	6.658	Mean	:	6.306	Mean	:	6.66
3rd Qu.	:	8.200	3rd Qu.	:	8.200	3rd Qu.	:	7.900	3rd Qu.	:	7.80
Max.	:	10.000	Max.	:	10.000	Max.	:	10.000	Max.	:	10.00

Fig. 1 Statistical measures related to the main components (student grades)

As part of the data pre-processing phase, duplicate records, records with null values in the PG1, PG2 and PG3 components, and course grade records from the general education area (English, academic writing, communication and language, etc.) were eliminated. A brief summary of statistical measures of the final dataset is shown in Fig. 1.

4 Results

Once the dataset is loaded into the weka tool, it is possible to see in Fig. 2 all the data graphically to more easily appreciate the correlation between all the columns with respect to the result (pass or fail the course, column "desSituation"; blue = "pass", red = "fail"). This information can be seen in the lower left corner of Fig. 2.

Figure 2 could be seen like dashboards where it is possible to measure the influence/relationship of every particular feature (PG1 = N1, PG2 = N2, PG3 = N3)

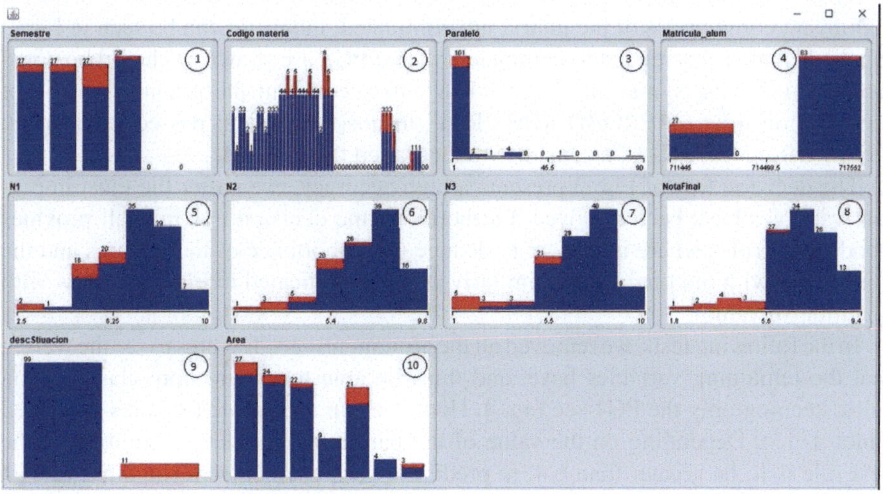

Fig. 2 Visualization and correlation with all data

regarding the final grade (descSituation). There are obvious cases where is clear to identify that correlation. The final grade (NotaFinal) clearly identify (almost with a perfect line) that up to 5.6 the final grade will be "fail", whereas over this value the final grade will be "pass".

Most of the remain dashboards are not so simple to interpret. They often show mixes of red & blue to misunderstand the correlation. Of course, there are general indications of these indicators like the partial grades which indicate a trend to blue when the value increases, and they are rather red when the value is low. In fact, this is the clear objective of an indicator, clear and concise. It is also worthwhile to mention the variable "Area". There will be always a majority of blue as the classes (pass and fail) are totally unbalanced.

Figure 2 shows interesting information regarding the number of students who passed or failed the subject according to each of the components (PG1, PG2 and PG3). Thus, for example: only 0.7 % of students approved the subject with a PG1 < 4; while 12.11 % of students passed with a 4 = < PG1 < 6; and 86.26 % of students passed the subject with a PG1 > = 6. In the case of PG2, the following data are available: only 0.5 % of students approved the subject with a PG1 < 4; 14.3 % of students passed with a 4 = < PG1 < 6; and 82.2 % of students passed the subject with a PG1 > = 6. In the case of PG3 the following data are available: only 1.9 % of students passed the subject with a PG1 < 4; 18.9 % of students passed with a 4 = < PG1 < 6; and 76.7 % of students passed the subject with a PG1 > = 6. Another aspect that can be observed from Fig. 2 is to analyze a group of 124 students who with partial grades (PG1 and PG2) > = 6 and with grades in PG3 <= 4 approved the subject.

In the first tests, obvious relations can be observed when analyzing all the columns of the matrix. For example, it can be seen that the final grade column is the one that determines if the student has passed the subject or not. However, nothing is observed in relation to the way PGs (PG1, PG2, PG3) influence FG. However, if the most significant components of the matrix are eliminated, the result can be seen in Fig. 3 in which some rules related to columns PG2 and PG3 are shown. It should be noted that the following syntax has been taken into account in the graphic scheme of the decision tree: ("APROBO MATERIA" means the student passed the subject, "REPROBO MATERIA" means the student failed the subject).

The decision tree of Fig. 3 presents a high accuracy, even after the most important variables have been removed. Furthermore, the decision tree in itself provides good visual rules where is evident to deduce the importance of the features and the correlation with the final grade (similarly as it was indicated in Fig. 2 but now with quantitative value).

In the following tests, we removed all the aforementioned features to see the weight that the remaining variables have, and thus, be able to obtain more classification rules, keeping only the PG1 see Fig. 4. Here, we can appreciate the rules generated (lines 4–13). Depending on the value of the partial grade N1 the condition of the first rule is to be greater than 6.4, to predict to pass any subject. The other rules 4 generated regard the subject (Codigo materia) and finally the last one (lines 12–13) combines these two features (codigo materia and when N1 is less or equal than 5.1.)

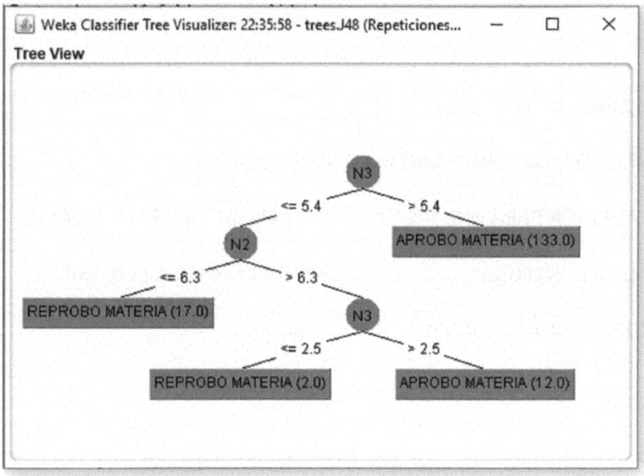


```
Correctly Classified Instances         157               95.7317 %
Incorrectly Classified Instances         7                4.2683 %
Kappa statistic                          0.805
Mean absolute error                      0.0434
Root mean squared error                  0.1988
Relative absolute error                 20.7941 %
Root relative squared error             62.0889 %
Total Number of Instances              164

=== Detailed Accuracy By Class ===

                 TP Rate  FP Rate  Precision  Recall   F-Measure  MCC      ROC Area  PRC Area  Class
                 0.966    0.105    0.986      0.966    0.976      0.808    0.961     0.991     APROBO MATERIA
                 0.895    0.034    0.773      0.895    0.829      0.808    0.961     0.814     REPROBO MATERIA
Weighted Avg.    0.957    0.097    0.961      0.957    0.959      0.808    0.961     0.970

=== Confusion Matrix ===

   a   b   <-- classified as
 140   5 |   a = APROBO MATERIA
   2  17 |   b = REPROBO MATERIA
```

Fig. 3 Experiments without the two more significant values. (PG2 = N2, PG3 = N3)

In this latter case the prediction is to fail. Lines 17–18 indicate the accuracy of the prediction. For example, the following measures are obtained: TP Rate (rate of true positives, instances correctly classified), FP Rate (rate of false positives, instances falsely classified), Kappa (measure of agreement between the classifications and the true classes), ROC Area (curve of probability, representing a degree or measure of separability. It indicates the ability to distinguishing between classes), just to mention some of these measures.

Finally the confusion matrix (lines 33–37) it is a working measurement for a classification problem that in our problem provides the accuracy in the final classification (fail or pass, “APROBO MATERIA” or “REPROBO MATERIA”, respectively).

```
PART decision list
------------------

N1 > 6.4: APROBO MATERIA (117.0/1.0)

Codigo materia  = ACI630-BASE DE DATOS II: REPROBO MATERIA (4.0/1.0)

Codigo materia  = AES300-ESTADISTICA PARA INGENIERIA          : APROBO MATERIA (3.0/1.0)

Codigo materia  = MAT310-CALCULO INTEGRAL                     : APROBO MATERIA (3.0/1.0)

Codigo materia  = IES541-ELECTRONICA ANALOGICA                AND
N1 <= 5.1: REPROBO MATERIA (3.0/1.0)

=== Summary ===

Correctly Classified Instances       143           87.1951 %
Incorrectly Classified Instances      21           12.8049 %
Kappa statistic                   -0.0226
Mean absolute error                 0.1778
Root mean squared error              0.3161
Relative absolute error             85.1027 %
Root relative squared error          98.7213 %
Total Number of Instances            164

=== Detailed Accuracy By Class ===

             TP Rate  FP Rate  Precision  Recall  F-Measure  MCC     ROC Area  PRC Area  Class
             0.986    1.000    0.883      0.986   0.932      -0.040  0.789     0.952     APROBO MATERIA
             0.000    0.014    0.000      0.000   0.000      -0.040  0.789     0.253     REPROBO MATERIA
Weighted Avg.  0.872  0.886    0.780      0.872   0.824      -0.040  0.789     0.871

=== Confusion Matrix ===

   a     b   <-- classified as
 143     2 |   a = APROBO MATERIA
  19     0 |   b = REPROBO MATERIA
```

Fig. 4 Experiments using only the first grade PG1

5 Conclusions and Future Works

The information presented in this article is part of a work in progress. In the presented work some results were obtained that, although they are not too conclusive, they allow us to be optimistic about future work when more data can be used to establish better conclusions. The results obtained in the experiment are quite encouraging, in relation to related works. A complete series of experiments have been carried out in order to establish the best correlations between the input variables and the result, which is the prediction of whether the student will pass a certain subject or not. These experiments have combined the different partial grades choice, as well as the follow-up of the students. The results obtained by the experiments allow concluding about the creation of action plans to avoid dropout in the classrooms and to personalize the student follow-up as much as possible.

As a result of the research carried out at the university on issues of educational data mining and learning analysis, together with the application of academic quality standards, the change in the percentage weights assigned to each partial grade was approved (PG1, PG2 and PG3). From the present work it was concluded that a significant number of students neglected their academic performance in PG3, because with a moderate performance in PG1 and PG2 they pass the subject. After this change in the percentage weights, a better academic performance of the students in the PG3 was observed, as well as a reduction in the rate of student absenteeism at the end of each academic period.

We believe that additional experiments are needed to obtain a better and more accurate prediction. For this, it is necessary to evaluate other attributes and other criteria that significantly influence the performance and quality of the prediction. In addition, there are numerous factors that hinder the predictive analysis of the academic performance of students, such as demographic factors, social factors, family factors among others.

References

1. Bogarín, A., Cerezo, R., Romero, C.: A survey on educational process mining. Wiley Interdiscip. Rev.: Data Min. Knowl. Discov. **8**, 1230 (2018)
2. Park, K., Ali, A., Kim, D., An, Y., Kim, M., Shin, H.: Robust predictive model for evaluating breast cancer survivability. Eng. Appl. Artif. Intell. **26**, 2194–2205 (2013)
3. Gil, D., Fernández-Alemán, J., Trujillo, J., García-Mateos, G., Luján-Mora, S., Toval, A.: The effect of green software: a study of impact factors on the correctness of software. Sustainability **10**, 3471 (2018)
4. Sweeney, M., Lester, J., Rangwala, H.: Next-term student grade prediction. In: Proceedings—IEEE International Conference on Big Data, pp. 970–975 (2015)
5. Rechkoski, L., Ajanovski, V. V., Mihova, M.: Evaluation of grade prediction using model-based collaborative filtering methods. In: IEEE Global Engineering Education Conference, pp. 1096–1103. EDUCON (2018)
6. Hwang, C.S., Su, Y.C.: Unified clustering locality preserving matrix factorization for student performance prediction. IAENG Int. J. Comput. Science. **42**, 1–9 (2015)
7. Hershkovitz, A., Nachmias, R.: Learning about online learning processes and students' motivation through web usage mining. Interdiscip. J. E-Learn. Learn. Objects. **5**, 197–214 (2009)
8. Mashiloane, L., Mchunu, M.: Mining for marks: a comparison of classification algorithms when predicting academic performance to identify "students at risk." In: Proceedings of: International Conference, Mining Intelligence and Knowledge Exploration, pp. 541–552. MIKE (2013)
9. Huete, J.F., Fernández-luna, J.M., Campos, L.M.De, Rueda-morales, M.A.: Using past-prediction accuracy in recommender systems. Inf. Sci. **199**, 78–92 (2012)
10. Bydžovská, H.: Are collaborative filtering methods suitable for student performance prediction? In: Proceedings of 17th Portuguese Conference on Artificial Intelligence, pp. 425–430. EPIA (2015)

11. Georgios, K., Sotiris, K., Panagiotis, P.: Predicting student performance in distance higher education using semi-supervised techniques. In: Proceedings of 5th International Conference, pp. 259–270. MEDI (2015)
12. Polyzou, A., Karypis, G.: Grade prediction with models specific to students and courses. Int. J. Data Sci. Analytics. **2**, 159–171 (2016)
13. Rendle, S.: Factorization machines. In: Proceedings—IEEE International Conference on Data Mining, pp. 995–1000. ICDM (2010)

Computational Thinking in Basic Education in a Developing Country Perspective

Daniel Chagas and Elizabeth Furtado

Abstract In a connected world, where information is the most valuable input, compulsory education in computational thinking, especially in early ages, had became an important topic for governments who aim in a economy based on technology. This brought initiatives for compulsive adoption on basic education in USA and EU, but few actions on developing countries. This article presents a systematic review of academic papers and commercial products that present the teaching of logic to young people, and that deal with the use of tangible devices, robots or specific software. From the analysis performed with the review, we define requirements for teaching. Thus, considering the factors of analysis, such as pricing and replicability, we generate a series of sub-requirements aimed at adopting a solution for public schools from developing countries. As preliminary results, an interactive robot and a set of tangible artifacts adhering to the identified requirements are presented as a proposal for the teaching of computational thinking.

1 Introduction

The questioning of the evolution of teaching methodologies in schools and the adoption of technologies in the classroom is not new. Already in the 80's, Papert [1] criticizes that the school environment has not changed much in 200 years of formal education, differently to the areas of medicine and engineering, for example.

At the same time a worldwide trend of STEM teaching—Science, Technology, Engineering and Mathematics is accelerating [2]. This trend aims to treat technological education in a more practical and applied form, from elementary to higher education. The White House document, The Nation of Makers [3], states the impor-

D. Chagas (✉) · E. Furtado
LUQS, Universidade de Fortaleza, W. Soares, 1321, Fortaleza 60821-910, Brazil
e-mail: prof.daniel.chagas@gmail.com
URL: http://www.unifor.br

E. Furtado
e-mail: elizabethsfur@gmail.com

A. Visvizi and M. D. Lytras (eds.), *Research & Innovation Forum 2019*,
Springer Proceedings in Complexity,
https://doi.org/10.1007/978-3-030-30809-4_14

tance of technological education to foster innovation and research in the United States of America, and compiles various initiatives from American educational institutions on the subject. Among them is the teaching of programming for ever younger ages, the application of robotics and automation in teaching, and the creation of specific tools to aid teaching.

The European Union [4] addresses the issue first by showing the importance of Computational Thinking (CT) not only for technological activities, but for all professions. It shows the strategies in Europe and around the world in regard to compulsory adoption of CT subjects in primary and secondary schools, but also addresses implementation difficulties linked to teacher training and the lack of educational objects that support the educational process for younger children.

Developing countries, such as Brazil, are behind when it comes to teaching computational thinking in primary and secondary schools (up to 15 years), mainly in the public schools. In addition, data [5, 6] show that university students of computing are among those that drop out the most (around 30%).

From this perspective, the present article aims to evaluate some academic and commercial approaches to teaching logic, and their applicability in the Brazilian-like reality. In addition to classifying approaches in principles of human-computer interaction (HCI), device types and software, it provides an empirical evaluation of the cost-of-implementation (often not addressed in academic solutions) and possibility of replication of the material (the use of free licenses, such as Creative Commons). This allows us to seek pragmatic solutions to a developing country reality, especially in relation to public schools and municipal governments.

The article is divided into five parts. The first four parts, before the conclusion, are as follows: Background on the teaching of computational thinking; Research method with the search and classification of articles and solutions in common characteristics, and the development and evaluation of artifacts; Objectives of this work; and Findings, presenting a set of requirements for computational thinking teaching solutions, the presentation of the final artifacts, with a preliminary assessment of the solution with users.

2 Background

2.1 A Multidisciplinary View of the Difficulties in Teaching Computational Thinking

Bosse and Gerosa [6], describe a description of the problems in the teaching of computational thinking and programming, as well as the difficulties of the students, when using the programming environments, and of the teachers. Students report problems with syntax errors (they do not know how to write commands), variables and types of variables (making many typing errors of variable names), lack of analytical thinking in the resolution (they do not understand the semantics and try to solve with trial and error). They also claim that error messages do not help in correcting problems. Among the difficulties for the teachers, the emphasis is on the heterogeneity of the knowledge in the class (students with a lot of knowledge or who "catch on quickly" dominate the class and leave other students behind), little participation and attendance at classes, disinterest and trauma with the subject.

We classify the main difficulties of the students and those reported by Bosse and Gerosa in regard to HCI, using the following design principles [7]: visibility, feedback, restrictions, consistency and affordance. Our classification was empirical, considering the characteristics of the programming environments and the implications on the mental model of the user, in this case, the student. In addition, the authors of this work have experience in the technological teaching of adolescents and university students, and in the evaluation of technology.

Student errors in syntax and semantics are linked to the visibility [7] and constraints [7] of the programming interface: The student, when viewing a development interface for the first time, has no way to advance alone. The interface does not show what can be done or how. The restriction is also not present, since the interface allows any command or symbol to be typed (it does not restrict the user when erroneously typing the syntax of an IF command, or a variable).

With feedback [7], we have the lack of information for error correction in the messages given by the system. Compiler software (who translate the source code to the machine) are known for their obscure messages, which leaves experts confused, let alone newbies.

We have an interface with good affordance [7] when its attributes allow users to know how to use it (such as a door handle, something easy to hold and move at hand height). The base programming interface is a free typing text box, such as a text editor. We can extrapolate that students make mistakes by comparing the act of programming with writing a letter to a colleague: the student has no restrictions on how to begin or end, nor will s/he be misunderstood if s/he writes some parts wrongly or differently. A software code works in a much more constraints way, as a movie script, which every sentence has a form to be write, and the text is not ambiguous.

Human beings are always seeking similarities of new things with what they already know. We call this consistency [7], and it is a good indication of the interface: to be consistent with what the user already knows or with the real world. We have consistency in programming as a program (abstract and mathematical) works similarly to something from the real world (concrete). Teachers often change the abstract character of logic programming to something concrete, seeking to improve student understanding. An example being when they exemplify IF-ELSE with natural language phrases: "If you score above 8 you will pass, otherwise you will have to do a final exam".

2.2 *Active Learning Strategies to Deal with Programing Abstraction*

Abstraction is a human trait. We judge more or less intelligent animals by their capacity for abstraction. In human development, from childhood to adulthood, we develop our ability to abstract from the concrete play of the baby (moving hands and objects), through to playing and imagining, until reaching the capability of association of non-tangible concepts. Vygotsky [8, 9] in his studies on development and learning, is celebrated for the relevance he gives to the role of language and thought in the formation of the human infant. For Vygotsky, the toy provokes the imagination in the child, which constitutes the fundamental basis for the consolidation of abstract thought and also for the process of internalization of language.

When developing interfaces, the designer of the interaction must be aware of the space of the problem [7] and the mental models [7] of its user [10]. In the case of children from 7 to 9 years old, the scope of this project, we have a user who is still not very skilled with written mathematical language. And only recently (for Vygotsky from the age of 4) is s/he exercising her/his mental models for abstract situations. The realization of the interface through a toy, taking advantage of tactile dexterity already consolidated at the age of 7–9 years, is a way to 'materialize' programming concepts: such as block programming and robotics [11]. Such approaches have shown promising data, especially when adopted at younger ages.

The characteristics of the tangible interaction and its role in the educational context of programming teaching are evidenced by van Gennip [12] when he says:

> While such screen-based GUI interfaces bring many benefits, chiefly among them the flexibility to show and manipulate a wide variety of interfaces, cognitive challenges remain. [...] For example, in an educational context, teaching children the basics of programming is often considered complex and dependent on the possession of basic language skills. However, when approaching this challenge with things children already know, such as building blocks, it turns out that young children are already able to reason about programmatic flow.

In fact, several studies [13, 14] deal with the importance of tangible interfaces in teaching, especially for teaching children [15, 16].

Tangible interfaces also reinforces collaboration and social activity between students, a non-common interaction in the solitary programming activity. Vygotsky [8, 9] in their studies about active learning strategies demonstrate the importance of speech in the mental process of development, delineating the process of children progressing, from external speech to egocentric speech to inner speech.

3 Objectives

The present research aims to define requirements for solutions to teaching of computational thinking for children using tangible solutions, which are validated through the prototyping of an interactive product.

The specific objectives of this proposal are:

(1) Identify requirements for the teaching of computational thinking based on systematic review of articles and analysis of products and solutions, especially aimed to developing countries;
(2) Design a solution that adheres to the requirements raised, based on tangible interfaces;
(3) Make a preliminary validation of the designed solution.

4 Research Method

The search for the state of the art was carried out by means of the methodology of systematic review of publications on the subject, and by an empirical search of commercial solutions to logic teaching and/or programming logic presented as products. After categorizing academic and commercial solutions, physical prototypes were created through the ideas of design thinking, culminating in an assessment of artifacts with users.

4.1 Systematic Review

The bibliographic research was based on the definition of certain investigative questions, which guided the definition of key elements for the research, from two databases. The questions were associated with the objective of the research, which was to acquire knowledge about other academic and scientific projects aimed at teaching logic programming, mainly for elementary and high school students with different educational objects, in Portuguese and English. In addition, the focus was to identify the processes of the execution of these projects, the target audience and the results that were obtained.

The research questions were: What educational object was used in the teaching of logic? How was the educational object used? With which public was the educational object used? The descriptors used were: education, hardware, programming, educational objects, learning. The search strings were: "education and hardware and programming", "educational objects and hardware and learning not software", and "educational objects and teaching logic". The criterion for the selection of search sources were digital article libraries that commonly publish scientific papers on HCI that were published between 2000 and 2017. The sources chosen were the ACM Digital Library and IEEE library because they contained articles from the main publications on human computer interaction.

For the filtering of the collected material, articles were selected where the title and/or abstract presented data about educational objects, or didactic resources, or hardware, applied to the teaching of logic, or mathematical logic, or programming logic. In addition, publications should answer at least two of the questions in this research. Thus, 10 articles were selected for analysis.

Commercial initiatives are also presented in the categorization below for illustration. The methodology for the survey of commercial solutions included common search engines, with the same key words from the systematic review, for products launched in the last 10 years. Products were chosen for the child to adolescent public only.

These works bring different approaches to technological teaching, which can be grouped as:

• **Interaction by manipulation**: the substitution of typed commands with command blocks has been a strategy for teaching programming for both children and adults. Tests with 100 students using block programming showed that most of them were able to predict, understand and apply block behavior, even though they had never seen it before [17]. When dragging and connecting pieces representing commands and parameters, several ideal interface characteristics are favored [10], such as visibility (The user can identify a series of commands that are possible to use, and plan the execution of the schedule), and constraints (the shapes of the blocks or pieces represent the combination options that may or may not be viable).

Virtual manipulation is present in papers: Junior Soccer Simulation [18] (competition between teams at a virtual football match), CodeSpells [19] (Role Playing game for teaching programming, where commands to be taught are seen as "spells"), and Azemi and Pauley [20] (new approach to programming logic with MATLAB). Real manipulation are present in Sarkar and Craig [21] (circuit board for teaching hardware), and Katterfeldt et al. [11] (physical computing with his Talkoo Kit- physical modules connected by small cables). Commercial solutions analysed are Cozmo robot [22] and Fischer-Price's Think and Learn [23] (activated by manipulating electronic blocks).

• **Feedback and use of robot**: Visualization of the results of the program in a playful way, with the use of a robot, robotic arm or electronic puppet, simulated or physical. Usually the teaching of programming is done with terminal commands for mathematical outputs, which is viewed as being dull by students. The vast majority of solutions are presented in something more visible or tangible. The movements (of head, body, arms or images) of physical or virtual robots, make feedback a playful response to programming.

Robots are present in the work of Hongjun et al. [24] (MountTai robot, platform for developing educational robots with viable cost), Aparicio et al. [25] (robot arm), robot development kit [26] (electronic and digital scrap for university students), Barreto et al. [27] (Teaching Programming for High School Students Using the Lego Mindstorms Robot), and Lalonde, Bartley, Nourbakhsh paper [28] (robots are notebooks with webcams on locomotion platforms). Mixed hardware (Desktop, Wearable, electronics and/or Robotics) are present in Member et al. [29] and Merkouris, Chorianopoulos and Kameas [30]. In the commercial solutions, Ozobot [31] (small line-following robot, programmed by drawing the paths using different colors), and Codipeia [23] (electronic centipede, where each segment of the centipede is a command block).

• **Pictographic commands**: Use of simple commands, in the form of pictographs, can be used to move robots, similar to the LOGO language [1]. The works also present the use of simple commands and pictographs. The act of programming a computer is an interaction of the instruction kind [10], typing previously defined commands. The beginner student's difficulty in not knowing what to type or how to organize these commands is what makes an IDE development screen the first obstacle. When replacing typed commands with blocks, physical or virtual, pictographic, and generalist in their functions, one uses the principle of design affordance [10], defined as "an attribute of an object that allows people to know how to use it". The manipulation of command blocks with connectors brings an affordance similar to that of card games, jigsaw puzzles, or dominoes, breaking the barrier of initial interaction, thus avoiding typing errors and syntax, which are so common in beginners.

The Adroit project [32] shows a pictographic visual language, based on blocks, used for children and pre-adolescents to program the movements and actions of a robot on the ground. A similar approach is found in Krishnamoorthy and Kapila [33] where a visual language is based on Google's Blockly. The project of Gupta, Tejovanth and Murthy [34] presents a discussion of the results obtained when introducing programming and a hardware interface using the platform Scratch/Arduino. The commercial product Osmo Coding Game [35] is a software set for tablets with plastic pieces representing program commands, which together can be used to manipulate a character on the tablet screen.

The uses of pedagogical methodologies for the researched solutions, and the influence of the physical environment on users is not yet part of the scope of this study.

4.2 Design Thinking

The development of the prototypes followed the cycle of design thinking which includes empathy, definitions, ideation, prototyping and evaluation. The whole process took place in the Research and Innovation Laboratory of the University of Fortaleza, which relies on the machinery and structure (FabLab) available, both for the manufacture of prototypes and for their evaluation with users. The structure allowed for speed in adopting new approaches as well as realistic cost evaluation.

Empathy. The empathy with the problem came from the support of a programming company teaching children's programming using free tools. Teachers and coordinators participated in meetings of ideation, as well as being part of the evaluation of the prototypes.

Definitions. The analysis of the works and products for the teaching of computational thinking for children assisted in the definition of requirements to be followed in the subsequent phases. In addition to the general requirements, the characteristic being pursued was always of artifacts being used in multiple forms, mainly associated with playful strategies for their exploration among students.

Ideate. Ideation meetings were held biweekly with researchers and sometimes with teachers.

Prototyping. The prototyping of new solutions allowed for the construction of several prototypes, which were evolved in each design cycle. The choice of laser-cut materials (rather than 3D) sped up the process of ideation-prototyping-evaluation, as well as the choice of inexpensive and easily adjustable materials (Medium-Density Fiberboard—MDF and Ethylene-vinyl acetate—EVA).

Evaluation. The evaluation of the prototypes occurred primarily in the laboratory, with researchers and teachers involved, and then in the school of programming for children, with the support of the local coordination.

5 Findings

5.1 Requirements

The elaboration of the requirements begins with the analysis of the references studied in the state of the art, and the empirical analysis of the authors on the aspects of HCI in Sect. 2.1. By design thinking meetings involving researchers and teachers, it was possible to evaluate the systematic review findings and commercial solutions presented. Teacher's experiences and empirical analysis, supported by discussions about active learning strategies, helped to define requirements related to teaching. Partner companies, related to the technological teaching business, has a important role in the sustainability of solutions, especially in a scenery of developing countries local production, and about the choice of Open Source solutions.

The basic requirements for teaching logic in schools were listed as:

- Presentation of programming commands not as text, but as real or virtual command blocks.
- Start with codes that generate playfulness (movement, sound, light, etc.) using interactive physical devices.
- Encourage collaboration among students to solve challenges.
- Support digital inclusion for everyone, and the accessibility of the solution is a relevant requirement.

Reality of underdeveloped countries. In addition to the set of basic requirements presented, the education context of developing countries (such as Brazil) needs a specific subset of requirements. This is due to the inadequate physical structure and laboratories in Brazilian schools in general, and the lack of teacher training in the adoption of technologies in the classroom. These include:

- Low implementation cost, with the use of software that allow virtual interaction with the elements of the proposal.

- Use of low cost materials for tangible objects such as M.D.F. (medium-density fiberboard) wood, E.V.A., acrylic and paper, interacting with software through computer vision, or sensors present in smartphones.
- When using hardware, choose Open Source projects [36], which can easily be replicated and are designed with accessible parts or parts obtained from scrap.

5.2 Proposed Solution Featuring Tangible Interactive Artifacts

Given the requirements it was possible to prototype a set of interactive artifacts for teaching computational thinking. The set of tangible interactive artifacts was called Code Domino. These are pieces of low-cost material (MDF) fitted with an adhesive (TAG) with an RFID chip, representing commands based on LOGO codes [1].

Programming is achieved using a domino game, where users place a sequence of commands to be executed online by a robot Fig. 1.

The use of tangible pieces allows for a collaborative interaction among users. The pieces, colored and designed to be connected, are aimed at exploration and learning in a playful way. The use of a set of pieces for programming with IoT (Internet of Things) RFID was successfully evaluated by Carbajal and Baranauskas [37], at a low cost and with a negligible number of reading errors. But, unlike Carbajal and Baranauskas where the code is executed via software, the Code Domino result from coding is visualized by the users through a robot, which registers and executes the steps that have been programmed into it Fig. 1.

Tangible Command Parts The set of pieces is organized into two sets:

- A basic domino piece, which is pictographic, representing only movements (front, right, left, etc.) and non-abstract actions (pause, musical notes);
- An extended domino piece representing, in addition to the basic actions, common commands in programming logic (otherwise, repeat, arithmetic and boolean operations, variables and functions).

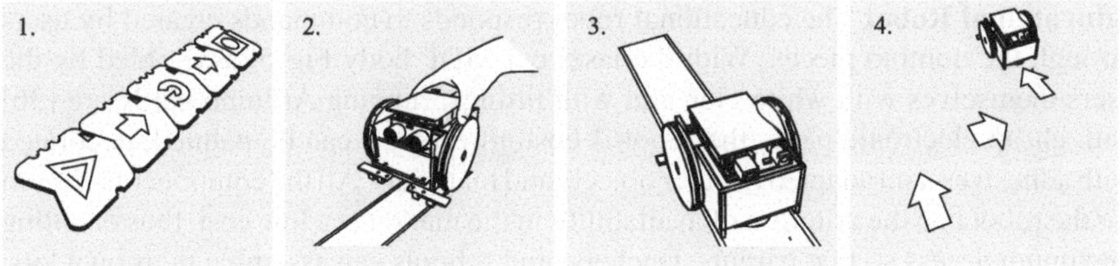

Fig. 1 Programming sequence of robot movements through physical parts: 1—Coding, 2—Positioning, 3—Loading, and 4—Execution

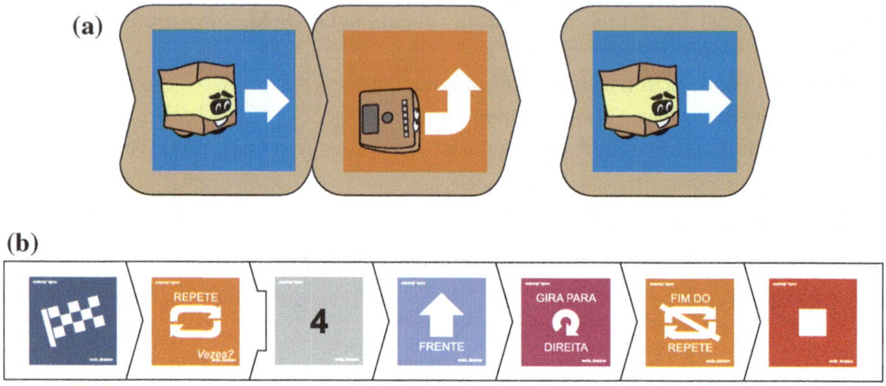

Fig. 2 Different set of pieces of the robot commands: **a** Basic pieces, and **b** Extended pieces

The two sets of pieces with different complexity levels were created to suit the different age ranges of users Fig. 2). The most basic set aims to serve children not yet literate up to 8 years old. There are three pieces, the first two being fitted together, are located at the top of Fig. 2. In this image, each piece is available in an MDF base for easy manipulation. The most complex set of pieces aims to teach the basics of computer programming, and focuses on users over 9 years of age. Seven pieces representing the robot commands, are located at the bottom of Fig. 2.

Pieces with accessibility support Given the tangibility of the pieces, it was possible in their design to think about their accessibility for blind users, adopting braille code in grooves on the side of the pieces Fig. 3. The approach is similar to Koushik's work [38] where tangible pieces receive tactile markings for storytelling with blind children.

To facilitate assembly for very new users or for those with limited movement, the parts can be mounted on the MDF guide bases, where they do not move out of position easily Fig. 4. The base also allows for a more comfortable reading of the assembled code for blind users, without undoing the order.

Educational Robot The educational robot responds to commands created by users through the domino pieces. With the laser cut MDF body Fig. 5, assembled by the users themselves with white glue and with fittings, running Arduino hardware [36] with cheap electronic parts, the robot is customizable, it can be painted, decorated with adhesives, and adapted to other objects and materials. All the components chosen for the robot had the criterion of availability in the market at a low cost, thus enabling maximum access so that parents, teachers, and schools can assemble their own kits.
Preliminary Analysis with Users The purpose of the evaluation described in this subsection was to validate the proposed solution with users (teachers and volunteer students). With teachers who teach programming for children in a private programming course, the applied technique was an interview, and we wanted to understand

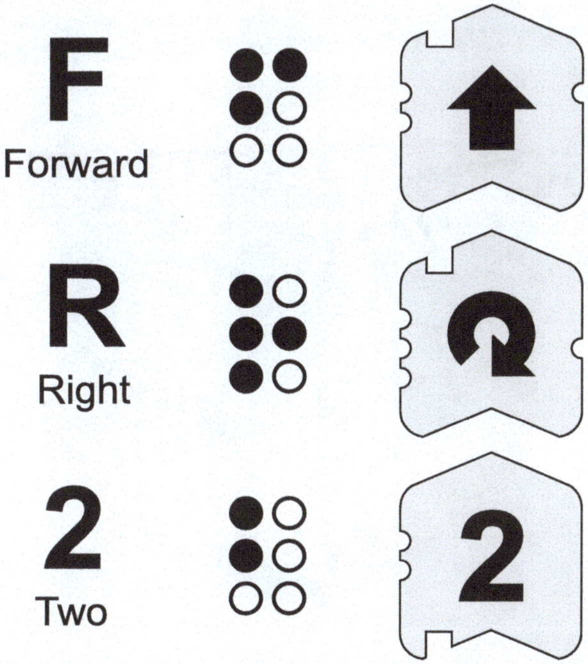

Fig. 3 Representation of the command in braille on the side of each piece

Fig. 4 Pieces representing commands and inserted into the guide base in MDF

Fig. 5 Robot with RFID reader to execute programming made with tangible commands

how they could explore the solution in the classroom. With the students, we held a session where they manipulated the pieces and the robot and investigated their acceptance and understanding.

5.3 Teachers

The three teachers interviewed said that the artifacts (robot, pieces and firmware) can support their teaching practice, as they allow the solutions to be fun and also activities to be performed in the form of challenges to students, and in competitions between teams. One of them pointed out that one of these competitions could be to burst a balloon: A balloon is positioned in a given position of the room (or stuck on a robot of another team), and teams must code their robots to reach the balloon and burst it first (the robot has a sharp stick attached to the front).

Another point of great relevance addressed by teachers is an increase in collaboration among students, since the pieces are distributed and are directly accessible to everyone. In a class using software and shared computers, it is common for a student in the group to master the machine, programming alone, and thus other students just end up watching the activity being performed.

5.4 Students

Usability tests were performed with three children (Fig. 6) to assess whether the solution of tangible parts would be comprehensible to the initial target audience (7–9 years). The students were children with very basic programming experience. Each session lasted on average 30 min, and was accompanied by two team participants (the evaluator and the developer). During the manipulation of the robot with the programming domino pieces by each student, their behavior was observed and questions were asked to investigate their reactions.

Results were as follows: The simplification of the programming process (without the use of devices and/or screens) was a point perceived by students; The students showed empathy with the educational robot. The fact that the robot appears to be approachable, to have a nice face, to have a shape that invites manipulation and exploration were comments reported by the children.

The commands were well understood by the pictograms, and pieces were enjoyable to touch and manipulate.

The positioning of the robot over the commands was found to be difficult, because at the time of the tests (as shown in Fig. 6), there was still no guide base, as shown in Fig. 4.

Fig. 6 Usability test with children

6 Discussion and Conclusion

The adoption of computational thinking in schools involves the adoption of technologies that adhere to school practices. These solutions should speak the "language" of the school, merge with the teaching methodologies, and converse with the school agent: the teacher. From the systematic review, it can be observed that most of the analyzed solutions focus on more technical aspects to the detriment of the questions and/or more pedagogical approaches, culminating in a perspective that views only the teaching of logic programming.

In regard to the requirements presented with the analysis of the solutions, it is shown that working with interaction by block manipulation (virtual or not) with results that are fun can be decisive factors in the success of an educational solution. Regarding the developing countries reality, issues such as cost of solutions and replicability are key factors for adoption, given that they support the teacher in seeking solutions that adhere to their reality.

This paper innovate by bringing requirements and a implemented solution related to developing countries, especially about sustainability and replicability of the solution. It reinforces the necessity of science to reach the poor population, and implies to researchers and professionals the importance of applied science focusing in developing countries and their most needed populations (in slums, ghettos, etc.).

The prototyping of the set of interactive artifacts presented, adhering to the requirements raised and in accordance with the many challenges of human computer interaction of the Brazilian Computer Society (SBC GrandIHC [39]), show that the adoption of new forms of interaction and elements of IoT (Internet of Things) can fill the existing gap between schools and technology, as well as fostering digital inclusion for everyone.

In order to expand the research started, the team of this study intends to carry out an experimental research, with control groups, with classes of children in a course of the same teaching project, but with different pedagogical tools: Code Domino (artifacts from this research) compared with Scratch, and Lego WeDo (other pedagogical teaching tools for children's programming). The evaluation will take into account the assimilation of the content, the engagement (frequency and activity) and the satisfaction with the tool (affection).

References

1. Papert, S.: The Childrens Machine: Rethinking School in the Age of the Computer. Basic Books Inc., New York (1993)
2. Hatch, M.: The maker movement manifesto. Mak. Mov. Manif. (2014). https://doi.org/10.1162/INOV_a_00135
3. White House: The Maker Movement. https://www.whitehouse.gov/nation-of-makers
4. Bocconi, S., Chioccariello, A., Dettori, G., Ferrari, A., Engelhardt, K., Kampylis, P., Punie, Y.: Developing computational thinking in compulsory education. Proc. EdMedia **2016** (2016). https://doi.org/10.2791/792158

5. Takahashi, F.: Matemtica e Cincia da Computao tem alta taxa de abandono. http://www1.folha.uol.com.br/educacao/2009/04/546576-matematica-e-ciencias-da-computacao-tem-alta-taxa-de-abandono.shtml
6. Bosse, Y., Gerosa, M.A.: Why is programming so difficult to learn?: Patterns of difficulties related to programming learning mid-stage. SIGSOFT Softw. Eng. Notes. **41**, 16 (2017). https://doi.org/10.1145/3011286.3011301
7. Rogers, Y., Sharp, H., Preece, J.: O que Design de Interao? In: Design de Interao. p. 2529. Bookman, Porto Alegre, Brazil (2013)
8. Rabello, E., Silveira, J.: Vygotsky e o desenvolvimento humano. **110** (2011)
9. Vygotsky, L.S.: Mind in society (1978)
10. Rogers, Y., Sharp, H., Preece, J.: Compreendendo e Conceitualizando a Interao. In: Design de Interao. p. 4647. Bookman, Porto Alegre, Brazil (2013)
11. Katterfeldt, E.-S., Cuartielles, D., Spikol, D., Ehrenberg, N.: Talkoo. A new paradigm for physical computing at school. Proceedings of the 15th International Conference on Interaction Design and Children-IDC pp. 512–517 (2016). https://doi.org/10.1145/2930674.2935990
12. van Gennip, D., Orth, D., Imtiaz, M.A., van den Hoven, E., Plimmer, B.: Tangible cognition: bringing together tangible interaction and cognition in HCI. In: Proceedings of the 28th Australian Conference on Computer-Human Interaction. pp. 662–665. ACM, New York (2016)
13. Horn, M.S., Solovey, E.T., Crouser, R.J., Jacob, R.J.K.: Comparing the use of tangible and graphical programming languages for informal science education. In: Proceedings of the 27th International Conference on Human Factors in Computing Systems-CHI. p. 975 (2009). https://doi.org/10.1145/1518701.1518851
14. McNerney, T.S.: From turtles to tangible programming bricks: explorations in physical language design. Pers. Ubiquitous Comput. **8**, 326337 (2004). https://doi.org/10.1007/s00779-004-0295-6
15. Futschek, G., Moschitz, J.: Learning algorithmic thinking with tangible objects eases transition to computer programming. Lect. Notes Comput. Sci. (including Subser. Lect. Notes Artif. Intell. Lect. Notes Bioinformatics). 7013 LNCS, 155164 (2011). https://doi.org/10.1007/978-3-642-24722-4_14
16. Posada, J.E.G., Baranauskas, M.C.C.: A socio-constructionist environment to create stories using tangible interfaces. In: Proceedings of the 14th Brazilian Symposium on Human Factors in Computing Systems. pp. 1:1–1:10. ACM, New York (2015)
17. Honig, W.L.: Teaching and assessing programming fundamentals for non majors with visual programming. In: Proceedings of the 18th ACM conference on Innovation and Technology in Computer Science Education-ITiCSE 13. p. 40. ACM, New York (2013)
18. Martins, F.N., Gomes, I.S., Santos, C.R.F.: Junior soccer simulation: providing all primary and secondary students access to educational robotics. In: Proceedings of the 12th LARS Latin American Robotics Symposium 3rd SBR Brazilian Symposium on Robotics LARS-SBR 2015—Part Robotics Conference 2015. pp. 61–66 (2016). https://doi.org/10.1109/LARS-SBR.2015.16
19. Esper, S., Foster, S.R., Griswold, W.G.: CodeSpells. In: Proceedings of the 18th ACM Conference on Innovation and Technology in Computer Science Education-ITiCSE 13. p. 249 (2013). https://doi.org/10.1145/2462476.2465593
20. Azemi, A., Pauley, L.L.: Teaching the introductory computer programming course for engineers using Matlab. In: 2008 38th Annual Frontiers in Education Conference (2008). https://doi.org/10.1109/FIE.2008.4720302
21. Sarkar, N.I., Craig, T.M.: A low-cost PIC unit for teaching computer hardware fundamentals to undergraduates. ACM SIGCSE Bull. **39**, 88 (2007). https://doi.org/10.1145/1272848.1272892
22. Anki: Cozmo. https://www.anki.com/en-us/cozmo (2015)
23. Fischer-Price: Think and learn. http://www.fisher-price.com/pt_BR/brands/think-and-learn/index.html (2017)
24. Hongjun, S., Xin, M., Fengyu, Z., Yibin, L.: The design and implementation of OpenGL-based comprehensive educational robot system. In: Proceedings of the IEEE ICIA 2006—2006 IEEE International Conference on Information Acquisition. pp. 522–527 (2006). https://doi.org/10.1109/ICIA.2006.305788

25. Garduno-Aparicio, M., Rodriguez-Resendiz, J., Macias-Bobadilla, G., Thenozhi, S.: A multidisciplinary industrial robot approach for teaching mechatronics-related courses. IEEE Trans. Educ. **61**, 5562 (2018). https://doi.org/10.1109/TE.2017.2741446
26. Lopes Filho, J.A.B., Almeida, W.R.M., Martins, S.G.: Development of a multitasking mobile robot for the construction of educational robotics kits. In: International Conference on Electronic Devices, Systems and Applications (ICEDSA). pp. 213–216 (2011). https://doi.org/10.1109/ICEDSA.2011.5959090
27. Barreto, V.B., LErario, A., Fabri, J.A.: Ensino de Programacao para Alunos do Ensino Mdio Utilizando o Robo Lego Mindstorms. In: 2015 10th Iberian Conference on Information Systems Technologies CISTI (2015). https://doi.org/10.1109/CISTI.2015.7170521
28. Lalonde, J.-F., Bartley, C.P., Nourbakhsh, I.: Mobile robot programming in education. In: Proceedings of the 2006 IEEE International Conference on Robotics and Automation pp. 345–350 (2006). https://doi.org/10.1109/ROBOT.2006.1641735
29. Member, M.R., Lysecky, S., Rozenblit, J.: Educational technologies for precollege engineers. **5**, 2037 (2011)
30. Merkouris, A., Chorianopoulos, K., Kameas, A.: Teaching programming in secondary education through embodied computing platforms. ACM Trans. Comput. Educ. **17**, 122 (2017). https://doi.org/10.1145/3025013
31. Ozobot, Evollve.: Ozobot, www.ozobot.com (2017)
32. Besari, A.R.A., Sukaridhoto, S., Wibowo, I.K., Berlian, M.H., Akbar, M.A.W., Yohanes Yohanie, F.P., Aldi Bayu, K.I.: Preliminary design of interactive visual mobile programming on educational robot ADROIT V1. In: Proceedings of the 2016 International Electronics Symposium. IES 2016. pp. 499–503 (2017). https://doi.org/10.1109/ELECSYM.2016.7861057
33. Krishnamoorthy, S.P., Kapila, V.: Using a visual programming environment and custom robots to learn C programming and K-12 STEM concepts. In: Proceedings of the 6th Annual Conference on Creativity and Fabrication in Education-FabLearn 16. pp. 41–48 (2016)
34. Gupta, N., Tejovanth, N., Murthy, P.: Learning by creating: interactive programming for Indian high schools. In: Proceedings of the 2012 IEEE International Conference on Technology Enhanced Education ICTEE 2012. p. 24 (2012). https://doi.org/10.1109/ICTEE.2012.6208643
35. Tangible Play Inc.: Osmo (2013)
36. Banzi, M.: Getting Started with Arduino (Make: Projects). Make Books (2008)
37. Carbajal, M.L., Baranauskas, M.C.C.: TaPrEC: Desenvolvendo um ambiente de programao tangvel de baixo custo para crianas. An. do XX Congr. Int. Informtica Educ.- TISE. **11**, 363–370 (2015)
38. Koushik, V., Kane, S.K.: Tangibles + Programming + Audio Stories = Fun. In: Proceedings of the 19th International ACM SIGACCESS Conference on Computers and Accessibility-ASSETS 17. pp. 341–342 (2017). https://doi.org/10.1145/3132525.3134769
39. Baranauskas, M.C.C., de Souza, C.S., Pereira, R.: GranDIHC-BR: Prospeco De Grandes Desafios De Pesquisa Em Interao Humano-computador No Brasil. In: Companion Proceedings of the 11th Brazilian Symposium on Human Factors in Computing Systems. p. 6364. Brazilian Computer Society, Porto Alegre, Brazil (2012)

Cognitive Computing and Social Networking

Cognitive Computing Approaches for Human Activity Recognition from Tweets—A Case Study of Twitter Marketing Campaign

Jari Jussila and Prashanth Madhala

Abstract Big data analytics is a growing area of research and has in the recent years introduced many applications for business sector to create value from data. Cognitive computing provides new approaches for analysing social media data, e.g. text analytics and machine vision, available via APIs and software libraries. This study focuses on cognitive computing approaches for recognizing and categorizing valence, arousal and human activity from tweets. The aim is to identify human activities from tweets and map these activities on pleasure and arousal dimensions using cognitive computing services in order to derive new insights for business. A case study of Twitter marketing campaign of a company providing stress management services is presented. In the marketing campaign participants submitted images and text on events that help them to recover from stress. These tweets are analysed with cognitive computing services to recognize activities that participants of the campaign performed to recover from stress and these are compared against qualitative analysis. The caption recognition was successful at detecting some of the activities of the users and overall results indicate that many participants cope with stress by e.g. performing outdoor activities. The two sentiment detection algorithms were also successful in detecting consumer sentiments from the text component of the tweets. Finally, implications for business are drawn from the results of analysis.

1 Introduction

Social media has transformed the ways in which companies and customers interact. A great deal of social media interaction are emotionally loaded. Furthermore, competition is increasingly based on a brand's ability to inspire emotional experiences [1]. According to a recent study, fully emotionally connected customers are 52% more valuable on average than just highly satisfied ones [2]. In fact, for more than a decade, emotions have been recognized as being essential in marketing and

J. Jussila (✉) · P. Madhala
Häme University of Applied Sciences, 13100 Hämeenlinna, Finland
e-mail: jari.jussila@hamk.fi

A. Visvizi and M. D. Lytras (eds.), *Research & Innovation Forum 2019*, Springer Proceedings in Complexity,
https://doi.org/10.1007/978-3-030-30809-4_15

consumer behavior studies [3]. These studies have highlighted the fact that, instead of rational decision making based on utilitarian product attributes and benefits, consumer decisions are "biased" by emotions [1]. Understanding customers' emotions can be further augmented by including information about the activity that the emotion is related to. Activity recognition is a significant research topic by itself with far-reaching applications in many areas such as healthcare, personal assistance [4] and marketing; and combined together with emotion recognition can provide novel solutions for several application areas.

Although there is a wealth of social media data available related to consumer and business-to-business customers' emotions and activities for companies to collect and analyze [5], however the sheer volume of the data makes manual analysis and interpretation challenging, time-consuming, and costly [6]. Consequently, social media monitoring and analytics tools have emerged that can provide a cost-effective way to gather relevant data and the means of processing the data into knowledge, enabling more accurate and valuable marketing decisions [7].

Recent studies [8, 9] emphasize that big data (BD) generated by social media should be used for understanding customers better. In the literature, some authors distinguish big social data (BSD) from the broader category of big data [10]. Big data has been, for instance, conceptualized as any data that is produced as a result of the quantification of the world that may include data from sensors, multiple industrial and domestic networks as well as financial markets, whereas big social data is produced as a result of the mediated communication practices of our everyday life, "whenever we go online, use our smartphone, use an app or make a purchase" [11]. In recent years, several approaches to big social data analytics (BSDA) [10, 12–14] have been introduced that make use of computational methods for detecting and analyzing human emotions and activities. Cognitive computing defined by Valiant [15] as *"a discipline that links together neurobiology, cognitive psychology and artificial intelligence"* was on the other hand introduced already in 1995—long before there was any buzz on social media and big data. The objective of cognitive computational models "*to endow computer systems with faculties of knowing, thinking, and feeling*" [16] is, however, as timely as in the nineties.

The current big data research has been also criticized. For instance, Boyd and Crawford [17] question the objectivity and accuracy of the data, and challenge whether big data is always more suited to the research task than small data and whether it can preserve the contexts of what it is intended to describe. Bruns [18] continues this by stating that the challenge is to improve the quality rather than (or at least the same time as) the quantity of the data gathered. Researchers must carefully consider the data they collect, and what the limitations of these sources are—rather than use the datasets that were the easiest to capture while still appearing to contain meaningful data at face value [18]. Bruns [18] also points out that discussion of research choices in big data research should not be limited to data gathering approaches, for it is just as important that the further steps in processing the data are documented in detail in order to ensure the replicability of results.

Taking note of the criticism towards big data research, this study explores how human emotions and activities can be recognized from data collected from Twitter.

The aim is to go beyond the typical sentiment and emotion detection algorithms, and relate [19] the emotions expressed to human activities that can be identified from the tweets.

A case study of Twitter marketing campaign is presented, where users participated in the Twitter campaign by posting pictures of their stress management practices on Twitter. The case study attempts to follow the vision of computing being the 5th utility, where computing services are commoditized to a degree that allows them to be delivered in a manner similar to traditional utilities such as water, electricity, gas and telephony [20], and where computational models can be effectively used to determine human emotions and activities to support decision-making [16].

2 Related Research

2.1 Recognizing Emotions from Social Media

Recognizing emotions from social media is typically done using one or more theoretical models. Some of these theoretical models have been implemented as software features and are available via API's or software libraries. According to a recent review [6], the most common emotion theories used in social media studies include: Ekman's [21] theory of basic emotions and Plutchik's [22] emotion theory. Some relevant research regarding recognizing emotions from social media is next introduced.

Jussila et al. [7] introduced a framework of affective experiences (emotions), where four pleasure-arousal combinations (elation, serenity, lethargy, and tension) [23–25] are further divided according to dominance to form eight "affective families". The framework of 8 affective families and 24 affective experiences [e.g. 3]: excited, astonished, curious, exuberant, elated, thrilled, satisfied, content, serene, sleepy, tranquil, calm, unconcerned, lazy, indolent, gloomy, droopy, dejected, distressed, anxious, bewildered, hostile, defiant, angry were used to categorize tweets by human experts collected of two software companies in Finland.

Zhao et al. [26] demonstrated an interactive visual analytic tool for understanding personal emotion style from social media. For evaluation of the tool they conducted a series of studies. First, they recruited 10 participants from a large IT company to label their own tweets with the following information: valence, arousal, and dominance level, each on 9-point scale (-4 most negative, 0 neutral, and $+4$ most positive) [27], the emotional category from the Plutchik's and Robert's categorical model for primary emotion pairs [22], such as anger-fear, anticipation-surprise, joy-sadness, and trust-disgust [28, 29] and emotional styles [30] on the above-mentioned emotions and dimensions. They received a total of 308 labeled tweets belonging to 60 mood segments. Second, they carried out two user studies on the use of PEARL for self-reflection, and a Mechanical Turk Study to evaluate PEARL's effectiveness in assessing the emotional style of others.

Zimmerman et al. [31] presented a prototype of automatic 'Feelings Meter' able to detect feelings from Facebook posts. They applied commonly accepted two-dimensional (valence and arousal) semantic space for feelings adapted from Russell [32] and Scherer [25]. They collected 18 months of data using Radian6 tool of public mentions of feelings based on 143 different combinations and analyzed and visualized the resulting dataset with Tableau Desktop. In their study, the authors selected 44 'Facebook Feelings' in the final analysis. The valence and arousal of the feelings were marked on a scale between (−0.5 and +0.5). The results showed that, for instance, 'sad' (valence—0.488) was considered more negative than 'angry' (0.349). However, 'angry' (0.247) was more aroused compared to 'sad' (−0.119). Meanwhile, 'wonderful' (−0.427) was considered more positive in valence than 'awesome' (−0.399) but 'awesome' (0.208) had a higher arousal value compared to 'wonderful' (0.121).

Wang et al. [33] introduced a social media analytics engine, which employs a social adaptive fuzzy similarity-based classification method to automatically classify text messages into four sentiment categories (positive, negative, neutral and mixed), with the ability to recognize also their underlying emotion categories (satisfaction, happiness, excitement, anger, sadness, and anxiety) based on combined emotion studies [21, 22, 34, 35].

Menon et al. [36] presented a text analytics methodology of Facebook posts and comments, where Natural Language Toolkit (NLTK) was employed for text extraction, preprocessing, word frequency analysis, collocation analysis as well as text classification. They used consumer psychology (hierarchy of effects) domain model [37] and Ekman's categorization of basic emotions [21] to classify Facebook posts of two companies to joy, anger, sadness, surprise, fear and disgust categories. Also [38–40] employed Ekman's model of basic emotions for classifying social media data.

Shukri et al. [41] presented a case study of polarity and emotion classification in the automotive industry. They collected 3000 tweets on three types of cars: Mercedes, Audi and BMW extracted from Twitter API using R. Wiebe's polarity lexicon [42] was used in polarity classification and Strappavara emotions lexicon [43] for emotions classification to categorize emotions into classes such as anger, disgust, fear, joy, sadness, surprise, and unknown.

Examples of related studies that did not employ any emotion theories in classification include Sun et al. [44] that classified emotion into neutral, happy, surprised, sad, and angry, without referring to any emotion theories and Xu et al. [45] that classified emotions without referring to any emotion theories, first, into two categories: emotional and informational, and second, into three additional categories: appropriateness, empathy, and helpfulness.

2.2 Recognizing Activities from Social Media

Zhu et al. [4] proposed a method for extracting, filtering, and semantically categorizing "tips" from Foursquare. Using Foursquare API they collected foursquare "tips" data from the geographical area of San Francisco that included tip text, posting time and associated venue information (e.g. name semantic category, GPS, coordinates). They applied Linear SVM of the scikitlearn package [46] to distinguish between activity/non-activity—related short texts. Then they made use of American Time-Use Study (ATUS) taxonomy [47] to categorize the 3343 activity-related text via Amazon Mechanical Turk into the following categories: Household; Socializing, Relaxing and Leisure; Eating and Drinking; Consumer Purchases; Sports, Exercise, and Recreation; Work-Related; Traveling; Personal Care; Education; and Professional and Personal Care Services.

In a later study Zhu et al. [48] investigated whether the activity of a user can be automatically estimated from social media posts. They collected 157,257 tweets between July 4, 2013 and July 30, 2013 limited to English tweets in the San Francisco metropolitan area. Using the categories introduced in the earlier article [4] they crowdsourced the task of manually inferring activities from instances by using the CrowdFlower [49] platform which distributes tasks to other crowdsourcing platforms where the workers manually labelled 6004 tasks. The task involved workers to look at the image and manually select an activity category based on different contexts such as time, location, type of venue and text. 50% of the instances received confidence score around 1 and labeling agreement between 0.8 and 1.

Beber [50] proposed in his thesis a new method to recognize multiple activities performed at a place, and to identify all the individuals involved in group activities, using people trajectories and knowledge extracted from social media. For categorizing the activities a dataset generated from ATUS dataset [47] was employed. Similar to Zhu et al. [48], Foursquare is used to gather information on where the tweet activity happened.

Recognizing activities from social media is yet a little investigated research area. The previous work conducted has applied existing taxonomies [47] to categorize activities, and the categorization work has be mostly crowdsourced via various micro-tasking platforms.

3 Methodology

3.1 Case Study

A single case study is conducted on Twitter marketing campaign related to a health tech company providing stress management products and services. Case study method was considered as appropriate for this research, because it allows empirical investigation of contemporary phenomenon within its real life context [51, 52].

The product of the case company allows to measure emotions by a simple emotional circumplex model [53]. The arousal dimension (from alert to lethargic) is measured by a smart ring sensor, which measures electrodermal activity (EDA) that is commonly used physiological measure of arousal [54] and the valence dimension (from happy to unhappy) is measured via self-reporting using diary function of mobile application [55]. The analytics function of the mobile application displays the circumplex model of experienced emotions categorized into the following activities self-reported by the user: Sport, Work, Family, Relax, TV/Web, Dining, Art, Travel, Sleep, and Other [55]. Consequently, the company is also interested to understand how these emotions and activities could be also recognized from social media discussions.

The company designed a Twitter marketing campaign, where the followers of the company and other users were given an opportunity to share their practices of managing stress. Originally, the call for participant's tweets was in Finnish language. The campaign ran for 22 days between 31st August 2018 and 21st of September. Figure 1 illustrates the tweet of the initial call for participation. Translated into English: "September brings challenges to everyday life. There is hustle, but also excitement! What is your best stress management practice?" Participate to Moodmetric Twitter-campaign—You can win a Moodmetric smart ring. Instructions of the campaign: http://www.moodmetric.com/fi/uutiset/.

In Fig. 1. The tweet contains the campaign hashtag "#moodmetricstressinhallintakeino" (moodmetricstressmanagementpractice). The participants were requested to tweet a message along with a picture using the campaign hashtag.

Fig. 1 Campaign tweet which translates to "What is your best stress management practice?"

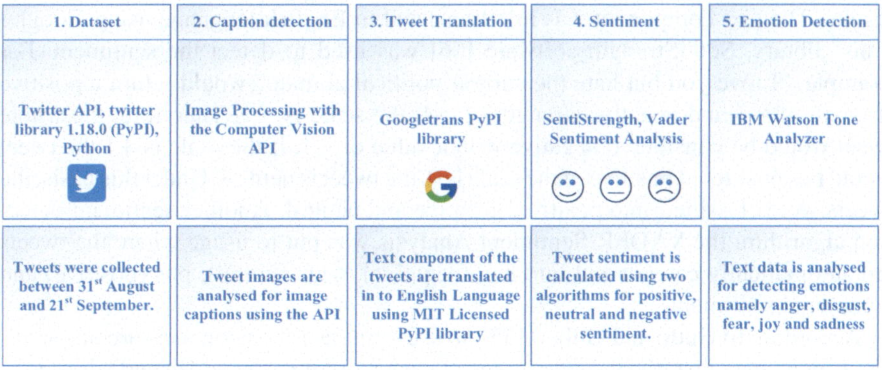

Fig. 2 APIs, libraries and algorithms employed in the analysis of the tweet content

3.2 Data Collection, Processing and Analysis

During the campaign, the tweets were collected on a weekly basis. The tweets were gathered using Python and Twitter API. A Python "twitter 1.18.0"[1] library from Python Package Index (PyPI), which is a repository for Python programming language was used in this process to extract the tweets. The API is limited to 100 tweets per request. Tweets can be gathered up to 1-week time window from the time of request. After gathering the data and excluding all the retweets, a total of 47 tweets with images were further processed and analyzed. Figure 2 illustrates the APIs, libraries and algorithms employed for this study. In addition to the computational approaches, arousal and valence dimension, as well as, activity was manually annotated for each tweet.

As shown in Fig. 2, for analyzing the images, Image Processing with Computer Vision API[2] was used. The Computer Vision API was primarily used to detect the captions from these images; The API analyses up to 20 images per minute. A single caption (sentence) is returned for each image along with tags. The Computer Vision API attempts to provide description (for example: "people waiting at a train station") and tags (for example: "waiting", "walking", "indoor", "holding", "train") for an image.

For analyzing the text sentiment from the tweets, SentiStrength[3] and VADER (Valence Aware Dictionary and sEntiment Reasoner) Sentiment Analysis[4] was used. The idea behind using two sentiment detection algorithm/software was to gather multiple viewpoints regarding the sentiment detection from the text. The sentiment analysis with two algorithms were carried out with a notion to understand the consequence of two algorithms and to figure out whether they produce similar or different

[1] https://pypi.org/project/twitter/.

[2] https://azure.microsoft.com/en-us/services/cognitive-services/computer-vision/.

[3] http://sentistrength.wlv.ac.uk/.

[4] https://github.com/cjhutto/vaderSentiment.

results. The text component was first translated to English language using Googletrans[5] library. SentiStrength software [56] was used to detect the sentiment. For example, "I love you but hate the current political climate" would return a positive strength of $+3$ and negative strength of -4, the scale is -1. Therefore, the whole tweet would be considered negative with a value of -1, if the scale is $+1$ the tweet would be considered positive, if the scale is 0, the tweet is neutral. Under this basis, the tweets were classified into positive, negative and neutral. Another sentiment detection algorithm, the VADER Sentiment Analysis was put to usage where the tweets are automatically categorized into neg (negative), neu (neutral), pos (positive) and compound sentiment.

According to Hutto and Gilbert [57], to use a single dimension measure of a sentiment for a given text, the valence scores of each word must be added and adjusted to be normalized between -1(most extreme negative) and $+1$(most extreme positive), it can also be called as 'normalized, weighted composite score'. For example, "I love you but hate the current political climate" would return as 'compound': -0.5346, 'neg': 0.37, 'neu': 0.44, 'pos': 0.19. Therefore, we can assume the statement to be negative. Under this premise, the sentiment for all the tweets were computed.

Additionally, to detect the emotions from the text other than positive, negative, and neutral, IBM Watson Tone Analyzer API[6] was utilized which categorizes emotional tones from text into anger, disgust, fear, joy and sadness. Since most of the tweets were in Finnish language, they were first translated into English language. By making use of the Googletrans library, a MIT licensed library from Python Package Index (PyPI), the tweets were translated from Finnish to English, if the language detected was in Finnish.

For the qualitative content and quantitative analysis, the tweets and images were manually annotated for its valence, arousal and activity category. The activity category is divided into 'Sport', 'Work', 'Family', 'Relax', 'TV/Web', 'Dining', 'Art', 'Travel', 'Sleep', and 'Other' following the categorization used in the mobile application of the case company [55]. The valence and arousal values were annotated based on the understanding derived from Zimmerman et al. [31]. Based on this understanding, the activity is mapped under the valence and arousal dimensions for qualitative data.

By looking at the image and tweet, a qualitative assessment was concluded on a scale between -5 and $+5$ on both the valence and arousal dimensions. Simultaneously, the activity of the participant was also evaluated. For instance, if the participant tweeted while he/she was in the woods or in the forest, then the activity will be considered as "Travel". Accordingly, the tweets were reviewed for activities in the context of "Travel", "Art", "Work", "Relax", "Dining", "Sport", "Other".

[5]https://pypi.org/project/googletrans/.

[6]https://www.ibm.com/watson/services/tone-analyzer/.

4 Results

4.1 Quantitative Analysis of Sentiment and Emotions

After gathering the tweets using the Twitter API and excluding retweets, the final image count from the tweets were 47. Every image of a tweet has multiple tags associated with it. From all the images, a total of 225 unique tags were extracted. To find out the sentiment associated with each tag, a loop was created for each unique tag to identify the corresponding sentiment. Thereby, tags could have multiple sentiment counts at the end of final analysis. For instance, ‘holding’ may have neutral, negative and positive counts based on its occurrence and related sentiment of the tweet. The following image shows the tags and its associated sentiment.

In Fig. 3, the tags and its sentiment score are shown in descending order. The tags with high sentiment scores are “sitting”, “man”, “standing”, “holding”, “large”, “table”, “white”, “old”, “red” and “outdoor”. Similarly, the analysis of tag sentiment using VADER Sentiment Analysis is shown in the following image.

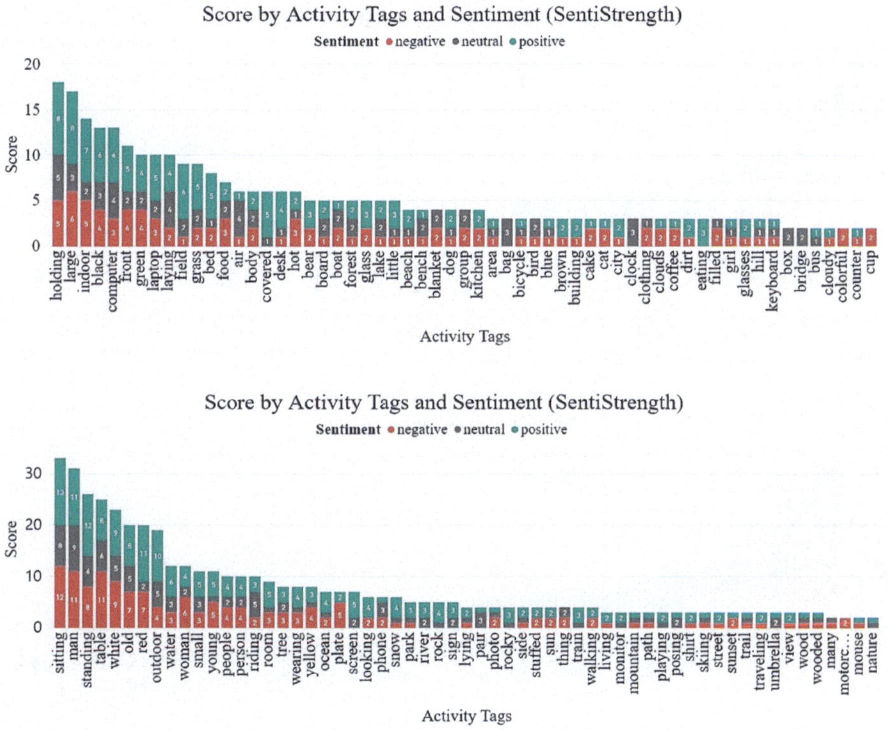

Fig. 3 Quantitative analysis of tags and sentiment (SentiStrength)

Figure 4 depicts the tags and its sentiment score from VADER Sentiment Analysis in descending order. The tags with high sentiment scores are "sitting", "man", "standing", "table", "white", "red", "old", "outdoor", "holding", "large" and "indoor". This is not a surprise as the above-mentioned tags are more prevalent throughout the images. The next table shows the most positive, neutral and negative analyzed by two algorithms.

In Table 1. Top 7 tags from each sentiment is displayed. The most common tags are "sitting", "standing", "man", "table", "white", "red". The emotions for tag which were analyzed using IBM Watson Tone Analyzer API are illustrated in the image below.

Figure 5 demonstrates the analysis of tags and emotions using IBM Watson Tone Analyzer API in descending order. The tags were analyzed to check for 5 different emotions namely "Anger", "Disgust", "Fear", "Joy" and "Sadness". The number of tags under "Anger" were 23, under "Disgust"—10. However, under "Fear" there were 102 tags. Top 10 "Fear" tags include ("bench"; "black"; "hot"; "indoor"; "outdoor"; "laptop"; "food"; "person"; "red"; "small", "white"; "woman"—3), ("computer"; "front"—4), ("holding"; "standing"; "table"—5), ("man"; "old", "sitting"—6).

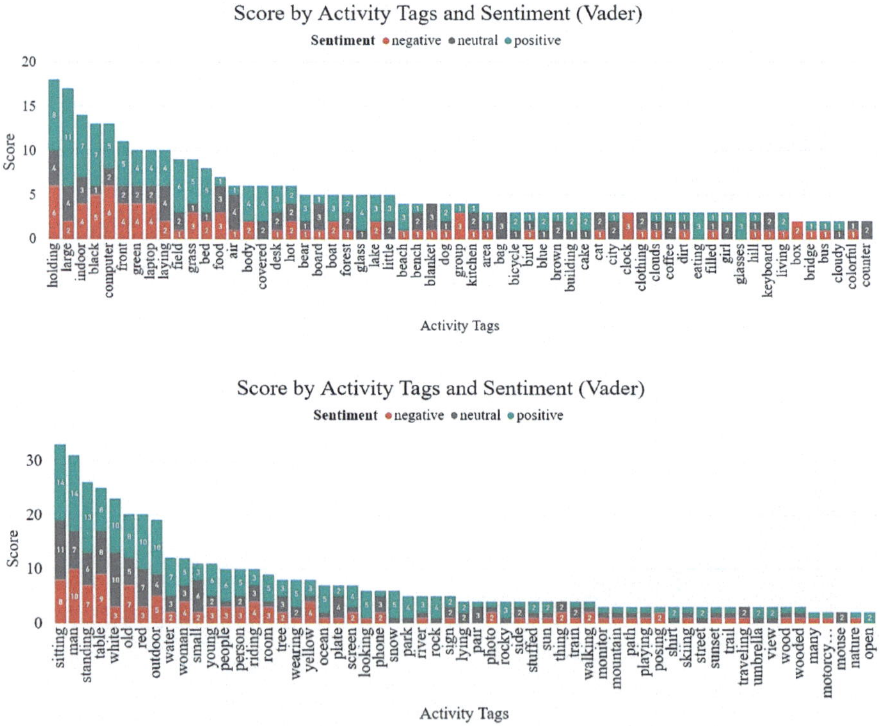

Fig. 4 Quantitative analysis of tags and sentiment (VADER Sentiment Analysis)

Table 1 Sentiment for activity tags from SentiStrength and VADER

SentiStrength analysis		
Positive	Neutral	Negative
sitting—13	man—9	sitting—12
standing—12	sitting—8	table—11
red/man—11	standing—6	man—11
outdoor—10	table—6	white—9
white—9	white/riding/old/outdoor/holding/air—5	standing—8
large/table/holding/old—8	computer/laying—4	red/old/large—7
indoor—7	bag/black/clock/large/pair/phone—3	woman—6
VADER sentiment analysis		
sitting/man—14	sitting—11	man—10
standing—13	white—10	table—9
large—11	table—8	sitting—8
outdoor/red/white—10	red/man—7	old/standing – 7
holding/old/table—8	small/standing—6	computer/holding—6
black/indoor/water—7	old—5	black/outdoor—5
field/people/young—6	air/holding/large/laying/outdoor/plate/—4	front/green/indoor/laptop/riding/woman/yellow—4

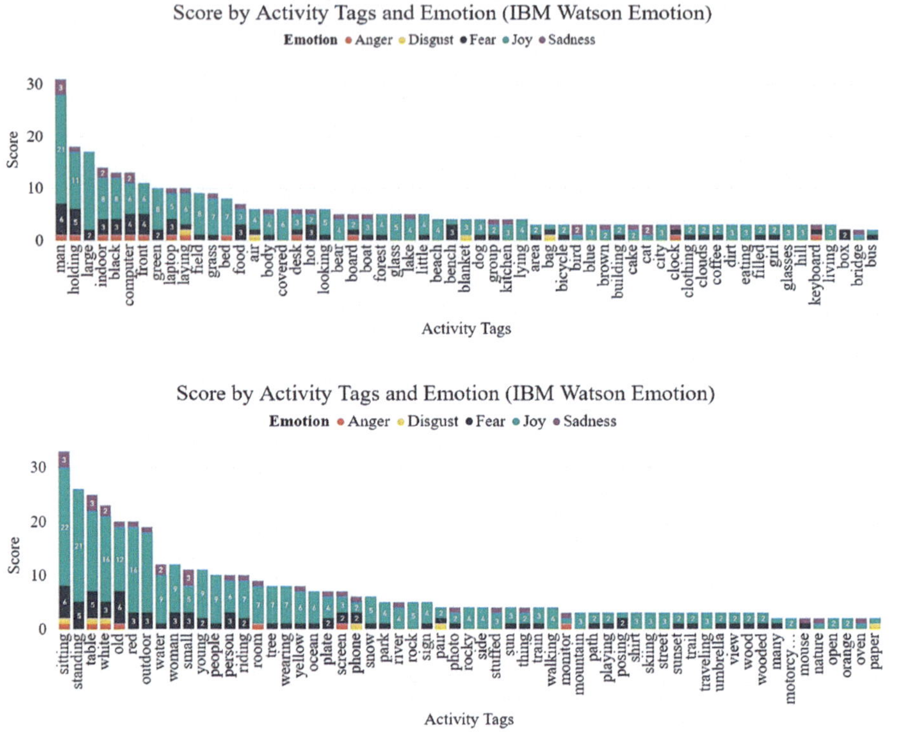

Fig. 5 Quantitative analysis of tags and emotions (IBM Watson)

"Joy" was the most prominent emotion that was evident from the tags. "sitting"—22, ("standing"; "man"—21), ("white"; "red"—16), ("table"; "outdoor"; "large"—15), "old"—12, "holding"—11 displayed the highest count. Whereas, "Sadness" included tags such as ("sitting"; "small"; "table"; "man"—3) and ("bird"; "cat"; "computer"; "indoor"; "water"; "white"—2).

4.2 Qualitative and Quantitative Analysis of Activities

The valence and arousal dimensions of all the tweets and image combination were annotated. The corresponding activity for the tweet and image combination was also annotated. The visualization of the obtained results are as shown in Fig. 6.

In Fig. 6 the four quadrants of the image depict the magnitude of valence and arousal values. In a clockwise manner, the second and third quadrant have the most values. All activities were classified to have positive valence. Lowest arousal scores were related to activities categorized as Relax(ing) and surprisingly there were also some Travel activities that were scored low on arousal. When investigating the tweets

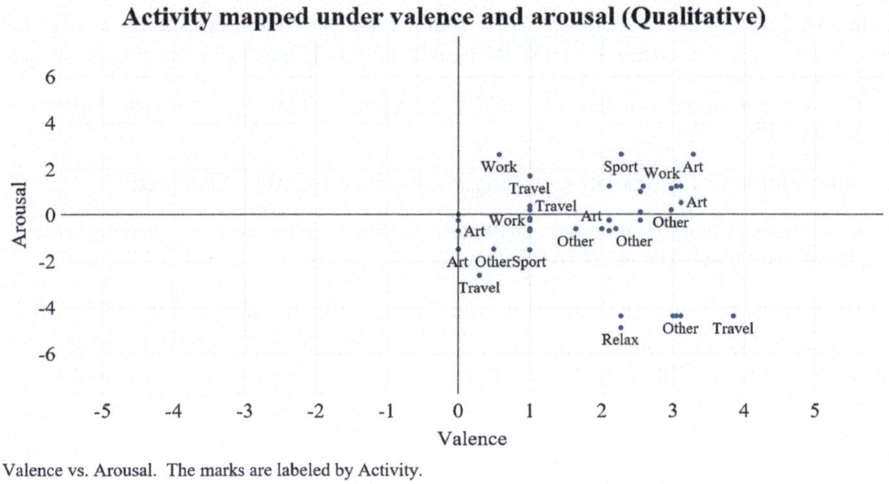

Fig. 6 Qualitative analysis of activities mapped into arousal and valence space

in detail, these turned out to be activities like wandering in the forest and traveling to the sea. The highest arousal scores were related to Sport, Work, and Art.

Using the same data points the emotions classified by IBM Watson Emotions were next mapped and included in Fig. 7.

From Fig. 7 it can be understood that most activities fall under a positive valence with 'Joy' as the most prominent emotion and 'Disgust' and 'Anger' as the least prominent emotion. When examining the emotion of tweets classified by IBM Watson

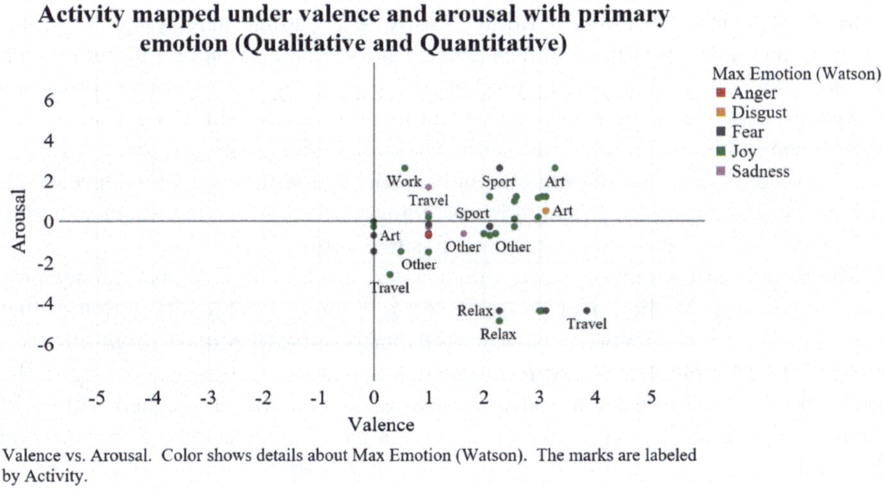

Fig. 7 Qualitative analysis of activity mapped into valence and arousal dimensions including quantitative analysis of emotions

Tone Analyzer, there turned out to be many false positives. An example of false positive of 'fear' classified by IBM Watson Tone Analyzer:

> How I relieve stress? Drinking a lot of NOT ALCOHOL (TM) tap water. #moodmetric-stressinhallintakeino

An example of false positive of 'anger' classified by Tone Analyzer:

> Play Paranoid #moodmetricstressinhallintakeino #moodmetricmuusikot @moodmetric @Larjovuori https://t.co/bFWF7AXKvY

IBM Watson Tone Analyzer was more successful in categorizing 'joy', but as indicated by the previous qualitative analysis (see Fig. 6) that showed no tweets with negative valence, 30% of the tweets were incorrectly categorized with emotion.

5 Discussion and Conclusions

The aim of the study was to explore cognitive computing approaches for recognizing human activities from social media, and relating detected emotions to the recognized human activities. In contrast to widely available computational approaches for sentiment and emotion classification, no libraries or API's were discovered for automatic activity recognition.

The closest approach to activity recognition available was via Computer Vision API, which is part of Cognitive Services provided by Microsoft Azure. The Computer Vision API enabled to tag all the pictures on the tweets based on recognizing features from the image. The 225 unique tags contained several relevant contextual factors, some of which were also activities, such as "sitting", "standing", "holding", "playing", "posing", "walking", "riding"; "looking", "eating", and "lying". Combining sentiment analysis with these tags (Figs. 3 and 4; Table 1) enables to infer which activities (and other contextual factors) were expressed by consumers and business-to-business customers' in most positive and most negative light. For instance, outdoor(s) and holding were predominantly associated with positive valence, whereas "old" and "computer" were predominantly associated with negative valence. Such insights could be used for example in designing marketing messages and selecting images that the audience associates most positive with.

The mapping of activities in two dimensional space of arousal and valence provided the most interesting insights (see Fig. 6). It allowed to identify activities that were associated as most relaxing (low arousal) and positive (medium to high valence). These were example Travel and Relax related activities, such as wandering in the forest and traveling to the seas. Also interesting are activities associated with most exciting (high arousal and positive (medium to high valence), such as Sport, Art and Work related. For a health tech company these are some examples of good stress management practices that could be in general recommend for the user (also referred to as weak personalization) [58]. A strong personalization of stress management service [58] would, however, use the data of individual user and his or her experiences to personalize recommendations that have worked for the particular user, e.g. in the event that a person has a long period of high arousal states the mobile application could feedback recommendations of activities that helped to relax and calm down that person in the past most effectively. When including the emotion classification (Fig. 7) even more customized recommendations could be given. For example, listening to which artist or song will generate the most joy for the person. Using computational models with a larger set of distinct motions, from five basic emotions [21] to 24 affective experiences [7] to 51 moods [59] would give a more holistic picture and opportunities to tailor more specific recommendations. The drawback of having more emotion classes, on the other hand, typically leads to more false positives and the likelihood that the given recommendations are wrong or misleading.

As to the "big data" and "small data" debate this study brings interesting viewpoints. To the question: "which is more valuable?" the study implies that it depends on the use case. Small data collected of individual user can be used to generate strong personalization, in this case, stress management recommendations that are based on the individual users past behavior and experiences. This can be highly valuable from the point of view of an individual user, and it could be implemented in the mobile application without the need to send personal data to cloud. Whereas, big data can be a source of weak personalization and used in marketing, e.g. to design marketing messages, and in research and development to e.g. discover most favorable mobile phone application features, use cases, etc.

In terms of enjoying computing as the 5th utility, there are still many limitations. The cognitive computing API's for instance support a limited number of languages. For example, Finnish language support is missing from most services. Although computing services and infrastructure are easily available—they are commodity only to those that have programming skills (such as Python), and various digital competences that are necessary in working with computational models. In current level of commodization, cognitive services like emotion detection and activity recognition are not as easy as flipping the light switch to turn on the lamp, or turning on the faucet to drink tap water.

References

1. Jalonen, H., Jussila, J.: Developing a Conceptual Model for the Relationship Between Social Media Behavior, Negative Consumer Emotions and Brand Disloyalty. Presented at the (2016)
2. Magids, S., Zorfas, A., Leemon, D.: The new science of customer emotions. Harv. Bus. Rev. November 2, (2015)
3. Laros, F., Steenkamp, J.: Emotions in consumer behavior: a hierarchical approach. J. Bus. Res. **58**, 1437–1445 (2005)
4. Zhu, Z., Blanke, U., Calatroni, A., Tröster, G.: Prior knowledge of human activities from social data. In: Proceedings of the 2013 International Symposium on Wearable Computers. pp. 141–142. ACM (2013)
5. Jussila, J.: Social media in business-to-business companies' innovation. http://urn.fi/URN:ISBN:978-952-15-3621-2 (2015)

6. Madhala, P., Jussila, J., Aramo-Immonen, H., Suominen, A.: Systematic literature review on customer emotions in social media. In: ECSM 2018 5th European Conference on Social Media (2018)
7. Jussila, J., Boedeker, M., Jalonen, H., Helander, N.: Social media analytics empowering customer experience insight. In: Kavoura, A., Sakas, D., Tomaras, P. (eds.) Strategic Innovative Marketing, pp. 25–30. Springer, Cham (2017)
8. Fosso Wamba, S., Akter, S., Edwards, A., Chopin, G., Gnanzou, D.: How "big data" can make big impact: Findings from a systematic review and a longitudinal case study. Int. J. Prod. Econ. **165**, 234–246 (2015)
9. Vatrapu, R.: Understanding social business. In: Emerging Dimensions of Technology Management. pp. 147–158. Springer India, India (2013)
10. Jussila, J., Menon, K., Gupta, J., Kärkkäinen, H.: Who is who in big social data? A bibliographic network analysis study. In: Proceedings of The 4th European Conference on Social Media. pp. 161–169. Academic Conferences and publishing limited, Vilnius (2017)
11. Coté, M.: The materiality of big social data. Cult. Stud. Rev. **20**, 121–149 (2014)
12. Bravo-Marquez, F., Mendoza, M., Poblete, B.: Meta-level sentiment models for big social data analysis. Knowl.-Based Syst. **69**, 86–99 (2014)
13. Batrinca, B., Treleaven, P.C.: Social media analytics: a survey of techniques, tools and platforms. AI Soc. **30**, 89–116 (2015)
14. Bello-Orgaz, G., Jung, J., Camacho, D.: Social big data: Recent achievements and new challenges. Inf. Fusion. **28**, 44–59 (2016)
15. Valiant, L.: Cognitive computation. In: Proceedings of IEEE 36th Annual Foundations of Computer Science. IEEE (1995)
16. Gutierrez-Garcia, J., López-Neri, E.: Cognitive computing: a brief survey and open research challenges. In: 3rd International Conference on Applied Computing and Information Technology/2nd International Conference on Computational Science and Intelligence. pp. 328–333. IEEE (2015)
17. Boyd, D., Crawford, K.: Critical Questions for Big Data. Information, Commun. Soc. **15**, 662–679 (2012)
18. Bruns, A.: Faster than the speed of print: reconciling 'big data'social media analysis and academic scholarship. First Monday. 18, (2013)
19. Davenport, T.H., Prusak, L.: Working knowledge : how organizations manage what they know. Harvard Business School Press (1998)
20. Buyya, R., Yeo, C., Venugopal, S., Broberg, J., Brandic, I.: Cloud computing and emerging IT platforms: vision, hype, and reality for delivering computing as the 5th utility. Futur. Gener. Comput. Syst. **25**, 599–616 (2009)
21. Ekman, P.: An argument for basic emotions. Cogn. Emot. **6**, 169–200 (1992)
22. Plutchik, R.: Robert: the nature of emotions. Am. Sci. **89**, 344 (2001)
23. Mehrabian, A., Russell, J.: An Approach to Environmental Psychology. The MIT Press, Cambridge (1974)
24. Seo, M., Barrett, L., Jin, S.: The structure of affect: history, theory, and implications for emotion research in organizations. In: Ashkanasy, N., Cooper, C. (eds.) New Horizons in Management. Research Companion to Emotion in Organizations. pp. 17–44. Edward Elgar Publishing, Northampton (2008)
25. Scherer, K.R.: What are emotions? And how can they be measured? Soc. Sci. Inf. **44**, 695–729 (2005)
26. Zhao, J., Gou, L., Wang, F., Zhou, M.: PEARL: an interactive visual analytic tool for understanding personal emotion style derived from social media. In: 2014 IEEE Conference on Visual Analytics Science and Technology (VAST). pp. 203–212. IEEE (2014)
27. Bradley, M., Lang, P.: Affective norms for English words (ANEW): Instruction manual and affective ratings. (1999)
28. Mehrabian, A.: Basic dimensions for a general psychological theory : implications for personality, social, environmental, and developmental studies. Oelgeschlager, Gunn & Hain (1980)
29. Russell, J.A.: A circumplex model of affect. J. Pers. Soc. Psychol. **39**, 1161–1178 (1980)

30. Davidson, R.J., Sharon Begley, W.: The emotional life of your brain: how its unique patterns affect the way you think, feel, and live-and how you can change them (2012)
31. Zimmerman, C., Stein, M.-K., Hardt, D., Vatrapu, R.: Emergence of things felt: harnessing the semantic space of Facebook feeling tags (2015)
32. Russell, J.: Pancultural aspects of the human conceptual organization of emotions. http://psycnet.apa.org/record/1984–14108-001
33. Wang, Z., Chong, C.S., Lan, L., Yang, Y., Beng Ho, S., Tong, J.C.: Fine-grained sentiment analysis of social media with emotion sensing. In: 2016 Future Technologies Conference (FTC). pp. 1361–1364. IEEE (2016)
34. Socher, R., Socher, R., Perelygin, A., Wu, J.Y., Chuang, J., Manning, C.D., Ng, A.Y., Potts, C.: Recursive deep models for semantic compositionality over a sentiment treebank. (2013)
35. Chafale, D., Pimpalkar, A.: Review on developing corpora for sentiment analysis using Plutchik's wheel of emotions with fuzzy logic. Int. J. Comput. Sci. Eng. (2014)
36. Menon, K., Karkkainen, H., Jussila, J., Huhtamaki, J., Mukkamala, R., Lasrado, L., Hussain, A.: Analysing the role of crowdfunding in entrepreneurial ecosystems: a social media event study of two competing product launches. Int. J. Entrep. Small Bus. **33**, 575–606 (2018)
37. Lavidge, R., Steiner, G.: A model for predictive measurements of advertising effectiveness. J. Mark. 59–62 (1961)
38. Larsen, H.H., Forsberg, J.M., Hemstad, S.V., Mukkamala, R.R., Hussain, A., Vatrapu, R.: TV ratings versus social media engagement: big social data analytics of the Scandinavian TV talk show Skavlan. In: 2016 IEEE International Conference on Big Data (Big Data). pp. 3849–3858. IEEE (2016)
39. Sarakit, P., Theeramunkong, T., Haruechaiyasak, C., Okumura, M.: Classifying emotion in Thai youtube comments. In: 2015 6th International Conference of Information and Communication Technology for Embedded Systems (IC-ICTES). pp. 1–5. IEEE (2015)
40. Li, W., Xu, H.: Text-based emotion classification using emotion cause extraction. Expert Syst. Appl. **41**, 1742–1749 (2014)
41. Shukri, S.E., Yaghi, R.I., Aljarah, I., Alsawalqah, H.: Twitter sentiment analysis: a case study in the automotive industry. In: 2015 IEEE Jordan Conference on Applied Electrical Engineering and Computing Technologies (AEECT). pp. 1–5. IEEE (2015)
42. Wilson, T., Wiebe, J., Hoffmann, P.: Recognizing contextual polarity in phrase-level sentiment analysis. In: Proceedings of the conference on Human Language Technology and Empirical Methods in Natural Language Processing—HLT '05. pp. 347–354. Association for Computational Linguistics, Morristown, NJ, USA (2005)
43. Strapparava, C., Valitutti, A.: WordNet-Affect: an affective extension of WordNet. In: Proceedings of the International Conference on Language Resources and Evaluation (2004)
44. Sun, X., Zhang, C., Li, G., Sun, D., Ren, F., Zomaya, A., Ranjan, R.: Detecting users' anomalous emotion using social media for business intelligence. J. Comput. Sci. (2017)
45. Xu, A., Liu, Z., Guo, Y., Sinha, V., Akkiraju, R.: A new chatbot for customer service on social media. In: Proceedings of the 2017 CHI Conference on Human Factors in Computing Systems—CHI '17, pp. 3506–3510. ACM Press, New York, USA (2017)
46. Pedregosa, F., Varoquaux, G., Gramfort, A., Michel, V., Thirion, B., Grisel, O., Vanderplas, J.: Scikit-learn: machine learning in Python. J. Mach. Learn. Res. **12**, 2825–2830 (2011)
47. Kristina, S., Shelley, K.: Developing the American time use survey activity classification system. Mon. Labor Rev. **3**, (2005)
48. Zhu, Z., Blanke, U., Calatroni, A., Tröster, G.: Human activity recognition using social media data. In: Proceedings of the 12th International Conference on Mobile and Ubiquitous Multimedia. p. 21 (2013)
49. Dearman, D., Sohn, T., Truong, K.: Opportunities exist: continuous discovery of places to perform activities. In: Proceedings of the SIGCHI Conference on Human Factors in Computing Systems. pp. 2429–2438. ACM (2011)
50. Beber, M.: Individual and group activity recognition in moving object trajectories (2017)
51. Saunders, M.N.K., Lewis, P., Thornhill, A.: Research methods for business students. Prentice Hall (2009)

52. Yin, R.K.: Case study research : design and methods. Sage Publications (2003)
53. Cowley, B.U., Torniainen, J.: A short review and primer on electrodermal activity in human computer interaction applications (2016)
54. Bradley, M., Lang, P.: Emotion and motivation (2000)
55. Jussila, J., Venho, N., Salonius, H., Moilanen, J., Liukkonen, J., Rinnetmäki, M.: Towards ecosystem for research and development of electrodermal activity applications. In: Proceedings of the 22nd International Academic Mindtrek Conference. pp. 79–87. ACM (2018)
56. Thelwall, M.: The Heart and soul of the web? Sentiment strength detection in the social web with SentiStrength. In: Holyst, J. (ed.) Cyberemotions. pp. 119–134. Springer (2017)
57. Hutto, C.J., Gilbert, E.: Vader: a parsimonious rule-based model for sentiment analysis of social media text. ICWSM. (2014)
58. Oinas-Kukkonen, H.: Personalization Myopia: a viewpoint to true personalization of information systems. In: Proceedings of the 22nd International Academic Mindtrek Conference. pp. 88–91. ACM (2018)
59. Bernabé-Moreno, J., Tejeda-Lorente, A., Porcel, C., Fujita, H., Herrera-Viedma, E.: Emotional profiling of locations based on social media. Procedia Comput. Sci. **55**, 960–969 (2015)

Assessing Strategies for Sampling Dynamic Social Networks

Paolo Ceravolo, Francesco Ciclosi, Emanuele Bellini and Ernesto Damiani

Abstract Social Networks represents an invaluable source of information to detect, understand and predict social trends and complex dynamics. Unfortunately, the presence of several constraints in data collections as costs, dimensions, time and so forth, requires the implementation of a sampling strategy able to maximize the information value for the analysis. The paper defines a number of parameters and thresholds defining a new strategy for data sampling in social network and compares samples obtained with different strategies. The test case has been conducted on the social network of the mayors of four Italian metropolitan areas. Results of the assessment reveal that the parameter designed for configuring a strategy impacts on the dimension, the extraction time and the quality of the generated network as expected. The best tradeoff between quality and execution time has been identified and discussed.

1 Introduction

Social Networks (SN) offer a fruitful and effective source of information as they track the users' social links and circles at an extensive detail level. This deep tracking detail is crucial for inferring advanced knowledge such as networks dynamics or latent potential but, at the same time, implies to get along with a large volume of data.

Datasets generated from SN are in fact classified as Big Data by Gartner, that state "*Big data is high-volume, high-velocity and/or high-variety information assets that demand cost-effective, innovative forms of information processing that enable enhanced insight, decision making, and process automation*" [1]. Nonetheless, the European Data Protection Supervisor (EDPS) highlight that the added value of Big

P. Ceravolo (✉)
Sesar Lab, Università degli Studi di Milano, Milan, Italy
e-mail: paolo.ceravolo@unimi.it

F. Ciclosi
Università di Macerata, Macerata, Italy

E. Bellini · E. Damiani
Center of Cyber-Physical System, Khalifa University, Abu Dhabi, UAE

A. Visvizi and M. D. Lytras (eds.), *Research & Innovation Forum 2019*,
Springer Proceedings in Complexity,
https://doi.org/10.1007/978-3-030-30809-4_16

Data "*is derived from the monitoring of human behaviour, collectively and individually, and resides in its predictive potential*" [2]. As it is well known "*social network is a term used to describe web-based services that allow individuals to create a public/semi-public profile within a domain such that they can communicatively connect with other users within the network*" [3], in facts "*social network is a graph consisting of nodes and links used to represent social relations on social network sites*" [4].

The underlined theory supporting analytics on SN is Graph theory where "*the nodes include entities and the relationships between them forms the links*" [3]. It offers metrics and algorithms to identify the centrality of nodes, connected or isolated components, it supports the characterization of networks in terms of topological properties such as density and assortativity.

2 The Problem of Sampling a Social Network

Coscia and Rossi [5] highlight that, even though "*online social media contain valuable quantitative and qualitative data, necessary to advance complex social systems studie*s" [5], this valuable source of information are not easily accessible to network scientists due to the limitations that social media companies impose on their API to calibrate the shareability of data on the business goals they target. Sharing is encouraged for a subset of data that can positively affect the diffusion of their product, rate limiting is applied on API requests to hinder collecting large volumes of data. For example, Twitter clearly states that "*our free, standard APIs are great for getting started, testing an integration, or validating a concept*" [6]. As a matter of fact, "*the vast majority of social media platforms recognize data as one of their primary source of revenue, and therefore put restrictions on the amount of data you can extract from them per unit of time*" [5].

To address these limits sampling strategies have been acknowledged as a relevant mean. The idea is working on a subset of the data selected to be representative of the entire dataset. A sampling procedure is as solid as the metrics that can be calculated on the sample are closed to the metrics calculated on the entire dataset. In general terms, a sampling strategy can be seen as a function that from one side reduce the budget required to mine data, from the other side introduce a bias on some data distribution property. An optimal strategy has to minimize both budget and bias.

Several works in the literature studied optimal strategies taking the assumption that the cost of access to data is homogeneous during the entire mining procedure. However, this assumption is artificial for real-world examples because, as said, vendors' API impose restrictions. Moreover, when defining a strategy, the budget available in terms of time or computational resources is a factor that concretely matters and should take a role. In fact, if we "*consider crawling cost: perhaps some methods do not represent the original network well on a large budget, but outperform the alternatives on a tighter one*" [5]. In this sense, the identification of a strategy requires to mediate between three goals:

1. statistical significance of the sampling;
2. consumed budget (number of request calls available);
3. fitness with the API of a vendor.

A way for addressing the challenge is to exploit multiple algorithms to create samplings, searching the best "*tradeoff between the cost of a network sampling strategy and the representativeness of the sample it creates*" [5]. The strategy proposed in this paper is aimed at simplifying the analysis of a social network by operating on a subset sufficiently representative of the main properties of the full network. The experimental activity carried out highlighted that mining strategies require to consider a greedy approach exploiting specific properties of a node to reduce the number of calls consumed during the mining procedure. The final goal is allowing users with a limited budget to analyze social network in a consistent way.

3 Comparison of the Main Graph Navigation Algorithms

The state of the art in terms of graph navigation algorithms propose two main approaches: *deterministic* and *probabilistic*. Our work starts by considering these algorithms as a baseline for comparing any other strategy. In particular, we focus here on two approaches. The Breadth-First Search (BFS), a well-known deterministic approach that visits all nodes of a graph by exploring the entire sequence of a path, and Forest Fire, a random approach that visits a subset of the nodes with equal probability for each different area of the graph.

Breadth-First Search (BFS) "*start from a seed v and we put its neighbors in a First-In, First-Out (FIFO) queue. In this case we call v a parent, and its neighbors are its children. Children of the same parent are siblings*" [5]. As a result, all the neighbors of a node will be explored before its children are included in the FIFO queue (cfr. Fig. 1). The algorithm explores first an entire level of the graph and passes then to the subsequent level. If a limited budget is available, for example, if a limited number of nodes can be visited, the obtained sample will be focused on the first levels of the graph.

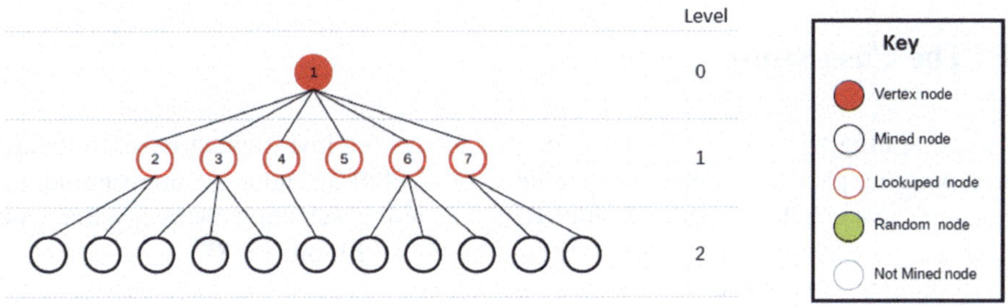

Fig. 1 Exploring a graph using the BFS algorithm (no filter is applied)

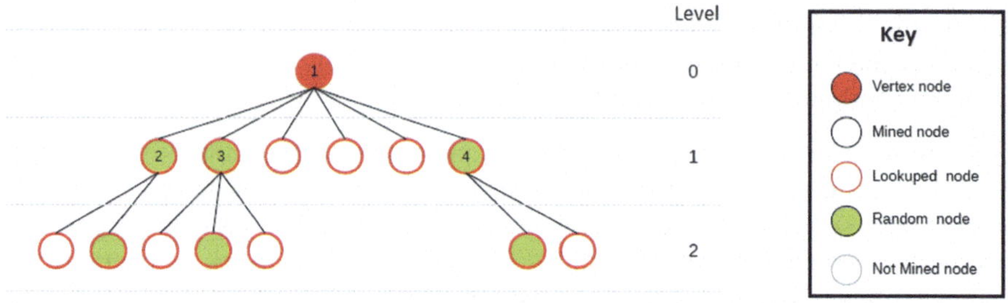

Fig. 2 Exploring a graph using the FF algorithm (filter: followersCount = 50%)

The main limit of this approach is that, if not parametrized, the growth in the research space may be exponential. Parameters on the dimension of the node set per level can be adopted in order to control the growth but this strategy typically produces an unbalanced sample.

On the other side, the probabilistic approach represented by the Forest Fire (FF), implements a "*sampling strategies which have a probabilistic component, but which do not base their core strategy on a random walk*" [5]. The idea is calibrating the effort of the research procedure on multiple areas of the addressed graph. In short, the procedure followed by FF is the following: "*one starts by performing a BFS exploration of the graph* (cfr. Fig. 2). *However, FF introduces a parameter: the «burning probability» . If a node passes the burning probability test, the node is «burned» and the BFS exploration will move on to its neighbors*" [5]. This can be obtained by the lookup of each node to obtain the attributes of the current node including its level, then the procedure executes the *burning probability* in order to define if the current node will be used for exploring new nodes by the list of its children or not.

Our approach, in order to avoid an exponential growth of the BFS algorithm, introduces restrictions to the degree of relationships that can be explored and to the nodes to be considered. This way we avoid the extraction of protected or potentially fake accounts. The experimental results presented in Sect. 5 confirms the correctness of this approach.

4 The Case Study

By elaborating from these two baseline procedures we investigated new strategies incorporating parameters that are specific to the API of a vendor. As an example, in this work, we use the standard Twitter API. The aim is selecting a strategy that can generate good samples not being penalized by the API limits [7, 8]

In order to assess the quality of a sample, we analyzed the social network of the mayors of four Italian metropolitan areas: Virginia Raggi (Roma), Beppe Sala (Milano), Chiara Appendino (Torino) e Leoluca Orlando (Palermo). We compare

Table 1 The Twitter endpoints used in our study

Endpoint	Resource family	Requests/Window (user auth)	Requests/Window (app auth)
GET application/rate_limit_status	Application	180	180
GET followers/ids	Followers	15	15
GET statuses/show/:id	Statuses	900	900
GET users/lookup	Users	900	300

samples obtained with different strategies imposing to each strategy a same budget. To assess the quality of a sample a set of metrics on its SN is compared with the same metrics computed on a sample generated using a significantly higher budget. The choice of politicians is motivated by the public interest on the domain and by the need of using a generating seed to construct the SN [9]. Each mayor is elected as the seed of the SN to be mined, from the seed we get the list of its followers and then apply our strategies to identify the nodes to be visited, recursively until the entire budget is consumed.

Each strategy has been designed to comply with the maximal number of requests that are allowed by the Twitter API. In Table 1 we list the endpoints exploited by our algorithms with the limits applied computed within a time window of 15 min. This implies that "*the maximum number of visits that can be implemented is a function of the running time and the rate_limit_status: 180 requests per 15 min, even if the requested endpoint has higher limits*" [10].

To improve the capacity of our algorithms, in terms of the number of nodes that can be visited in a time window, we adopted in our code a set of countermeasures. For instance, we applied the Object Pool pattern design, creating a pooler of Twitter credentials the algorithm can access when issuing a request. This object will return an instance of twitter4j [11] built to exploit a specific credential of the pool and checking the availability, in the time window, of requests for this credential. Another schema we adopted is electing nodes to be visited only after having executed a lookup procedure, i.e. having all the information required to access a node without the need of issuing a new request. An internal cache is exploited to faster the reading procedures. Finally, a set of filters has been adopted. Filters apply on the minimum and maximum number of followers a node has to list to be visited and on the maximum percentage of children that can be selected from a node. The rationale behind these filters is to equally distribute the budget available in different areas of the network. In order to avoid visiting nodes of little interest the availability of specific attributes, such as "$lang = it$", has been considered a pre-condition to visit a node.

Table 2 lists the strategies we adopted. Each strategy generated an SN for each mayor. Comparing the generated SNs we tested the quality of the samples generated by these different strategies.

Table 2 List of strategies adopted with filters and thresholds applied

Network name (generation algorithm)	Filters and thresholds of level 1 (min. followers, max followers, % nodes)	Filters and thresholds of level 2 (min. followers, max followers, % nodes)
NET-A (Mining RANDOM)	100, 30,000, 2%	100, 30,000, 10%
NET-B1 (Mining RANDOM)	100, 30,000, 0,2%	100, 30,000, 1%
NET-B2 (Mining RANDOM)	200, 7000, 2%	200, 7000, 10%
NET-B3 (Mining RANDOM)	300, 3000, 2%	300, 3000, 10%
NET-B4 (Mining BFS)	n.a., n.a., 0, 2%	n.a., n.a., 10%

5 Achieved Results

In our tests, NET-A has been identified as the benchmark network as it represents the sampling generated using the highest budget. The other networks differ for the budget available and the filter applied.

NET-B1 uses the same parameters of NET-A but visiting 10% of the nodes visited in NET-A. NET-B2 use the same percentage of nodes than NET-A but imposing stronger restrictions with filters that, in fact, reaching around 65% of the nodes reached by NET-A. NET-B3 uses similar parameters but imposes even more restricted filters, resulting in a number of visited nodes around 50% of NET-A. Finally, NET-B4 uses the same percentage of nodes of NET-B1 but apply BFS as a research algorithm; so, selecting nodes it does not implement any randomization but simply follow the order provided by the Twitter API, that is related to the age of the link connecting two nodes.

5.1 Analysis of the Results

An initial comparison between the generated networks can be provided by observing the dimension of the networks. Table 3 shows for each network, the total number of nodes, the total number of nodes at level 1 and the nodes at level 1 that was randomly selected, the total number of nodes at level 2 and the nodes at level 2 that was randomly selected. We can observe that the percentage of nodes randomly selected at level 1 is proportional to the parameter fixing the percentage of child nodes to select. This is not true at level 2 due to the intersections between child nodes that can exist.

To compare the different strategies implemented we considered a set of SN metrics capturing the topology of a graph. One of the metrics used is the PageRank of the seed nodes.[1] Another metric considered is the number of communities in the network.[2]

[1]The PageRank values have been normalized dividing by the max value in the network.

[2]Communities are detected using the Louvain's community detection algorithm.

Table 3 Comparing the dimension of the generated networks

Metrics[a]	NET-A	NET-B1	NET-B2	NET-B3	NET-B4
CNT.ALL	1.007.788	558.038	877.875	767.094	45.460
CNT.L1	486.590	486.675	486.642	486.678	1.125
CNT.L1.RND	1.374	137	925	657	na ≈ 1.125
CNT.L2	521.194	71.419	391.229	280.412	44.331
CNT.L2.RND	31.222	369	18.046	9.520	na ≈ 44.331

[a]The rows have the following values. CNT.ALL is the total number of nodes in a network; CNT.L1 and CNT.L2, the total number of rows at level 1 and 2; finally, CNT.L1.RND e CNT.L2.RND, refer to the node randomly selected in level 1 and 2

The density is correlated with the dimension of the network but does not seem correlated with the strategy adopted. The same is with the number of communities detected. The extraction time is correlated with the dimension of the network with NET-B4 as the faster option, due to the fact that lookup operation is not implemented for this strategy (cfr. Table 4).

The PageRank values for the seed nodes of the generated networks are compared in Table 5. Comparing the normalized values, we observe that NET-B3 provide the results that best fit with the benchmark (NET-A) while NET-B4 provide results largely misaligned with the other networks.

Table 4 Comparing the generated networks based on density, number of communities and extraction time

Metrics	NET-A	NET-B1	NET-B2	NET-B3	NET-B4
Density (10^{-4})	1.54	35.17	2.12	2.88	2253
Com. number	16	14	17	23	38
Extraction time (min)	1.126	332	839	664	258

Table 5 Comparing normalized PageRank values for seed nodes in generated networks

Seed name	NET-A	NET-B1	NET-B2	NET-B3	NET-B4
Raggi V.	1.00	1.00	1.00	1.00	0.35
Appendino C.	0.52	0.82	0.57	0.45	0.93
Orlando L.	0.40	0.28	0.36	0.39	0.07
Sala B.	0.29	0.33	0.38	0.26	0.13

6 Conclusions and Future Works

The results obtained underlined that the parameter set we designed for configuring a strategy impacts on the dimension, the extraction time and the quality of the generated network. Quality was assessed by observing the consistency of the PageRank value of a network with a benchmark network and with the other generated networks. We then observe that NET-B3 offer the best tradeoff between quality and execution time. Our observation could probably be generalized to the networks generated using the Mining RANDOM algorithm; this is justified considering that in case of selection performed with the Mining BFS algorithm the selection only depends on how Twitter returns the data.

Considering that in the context of Big Data, "one of the greatest challenges, for the purposes of scientific investigation, is represented by the complexity in analyzing and interpreting big and varying datasets" [12], our strategy can be effectively used by small research groups, allowing to analyze social networks of considerable size with low computational budget. Some authors have pointed out that "although denied to the general public, in the era of Big Data three types of objectives are emerging at incremental risk for democracy: commercial influence by companies; socio-political influence by the parties; social control up to mass surveillance by some governments" [13]. This conclusion is emerging also from experimental studies, such as the study developed by Swansea University and the University of California, Berkeley [14], in which clues were found on the attempt to condition the referendum on Brexit held on 23 June 2016, conveying a massive quantity of messages with an orientation towards "leave". These are elements that should not be overlooked considering that although "where the majorities are large, the social media seem to have a significant impact; however, they could be determinant in uncertainty conditions" [13], such as those characterized by strong polarization. Therefore, our strategy can be of particular utility when used to detect in a short time the changes occurring in a reference community or to identify which are the reference communities "most suitable for to resolve particular needs or to deepen certain interests" [15].

References

1. Gartner, I.: Gartner IT Glossary > Big Data . [In linea]. Available at: https://www.gartner.com/it-glossary/big-data/. [Consultato: 28-gen-2019]
2. Buttarelli, G.: European data protection supervisor. In: Opinion 7/2015—Meeting the Challenges of Big Data, pp. 1–21. European Data Protection Supervisor, Brussels (2015)
3. Adedoyin-Olowe, M., Gaber, M.M. Stahl, F.: Survey of data mining techniques for social media analysis. J. Data Min. Digit. Humanit. Digit. Humanit. **2014** (2014)
4. Borgatti, S.P.: Social network analysis, two-mode concepts in. In: Computational Complexity: Theory, Techniques, and Applications, pp. 2912–2924. Springer New York, R. A. Meyers, A c. di New York, NY (2012)
5. Coscia, M., Rossi, L.: Benchmarking API costs of network sampling strategies. In: 2018 IEEE International Conference on Big Data (Big Data), pp. 663–672 (2018)

6. (Twitter Inc.).: Twitter—API access that scales with you and your solution. [In linea]. Available at: https://developer.twitter.com/en/pricing. [Consultato: 28-gen-2019]
7. Twitter—Rate limits. [In linea]. Available at: https://developer.twitter.com/en/docs/basics/rate-limits. [Consultato: 28-gen-2019]
8. Franchin, M.: Strategie di sampling di social network. Università degli Studi di Milano (2018)
9. Ceravolo, P., Guerretti, S.: Testing social network metrics for measuring electoral success in the Italian municipal campaign of 2011. In: 2013 International Conference on Cloud and Green Computing, pp. 342–347 (2013)
10. Gambino, G.: Studio sperimentale di algoritmi di social media mining applicati a Twitter. Università degli Studi di Milano (2018)
11. twitter4J. [In linea]. Available at: http://twitter4j.org/en/index.html. [Consultato: 28-gen-2019]
12. Martelli, F., Navigli, R.: Disambiguare le reti sociali. Gnos. - Riv. Ital. di Intell. **4**, 21–27 (2018)
13. Colajanni, M.: Social: raccomandazioni per l'uso . Gnos. - Riv. Ital. di Intell. **4**, 29–37 (2018)
14. Gorodnichenko, Y., Pham, T., Talavera, O.: Social media, sentiment and public opinions: evidence from #Brexit and #USElection (2018)
15. Iacus, S.M.: Social network, data science e intelligence. Gnos. - Riv. Ital. di Intell. **4**, 39–45 (2018)

Facebook Engagement—Motivational Drivers and the Moderating Effect of Flow Episodes and Age Differences

Inma Rodríguez-Ardura and Antoni Meseguer-Artola

Abstract The rise of Facebook creates opportunities for consumers to entertain, present themselves and interact socially as well as for organizations and brands, who can utilize this social networking site (SNS) as a strategic tool in their integrated marketing communication. Despite engagement being considered a key success factor for content and services on Facebook, an issue still to address is how to trigger a Facebook user's engagement. This paper is one of the few to theoretically and empirically investigate Facebook engagement and its psychological, motivational drivers. Drawing theoretical insights from uses and gratifications theory, theoretical accounts of flow and the socioemotional selectivity theory, we build and validate a model that connects the motivations that underlie Facebook behavior to involvement-related, emotional and conative facets of Facebook engagement. We test the model with the partially least squares approach on a sample of active Facebook users. The results strongly support the causal, mediating and moderating relationships included in the model. They show that three distinct types of motivational orientation (toward enjoyment, self-disclosure and community identification) contribute to a Facebooker's engagement behavior. Importantly, the findings reveal that flow episodes strengthen the causal path from enjoyment-related motivation towards engagement, and that the impact of engagement on continued use of Facebook is greater for older Facebookers than among younger users. Based on these findings, the paper provides practical knowledge for organizations making use of Facebook and bears implications for the managers of brands that have active Facebook pages.

I. Rodríguez-Ardura (✉) · A. Meseguer-Artola
Open University of Catalonia, Universitat Oberta de Catalunya, Ave. Tibidabo 39-43, 08035 Barcelona, Spain
e-mail: irodriguez@uoc.edu

A. Meseguer-Artola
e-mail: ameseguer@uoc.edu

I. Rodríguez-Ardura
University of Miami, 5250 University Dr, Coral Gables, FL 33146, USA

A. Visvizi and M. D. Lytras (eds.), *Research & Innovation Forum 2019*, Springer Proceedings in Complexity,
https://doi.org/10.1007/978-3-030-30809-4_17

1 Introduction

Despite Facebook being the most popular and widely used social networking service (SNS), little is known about what makes Facebook so engaging for many users [1, 2], to such an extent that lots of them regularly patronize Facebook. Based on the uses and gratifications theory [3–6], studies that have explored the users' motivational forces on Facebook have shed a great amount of light about the psychological benefits that Facebook usage satisfies, or the gratifications it provides (see e.g. [7–9]). However, they have not connected the dots between the motivational forces leading to Facebook usage and experiencing engagement toward Facebook.

For their part, research on engagement has mainly focused on defining the nature and scope of engagement [10, 11], reliably categorizing engagement behaviors [12, 13], and establishing how engagement contributes to value co-creation brand initiatives [14, 15] and business performance [16, 17]. Only a handful of studies have bridged engagement with the benefits sought by SNS users, but they only do so in specific contexts—usually a particular brand community or a social initiative in an SNS (i.e. [16, 18–20]).

Engagement is a key driver for a brand's or an organization's success within their integrated communication programs online. Because of this, of utmost importance is the unveiling and examining of the inner motives that lead people to engage on Facebook. This is precisely one of the main goals of this paper. We attempt to achieve this goal by considering not only brand contexts on Facebook but also non-marketing content and resources, since this will allow us to offer a wider and more comprehensive outlook of Facebook engagement experiences.

Furthermore, we aim to contribute to the emerging academic conversation about the interplay of engagement experiences with flow episodes and users' age differences. A couple of previous studies have empirically shown the distinct nature of flow and engagement online [21, 22], but none of them have considered their interplay with the users' motivations. Also, prior investigations have not yielded concluding evidence about the role of age differences in creating and processing SNS content [23–25], so there is a research need to clarify the role of age differences in the patronage of Facebook. Based on flow theory [26, 27] and the socioemotional selectivity theory [28, 29], we endeavor to find empirical evidence about the moderating role of flow episodes on engagement formation and the interaction effects of age difference in the impact of engagement.

2 Theoretical Framework

We adopt an integrative perspective to conceptually delimitate and define engagement experiences. Accordingly, we conceive engagement as a multidimensional mechanism [30] that embraces cognitive (i.e. mental involvement), emotional (i.e. positive

Table 1 Hypothesized paths of the model, theoretical backbones, and contexts of previous testing

	Hypothesized pathways	Supporting theories	Contexts of prior testing
H1 (+)	Enjoyment → Engagement	Uses and gratifications theory [3–5]	Social ventures in SNSs [18]
H2 (+)	Self-disclosure → Engagement	Uses and gratifications theory [3–5]	No prior testing
H3 (+)	Community identification → Engagement	Uses and gratifications theory [3–5]	Online brand communities [20]
H4 (+)	Engagement → Patronage	Social exchange theory [34–36]	Brand fan pages on Facebook [16]
H5 (+)	Flow × Enjoyment → Engagement	Flow theory [37, 38]	No prior testing
H6 (+)	Age × Enjoyment → Patronage	Socioemotional selectivity theory [39–41]	No prior testing

sensations) and conative facets (i.e. participation and socialization behaviors) and is triggered by motivational drivers that lead the user to interact online [12, 31–33].

Our integrated model of user's engagement experiences in Facebook includes a variety of pathways founded on four research theories: causal pathways that stem from uses and gratifications theory (H1, H3) [3–5], a causal pathway rooted in social exchange theory (H4) [34–36], a moderating pathway from flow research (H5) [37, 38], and a moderating path derived from the socioemotional selectivity theory (H6) [39–41]—see Table 1.

3 Research Method

Data was collected from active adult users of Facebook Spain. Participants were recruited by snowball sampling on Facebook. We employed an online survey to gather the data. For our analyses, we only used the questionnaires that were fully completed, which resulted in a final sample of 407 questionnaires. Because the differences between the demographic characteristics of the final sample and those of the target population were not statistically significant, we discarded the existence of non-respondent biases and confirmed the representativeness of the sample.

We measured the constructs with multi-item scales—all earlier validated in relevant studies and adapted to the context of Facebook.

4 Analysis and Results

We tested the hypotheses with the partial least squares (PLS) technique. PLS allowed us to assess the measurement of latent variables in the model and to determine the significance of the (direct, moderating and indirect) relationships between them.

4.1 Measurement Model

The internal consistency reliability, the individual item reliability, the convergent validity and the discriminant validity of the measurement model with reflective indicators were all satisfied.

4.2 Common Method Variance

In the design of the questionnaire, we applied preventative procedures of common method variance (CMV). In addition to this, we performed two post hoc tests to definitively discard the existence of any significant CMV affecting data analysis.

4.3 Structural Model

We created a second-order molar construct for engagement, reflectively related to its three facets (cognitive engagement, emotional engagement and conative engagement); and we introduced this molar construct in the model estimation by using the repeated-indicators approach. To assess the moderating relationships included in hypotheses H5 and H6, we applied the product-indicator approach.

We tested and confirmed the model's quality by means of the coefficient of determination (R^2) associated to the two regressions in the model and the significance of the path coefficients.

All hypotheses were supported based on the PLS analysis. Enjoyment, self-disclosure and community identification all had a positive, direct and significant effect on engagement ($\beta = +015$, $\beta = +0.25$, $\beta = +0.26$, respectively). The impact of community identification on engagement showed to be significantly moderated by flow ($\beta = +0.34$). Furthermore, the higher-order construct engagement (measured as a compound perception of cognitive engagement, affective engagement and behavioral engagement) had a positive and significant effect on patronage ($\beta = +\ 0.91$), which indeed is moderated by age ($\beta = -0.32$).

5 Concluding Discussion

The findings in this study improve our knowledge about engagement formation, the operationalization of engagement as a multidimensional construct and the mediating effects of engagement on Facebook patronage.

Our research framework (motivation-engagement-patronage) results from the integration, for the first time in the literature, of the research stream of engagement with the uses and gratifications theory. This novel framework has led us to conceive engagement as a mediating psychological mechanism rather than an outcome and to predict three different ways users might experience engagement in Facebook.

The study has also unearthed new evidence which shows that flow interacts with enjoyment motivation to prompt engagement, and that differences in age moderate the effect of engagement on patronage.

Furthermore, this study shows that the socioemotional selectivity theory might be applied to supplement the accounts offered by the social exchange theory. On the basis of this integrative theoretical lens, the model predicts age-related changes within the impact of an individual's engagement on their willingness to patronize.

Our future work will adopt a multi-analytical approach and use the results from PLS modelling as input and output nodes for two artificial neural network analyses.

References

1. Verhagen, T., Swen, E., Feldberg, F., Merikivi, J.: Benefitting from virtual customer environments: an empirical study of customer engagement. Comput. Human Behav. **48**, 340–357 (2015). https://doi.org/10.1016/j.chb.2015.01.061
2. Pansari, A., Kumar, V.: Customer engagement: the construct, antecedents, and consequences. J. Acad. Mark. Sci. **45**, 294–311 (2017). https://doi.org/10.1007/s11747-016-0485-6
3. Katz, E., Blumler, J.G., Gurevitch, M.: Uses and gratifications research. Public Opin. Q. **37**, 509–523 (1973)
4. McQuail, D.: McQuail's Mass Communication Theory. Sage Publications, London (2012)
5. Ruggiero, T.E.: Uses and Gratifications Theory in the 21st Century. Mass Commun. Soc. **30**, 3–37 (2000). https://doi.org/10.1207/S15327825MCS0301_02
6. Luo, M.M., Remus, W.: Uses and gratifications and acceptance of web-based information services: an integrated model. Comput. Human Behav. **38**, 281–295 (2014). https://doi.org/10.1016/j.chb.2014.05.042
7. Nadkarni, A., Hofmann, S.G.: Why do people use Facebook? Pers. Individ. Dif. **52**, 243–249 (2012). https://doi.org/10.1016/j.paid.2011.11.007
8. van Dijck, J.: 'You have one identity': performing the self on Facebook and LinkedIn. Media, Cult. Soc. **35**, 199–215 (2013). https://doi.org/10.1177/0163443712468605
9. Seidman, G.: Self-presentation and belonging on Facebook: how personality influences social media use and motivations. Pers. Individ. Dif. **54**, 402–407 (2013). https://doi.org/10.1016/j.paid.2012.10.009
10. Baldus, B.J., Voorhees, C., Calantone, R.: Online brand community engagement: scale development and validation. J. Bus. Res. **68**, 978–985 (2015). https://doi.org/10.1016/j.jbusres.2014.09.035

11. Hollebeek, L.D., Glynn, M.S., Brodie, R.J.: Consumer brand engagement in social media: conceptualization, scale development and validation. J. Interact. Mark. **28**, 149–165 (2014). https://doi.org/10.1016/j.intmar.2013.12.002
12. Eigenraam, A.W., Eelen, J., van Lin, A., Verlegh, P.W.J.: A consumer-based taxonomy of digital customer engagement practices. J. Interact. Mark. **44**, 102–121 (2018). https://doi.org/10.1016/j.intmar.2018.07.002
13. Hollebeek, L.D., Juric, B., Tang, W.: Virtual brand community engagement practices: a refined typology and model. J. Serv. Mark. **31**, 204–217 (2017). https://doi.org/10.1108/JSM-01-2016-0006
14. Hollebeek, L.D., Srivastava, R.K., Chen, T.: S-D logic–informed customer engagement: integrative framework, revised fundamental propositions, and application to CRM. J. Acad. Mark. Sci. (2016). https://doi.org/10.1007/s11747-016-0494-5
15. Jaakkola, E., Alexander, M.: The role of customer engagement behavior in value co-creation: a service system perspective. J. Serv. Res. **17**, 247–261 (2014). https://doi.org/10.1177/1094670514529187
16. Huang, S.-L., Chen, C.-T.: How consumers become loyal fans on Facebook. Comput. Human Behav. **82**, 124–135 (2018). https://doi.org/10.1016/j.chb.2018.01.006
17. Harmeling, C.M., Moffett, J.W., Arnold, M.J., Carlson, B.D.: Toward a theory of customer engagement marketing. J. Acad. Mark. Sci. **45**, 312–335 (2017). https://doi.org/10.1007/s11747-016-0509-2
18. Hall-Phillips, A., Park, J., Chung, T.L., Anaza, N.A., Rathod, S.R.: I (heart) social ventures: identification and social media engagement. J. Bus. Res. **69**, 484–491 (2016). https://doi.org/10.1016/j.jbusres.2015.05.005
19. Zheng, X., Cheung, C.M.K., Lee, M.K.O., Liang, L.: Building brand loyalty through user engagement in online brand communities in social networking sites. Inf. Technol. People. **28**, 90–106 (2015). https://doi.org/10.1108/ITP-08-2013-0144
20. Lee, D., Kim, H.S., Kim, J.K.: The impact of online brand community type on consumer's community engagement behaviors: consumer-created vs. marketer-created online brand community in online social-networking web sites. Cyberpsychol. Behav. Soc. Netw. **14**, 59–63 (2011). https://doi.org/10.1089/cyber.2009.0397
21. Rodríguez-Ardura, I., Meseguer-Artola, A.: Imagine, feel "there", and flow! Immersive experiences on m-Facebook, and their affective and behavioural effects. Inf. Technol. People. In press (2018). https://doi.org/10.1108/itp-10-2017-0358
22. Shin, D.: Empathy and embodied experience in virtual environment: to what extent can virtual reality stimulate empathy and embodied experience? Comput. Human Behav. **78**, 64–73 (2018). https://doi.org/10.1016/j.chb.2017.09.012
23. Hayes, M., Van Stolk-Cooke, K., Muench, F.: Understanding Facebook use and the psychological affects of use across generations. Comput. Human Behav. **49**, 507–511 (2015). https://doi.org/10.1016/j.chb.2015.03.040
24. Leung, L.: Generational differences in content generation in social media: the roles of the gratifications sought and of narcissism. Comput. Human Behav. **29**, 997–1006 (2013). https://doi.org/10.1016/j.chb.2012.12.028
25. Manzi, C., Coen, S., Regalia, C., Yévenes, A.M., Giuliani, C., Vignoles, V.L.: Being in the social: a cross-cultural and cross-generational study on identity processes related to Facebook use. Comput. Human Behav. **80**, 81–87 (2018). https://doi.org/10.1016/j.chb.2017.10.046
26. Csikszentmihalyi, M.: Play and intrinsic rewards. J. Humanist. Psychol. **15**, 41–63 (1975). https://doi.org/10.1177/002216787501500306
27. Csikszentmihalyi, M.: Flow: The Psychology of Optimal Experience. Harper and Row, New York, NY (2008)
28. Carstensen, L.L.: The Influence of a Sense of Time on Human Development. Science (80-.). **312**, 1913–1915 (2006). https://doi.org/10.1126/science.1127488
29. Carstensen, L.L.: Integrating cognitive and emotion paradigms to address the paradox of aging. Cogn. Emot. In press (2018). https://doi.org/10.1080/02699931.2018.1543181

30. Brodie, R.J., Hollebeek, L.D., Juric, B., Ilic, A.: Customer engagement: conceptual domain, fundamental propositions, and implications for research. J. Serv. Res. **14**, 252–271 (2011). https://doi.org/10.1177/1094670511411703
31. Bowden, J.L.-H., Conduit, J., Hollebeek, L.D., Luoma-aho, V., Solem, B.A.: Engagement valence duality and spillover effects in online brand communities. J. Serv. Theor. Pract. **27**, 877–897 (2017). https://doi.org/10.1108/JSTP-04-2016-0072
32. Harrigan, P., Evers, U., Miles, M.P., Daly, T.: Customer engagement and the relationship between involvement, engagement, self-brand connection and brand usage intent. J. Bus. Res. **88**, 388–396 (2018). https://doi.org/10.1016/j.jbusres.2017.11.046
33. Dessart, L., Veloutsou, C., Morgan-Thomas, A.: Capturing consumer engagement: duality, dimensionality and measurement. J. Mark. Manag. **32**, 399–426 (2016). https://doi.org/10.1080/0267257X.2015.1130738
34. Blau, P.M.: Exchange and power in social life. Transaction Publishers, New Brunswick, NJ (2009)
35. Emerson, R.M.: Social exchange theory. Annu. Rev. Sociol. **2**, 335–362 (2008). https://doi.org/10.3982/ECTA7195
36. Homans, G.C.: Social behavior as exchange. Am. J. Sociol. **63**, 597–606 (1958)
37. Fang, X., Zhang, J., Chan, S.S.: Development of an instrument for studying flow in computer game play. Int. J. Hum. Comput. Interact. **29**, 456–470 (2013)
38. Pelet, J.-É., Ettis, S., Cowart, K.: Optimal experience of flow enhanced by telepresence: evidence from social media use. Inf. Manag. **54**, 115–128 (2017). https://doi.org/10.1016/j.im.2016.05.001
39. Martins, B., Sheppes, G., Gross, J.J., Mather, M.: Age differences in emotion regulation choice: older adults use distraction less than younger adults in high-Intensity positive contexts. J. Gerontol. Ser. B Psychol. Sci. Soc. Sci. **73**, 603–611 (2018). https://doi.org/10.1093/geronb/gbw028
40. Mather, M., Carstensen, L.L.: Aging and motivated cognition: the positivity effect in attention and memory. Trends Cogn. Sci. **9**, 496–502 (2005). https://doi.org/10.1016/j.tics.2005.08.005
41. Chang, P.F., Choi, Y.H., Bazarova, N.N., Löckenhoff, C.E.: Age differences in online social networking: extending socioemotional selectivity theory to social network sites. J. Broadcast. Electron. Media. **59**, 221–239 (2015). https://doi.org/10.1080/08838151.2015.1029126

Temporal Trend Analysis on Virtual Reality Using Social Media Mining

Chen-wen Shen, Jung-tsung Ho and Hung-wen Ma

Abstract Many studies have discussed the widespread use of virtual reality (VR). However, few studies have investigated VR from the perspective of social media, even though social media has changed how people communicate and emerged as an essential marketing channel. An approach of two-layer hierarchical concept decomposition structure was proposed to investigate the temporal trend of VR development from the perspective of the public. Accordingly, Twitter posts related to VR in 2015 and 2016 were crawled and analyzed by our proposed approach. The mining results determined that public focus shifted from VR headsets in 2015 to content in 2016. This suggests that VR devices are perceived as having gradually developed and that the next challenge and business opportunity is VR content and applications. In the era of big data and artificial intelligence, our concept decomposition approach contributes to the content analysis of acquiring insight from massive user-generated content, which extracts temporal trends in a holistic view and individual insights on a detailed scale.

1 Introduction

Virtual reality (VR) headset sales are expected to grow at a 5-year compound annual growth rate of 48.7% from 2016 (9.2 million units) to 2021 (67.1 million units) [1]. In this burgeoning market, either its developers or marketing personnel are eager for having a thorough comprehension of what VR users and practitioners experience and expect the technology of VR development. It is critical for an enterprise to gain a thorough knowledge of what their target audients care about, and to response their expectation respectively. Therefore, the VR status as a whole and its trends of discussion from a public perspective evolve into a crucial subject to be addressed.

C. Shen (✉) · J. Ho · H. Ma
Department of Business Administration, National Central University, Jhongli District, Taoyuan City 32001, Taiwan, Republic of China
e-mail: cwshen@ncu.edu.tw

A. Visvizi and M. D. Lytras (eds.), *Research & Innovation Forum 2019*, Springer Proceedings in Complexity,
https://doi.org/10.1007/978-3-030-30809-4_18

Meanwhile, the maturity of social networking from past decades has changed how netizens share their experiences, manage their relationships, and access information [2]. In this context, social media has emerged as a critical marketing channel for disseminating information, promoting products, and managing public relationship [3]. The massive user-generated content enables researchers to analyze social patterns and trends [4]. As for the research of VR, numerous studies have focused on the engineering development [5, 6] and its applications in fields such as human learning [7], purchasing behavior [8], and medical rehabilitation [9]; however, studies regarding VR from the perspectives of social media are scant.

Accordingly, we propose a two-layer hierarchical concept decomposition approach to condense the information from a huge corpus, which would otherwise be a very challenging task for researchers to have a close reading of all the textual information. In this study, VR-related tweets from 2015 and 2016 were retrieved and the collective opinions of how the public perceives VR were analyzed. Our goals were to elucidate the status of VR development and compare the temporal trends of VR through the user posts of Twitter between 2015 and 2016. The mining results indicated that VR devices for consumer markets have gradually matured, and the public focus has moved from hardware details to VR applications and contents. The study contributes a complementary perspective that differs from that of case studies, surveys, and statistical analyses. To the best of our knowledge, this was the first study to investigate the status of VR development from the public's perspective through social media mining.

2 Literature Review

The modern meaning of VR was popularized by Jaron Lanier in 1989. VR is a technology that creates a three-dimensional (3D) computer-generated world [10], and it provides a user experience of immersion, interaction, and imagination [11]. Additionally, advanced VR detects user actions and responds simultaneously. In this context, it provides an immersive experience that can enhance human sensations [12]. In addition, VR technology can be used as a cognitive means to expand the user's thoughts and substantially enhance learning [13]. Therefore, an immersive VR environment initiates user perception and imagination regarding the computer-generated content [14]. Because VR creates abstract concepts [11], it is not only a type of user interface but also a technology for solving problems in our daily lives [15].

Since the early 1920s, VR technology has gradually improved for application in medicine, education, entertainment, and science [16], and its use for training in myriad industries and fields has been utilized and studied [17]. In medicine, VR has many applications, including medical training and diagnostic assistance [18]. The first VR application of on-site medical training, the Plattsburgh Procedural Simulator for Nursing Education, was designed to allow nursing students and other health personnel to practice venipuncture [19]. VR is utilized in medical training primarily because learners can practice detailed procedures in a VR environment without risking the well-being of a real patient [15, 20]. Also, VR technologies are an influential and promising tool in education [21]. Three-dimensional virtual worlds are used in both conventional classrooms and distance education [22]. They activate children's digital literacy in a playful but meaningful context [23]. In industry, VR is widely used to construct preliminary designs and virtual prototypes, which shortens the design cycle and lowers the research and design costs. VR-simulated evaluation of product functions and product interactions with humans and other components has become the most crucial design phase of a product, especially for evaluating assembly tasks [24].

3 Data Collection and Methodology

We utilized a concept decomposition approach to determine the status of VR in 2015 and 2016 based on mentions on Twitter, which provides a unique overview of public opinion. In this study, we first defined the VR concept-related tweets as tweets containing either "virtual reality" or "VR." During the data collection process, we retrieved tweets containing "virtual reality" or "VR" published between January 2015 and December 2016 from the advanced search page of Twitter. Accordingly, the corpora for 2015 and 2016 were 110,385 and 386,572 VR-related tweets, respectively. For both corpora, each tweet was considered as a single document, and text-mining analysis was performed separately.

We performed the hierarchical concept decomposition on the basis of the strength of the concept links between terms. Each tweet in a corpus was parsed into a collection of terms, and each corpus was parsed into a bag-of-words. Following this procedure, the attributes of terms, such as parts of speech, frequency, and synonyms, were determined by applying linguistic rules. Irrelevant characters, such as controls, digits, and graphics, were excluded. To simplify the analysis, we retained both nouns and proper nouns; all other parts of speech were excluded. After parsing, the term-frequency statistics were manually examined to filter out irrelevant terms, such as "we" and hyperlinks, and to combine synonyms, such as Pok and Pokémon, if necessary. Subsequently, we calculated the strength of the concept links between terms. The strength of association is defined as the statistical probability between two concepts based on their co-occurrence in the corpus. For a given pair of terms, the strength of association between a $term_a$ and the associated $term_b$ for r given tweets is calculated on the basis of the following binomial distribution:

$$\text{Strength of Assocation} = \log_e\left(\frac{1}{prob_k}\right) \quad (1)$$

and the sum of probabilities is calculated as follows:

$$prob_k = \sum_{r=k}^{r=n} \frac{n!}{r!(n-r)!} p^r (1-p)^{(n-r)} \tag{2}$$

where n is the number of tweets that contain $term_b$, k is the number of tweets that contain both $term_a$ and $term_b$, and $p = k/n$ is the probability that $term_a$ and $term_b$ co-occur, assuming that they are independent of each other [25]. We decomposed the concept of VR development into two layers with the eight most associated first-layer keywords and their associated second-layer keywords. In the two-layer concept decomposition analysis, we identified the first-layer keywords that were strongly associated with the concept of VR by using (1). The first-layer keywords represent the most significant concepts regarding VR that individuals had mentioned on Twitter. However, to more thoroughly understand the status of VR, we required more information regarding how users perceived the first-layer keywords. Therefore, the second-layer keywords, which strongly exhibited the concepts of the first-layer keywords, were also identified using (1). Accordingly, VR status was outlined by the first-layer keywords, and the concepts of the first-layer keywords were further elucidated by the second-layer keywords. In the following section, we refer to the first- and second-layer keywords as *keywords* and *terms*, respectively.

4 Mining Results

4.1 Similar Keywords in 2015 and 2016

The mining results revealed four similar keywords regarding VR-related tweets in 2015 and 2016; however, the different terms associated with these keywords indicate the subtle differences between them (Table 1).

Table 1 Keywords and terms

2015		2016	
AR	leap, Hololens, Sulon, cardboardvirtualreality, aid, TV, trend, difference	AR	leap, mixedreality, Pok, Manchester, unity, report, wall, AI
tech	innovation, technews, dk2, consumer, aid, support, ramification	tech	innovation, technews, artificialintelligence, bigdata, GameDev, challenge, times, film
future	journalism, godmother, storytelling, glimpse, investor, window, anthonye_vr, takeaway	future	network, entertainment, eye, prediction, pioneer, fad, Forbes, innovation
headset	vive, razor, galaxy, LG, IncrediSonic, magnet, fall, air, telecommunications	headset	vive, razor, galaxy, Asus, hadband, MergeVR, QualComm, phones, million

AR is a keyword identified for both 2015 and 2016, and AR technology integrates partially computer-generated virtual content with the physical world, whereas VR creates a fully simulated reality. The term *leap* appeared in 2015 and 2016 which identifies an AR startup, Magic Leap. Most tweets related to *leap* concerned the fundraising status of Magic Leap, suggesting that AR technology gained favor among investors. *HoloLens* refers to smart glasses developed by Microsoft that provide vivid user interaction with holograms. The term *mixedreality* refers to mixed reality (MR), which is a concept created by Microsoft as a marketing strategy to differentiate HoloLens from VR and AR technology. The term *Sulon* refers to an AR technology startup based in Toronto and the term *cardboardvirtualreality* is a hashtag created for the Google Cardboard VR, which is a stereoscope-style of the head-mounted enclosure (HME) enabling affordable VR gaming and video experiences on smartphones. In 2016, the term *Pok* refers to Pokémon Go, which is an AR mobile game for iOS and Android. The relevant tweets discuss whether Pokémon Go's popularity proved the market viability of AR technology (e.g. "If the recent Pokmon Go craze is any indication AR has arrived #VirtualReality #AugmentedReality"). In 2015, the terms *difference* and *trend* were used to discuss the differences between VR and AR. In 2016, the term *report* refers to a technical review by *Verge*, and *wall* refers to *The Wall Street Journal*.

The keyword *tech* typically refers to different aspects of its technological development updates. The terms *innovation* and *technews* were identified in both 2015 and 2016. The term *innovation* typically refers to the discussion of VR technology and its future development, whereas *technews* was a hashtag created to increase the readership of technology news posted to Twitter. In 2015, *dk2* referred to the second-generation developer kit for Oculus Rift, which was launched with HMDs. Most *consumer*-related tweets (e.g., "#tech Vive VR Headset Full Consumer Launch Slips to Q1 2016 ? You wait years and years for virtual reality to …") refer to releases of consumer-edition HMDs by HTC Vive VR. In 2016, *bigdata* and *artificialintelligence* were hashtags for big data and AI, respectively. *GameDev* refers to Amazon GameDev, which is a cross-platform building block for game developers. The term *challenge* is used to discuss whether the challenges VR faces derive not from the technology but from what the content it provides to customers.

When people discuss the *future* of VR, they usually contemplate how VR technology will change lifestyles and whether the public will accept VR. In 2015, the keyword *journalism* was used to discuss how placing viewers within VR immersive environment would remodel journalism and the term *godmother* refers to a VR pioneer who encouraged journalism to adopt VR technology(e.g., "Godmother of VR sees #journalism as the future of virtual reality | Technology | The Guardian #VR"). The term *storytelling* associated tweets were the discussions of how VR as a medium would change the way we communicate our ideas. Similar discussions involved different fields in 2016, including social networks, the civil aviation industry, and real estate. The term *network* identified emerging social phenomena and business opportunities that VR users might use to improve their social networking. The term *entertainment* was used to suggest new applications for VR in airplane entertainment. However, the future of the VR market was debated on Twitter. In

2015, *glimpse* and *investor* were used to reveal that investors experienced considerable uncertainty regarding the development of VR. In 2016, these concerns lingered. The term *prediction* was used in tweets to convey promising predictions for the VR market in 2017. However, opinions different from those expressed by a VR *pioneer* caused controversy in the Twitterverse (e.g., "Virtual Reality Pioneer Worries About the Future of the Technology—Sci-Tech Today Sci-Tech #realitevirtuelle, #VR"). Furthermore, *fad* was used to discuss whether VR is a temporary phenomenon or a durable technology.

A headset is related to an HMD. It is a wearable device that displays projected images and provides an immersive user experience. As suggested by Table 1, most terms regarded HMD manufacturers: HTC, Razor, Samsung, LG, and IncrediSonic in 2015 and HTC, Razor, Samsung, Asus, CloudSeller, Merge VR, and QualComm in 2016. VR is a highly unpredictable and competitive business opportunity; however, the new entrants are willing to invest in VR technology because of its promising market potential. The term *vive* refers to HTC Vive HMDs and their facilities. *Razer* is a personal computer (PC) peripheral-maker that introduced a framework for open-source VR to compete in the market. *QualComm* was used to report that despite being a semiconductor giant, QualComm competes in the VR market. The terms *IncrediSonic*, *LG*, and *MergeVR* refer to stereoscope-style enclosures similar to Google Cardboard. The term *LG* refers to a marketing promotion that included a free VR headset as part of an LG mobile phone package. In 2016, *phone* generally refers to VR applications for smartphones. The relevant tweets described that the largest VR market segment was smartphones (e.g., "RT @ProblemVR: Report: 98% of VR headsets sold this year are for mobile phones #VR #VirtualReality").

4.2 Different Keywords in 2015 and 2016

Of the mining results for 2015, the keywords *Oculus*, *rift*, *Samsung*, and *gear* refer to the companies Oculus and Samsung and their VR products, Rift and Gear VR, respectively. In 2016, the keywords *game*, *vrporn*, and *video* refer to the applications of VR and *domain* related to domain names for sale (Table 2). This indicates that VR hardware has gradually become ready for customers and that the digital content and its applications have gained public attention.

In 2015, people frequently tweeted reviews of VR equipment use, messages regarding preorders of the consumer edition, and about the release of the developer kit, making *Oculus*, *Rift*, *Samsung*, and *Gear* the keywords most strongly associated with VR. Oculus VR is a major VR developer based in the United States, and Oculus Rift is the brand name of its HMD. The *Oculus Rift*–associated term *Iribe* is the name of the company's chief executive officer, who announces company updates and reveals VR trends. *GTA* refers to the game Grand Theft Auto, and *gamermuscle* refers to the game GamerMuscle. Additionally, *UE4* was the hashtag created for discussing the game engine Unreal Engine 4, and *gameart* refers to content art design. These terms indicate that people frequently tweeted gaming-related messages and

Table 2 Keywords and terms

2015		2016	
Oculus	dk2, development, gameart, jam, Iribe, story, UE4, viewer	game	indiedev, MergeVr, phones, PS4, report, unity, victusvincimus, vive, sword
Rift	consumer, dk2, development, gamermuscle, GTA, PC, TV, wind	vrporn	episode, girl, Kate, Onix, Pax, Rachel, Rose, sister, stroke
Samsung	edge, insight, fall, telecommunications, galaxy, edition, preorder, review, web	video	chair, iphone7, GoPro, music, player, portal, television, unboxing, vive
Gear	edge, edition, fall, galaxy, insight, preorder, solitaire, telecommunications, web	domain	Bitcoin, dm, Gamedev, internetofthings, offer, opportunity, startup, vrgame

that Oculus Rift fans were interested in playing VR games. The term *story* refers to Oculus Story Studio, which was a subdivision of Oculus VR providing short narrative VR films. Keeping the developer kit up to date, Oculus Rift established a studio for creating VR videos for storytelling and even allowed their HMD to be used for adult entertainment (e.g., "Oculus to allow virtual reality porn: People try out Oculus VR's headset Oculus Rift development kit 2 at its…").

The keyword *Samsung* and the term *telecommunications* refer to Samsung Telecommunications, the subgroup of Samsung specializing in mobile communications. The term *galaxy* refers to the company's Galaxy S series smartphone, and *edge* refers to its upgrade. The keyword *gear* refers to the Samsung Gear VR, which is an HME used in combination with a compatible Galaxy smartphone to simulate a 3D video or gaming experience, and *edition* refers to the release of the upgraded Gear VR. The term *preorder* refers to preorders of the Gear VR (e.g., "Samsung's virtual reality headset Gear VR is now available for pre-order via @Techland"). Additionally, the term *web* refers to a VR web browser developed by Samsung (e.g., "Samsung is releasing a virtual reality Web browser for the Gear VR via @thenextweb"). These examples imply Samsung planned that its potential customers would consider the resources of VR content as a crucial factor in their purchasing decisions.

In 2016, after VR technology had developed sufficiently and customer-edition HMDs were released, people's focus had shifted from the technology to its applications. Thus, the keywords *game*, *video*, and *vrporn* were identified as first-layer keywords, which are associated with VR content in different applications. VR technology began to penetrate industries, such as the smartphone, music, TV, gaming, and adult entertainment industries. The keyword *game* represents one of the major market segments of VR. The terms *PS4*, *vive*, and *MergeVR* refer to Sony PlayStation VR (PS VR), HTC Vive, and MergeVR respectively. The term *indie* refers to indie developers who are independent from a publisher. The tweet "#vr #virtualreality #virtualrealityworld #game #videogames #cardboard #oculus #htcvive #oculusrift #indie #indiega? http://ift.tt/1WQa3aP?" was an extreme example in which an experienced user created hashtags of each word to maximize exposure.

The keyword *video* mostly refers to VR video content. The term *television* was used to discuss how VR video affected TV broadcasting. The terms *music*, *GoPro*, and *player* refer to 360° music videos, the camera used to shoot them in 3D, and the multimedia player used for watching them, respectively. The music industry is a pioneer in this technology. *Music*-related tweets (e.g., "OH. - Love of Avalanches (360? #virtualreality video) via @YouTube #musicvideo #vr #music") link to 360° music videos hosted on YouTube. The term *Gopro* refers to the GoPro camera, which can be used to shoot footage from different angles simultaneously. A tweet (e.g., "How to Stitch @GoPro Footage into 360? Spherical Video #VirtualReality #VR") provided instructions on how to operate GoPro cameras to obtain high-quality 360° video. *Player*-related tweets indicate that multimedia players support 360° video playback, suggesting that 360° videos are becoming increasingly widely viewed. The term *portal* refers to VAR-Port's Portal headset, which can be used in tandem with a smartphone to create VR experiences.

Because the major appeal of VR is its immersive user experience, the adult entertainment industry has embraced this new technology. The keyword *vrporn* refers to adult VR videos. Most terms associated with it, such as *Onix* and *Rose*, are the names of adult VR video actresses, and the terms *stroke* and *girl* refer to the videos' content. Numerous tweets were mined from an adult VR video company that uses Twitter to announce new video releases. Furthermore, the keyword *domain* refers to the domain name of the website. Relevant tweets (e.g., "VRIoT, Trade on #sales #domain #vr #VirtualReality #IoT #app #InternetOfThings #tech #bigdata") contained information regarding VR-related domain names for sale. The term *internetofthings* is a hashtag created to refer to the internet of things. Experienced users combined different widely used hashtags (e.g., "Domain Name http://VRBitcoinGames.com? is for sale #virtualreality #bitcoin #btc #bitcoins #VRgames #vrgame #vr #domains #videogames") to maximize exposure. Most such terms, such as *opportunity*, *startup*, *offer*, *Bitcoin*, *vrgame*, *dm*, and *Gamedev*, refer to sales of registered domain names.

5 Conclusion

We proposed a two-layer concept decomposition approach for using social media posts to track VR status and implement the approach for 500,000 tweets from 2015 and 2016. From our findings, the temporal trend indicates that people's concerns shifted from details of HMDs in 2015 to VR content (e.g., gaming, video, and adult entertainment) in 2016. This suggests that HMDs have gradually improved in users' opinions, and the next challenge and business opportunity is VR content and its applications. Additionally, AR as a key concept in both years and the popularity of Pokémon Go suggests that the AR market has been accepted by the public because of its more straightforward and cheaper technology requirements compared with VR. The other mining results from the keyword headset identified new entrants in 2016, such as Asus and QualComm, who are the PC and telecom chip industry leaders, respectively. Because of their investments and endorsement, VR's future appears more secure than it did in 2015. The other new direction is suggested by the term storytelling, journalism and network, regarding how the VR technology and usage will reshape the way the media sharing their storyline. With a VR immersive and interactive environment that differs from the text, sound, and video, an abstract concept that is difficult to envision can now be represented into a concrete one.

Furthermore, our hierarchical concept decomposition approach identified vital events from the aspect of marketing, such as Microsoft's alliance with Autodesk, Oculus VR's establishment of a VR studio, and Samsung's development of a VR web browser. These actions indicate that most VR developers acknowledge that developing adequate VR content is necessary for the early stage of VR development lifecycle to increase their market share and beat the competition. As one of the world's most widely used marketing channels among enterprises, consumers, and professional bloggers, Twitter offers a platform for promoting users' activities, ideas, and products. From our mining result, we notice that the executive of VR developer (e.g., the term *Iribe*) was using social media to communicate with the company fans. These activities of sharing ideas through social media are not only trying to build his reputation but also managing customer relationship. Moreover, the question is not whether the aforementioned actors should use Twitter as an advertising channel but how target audiences can access information related to these agents.

As a proof-of-concept approach, we proposed a two-layer decomposition system to determine VR status by using social media posts. Our results provide only major status milestones from this large dataset of half a million tweets. If a detailed status report were required, we would suggest performing multilayer hierarchical analysis to collect more key concepts, from which it would be possible to extract more insights. Furthermore, although we examined our approach's usage for social media mining, some limitations must be acknowledged regarding the corpus we analyzed. In our study, we selected Twitter for VR status analysis, but other major social media platforms were not considered. Because social media and social networking platform users differ in specific demographic attributes, we believe that using our approach to mine the content of other social media platforms would extend our study.

Acknowledgements This research was supported by the Ministry of Science and Technology, Taiwan, under contract number MOST 107-2410-H-008-042.

References

1. IDC.: Worldwide shipments of augmented reality and virtual reality headsets expected to grow at 58% CAGR with low-cost smartphone VR devices being short-term catalyst, according to IDC 2017 [cited 2018 June 05]; Available from: http://www.idc.com/getdoc.jsp?containerId=prUS42807717
2. Kreiss, D.: Seizing the moment: the presidential campaigns' use of Twitter during the 2012 electoral cycle. New Med. Soc. **18**(8), 1473–1490 (2016)
3. Hays, S., Page, S.J., Buhalis, D.: Social media as a destination marketing tool: its use by national tourism organisations. Curr. Issues Tourism **16**(3), 211–239 (2013)

4. Neubaum, G., Krämer, N.C.: Opinion climates in social media: blending mass and interpersonal communication. Hum. Commun. Res. **43**(4), 464–476 (2017)
5. Turner, C.J., et al.: Discrete event simulation and virtual reality use in industry: new opportunities and future trends. IEEE Trans. Hum.-Mach. Syst. **46**(6), 882–894 (2016)
6. Bovet, S., et al.: The critical rôle of self-contact for embodiment in virtual reality. IEEE Trans. Vis. Comput. Graph. **24**(4), 1428–1436 (2018)
7. Shen, C.-W., et al.: Behavioural intentions of using virtual reality in learning: perspectives of acceptance of information technology and learning style. Virtual Reality (2018)
8. Zhao, X., et al.: Analysis of mental workload in online shopping: are augmented and virtual reality consistent? Front. Psychol. **8**, 71 (2017)
9. Gerber, S.M., et al.: Visuo-acoustic stimulation that helps you to relax: a virtual reality setup for patients in the intensive care unit. Sci. Rep. **7**(1), 13328 (2017)
10. Vince, J.: Introduction to virtual reality. Springer Science & Business Media (2004)
11. Burdea, G.C., Coiffet, P.: Virtual Reality Technology, vol. 1. John Wiley & Sons (2003)
12. Burdea, G.C.: Keynote address: haptics feedback for virtual reality. In: Proceedings of International Workshop on Virtual Prototyping. Laval, France (1999)
13. Jonassen, D.H.: Transforming learning with technology: beyond modernism and post-modernism or whoever controls the technology creates the reality. Educ. Technol. **40**(2), 21–25 (2000)
14. Huang, H.-M., Rauch, U., Liaw, S.-S.: Investigating learners' attitudes toward virtual reality learning environments: Based on a constructivist approach. Comput. Educ. **55**(3), 1171–1182 (2010)
15. Huang, H.-M., Liaw, S.-S., Lai, C.-M.: Exploring learner acceptance of the use of virtual reality in medical education: a case study of desktop and projection-based display systems. Interact. Learn. Environ. **24**(1), 3–19 (2016)
16. de Strulle, A.: Differentiation of the causal characteristics and influences of virtual reality and the effects on learning at a science exhibit. University of San Diego (2004)
17. Van Wyk, E., De Villiers, R.: Virtual reality training applications for the mining industry. In: Proceedings of the 6th International Conference on Computer Graphics, Virtual Reality, Visualisation and Interaction in Africa. ACM (2009)
18. Lu, J., et al.: Virtual learning environment for medical education based on VRML and VTK. Comput. Graph. **29**(2), 283–288 (2005)
19. Higgins, G.A., et al.: Teleos™: development of a software toolkit for authoring virtual medical environments. Presence: Teleoperators and Virtual Environ. **6**(2), 241–252 (1997)
20. Temkin, B., et al.: Web-based three-dimensional virtual body structures: W3D-VBS. J. Am. Med. Inform. Assoc. **9**(5), 425–436 (2002)
21. Mikropoulos, T.A., Natsis, A.: Educational virtual environments: a ten-year review of empirical research (1999–2009). Comput. Educ. **56**(3), 769–780 (2011)
22. Dickey, M.D.: Brave new (interactive) worlds: a review of the design affordances and constraints of two 3D virtual worlds as interactive learning environments. Interact. Learn. Environ. **13**(1–2), 121–137 (2005)
23. Merchant, G.: Literacy in virtual worlds. J. Res. Reading **32**(1), 38–56 (2009)
24. Pontonnier, C., et al.: Designing and evaluating a workstation in real and virtual environment: toward virtual reality based ergonomic design sessions. J. Multimodal User Interfaces **8**(2), 199–208 (2014)
25. Chakraborty, G., Pagolu, M., Garla, S.: Text Mining and Analysis: Practical Methods, Examples, and Case Studies Using SAS. 2014: SAS Institute

Using User Contextual Profile for Recommendation in Collaborations

Siying Li, Marie-Hélène Abel and Elsa Negre

Abstract Nowadays, many digital technologies are developed to support collaboration and facilitate its efficiency. When using them, users will leave contextual information explicitly and implicitly, which could contribute to identifying users' situations and thus enabling systems to generate corresponding recommendations. In the framework of collaborations, we are interested in considering user context with user contextual profile to suggest appropriate collaborators. In this article, we present the user contextual profile that we established and how it can be used to generate recommendations for collaborations in digital environments.

1 Introduction

Nowadays, more and more people consider collaborations as an effective way to work [1], particularly with the rapid development of digital technologies, such as blogs [2], wiki [2] and digital ecosystems [3]. These bring great convenience to collaborations from the term of tools [4] and environments [1]. However, at the same time, these technologies also pose problems in acquiring relevant information about collaborators, which are caused by information overload [5]. Thus, finding appropriate collaborators remains an urgent problem to be solved [6]. In order to tackle it, many researchers have dedicated great effort to recommend collaborators [5, 7, 6]. As part of our research, we focus on recommending collaborators to a user

S. Li · M.-H. Abel (✉)
Sorbonne Universités, Université de technologie de Compiègne CNRS UMR 7253, HEUDIASYC, 60203 Compiègne, France
e-mail: marie-helene.abel@utc.fr

S. Li
e-mail: siying.li@utc.fr

E. Negre
Paris-Dauphine University, PSL Research University, CNRS UMR 7243, LAMSADE, 75016 Paris, France
e-mail: elsa.negre@dauphine.fr

A. Visvizi and M. D. Lytras (eds.), *Research & Innovation Forum 2019*, Springer Proceedings in Complexity,
https://doi.org/10.1007/978-3-030-30809-4_19

of a digital collaborative environment. We aim at developing a recommender system based on the information collected from this digital environment.

In order to generate recommendations, many traditional approaches have been already used in many applications and have excellent performances, such as content-based, collaborative filtering and hybrid approaches [8]. Others have also been proposed and studied recently, such as link prediction in graphs/networks [5] and topic modeling [7]. Among them, what attract our attention are context-aware approaches, which allows to generate more relevant recommendations through considering the specific contextual situations of the user [9]. Such situations are described by user context, which could be explicit or implicit [10]. To record user's contextual information, the user profile has been introduced and applied [11]. Therefore, we are interested in analyzing user contextual information and integrate it with user profile into user contextual profile to propose collaborator recommendations in collaborations.

The remainder of this paper is constructed as follows. Section 2 discusses user context and user profile, and studies context-aware recommender systems. Our user contextual profile is presented in Sect. 3. We apply a scenario to illustrate how to use our user contextual profile on recommending collaborators to a given user in the framework of collaborations in Sect. 4. We then discuss our collaborator recommendations and the proposed user contextual profile in Sect. 5. Finally, some conclusions and future work are put forward in Sect. 6.

2 Related Work

In this section, we analyze two notions: user context and user profile. Moreover, we conduct a survey on context-aware recommender systems (CARSs).

2.1 User Context and User Profile

User context is a widely addressed concept in CARS since it matters in better predicting users' behaviors [12]. Usually, such contextual information can be collected from two approaches: requiring explicitly to the user and being learned implicitly from the user's behaviors [13]. However, user context remains dependent on applied domains and applications, such as user's intent [12], user's research topics [7]. According to the definition of context proposed by Dey [14], user context can be defined as any information that can be used to characterize the situation of a user who is considered relevant to the interaction with an application, including the user and applications themselves. Based on the properties of contextual information, it can be divided into static/dynamic context, long-term/short-term context [10].

As for user profile, its main features are user interests and preferences [15]. It depends on demographics or online user behaviors [13]. Based on the properties of different information in user profile, it is divided into two parts: static (e.g. user identity) and dynamic (e.g. current activities) [16]. While in [17], it is separated into short-term and long-term user profile depending on whether such information deals with users' current behaviors or not. In addition, the two main ways of collecting information for user profiles are also explicit and implicit [17, 15].

From these researches, it is obvious that user context and user profile have many features in common, such as collecting approaches, classifications. However, not any user contextual information could be included in user profile to express users' interests or preferences. For example, in a movie recommender system, user's companions [18] is critical contextual information that does not need to be recorded in the user's profile. In other words, user context is a larger concept than user profile. Conversely, user profile is included in user context. Since any information in user profile focuses only on users themselves, which definitely also belongs to user context.

Furthermore, a joint analysis of user context and user profile would provide a stronger comprehension of users. Palmisano et al. [12] uses a dataset of demographic, transactional and contextual information about the users to analyze users' behaviors. According to our discussions above, the first two parts belong to user profile, while the last one is under user context. Therefore, in our research, we are interested in integrating user context with user profile in order to construct user contextual profile, which could be used to predict users' behaviors in the framework of collaborations and thus generate corresponding recommendations.

2.2 Context-Aware Recommender System

Traditionally, recommender system (RS) deals with two types of entities: users and items [19]. Using a known set of users' ratings about items, these RSs are capable to predict those unknown ratings of users and thus recommend items based on the forecasted ratings. For these RSs, their rating function R_{RS} is

$$R_{RS}\text{: Users} \times \text{Items} \rightarrow \text{Ratings} \quad [9] \tag{1}$$

As for CARSs, they apply at least three types of information: users, items and context. Sometimes, it is even possible to construct a multi-dimensional CARS. In [19], context of a movie CARS is separated into 3 dimensions: Place, Time, and Companion. Thus, for an n-dimensional CARS, its rating function R_{CARS} is

$$R_{CARS}\colon D_1 \times D_2 \times \cdots \times D_n \rightarrow \text{Ratings} \quad [19] \tag{2}$$

where $D_1, D_2, \ldots, D_n$ represent its n dimensions (including Users, Items, …).

To compare the different performances of CARS and RS, Palmisano et al. [12] demonstrates that knowing the context could help RS to perform better. Therefore, in order to generate more relevant recommendations, we are interested in constructing a CARS to recommend collaborators to a given user in the framework of collaborations. A similar research has been carried out in [7]: a first context-aware academic collaborator recommender system is built, where the context is referred to a set of topics that users will jointly work on. However, contextual information collected from users' historical collaborations is not utilized, such as users' historical interactions with others. Therefore, in our research, the context is referred to the contextual information obtained from users' historical collaborations in digital environments. We focus on utilizing such user context and integrating it into user contextual profile to predict unknown ratings of users on collaborators to generate collaborator recommendations in collaborations.

3 User Contextual Profile in Our CARS

In this section, we introduce user contextual profile we built for collaborations in digital environments. Then the formalized elements of our CARS are presented.

3.1 User Contextual Profile

We first introduce the relations between user profile and user context in collaborations. We discuss how to express users' interests and preferences properly and what contextual information could be collected from their historical collaborations. Finally, we propose a user contextual profile in a digital collaborative environment. In this profile, we retain demographic information and activity traces related to collaborations and discuss privacy protection of users' personal data.

3.1.1 Relations Between User Profile and User Context

Based on the discussions in Sect. 2.1, user profile belongs to user context (as shown in Fig. 1). Besides, some of contextual information of users could be derived from their historical collaborations, as we discussed before in Sect. 2.2. In other words, user context is crossing with collaborations (as shown in Fig. 1).

Therefore, our user contextual profile is constructed based on user profile (represented by triangles) and user contextual information obtained from users' historical collaborations (represented by the intersected parts of circles and squares).

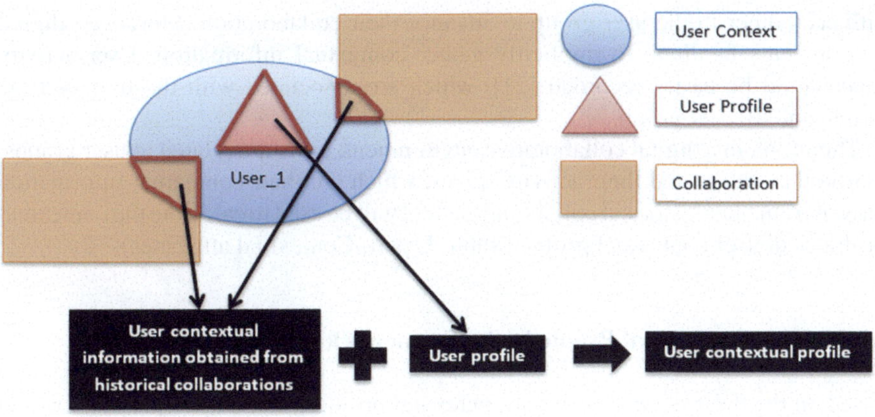

Fig. 1 User contextual profile in collaborations

Table 1 User contextual profile

	Attributes
Demographic attributes	Login; Gender; Competencies; Age; Educational level; Spoken languages; Employment; Home Country; Home City
Contextual attributes	Historical collaborations; Diversity of collaborators; Activeness

3.1.2 User Profile

In our research, a user is a person who holds a user account in a digital environment that could provide digital support (such as tools and resources) to facilitate collaborations. According to [13], we apply demographic attributes to express user's personal characteristics in user profile (Table 1, part: Demographic attributes). Based on the values of these attributes, our CARS could predict users' possible collaborators. For example, by checking whether there is a common value of 'Spoken languages' between users, our CARS could determine whether they have adequate communication skills in collaborations. Besides, in order to identify users, we use 'Login' as a unique authentication.

3.1.3 User Context

According to [20], a collaboration involves at least two persons and comprises a set of human actors' actions on behalf of the corresponding collaborator in order to achieve a shared goal. Thus, in a digital environment, collaboration necessities at least two members and a shared goal among all the members. To represent properly collaborations in digital environments, we use 'user group'. A user group consists of at least two users with a shared objective. These users could communicate and interact with each other in the user group to advance their collaboration. Moreover, digital environments facilitate to implicitly gather contextual information. User activity traces could be easily recorded [21], which are associated with the user and the corresponding user group.

Therefore, in a digital collaborative environment, users are related to user groups, their collaborators and their activity traces, which are their contextual information derived from their historical collaborations and will be used to calculate the contextual attributes in user contextual profile (Table 1, part: Contextual attributes).

3.1.4 User Contextual Profile in the Framework of Collaboration

Based on the discussion above, our contextual profile contains two parts (shown in Table 1): demographics attributes, and contextual attributes.

In particular, the contextual profile require collecting users' personal information and activity traces, which raises our concerns about users' personal data[1] privacy protection. Europe has enforced General Data Protection Regulation (GDPR) since 25 May 2018.[2] Thus, following GDPR, we ask users for their consents to collect and process their personal data in our CARS. Under the authorizations of users, they allow us to use their personal information and activity traces gathered through their historical collaborations to generate collaborator recommendations. All the collected data is only kept for users themselves to improve their collaborations and will not be shared with any third party.

3.2 *Formalized Elements*

This section presents the formalized terminologies that are used to formulate our collaborator recommendation problem. Then the definition and calculation of contextual attributes in user contextual profile are also explained.

3.2.1 Terminology

For a user u, who is in a user group $g(u \in g)$ at time t, the clarifications of terminologies are: (1) Collaborator: a collaborator c of u at time t is another user $c(\neq u)$ who is in the same user group $g(c \in g)$ at time t; (2) Contact (possible collaborator): a contact c_p of u at time t is another user $c_p(\neq u)$ who is not in the user group $g(c_p \notin g)$ at time t.

[1] According to [22], personal data in GDPR indicates information that can identify an individual directly or indirectly, specifically including online identifiers (e.g. IP addresses, cookies and digital fingerprinting, and location data).

[2] https://eur-lex.europa.eu/legal-content/EN/TXT/PDF/?uri=CELEX:32016R0679&from=EN.

Considering that collaborators and contacts are definitely users, we consider profile and context of collaborators and contacts are equal to user profile and user context. Besides, we use CCU to represent Contact/Collaborator/User in the following: (1) CCU profile consists of CCU's demographic information; (2) CCU context includes all the information that can be used to characterize the situation of a CCU. For the time being, in our research, CCU context is limited to the contextual information that is obtained from historical collaborations and CCU profile (shown in Fig. 1); (3) CCU contextual profile merges the information in CCU profile and CCU context derived from historical collaborations, including demographic attributes and contextual attributes (shown in Table 1).

Thus, our context-aware collaborator recommendation problem is formulated as: **Given a user u and his/her current user groups[3] $G = \{g_1, g_2, \ldots, g_{m_u}\}$, K (specified by user u) contacts (possible collaborators) c_p from all candidates $\mathrm{R_p} = \{\mathrm{c_p} \notin g_i, c_p \neq u\}$ $(i \in \{1, 2, \ldots, m_u\})$, who will collaborate with u with the highest probabilities. m_u represents the number of current user groups that user u is in**. Besides, in our CARS, the recommended items to a given user are his/her contacts (possible collaborators). Then the rating function R_{col} of our CARS is:

$$R_{col}\text{: User contextual profile} \times \text{Contact contextual profiles} \rightarrow \text{Ratings} \tag{3}$$

3.2.2 Contextual Attributes

The three contextual attributes are defined and calculated using contextual information derived from users' historical collaborations.

For the first attribute 'historical collaborations', its values are a list of all historical user groups in which the user has participated. Here, historical user groups mean that the represented collaborations are already terminated, no longer in progress.

Then, the diversity of collaborators of a user is measured by the entropy of historical collaborators[4] (denoted by H_u). According to [7], a large H_u indicates a user prefers to collaborate with different collaborators. In our research, H_u represents a logarithmic measure of the number of historical collaborators with a significant probability of working together, which follows:

$$H_u = -\sum_{u'} p_{u'} \ln p_{u'} \quad \left(\sum_{u'} p_{u'} = 1,\ u' \neq u\right) \tag{4}$$

---

[3]Current user groups mean that represented collaborations have not finished yet, still in progress.

[4]Historical collaborators represent the contacts that user u has collaborated with in his/her historical collaborations.

where $p_{u'} = \frac{N_{u,u'}}{N_u}$ represents the probability of collaborating with u' according to user u's historical collaborations records; $N_{u,u'}$ is the number of collaborations that user u collaborated with user u'; N_u indicates the number of collaborations that user u collaborated with others in all historical collaborations; the interval of H_u is $(0, \ln N_u]$.

Finally, the activeness of a user in collaborations aims at analyzing the willingness of the user to take actions and contribute to the collaboration itself. The activeness is measured by the average activity rate of the user in historical user groups (denoted by $\overline{A_u}$):

$$\overline{A_u} = \frac{1}{n_u} \sum_{j=1}^{n_u} A_{u,g_j} \tag{5}$$

where A_{u,g_j} is the number of activities that user u effected in the historical group g_j; n_u indicates the number of historical user groups that user u was in; the interval of $\overline{A_u}$ is $[0, +\infty)$.

4 Using User Contextual Profile for Collaborator Recommendations

In this section, we illustrate a scenario of Emma to discuss how to predict users' preferences on collaborators through user contextual profile: *Emma is a Ph.D. student in X laboratory. Her thesis is about data mining and user modeling, supervised by Elsa and Marie. During her thesis, Emma has collaborated with several people on three projects. Moreover, there are two other persons in X laboratory: Marinela and Jack. However, both of them never collaborate with Emma before. Now, Emma is looking for several partners to develop an application in order to analyze users' behaviors in her laboratory. However, she doesn't know who she will collaborate.* To help Emma, our CARS uses her contextual profile to predict a ranked list of all possible collaborators and recommend the top 3 ($K = 3$, specified by herself) collaborators to her. She can choose appropriate ones from the recommended collaborators to collaborate with.

In our scenario, suppose Emma's historical collaborations are represented by three user groups: g_1, g_2 and g_3. According to the recorded traces in the groups, she has acted 7 times in g_1, 6 times in g_2 and 5 times in g_3.

Then, Emma gets g_1, g_2, g_3 for the first contextual attribute. Besides, Emma has collaborated 2 times with Marie (in g_1, g_2), 1 time with Oriane (in g_1), 1 time with Nathalie (in g_2), and 1 time with John (in g_3), then we have $N_{Emma} = 2 + 1 + 1 + 1 = 5$ and $H_{Emma} = (\frac{2}{5} \ln \frac{2}{5} + \frac{1}{5} \ln \frac{1}{5} + \frac{1}{5} \ln \frac{1}{5} + \frac{1}{5} \ln \frac{1}{5}) = 1.332$. Considering $\ln N_{Emma} = \ln 5 = 1.609$, Emma gets a large $H_{Emma} (0 \ll 1.332 < 1.609)$. As for the third one, Emma has acted separately 7 times (in g_1), 6 times (in g_2) and 5 times (in g_3), then her activeness is $\overline{A_{Emma}} = \frac{7+6+5}{3} = 6$, which means that Emma

Table 2 Emma's contextual profile

Attributes	Example of Emma
Login	Emma_account
Gender	Female
Competencies	User modeling, Data mining
Age	25
Educational level	Ph.D. student
Spoken languages	French, English, Chinese
Employment	X laboratory
Home country	France
Home city	Compiègne
Historical collaborations (IDs)	g_1, g_2, g_3
Diversity of collaborators $H_u(H_u \in (0, \ln N_u])$	1.332
Activeness $\overline{A_u}(\overline{A_u} \in [0, +\infty))$	6

Table 3 Information from Emma's contacts' contextual profiles

Contact	Information
Jack	Speaks French and English; $\overline{A_{Jack}} = 10$, $H_{Jack} = 1.040$
Marinela	Speaks English; $\overline{A_{Marinela}} = 1$, $H_{Marinela} = 0.040$
John	Speaks French, English and Spanish; Has collaborated with Emma once in g_3; $\overline{A_{John}} = 5$, $H_{John} = 1.002$

performs 6 activities in a single collaboration on average. Finally, Emma's contextual profile is shown in Table 2.

Once Emma's contextual profile is complete, we need her contacts' contextual profiles. All her contacts work in X laboratory. And suppose that from their contextual profiles, we can know the following information (presented in Table 3).

Based on Table 3, all Emma's contacts are possible to collaborate with her since they all work in a same place and speak at least one common language. However, according to Emma's contextual profile (shown in Table 2), she gets a large H_{Emma} which indicates that when compared to contacts that she has already collaborated with, Emma prefers to work with other contacts. Thus, comparing with John, Emma would prefer to work together with Jack or Marinela. Besides, Emma is an active collaborator since she is willing to take actions in collaborations, averagely 6 times in a single one. However, Marinela is not an active collaborator based on the value of $\overline{A_{Marinela}}$. Then, Emma would rather not collaborate with Marinela. Therefore, Emma's top 3 possible collaborators is $R_{col}(Jack) > R_{col}(John) > R_{col}(Marinela)$.

In our scenario, user contextual profile contributes to generating collaborator recommendations. On the one hand, Emma's contextual profile concentrates on her personal preferences in collaborations. On the other hand, her contacts' contextual profiles (Jack, John, Marinela) help our CARS to calculate Emma's possible ratings for them.

5 Discussion

The generated recommendations are directly related to the user contextual profile. More information we have in the user contextual profile, more accurate the recommendation will be. We are therefore facing to two critical issues: how to improve the record of users' activity traces from a digital collaborative environment; how to let users agree to share their personal information and activity traces in the environment according to GDPR.

Using user contextual profile to generate recommendations offers diverse possibilities at research and practice level in future. The proposed user contextual profile could be broadened by widening the scope of collecting information in digital environments, such as specifying details of activity traces. Currently, we only apply the number of activities that user effected in historical collaborations, without any detailed information of an activity. Thus, we could enrich the user contextual profile by specifying the details of an activity in a user group, such as: one or multiple actor(s), time, resources used, and type of activity. Such profile could also be filled by the information coming from different software that users used for their collaborations. In addition, users need to give their agreements to apply this information to generate collaborator recommendations.

6 Conclusion and Future Work

In this paper, we focus on recommending possible collaborators to a given user in the framework of collaborations and propose a user contextual profile to generate recommendations. Based on the related literature in Sect. 2, we have explained why user contextual profile is needed to provide collaborator recommendations and justified the choice of building a CARS. We then presented the user contextual profile built for collaborations in digital environments and demonstrated how to use it in our CARS by illustrating a scenario. We finally discuss the issues of our recommendations and explore possibilities of the proposed user contextual profile in future.

Our research perspectives include the exploration of the relations between user context and collaboration context, as well as the implementation of the recommendation algorithms for collaborations in digital environments.

References

1. Mattessich, P.W., Monsey, B.R.: Collaboration: What Makes It Work. A Review of Research Literature on Factors Influencing Successful Collaboration. Amherst H. Wilder Foundation, 919 Lafond, St. Paul, MN 55104 (1992)
2. Godwin-Jones, R.: Blogs and wikis: environments for online collaboration (2003)
3. Boley, H., Chang, E.: Digital ecosystems: principles and semantics. In: Digital EcoSystems and Technologies Conference, pp. 398–403 (2007)
4. Patel, H., Pettitt, M., Wilson, J.R.: Factors of collaborative working: a framework for a collaboration model. Applied Ergonomics **43**(1), 1–26 (2012)
5. Li, J., Xia, F., Wang, W., Chen, Z., Asabere, N.Y., Jiang, H.: Acrec: a co-authorship based random walk model for academic collaboration recommendation. In: Proceedings of the 23rd International Conference on World Wide Web, pp. 1209–1214. ACM, New York (2014)
6. Yang, C., Liu, T., Liu, L., Chen, X.: A nearest neighbor based personal rank algorithm for collaborator recommendation. In: 2018 15th International Conference on Service Systems and Service Management (ICSSSM), pp. 1–5. IEEE, New York (2018)
7. Liu, Z., Xie, X., Chen, L.: Context-aware academic collaborator recommendation. In: Proceedings of the 24th ACM SIGKDD International Conference on Knowledge Discovery & Data Mining, pp. 1870–1879. ACM, New York (2018)
8. Resnick, P., Varian, H.R.: Recommender systems. Commun. ACM **40**(3), 56–58 (1997)
9. Adomavicius, G., Tuzhilin, A.: Context-aware recommender systems. In: Recommender Systems Handbook, pp. 217–253. Springer, Boston, MA (2011)
10. Shen, X., Tan, B., Zhai, C.: Context-sensitive information retrieval using implicit feedback. In: Proceedings of the 28th Annual International ACM SIGIR Conference on Research and Development in Information Retrieval, pp. 43–50. ACM, New York (2005)
11. Golemati, M., Katifori, A., Vassilakis, C., Lepouras, G., Halatsis, C.: Creating an ontology for the user profile: method and applications. In: Proceedings of the First RCIS Conference, pp. 407–412 (2007)
12. Palmisano, C., Tuzhilin, A., Gorgoglione, M.: Using context to improve predictive modeling of customers in personalization applications. IEEE Trans. Knowl. Data Eng. **20**(11), 1535–1549 (2008)
13. Anand, S.S., Mobasher, B.: Introduction to intelligent techniques for web personalization. ACM Trans. Internet Technol. (TOIT) **7**(4), 18 (2007)
14. Dey, A.K.: Understanding and using context. Pers. Ubiquit. Comput. **5**(1), 4–7 (2001)
15. Tamine-Lechani, L., Boughanem, M., Daoud, M.: Evaluation of contextual information retrieval effectiveness: overview of issues and research. Knowl. Inf. Syst. **24**(1), 1–34 (2010)
16. Tazari, M.R., Grimm, M., Finke, M.: Modeling user context. In: Proceedings of the 10th International Conference on Human-Computer Interaction, Crete (2003)
17. Shen, X., Tan, B., Zhai, C.: Implicit user modeling for personalized search. In: Proceedings of the 14th ACM International Conference on Information and Knowledge Management, pp. 824–831. ACM, New York (2005)
18. Ostuni, V.C., Di Noia, T., Mirizzi, R., Romito, D., Di Sciascio, E.: Cinemappy: a contextaware mobile app for movie recommendations boosted by DBpedia. SeRSy **919**, 37–48 (2012)
19. Adomavicius, G., Sankaranarayanan, R., Sen, S., Tuzhilin, A.: Incorporating contextual information in recommender systems using a multidimensional approach. ACM Trans. Inf. Syst. (TOIS) **23**(1), 103–145 (2005)
20. Li, S., Abel, M.H., Negre, E.: Contact and collaboration context model. In: 2018 IEEE 4th International Forum on Research and Technology for Society and Industry. IEEE, New York (2018)
21. Wang, N., Abel, M. H., Barthès, J. P., Negre, E.: Recommending competent person in a digital ecosystem. In: 2016 International Conference on Industrial Informatics and Computer Systems (CIICS), pp. 1–6. IEEE, New York (2016)
22. Goddard, M.: The EU General Data Protection Regulation (GDPR): European regulation that has a global impact. Int. J. Mark. Res. **59**(6), 703–705 (2017)

A Step Further in Sentiment Analysis Application in Marketing Decision-Making

Erick Kauffmann, David Gil, Jesús Peral, Antonio Ferrández and Ricardo Sellers

Abstract Nowadays, firms have realized the importance of Big Data, highlighting the need for understanding the current state of marketing practice with respect to Big Data analytics. Among the different sources of Big Data, User-Generated Content (UGC) is one of the most important ones. From blogs to social media and online reviews, consumers generate huge amounts of brand related information that have a decisive potential business value in targeted advertising, customer engagement or brand communication, among others. In the same line, previous empirical findings show that UGC has significant effects on brand images, purchase intentions, and sales. It plays an important role for customers' potential buying decisions. Thus, mining and analysing UGC data such as comments and sentiments might be useful for firms. Particularly, brand management can be one area of interest, as online reviews might have an influence on brand image and brand positioning. Within this context, as well as the quantitative star score usual in this UGC, in which the buyers rate the product, a recent stream of research employs Sentiment Analysis (SA) tools with the aim of examining the textual content of the review and categorizing buyers' opinions. While certain SA split the comments into two classes (negative or positive), other incorporate more sentiment classes. However, the review can have phrases with different polarities because the user can have different experiences and

E. Kauffmann · D. Gil · J. Peral (✉) · A. Ferrández · R. Sellers
University of Alicante, Crtra. San Vicente, 03690 San Vicente del Raspeig, Alicante, Spain
e-mail: jperal@dlsi.ua.es

E. Kauffmann
e-mail: eki@alu.ua.es

D. Gil
e-mail: dgil@dtic.ua.es

A. Ferrández
e-mail: antonio@dlsi.ua.es

R. Sellers
e-mail: Ricardo.Sellers@ua.es

E. Kauffmann
University of Costa Rica, San José 11501-2060, Costa Rica

A. Visvizi and M. D. Lytras (eds.), *Research & Innovation Forum 2019*,
Springer Proceedings in Complexity,
https://doi.org/10.1007/978-3-030-30809-4_20

sentiments about each feature of the product. Finding the polarity of each feature can be interesting for the decision makers of a product. In this paper, we consider that although these two scores (star and sentiment) are related, the sentiment score highlights extra information not detailed in the star score, which is crucial to be extracted in order to have better criteria of comparison between products. Moreover, we mine the positive and negative features of the products analysing the sentiment.

1 Introduction

Nowadays, with the globalized access of the Internet, a large amount of data is generated. Organizations want to take advantage of these data and convert them into relevant information that allows them to take better decisions and it is possible by analysing all this big data. The Internet user, on a daily basis, collaborates to generate huge amounts of data, being one of the most important sources of big data. By writing in blogs, participating on social media, or buying and reviewing products online is generating content. Thus, mining and analysing user generated content (UGC) such as comments and sentiments might be useful for firms. Particularly, brand management can be one area of interest, as online reviews might have an influence on brand image and brand positioning, including design decisions.

Within this context, sentiment analysis techniques are a useful way to examine opinionated text, which contains consumer opinions toward firms, products, brands, or events. Sentiment analysis is a subfield in Natural Language Processing (NLP) to automatically classify text by valence [16] extracting information from user opinions. [12]. Certain techniques split the comments into two classes (negative or positive), others incorporate more sentiment classes [7]. Generally, sentiment analysis is classification of the given text polarity at three levels: document level, sentence level or aspect level [12].

In addition to doing a global sentiment analysis in the review (document level) that allows us to measure whether the product is liked by people, we can also analyse each of the different phrases (sentence level) to find out what buyers like and dislike about that product. With this information, the decision-maker can make important decisions and for example, highlight the positive features of your product, and can also try to improve the negative features of it (aspect level).

In this paper, we study sentiment analysis techniques and PLN tools to apply in the marketing decision-making, firstly analysing the customer preferences according to star and sentiment scores, then splitting the negative and positive sections of the review, and finally discovering what are the main features of the products that provoke positive and negative feelings in the clients.

For this, we use a big data of online reviews on Amazon, one of the most important online market places to buy products. These reviews are written by the buyers for the products purchased and these are used by new consumers as a source of electronic word-of-mouth to make decisions of their own purchases. In this sense, brand image is derived not only by signals sent by firms, but also by online reviews written by consumers.

This research is not focused on proposing a new method of sentiment analysis or a new technique to discover product features; rather, it uses existing tools together for decision making.

Some previous studies qualify products in different ways and discover features considered by the consumers as positive or negative for use in decision-making. However, we do not find any research that gives a rating based on the qualification of the features and combines them with other scores that serve to classify the products, including the price and the global sentimental score. Our work develops a ranking for the product that classifies its features along with other indicators, such as the price and the sentimental score of the review.

Consequently, the aim of this paper is twofold:

i. to use the mined product features and the polarity of consumer opinions about each feature to obtain a product score.
ii. to combine star score, sentiment score of the review and sentiment score based on product features to rank each product and to assist marketing managers and consumers in their decision-making process.

This paper is organized as follows, Sect. 1 gives the introduction of the uses of sentiment analysis and the value of consumer's reviews in ecommerce for branding. Section 2 briefs the related work presented in the literature about big data techniques applied to marketing, sentiment analysis and product feature selection methods. In Sect. 3, we show the proposed tools to select and classify important product features. Section 4 details the data collection and tool setup, and the experimentation and results, and Sect. 5 gives the conclusion and future works.

2 Background

This paper deals with automatic Sentiment Analysis and Product Features Selection on product reviews and the benefits of applying Big Data techniques to Marketing, so we summarize their previous works. This section ends with an overview of the findings extracted from this previous work, which justifies our contributions to the state-of-the-art.

Given the current state of information technology, consumers can easily make online purchases and post reviews on social media. This user-generated content (UGC) may be relied upon by potential customers, thereby influencing future purchasing decisions [24]. Everybody can easily access these online reviews [6] in real time. It should be remembered that one of the factors consumers consider in their decision-making processes is word-of-mouth (WOM) [23]. Thus, measures of eWOM [9] or online reviews [5] can be included in marketing-mix models to provide better explanations and predictions of consumer choice and sales. These ratings and comments summarize the individual consumers' evaluations and act as indicators of product quality [14, 22]. Furthermore, and even more important, they act as a cue to help future consumers to determine product or brand attributes [21]. Such a large volume of constantly generated data is increasingly a big data challenge for businesses [19].

To analyse all this textual data on reviews, Sentimental Analysis can be used. Sentiment Analysis in product reviews is the process of exploring these reviews to determine the overall opinion or feeling about a product [10]. This information is unstructured and is not something that is "machine processable" [4]. Cambria also exposes that the challenge is huge because it is necessary an understanding of the explicit and implicit, regular and irregular, and syntactical and semantic language rules. Ratan et al. [18] shows six kinds of issues in Sentiment Analysis: (1) Opposite meaning in particular domains. (2) An interrogative sentence or conditional sentence may not have positive or negative sentiment. (3) The sarcastic sentences may have the opposite sentiment. (4) Sentiment information without using sentiment words. (5) A word can change the feeling polarization in two similar sentences, as well as the fact that for different person, a sentence may have a different sentiment. (6) Natural Language Issues Change Place to Place.

Firstly, sentiment analysis classifies product reviews as positive or negative; polarity classification is the basic task. In recent work, they are focused in applications that are more specialized. One such application is to use opinion mining to determine areas of a product that need to be improved by summarizing product reviews to see what parts of the product are generally considered good or bad by users [18]. The general opinion about a topic is useful, but also it is important to detect sentiment about individual aspects of the topic [25]. Also classifying a people based on your opinions or improving recommender systems using the positive and negative customer feedback.

Cambria et al. [3] classified the main existing approaches in four categories: Keyword spotting, lexical affinity, statistical methods and Concept-based approaches. The keyword spotting approach classifies text by affect categories based on the presence of unambiguous affect words. The lexical affinity approach detects obvious affect words and assigns arbitrary words a probable "affinity" to particular emotions. The Statistical methods include Bayesian inference and support vector machines; it is popular for affect text classification. It uses machine-learning algorithms that are trained with a large corpus of affectively annotated texts and the system learns the affective valence of keywords. The concept-based approaches use Web ontologies or semantic networks to accomplish semantic text analysis. All these techniques need to use a sentiment lexicon.

Many researches are focused in analysing product reviews to get feedback about the product and make decisions. García-Moya et al. [8] propose a new methodology for the retrieval of product features and opinions from a collection of free-text customer reviews about a product or service. Also, Singla [20] classifies the text in positive, negative and includes other sentiment classification. Paknejad [15] studies different machine learning approaches to determine the better options for sentiment classification problem for online reviews using product reviews from Amazon. Abbasi et al. [1] use SVM classifiers for sentiment analysis with several univariate and multivariate methods for feature selection, reaching 85–88% accuracies after using chi-squared for selecting the relevant attributes in the texts. A network-based feature selection method that is feature relation networks (FRN) helped improve the performance of the classifier.

There are several methods that are used in feature selection, where some are syntactic, based on the syntactic position of the word such as adjectives and some are univariate, based on each feature's relation to a specific category, and some are multivariate based on features subsets [10]. Archak et al. [2] use techniques that decompose the reviews into segments that evaluate the individual characteristics of a product (e.g., image quality and battery life for a digital camera). Then, as a major contribution of this paper, they adapt methods from the econometrics literature, specifically the hedonic regression concept.

3 The Proposed Architecture

In this section, we present our proposal, based on the use of Sentiment Analysis tools and product feature detection. Figure 1 illustrates the proposal.

As shown in the figure, we can distinguish 4 stages: (1) reviews NLP preprocessing, (2) Sentiment Analysis, (3) Product Feature Selection, and (4) dashboards.

We have a big collection of product reviews and the relevant information of each product, for example, the price, the brand and the categories in which the product can be classified. Our proposal analyses this data and discovers new information that will help managers and users to make decisions regarding the products.

These reviews usually contain an explicit star score assigned by the reviewer in a specific range, for example, from 1 (bad) to 5 (very good), and a comment in unstructured text. This numerical qualification is global with the product or the experience

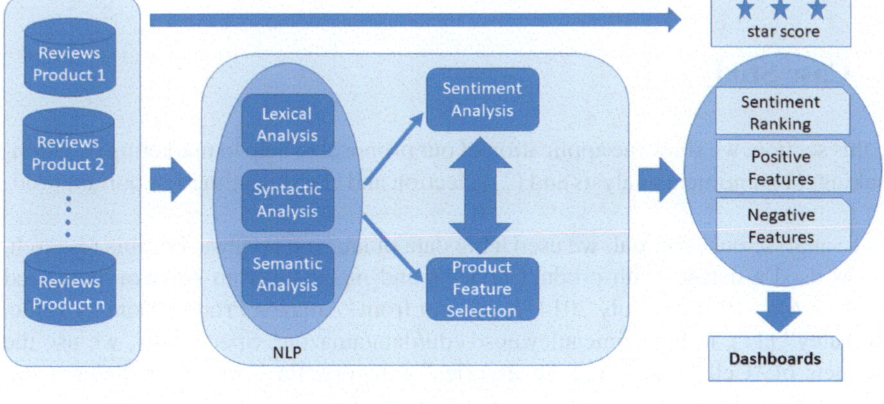

Fig. 1 Schema representing the proposal

of using it, although the user does not like some specific product characteristics. It may happen that the user qualified a product with 4 or 5 stars, but he/she criticized some aspects of it.

The textual comment has positive and negative opinions of different aspects of the product. On these comments, we have applied the following algorithm: (a) Calculate a global sentiment score for each review. (b) Split the review in phrases and calculate a phrase sentiment score. (c) Select the main features in all reviews of each product. (d) Assign a sentiment score for each feature based on the phrase sentiment score where each feature is. (e) Classify the opinions in positive and negative. (f) Show dashboards for decision-making, for example, wordcloud for positive and negative opinions for a product.

The sentiment qualification can be used as an additional criterion to search for the best products within a product category or within a brand of products. Also, the sentiment score of the features can be used to determine which product is the best according to the customer needs.

This proposal allows to use different NLP tools to do the preprocessing and to calculate the sentiment scores.

To identify the product features, we used the methods used by Archak et al. [2] because this research is not aimed at proposing a new method for the selection of features. They use a part-of-speech tagger to annotate each review word with its part-of-speech (POS), identifying whether the word is a noun, an adjective, a verb and so on. Nouns and noun phrases are popular candidates for product features, though other constructs (like verb phrases) can be used as well. Alternative techniques search for statistical patterns in the text, for example, words and phrases that appear frequently in the reviews. We select all nouns that appear in the reviews and then select the frequent nouns in all reviews of the same product.

To evaluate the sentiment polarity of product feature, we evaluate the sentiment score on each phrase that the feature appears. If the phrase sentiment score is positive, a point is added to sentiment of the feature. This is done for each feature. Then the features will be classified on positive scored feature or negative scored feature.

4 Case Study

In this section, we show the application of our proposal to help in marketing decision-making: (1) sentiment analysis and (2) selection and classification of important product features.

To analyse our proposal, we used a big data of amazon reviews. For this research, it was used a dataset with product reviews and metadata from Amazon compiled between May 1996 and July 2014. It was got from "Amazon Product Data by Julian McAuley" [11], in http://jmcauley.ucsd.edu/data/amazon/. Specifically, we use the products of "Cell Phones and Accessories" category. Each review includes a star score in range [1, 5] and a comment given by the user.

The dataset of these product reviews has the follow structure shown with a sample:

{
"*reviewerID*": "A6FGO4TBZ3QFZ",
"*asin*": "3998899561",
"*reviewerName*": "Abdullah Albyati",
"*helpful*": [1, 2],
"*reviewText*": "it worked for the first week then it only charge my phone to 20%. it is a waste of money.",
"*overall*": 1.0,
"*summary*": "not a good Idea",
"*unixReviewTime*": 1384992000,
"*reviewTime*": "11 21, 2013"
}

To get a quantitative sentiment score expressed in the reviews-text, we have used two tools. The first one, we use the approach that was used by other researches [13, 17]. Using the R language, the review-text is divided in individual words (Review_Words). We remove the words that belongs to stop word list. Each word in Review_Words is searched into Afinn Word List. If the word is in the list, add the emotional rating of this word. The sentiment of review is the mean of all these emotional ratings. The expression 1 shows the overall mathematical formula to obtain the sentiment value of a review in the range [−5, 5]. The second one, we use a natural language processing tool named Standford CoreNLP (https://stanfordnlp.github.io/CoreNLP/). Stanford CoreNLP provides a set of human language technology tools, including Sentiment Analysis. With this tool, we obtain a sentence classification of Very Negative = 0, Negative, Neutral, Positive or Very Positive = 5.

Expression 1. Sentiment Value of a Review

$$sentiment(Review) = \frac{\sum_{w \in R} emotional\_rating(w)}{|R|}$$

where $R = Review\_Words \cap Afinn\_Words$

With both tools, we get sentiment values that we use to score to the reviews and the products, and also to deduce which are the main positive and negative features of a product.

To normalize the values of sentiment analysis and star rating by the user, the following mapping is done in expression 2:

Expression 2. Review Sentiment Score Normalization

$$norm(x) = \begin{cases} 5, \; x \geq 3 \\ 4, \; 3 < x \leq 1 \\ 3, \; 1 < x \leq -0.5 \\ 2, \; -0.5 < x \leq -3 \\ 1, \; x < -3 \end{cases}$$

where $x = review\,sentiment\,in[-5, 5]calculated\,in\,expression$ 1

Analyzing a sample review phrase: "*Overall, it's a great headset for the price. Pretty amazing actually. You'll need to get a wired converter to convert the 2.5 mm to 3.5 mm like this one:Headset Adapter 3.5 mm to 2.5 mm Audio Adapter and Converter (Black) - Non-Retail Packaging. No biggie.But the curved rubber piece is really stiff and hurts my ear after about 30 min or less. Sometimes quite painfully. I have a dull voice and this is by far the best headset I have that helps people hear me clearly (I actually have an older version of the same headset, so I have two). No one complains they can't hear me or do they need to turn up their volume. Noise cancellation is also pretty good. Not as good as my cheap jawbone, but then people usually complain they can't hear me as well on that one. So, stay away from this if you plan to talk on the phone a while or have larger ears. Small ears may work just fine. Mine are probably about average for males. Otherwise, worth the price.*".

In this review, we found positive features like headset, price, volume and noise cancellation. Also, negative feature like rubber piece and ear. The sentiment score for the phrase "Noise cancellation is also pretty good" is 4. It is a positive sentiment. The feature "noise cancellation" will be classified as positive feature.

Following the algorithm proposed:

(a) We calculate the global sentiment score for each review. For this review, its sentiment score is 0.94, which is a neutral review. The average for all review sentiment scores for this product is 0.74, which is a neutral score.
(b) Split the review in phrases and calculate a phrase sentiment score. This review is separated in 13 sentences. For example, a positive sentence is *"Overall, it's a great headset for the price."* with a sentiment score of 3.0. Other example but a negative sentence is *"But the curved rubber piece is really stiff and hurts my ear after about 30 min or less."* with a sentiment score of −1.5.
(c) Select the main features in all reviews of each product. All collected features are 63 features in positive and 51 features in negative. These lists can be filtered selecting the more "important". For example, using tf-idf or semantic similarity between a feature and the product using WordNet.
(d) Assign a sentiment score for each feature based on the phrase sentiment score where each feature is.
(e) Classify the opinions in positive and negative and calculate the feature score and then calculate a semantic score of products based on the feature scores.
(f) Show dashboards for decision-making.

After recollecting the positive and negative features, we propose to show in the dashboard of a product a word cloud to represent the main liked and disliked features. For example, see Fig. 2.

Fig. 2 Positive and negative wordcloud in dashboard

5 Conclusions

In this paper, we proposed a step further in Sentiment Analysis application in Marketing Decision-Making, adding product feature selection and calculating additional information mined from online reviews. With these additional elements, we help to managers and users to take better decisions to analyse their products and purchases.

Our proposal uses NLP technology to analyse the reviews, to get sentiment values and to extract the main positive and negative features of the reviewed products. We have reviewed some algorithms to detect features of products and they were adapted to improve the information that is shown in the dashboards, including word-clouds to visualize the most important features mentioned by the consumers.

We put into practice the proposal on a corpus of reviews of cell-phone and accessories products extracted from Amazon. We have analysed that the star scores represent a general user's viewpoint about the product, but into a review the user can highlight specific features that they liked or not.

We propose for future works to use other types of corpus to validate our proposal. Furthermore, we must consider n-gram features and the positive as well as negative qualifiers of the features. Additionally, it is crucial to work on reducing the list of features considering only the relevant ones to the specific product discarding general features which are not related to the product.

Acknowledgements This work was supported in part by the Spanish Ministry of Science, Innovation and Universities through the Project ECLIPSE-UA under Grant RTI2018-094283-B-C32, the Project RESCATA under Grant TIN2015-65100-R, the Project PROMETEO/2018/089, and the Lucentia AGI Grant.

References

1. Abbasi, A., France, S., Zhang, Z., Chen, H.: Selecting attributes for sentiment classification using feature relation networks. IEEE Trans. Knowl. Data Eng. **23**(3), 447 462 (2011)
2. Archak, N., Ghose, A., Ipeirotis, P.: Show me the Money! Deriving the pricing power of product features by mining consumer reviews. ACM (2007). 978-1-59593-609-7/07/0008
3. Cambria, E., Das, D., Bandyopadhyay, S., Feraco, A.: Affective computing and sentiment analysis. IEEE Intell. Syst. **31**(2), 102–107 (2016 Mar–Apr)
4. Cambria, E., Schuller, B., Xia, Y., Havasi, C.: New avenues in opinion mining and sentiment analysis. Intell. Syst. IEEE. **28**, 15–21 (2013). https://doi.org/10.1109/MIS.2013.30
5. Chevalier, J., Mayzlin, D.: The effect of word of mouth on sales: online book reviews. J. Mark. Res. **43**(August), 345–354 (2006)
6. Dellarocas, C.: The digitization of word of mouth: promise and challenges of online feedback mechanisms. Manage. Sci. **49**, 1407–1424 (2003)
7. Feldman, R.: Techniques and applications for sentiment analysis. Commun. ACM **56**(4), 82–89 (2013)
8. García-Moya, L., Anaya-Sánchez, H., Berlanga-Llavori, R.: Retrieving product features and opinions from customer reviews. Intell. Syst. IEEE **28**, 19–27 (2013). https://doi.org/10.1109/MIS.2013.37
9. Godes, David, Mayzlin, Dina: Using On-line conversations to study word-of-mouth communication. Mark. Sci. **23**(4), 545–560 (2004)
10. Haddi, E., Liu, X., & Shi, Y.: The role of text pre-processing in sentiment analysis. ITQM (2013)
11. He, R., McAuley, J.: Ups and downs: modeling the visual evolution of fashion trends with one-class collaborative filtering (2016). https://doi.org/10.1145/2872427.2883037
12. Joshi, M., Prajapati, P., Shaikh, A., Vala, V.: A survey on Sentiment Analysis. Int. J. Comput. Appl. **163**(6), 34–39 (2017). https://doi.org/10.5120/ijca2017913552
13. Liske, D.: Tidy sentiment analysis in R (2018). https://www.datacamp.com/community/tutorials/sentiment-analysis-R
14. Noone, B.M., McGuire, K.A.: Effects of price and user-generated content on consumers' prepurchase evaluations of variably priced services. J. Hosp. Tour. Res. **38**(4), 562–581 (2014)
15. Paknejad, S.: Sentiment classification on Amazon reviews using machine learning approaches. Dissertation (2018)
16. Pang, B., Lee, L.: Opinion Mining and Sentiment Analysis. Found. Trends Inf. Retr. **2**(1–2), 1–135 (2008)
17. Paracchini, P.: "Sentiment analysis using the tidytext package (2016). https://rpubs.com/pparacch/236096
18. Ravendra, R.S.J.: A proposed novel approach for sentiment analysis and opinion mining. Int. J. UbiComp. **5**, 1–10 (2014)
19. Singh, J., Irani, S., Rana, N., Dwivedi, Y., Saumya, S., Roy, P.: Predicting the helpfulness of online consumer reviews. J. Bus. Res. **70**, 345–355 (2017). https://doi.org/10.1016/j.jbusres.2016.08.008
20. Singla, Z., Randhawa, S., Jain, S.: Sentiment analysis of customer product reviews using machine learning. In: 2017 International Conference on Intelligent Computing and Control (I2C2), 1–5 (2017)
21. Sun, T., Youn, S., Wu, G., Kuntaraporn, M.: Online word-of-mouth: an exploration of its antecedents and consequences. J. Comput. Mediat. Commun. **11**(4), 1104–1127 (2006)
22. Tsang, A.S.L., Prendergast, G.: Is a star worth a thousand words? The interplay between product-review texts and rating valences. Eur. J. Mark. **43**(11/12), 1269–1280 (2009)
23. Wang, J., Wang, L., Wang, M.: Understanding the effects of eWOM social ties on purchase intentions: a moderated mediation investigation. Electron. Commer. Res. Appl. **28**, 54–62 (2018). https://doi.org/10.1016/j.elerap.2018.01.011

24. Xu, X., Wang, X., Li, Y., Haghighi, M.: Business intelligence in online customer textual reviews: understanding consumer perceptions and influential factors. Int. J. Inf. Manage. **37**(6), 673–683 (2017)
25. Yi, J., Nasukawa, T., Bunescu, R., Niblack, W.: Sentiment analyzer: extracting sentiments about a given topic using natural language processing techniques. In: Proceedings of the 3rd IEEE International Conference on Data Mining ICDM-2003, pp. 427–434 (2003)

Knowledge Integration in Personalised Dietary Suggestion System Using Semantic Web Technologies

Antonella Carbonaro and Roberto Reda

Abstract In knowledge intensive nutrition-related contexts, such as personalised dietary, diet-sensitive diseases management and sport supplementation, ontologies play an important role. In this paper, we propose an ontology-based consultation system which aims to improve the life quality of both healthy people and individuals affected by chronic diet-related diseases. We developed a system which is capable of transferring human dietary and nutrition expertise into machine understandable knowledge through a set of semantic rules in order to better assist users in making the correct nutritional choices for their particular health status, age, lifestyle and food preferences. Our system makes use of open data, published ontologies, domain knowledge and IoT data to construct a domain representation consisting of unified concepts and instances suitable for reasoning processes. We described how several knowledge bases in knowledge-intensive contexts can be integrated to provide a unified structured and precise representation of heterogeneous information to provide better diet recommendation to individuals.

1 Introduction

Nowadays, there is a growing demand for wellbeing services to help people improve their eating behaviour and make informed choices about the food they consume. Indeed, improper nutrition can contribute to diseases development such as diabetes, hypertension, some types of allergies and even cancer [14].

The intrinsic potential of the available online nutrition data and IoT physical activity datasets can be exploited using sophisticated data analysis techniques such as

A. Carbonaro (✉) · R. Reda
Department of Computer Science and Engineering, University of Bologna, Bologna, Italy
e-mail: antonella.carbonaro@unibo.it

R. Reda
e-mail: roberto.reda@unibo.it

A. Visvizi and M. D. Lytras (eds.), *Research & Innovation Forum 2019*,
Springer Proceedings in Complexity,
https://doi.org/10.1007/978-3-030-30809-4_21

automatic reasoning to find patterns and extract information and knowledge in order to enhance decision-making and deliver better healthcare to the population. However, due to the high heterogeneity of data representation and serialisation formats, and a lack of common accepted standards, the wellbeing landscape is characterised by an ubiquitous presence of data silos which prevents domain experts from obtaining a consistent representation of the whole knowledge [17].

Semantic Web (SW) technologies are a promising solution for the integration and exploitation of data about food, health, nutrition and physical activity. SW technologies describe a new way to make resources content more meaningful to machines, whereas the meaning of data is provided by the use of ontologies. In the context of food and other relevant related contexts (e.g., nutrition, dietetics, individual diet-sensitive disease conditions, etc.), such a data model could be employed to analyse information from multiple data sources, like generic or domain specific dataset, and unify them in an interlinked data processing area. This scenario could promote interoperable communication among various information technology systems and can be used for automatic reasoning processes. Indeed, on the basis of the asserted knowledge, it is possible to automatically derive new knowledge about the current context and detect possible inconsistencies among the asserted information.

In this paper, we propose Food Data Manager (FDM), a dietary consultation system which aims to improve the life quality of both healthy people and individuals affected by chronic diet-related diseases. Food Data Manager is an ontology-based system whereby we can automatically reason about the food products and their characteristics (quantities, ingredients and composition) by inference engines in order to better assist users in making the correct diet choices for their particular health condition such as age, lifestyle and culinary preferences.Our system makes use of open data, published ontologies, domain knowledge and sensors data to achieve its goal. In addition, we developed semantic rules to transfer human expertise into machine understandable knowledge using the Semantic Web Rule Language (SWRL) [13] commonly used for building inference mechanisms in OWL-based knowledge systems.

The novelty of the proposed approach lies in using SW technologies to explicitly model all relevant information from user-generated data to domain knowledge and for implementing reasoning activities. Moreover, the integrated ontologies support the creation of the interactive semantic queries which allow us to bring the result sets closer to user's needs.

The remainder of this paper is organised as follows. Section 2 highlights the main open-issues in the field and introduces the previous works. Section 3 describes the technological aspects of the Semantic Web and how it fits into the context of dietary and nutrition. Section 4 shows the overall architecture of the proposed system. Section 5 describes the Food Data Manager user interface in details and provides several insights about the working principles of the system. Finally, some considerations close the paper.

2 Related Works

In recent years, there has been a great deal of interest in the development of semantic-based systems to facilitate data integration and knowledge representation of heterogeneous data [1–8, 18].

Wellbeing domain is characterised by the presence of a huge amount of information resources, and the knowledge formation process is often associated with multiple data sources. In this regard, interoperability and heterogeneous data integration are two vital issues still unsettled.

In [12] Helmy et al. focus on on the integration of different domain ontologies, like health, food, nutrition, and the user's profile ontology in order to help personalized information systems to retrieve food and health recommendations based on the users health conditions and food preferences.

FoodWiki [10] is an ontology-driven mobile safe food consumption system which performs its own inferencing rules within its own knowledge base. The system is equipped with an advanced semantic knowledge base and it can provide recommendations of appropriate foods before consumption by individuals.

The work presented in [9] aims to integrate multiple knowledge sources for problem solving modelling and knowledge-based system (KBS) development. A dietary consultation system for chronic kidney disease is constructed by using OWL and SWRL to demonstrate how a KBS approach can achieve sound problem solving modelling and effective knowledge inference.

SHADE (Semantic Healthcare Assistant for Diet and Exercise) [11] is a diet and exercise program recommender system for diabetic patients. The suggestion mechanism makes use of concepts and information from various domains knowledge base defined in separate OWL based ontologies.

From the aforementioned situation, it is possible to underline the main issues which affects nowadays health diet recommendation systems: the semantic interoperability of e-health systems and their data is crucial but still a problem poorly solved when it comes to representing disjointed contexts.

3 Knowledge-Intensive Contexts, Rules and Reasoning

Semantic Web (SW) describes a new way to make web content more meaningful to machines. Ontologies, as a source of formally defined terms, play an important role within knowledge-intensive contexts such as the one described in this article. Ontologies can be reused, shared, and integrated across applications, and aim at capturing domain knowledge in a generic fashion and provide a common agreed understanding of the domain. Rule languages allow writing inference rules in a standard way which can be used for automatic reasoning. SW technologies are a promising way for the integration and exploitation of food, nutrition, activity and personal data. In this field, ontologies enable the formal representation of the entities and their relationships and the associated background knowledge. While logic provides the theoretical underpinning required for reasoning and deduction. Reasoning is the process of extracting new knowledge from an ontology and its instance base and represents one of the most powerful features of SW technologies especially for dynamic and heterogeneous environments. A semantic reasoner is a software system whose primary goal is to infer knowledge which is implicitly stated by reasoning upon the knowledge explicitly stated, according to the rules that have been defined.

Rule languages designed for the SW are typically introduced by W3C, such as the Semantic Web Rule Language (SWRL) [13, 15]. SWRL language can enhance the ontology language by allowing us to describe relations that cannot be described using Description Logic (DL).

Linked Open Data (LOD) propose recommended best practices for exposing, sharing, connecting pieces of data, information and knowledge on the SW, using the Uniform Resource Identifiers (URI) mechanism to identify resources, the Resource Description Framework (RDF) and OWL. Therefore, the use of LOD as the data representation formalism enables the creation of a common model thus interconnecting a variety of heterogeneous data sources. In the context of food and wellbeing domain, such a data model could be employed to analyse information from unstructured data sources (e.g., data collected by IoT devices) along with generic or domain specific information, and unify them in an interlinked data processing area.

4 FDM System Architecture

The two main components of the proposed system architecture are the *Input Layer* and the *Knowledge Layer*.

The Input Layer is responsible for receiving data from users or IoT sensors. This layer includes the possibility of getting input data both from users using computer based solutions (e.g., web applications) and through automatic data transfer by connecting the platform directly to wearable devices. We wanted to reduce as much as possible the time consuming activities on the user side, for example, the input of all consumed foods during a week. We reduced the time required for the user to enter the data because the system partially retrieves the basic components from domain ontologies. Each user can also customise the system by inserting his own dishes and recipes in order to be able to use them whenever necessary, simply by choosing them through the graphical interface.

With the Knowledge Layer we wanted to define new methods for representing knowledge and for reasoning on it. For example, the percentage of cholesterol in a given food is a fact described in the food ontology. Inferring the fact that the percentage of the cholesterol is damaging for a specific patient, or is associated with another food, or is inserted in a rich menu or the consumption is protracted for days and is associated with lack of physical movement is possible by setting a series of rules in the system. The Knowledge Layer must therefore include both the management of the domain ontologies and the choice of rules to execute.

A recommendation system for food, nutrition and health habits also needs to take into consideration and combine information from different external domains. The *Personal Profile Ontology*, included in our project, models basic input parameters (i.e., gender, age, weight, height, physical lifestyle) and derivable information from input parameters (e.g., BMI, ideal weight, number of calories and nutrients for each meal type). While, the *Food Ontology* describes classes which represent specific concepts from the food product domain such as food Categories, recipes, nutrients and ingredients, and properties such as energy per 100 g or carbohydrates per 100 g.

The *Activity Ontology*, which is a re-adaptation of the IFO ontology [16, 17], is built around the notion of Episode. An episode represents the set of the all possible events that can be measured by an IoT fitness device or a wellness appliance. For example, an instance of the class Episode could be the heart rate measured during a running training session by a wearable wrist worn heart rate monitor or the person's body weight measured by a smart scale. To each episode is associated a time reference and a numeric measurement value with the related unit of measurement.

5 FDM User Interface

The proposed system can be used directly by single users interested in establishing healthy eating habits or by the nutritionists who wants to set a diet for their patient. Moreover, it can also be used for consulting the stored data about nutrition and food. The user interface can be used in two different modalities. The first one is used by the nutritionist, it allows the expert to enter details about foods and their nutrients, diseases related to eating disorders, and to create the patient's personalised profile. While, the second one allow users, to add preferred foods and recipes to the knowledge base, and enter the carried out activities and food eaten, like in a diary, and view the suggested foods according to the specific condition. The system is also able to represent the contribution of the nutrients present in a single food or in an entire recipe, offering wider useful features (Figs. 1 and 2).

Furthermore, the system interacts with external LODs portal, such as DBPedia, to get information related to diseases that may be significant for the definition of the diet, for instance, in presence of endocrinology and gastroenterology problems (Fig. 3).

Given the complexity and the heterogeneous nature of the components integrated within FDM, the validation of its effectiveness should be performed both from domain experts and user perspectives. Some interviews were conducted to evaluate the system usability and also the intuitiveness and simplicity have been taken into consideration. In general, users appreciated the system and considered FDM a useful tool, especially for increasing the awareness of their eating habits. However, domain experts should provide the correctness and appropriateness of the specific content representation. Nutritionist should validate the consultations and if any enhancements are needed, rules could be modified with coordination of domain and system experts.

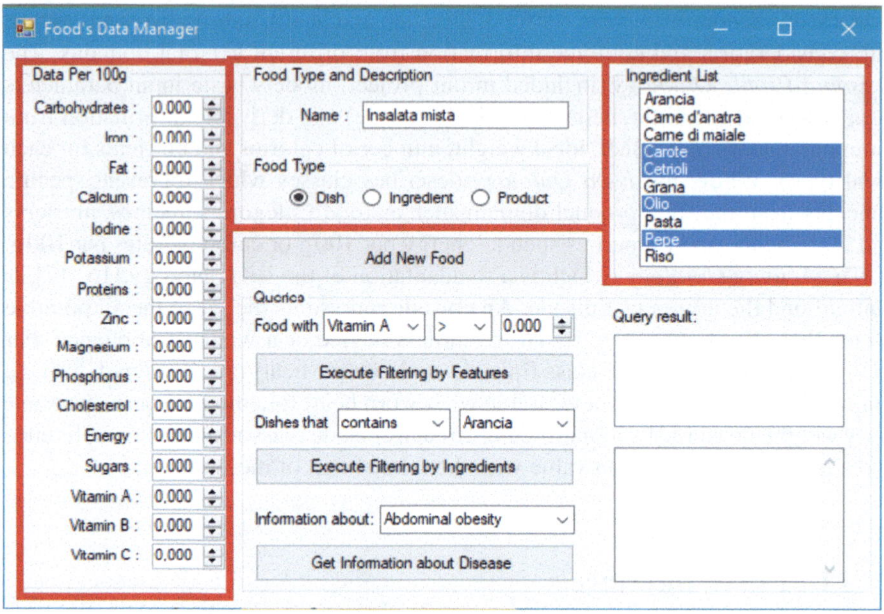

Fig. 1 An example of how to insert a dish into the system by choosing its ingredients and products automatically extracted using SPARQL queries

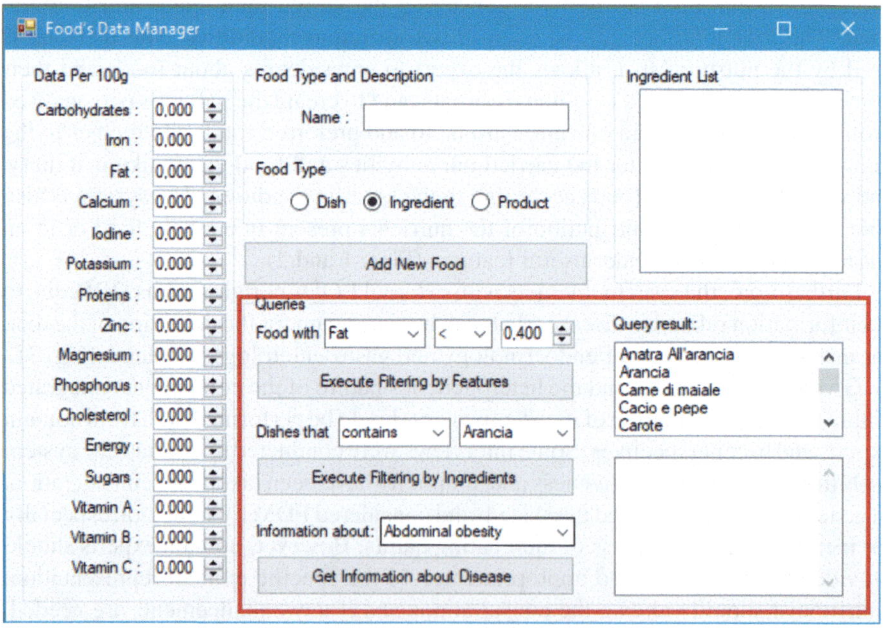

Fig. 2 The results of the search for foods that have a specific content, in this case lower fats

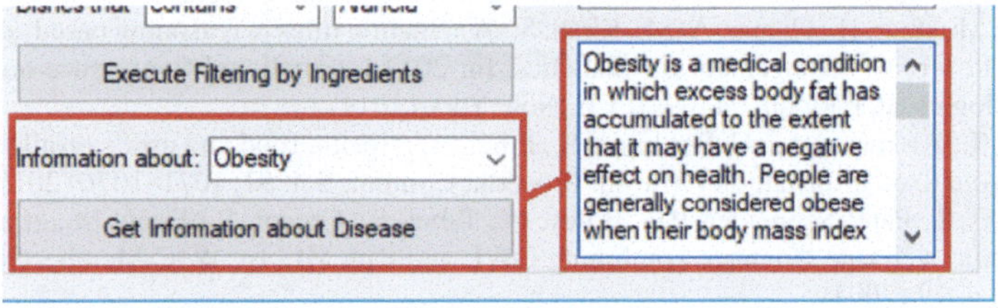

Fig. 3 The results of the search for foods that have a specific content, in this case lower fats

6 Conclusion

The availability of data on the health status of users and those that can be obtained by monitoring his physical activity allow us to develop increasingly sophisticated tools for the personalised suggestion of healthy eating habits in the context of nutrition. This paper presents Food Data Manager, a personalised dietary consultation system, focusing on design and implementation aspects. The system receives input data from both users and IoT fitness devices and is able to support users in their nutritional choices. Food Data Manager models the contribution of individual nutrients present in food or in a recipe, offering a wide range of useful features.

References

1. Carbonaro, A.: Collaborative and semantic information retrieval for technology-enhanced learning. In: Proceedings of the 3rd International Workshop on Social Information Retrieval for Technology-Enhanced Learning (SIRTEL 2009), Aachen, Germany (2009)
2. Carbonaro, A.: Improving web search and navigation using summarization process. In: World Summit on Knowledge Society, pp. 131–138. Springer, Berlin (2010)
3. Carbonaro, A.: Wordnet-based summarization to enhance learning interaction tutoring. J. e-Learn. Knowl. Soc. **6**(2), 67–74 (2010)
4. Carbonaro, A.: Interlinking e-learning resources and the web of data for improving student experience. J. e-Learn. Knowl. Soc. **8**(2), 33–44 (2012)
5. Carbonaro, A., Ferrini, R.: Ontology-based video annotation in multimedia entertainment. In: Consumer Communications and Networking Conference, 2007. CCNC 2007. 4th IEEE, pp. 1087–1091. IEEE, New York (2007)
6. Carbonaro, A., Ferrini, R.: Personalized information retrieval in a semantic-based learning environment. In: Social Information Retrieval Systems: Emerging Technologies and Applications for Searching the Web Effectively, pp. 270–288. IGI Global (2008)
7. Carbonaro, A., Ravaioli, M.: Peer assessment to promote deep learning and to reduce a gender gap in the traditional introductory programming course. J. E-Learn. Knowl. Soc. **13**(3) (2017)
8. Carbonaro, A., Reda, R.: A dietary consultation system using semantic rules and reasoning based approach. In: EAI International Conference on Smart Objects and Technologies for Social Good, pp. 314–315 (2018). https://doi.org/10.1145/3284869.3284929, ACM, New York
9. Chi, Y.-L., Chen, T.-Y., Tsai, W.-T.: A chronic disease dietary consultation system using owl-based ontologies and semantic rules. J. Biomed. Inf. **53**, 208–219 (2015)
10. Çelik, E.D.: Foodwiki: a mobile app examines side effects of food additives via semantic web. J. Med. Syst. **40**(2), 41 (2016)

11. Faiz, I., Mukhtar, H., Qamar, A. M., Khan, S.: A semantic rules & reasoning based approach for diet and exercise management for diabetics. In: 2014 International Conference on Emerging Technologies (ICET), pp. 94–99. IEEE, New York (2014)
12. Helmy, T., Al-Nazer, A., Al-Bukhitan, S., Iqbal, A.: Health, food and user's profile ontologies for personalized information retrieval. Procedia Comput. Sci. **52**, 1071–1076 (2015)
13. Horrocks, I., Patel-Schneider, P.F., Boley, H., Tabet, S., Grosof, B., Dean, M., et al.: SWRL: a semantic web rule language combining OWL and RuleML. In: W3C Member Submission, vol. 21, p. 79 (2004)
14. Kushi, L.H., Byers, T., Doyle, C., Bandera, E.V., McCullough, M., Gansler, T., Andrews, K.S., Thun, M.J.: American Cancer Society guidelines on nutrition and physical activity for cancer prevention: reducing the risk of cancer with healthy food choices and physical activity. CA Cancer J Clin. **56**(5):254–281, 2006
15. Rattanasawad, T., Saikaew, K.R., Buranarach, M., Supnithi, T.: A review and comparison of rule languages and rule-based inference engines for the semantic web. In: Computer Science and Engineering Conference (ICSEC), 2013 International, pp. 1–6. IEEE, New York (2013)
16. Reda, R., Piccinini, F., Carbonaro, A.: Semantic modelling of smart healthcare data. In: Proceedings of SAI Intelligent Systems Conference, pp. 399–411. Springer, Berlin (2018)
17. Reda, R., Piccinini, F., Carbonaro, A.: Towards consistent data representation in the iot healthcare landscape. In: Proceedings of the 2018 International Conference on Digital Health, pp. 5–10. ACM, New York (2018)
18. Riccucci, S., Carbonaro, A., Casadei, G.: Knowledge acquisition in intelligent tutoring system: a data mining approach. In: Mexican International Conference on Artificial Intelligence, pp. 1195–1205. Springer, Berlin (2007)

Using Physical Activity Monitors in Smart Environments and Social Networks: Applications and Challenges

Jose-Luis Sanchez-Romero, Antonio Jimeno-Morenilla, Higinio Mora and Francisco Pujol-Lopez

Abstract The use of smart watches and fitness wrists has been increasing in recent years. On the one hand, their cost has become cheaper and their performance has improved. On the other hand, the increase in the number of people who practice sports such as running and cycling is another factor to consider. The increase in the number of popular athletes has meant that these devices are no longer considered intended for a minority or an elite, but are even used in people's daily movements or in simpler activities such as walking. Some of these devices simply count the number of steps taken, while more advanced devices include energy consumption, distance travelled, speed, GPS tracking position, altimetry or heart rate. Moreover, there are social networks that allow athletes to share information gathered by their own activities, especially track and altimetry. This opens up a wider range of possibilities in sports training. However, the gathering of this rich information makes many more applications possible. For example, city planners can analyze the movements of people to detect possible shortcomings in public transport systems, deficiencies in urban pathways, and so on. This research work shows a taxonomy of different applications that use the data gathered by physical activity monitors and shared in social networks. The opportunities and drawbacks regarding the use of such applications in intelligent environments are also discussed.

J.-L. Sanchez-Romero (✉) · A. Jimeno-Morenilla · H. Mora · F. Pujol-Lopez
University of Alicante, 03690 San Vicente del Raspeig, Spain
e-mail: sanchez@dtic.ua.es

A. Jimeno-Morenilla
e-mail: jimeno@dtic.ua.es

H. Mora
e-mail: hmora@dtic.ua.es

F. Pujol-Lopez
e-mail: fpujol@dtic.ua.es

A. Visvizi and M. D. Lytras (eds.), *Research & Innovation Forum 2019*,
Springer Proceedings in Complexity,
https://doi.org/10.1007/978-3-030-30809-4_22

1 Introduction

People's concern for their health has evolved greatly in recent years, and has now become a fundamental element that should be part of each individual's daily or weekly routine. This concern for health must be seen as a series of long-term care and habits that must extend throughout an individual's life. It is clear that these habits must be adapted to the age and the specific conditions in which the individual finds himself. Physical exercise is fundamental within this set of habits, and is considered a fundamental pillar that surpasses the advantages of any pharmacological therapy. Experts argue that physical activity should be inspired by three pillars: impact aerobics, non-impact aerobics and strength training. This has led to the recent predominance of aerobic sports such as cycling, running or swimming, which combined with the other two factors give rise to sporting disciplines such as triathlon.

Nowadays, the use of smartwatches, activity wrists and other wearables is common, widespread among athletes and people concerned about their health. Technology has evolved in recent years and costs have been reduced, which has made it possible for many athletes, even those who only practice walking, to use these devices to objectify in some way the results of their training. The most basic devices simply measure aspects such as the number of steps, but the most common is to have models that provide much more appropriate parameters for proper monitoring of physical activity, such as distance traveled based on a GPS positioning system, altimetry or changes in route altitude, heart rate, maximum oxygen volume (VO2 max), energy expenditure, and others. There are even smartwatches that incorporate a count of the calories ingested by the individual based on the food consumed.

In this paper, some issues about monitoring physical activity by means of wearables are discussed. Section 2 deals with the concept of physical activity and the use of wearables for its measurement, as well as the incorporation of the resulting data into social platforms. In Sect. 3, an approach is proposed for the application of physical activity data at the level of smart cities. Section 4 summarizes the conclusions of the study carried out.

2 Health and Physical Activity Monitoring

Physical activity is the practice of healthy exercise. Some more precise definition on physical activity/inactivity can be found in [1]. There is wide evidence that the practice of PA helps improve people's social, psychological and physical well-being [1, 2].

In [3], the effects of physical activity on the heart function are studied. The effects of training on the myocardium are the reduction of oxygen consumption (VO2) and improvement of the maximum VO2 due to the decrease of the heart rate and the submaximum systolic blood pressure. As a consequence, a reduction in the energy expenditure of the myocardium and an improvement in work capacity are achieved.

In [4], it was observed that the systematic practice of physical exercise as part of treatment in a group of participants suffering from ischemic cardiopathy had a positive influence on the average values of body weight, blood glucose, cholesterol, and the values of systolic-diastolic blood pressure. A significant decrease in the number of cigarettes per day was also measured.

Some benefits of physical activity can be summarized [5]: lowers blood pressure, LDL cholesterol and triglycerides; increases HDL (*good*) cholesterol; lowers blood sugar; improves metabolic control; decreases fat mass and increases muscle mass; increases aerobic or functional capacity; decreases anxiety and stress; improves endothelial function; decreases myocardial oxygen consumption; has an antithrombotic effect; increases the vagal tone; decreases catecholamine release; has an antiarrhythmic effect; helps combat smoking; modifies sedentarism so as to adopt a more active life style; promotes early reintegration into the labour market after a serious illness; improves the socio-familiar and sexual relationships.

It is obvious that health is an essential element associated with the wellbeing of an individuals' community. Physical activity is a key element in maintaining the health of the inhabitants of a city, region or country. In recent years, activities such as cycling, running, and walking have experienced an important rise among common people. These are aerobic activities that promote health care as well as social relationships, which, in turn, results in greater psychological wellbeing.

Nowadays, it is assumed that wellness must be the concern of the government in a local, national, or even supranational scope, as in the case of the European Union programs [6]. Indeed, governors must promote the adequate actions and programmes so as to facilitate mobility of citizens. These actions should include: combining road traffic with the routes and training of pedestrians and athletes, create new routes and even modify existing ones so as to favour and encourage pedestrians and cyclers; with regard to safety, provide adequate means and city planning so as to avoid traffic accidents and decrease pollution indices (promoting the use the electric/hybrid car and public transport).

The term wearable (or *wearable device*) is applied to any electronic device portable by the individual integrated in his/her clothing or worn, with specific characteristics such as the inclusion of microprocessors or microcontrollers and with different connectivity and sensing facilities.

Wearables are incorporated into the user's personal space with which constant interaction takes place [7, 8]. The main idea is that the user can always carry the device with him/her, even when he/she is in motion or performing his/her daily activities [9], allowing the user to have his/her hands free. It is estimated that, by 2020, wearables sales will reach 237 million devices [10].

The increase in the number of popular athletes and the decrease in the costs of wearable devices has meant that these devices are no longer considered to be intended for elite athletes. Initially, the smartphone was the wearable device that popular athletes could afford. However, it has several limitations, such as size or weight.

With regard to that topic, devices such as smartwatches and fitness wrists are considered now the most common wearables [10, 11]. In fact, the characteristics of these devices, especially in terms of size and weight, ease of use and direct contact with the athlete, make them very suitable for sports practice [12, 13].

In [14], it is pointed out that wearables shouldn't be categorized by their technical features, but by their purpose. Some issues related to this topic are commented in the mentioned paper, among which it is worthwhile remarking:

- Biotech fusion: technologies evolve fast, promoting a closer relationship between wearable devices and the human body.
- Synchronized lifestyle: ability to synchronize with a wider ecosystem of connected technologies. Facilitating the connection between people and technology.
- Human enhancement: assistive technologies capable of restoring and enhancing senses and skills. This means focusing these technologies on improving people's safety and life quality.
- Health empowerment: empowering people to play a more active role in managing their personal well-being and health.

The performance and capabilities of wearables have been increasing while their cost has been decreasing [15]. Indeed, they are capable of measuring tracking (by GPS), heart rate (pectoral band or on the wrist), VO2 max, changes in altimetry, and so on [16]. Connectivity has also been improved, with WiFi, Bluetooth, NFC and even 4G communication capability. This allows, in some, cases the direct connection with a smartphone which, in turn, creates a global system connected to a sports monitoring application. Usually, athletes prefer to connect the wearable to the computer once they have finished the activity, in order to download the data collected in a secure way and incorporate them into a sports activity management platform.

In today's connected world, it is obvious that the information on the physical activity of individuals not only remains in the personal sphere of the individual [17], but also transcends personal limits and is incorporated, at the will of the user, in some specific social networks of athletes [18]. These social networks make it possible to interconnect athletes, sharing activities and forming groups to take part in joint training sessions [19]. There are different social environments or platforms, such as Endomondo and MapMyRun (property of Under Armour sportswear company), Runtastic, RunKeeper (Asics), and others [20]. There exist also some sport monitoring platforms with a more oriented approach to the management of sports teams with a professional, semi-professional, or advanced profile, which includes supervision by a coach/trainer, such as Sportlyzer, Team Snap, Game Time, Sports Engine, and so on. In this case, information is usually private and only shared with other athletes in the team or with the trainer.

Strava has probably become the most widespread platform or social network for sharing physical activities (especially cycling and running), with several tens of millions of cyclists, runners and swimmers sharing their activities [21, 22]. Although the exact number of athletes using Strava is unknown, it is claimed that the platform is adding one million new users every 45 days, with 8 million activities uploaded each day.

Strava has transcended the limits of a mere sports application and has become a valuable ally for city planners, as the large amount of data on hiking, running and

Fig. 1 A Strava colour map showing the tracks covered by cyclists in the city of Elche (Spain) and its surroundings. A more intense color indicates a greater training frequency on the corresponding track

cycling activities allows new itineraries to be planned in cities, as well as improving existing ones or adapting them to the uses of sportspeople [23]. This even means encouraging motorised vehicles to lose preference on these roads over pedestrians and cyclists and therefore, if necessary, to trace new itineraries for motorised vehicles in such a way that the physical activity of walkers, cyclists and runners is not damaged or endangered [24–26]. In this sense, Strava Metro is an application aimed at city planners using cycling and running data from Strava users in the cities and regions supported (see Figs. 1,2, and 3). Strava Metro partners with transport departments and city planning groups to plan, evaluate and improve infrastructure for cyclists and pedestrians [27]. Strava Metro provides data from more than 90 million bicycle rides and 24 million races to help city planners understand how and where cyclists and runners use public routes. Around a hundred transport planning departments worldwide use Strava Metro to improve their bicycle and footpath infrastructure.

3 Physical Activity Monitoring in Smart Cities

The increase in physical activity in a population with insufficient preparation or training results in an elevated risk of suffering injuries and cardiovascular problems, some of which could be serious. An index of physical suffering is theoretically given when the maximum heart rate is exceeded [28]. This maximum value is mainly related to the age of the athlete. On the other hand, the VO2 maximum defines the

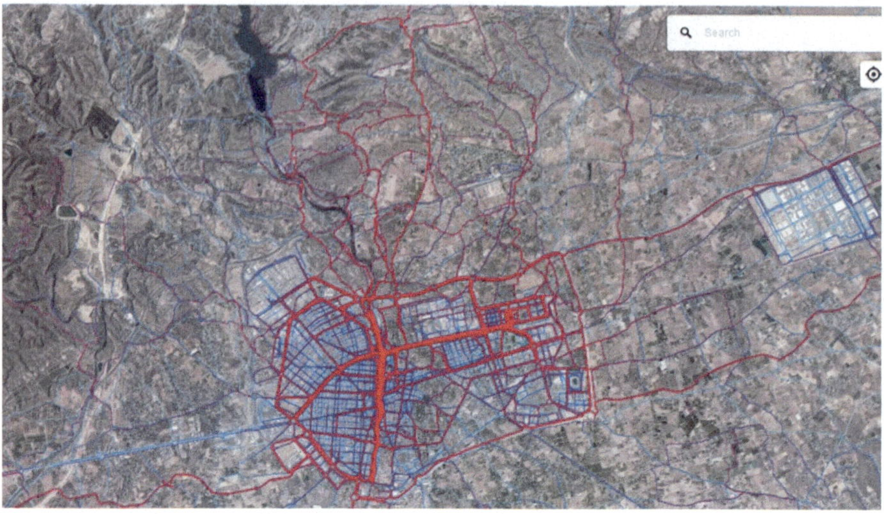

Fig. 2 A Strava colour map showing the tracks covered by runners in the city of Elche and its surroundings

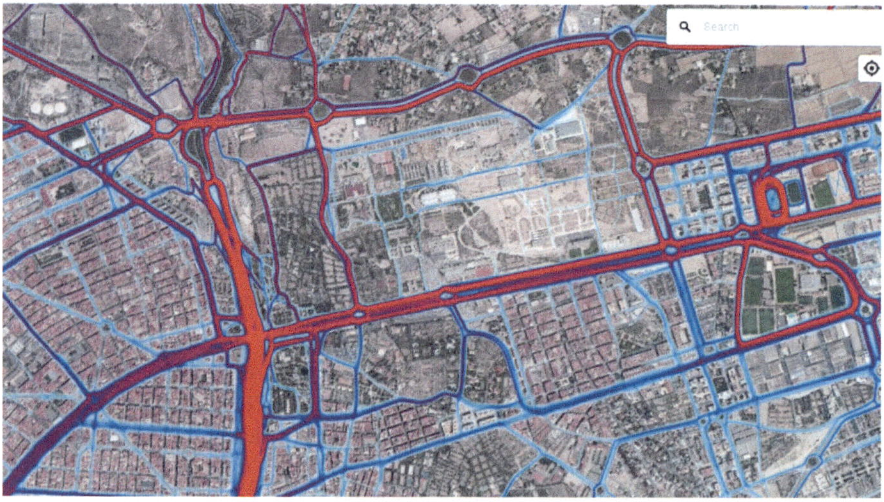

Fig. 3 The Strava colour map of the tracks covered by runners in the city of Elche shown in a greater detail

aerobic capability of the organism and is an index that can usually be improved with the proper sport practice [29].

The original application of smartwatches and wrists was to capture data about the physical activities of users [30, 31]. The approach proposed in this paper is to return to the fundamentals of such devices to reinforce the aspect of health and well-being supported by technology [32], an aspect that must be taken care of by city governors [33].

One first proposal consists in the development of public health indexes based on physical activity. This fact would imply the creation of a local medical team, possibly specialized or with a specific orientation towards sports medicine, that would be helped by tools of analysis of massive data so that an intensive study can be made of the public information on the physical activities carried out by the citizens.

Data about physical activity would be collected by platforms such as the ones mentioned above (Strava or similar, see Fig. 4). In this way, since the need for tags, RFID detectors and other similar systems to track athletes would not be indispensable, the infrastructure that the city should provide would be minimal. Indeed, the athletes themselves register their physical activities with their own wearables and then incorporate them into the platform. Each city would create a local group of athletes within the platform so as the collect the local athletes' activities. The formation

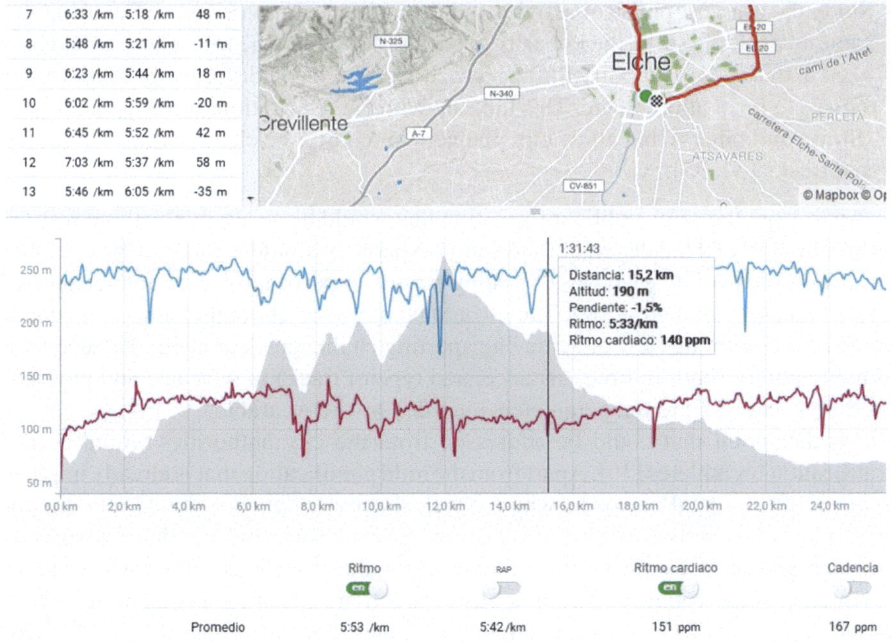

Fig. 4 Data corresponding to a physical activity obtained from Strava. It can observed that changes in heart rate or speed can be analyzed with regard to the slope of the track, as well as mean values of the different indices

of this local group would be promoted in collaboration with the different athletes' teams. These data would be based on the following parameters:

- Number of athletes (percentage of the total city population).
- Kilometres covered weekly (total and Km per athlete).
- Training frequency (days of practice). A low frequency could even be a detriment to health, especially when the training is very intense.
- Heart rates related to the age of athletes, speed, and slope of the tracks. Elevated heart rates could be an index of inadequate physical activities. Moreover, hear rate zones are commonly defined for each athlete (four zones being the most usual) so as to relate them with different training effort [28]. It would be preferable to handle this parameter individually (or by age) rather than as the global mean value.
- VO2 max. As previously mentioned, this is a health index. The higher the index, the greater the aerobic capacity of the athlete. It would be interesting to calculate statistical values (mean, standard deviation) related to this index, as well as its evolution.
- Most frequent training time zone. This could be a work-life balance index. Running very early or very late within the day in winter may be an indication that work-life balance is not fully achieved.
- Analyze the areas of frequent physical activity with regard to the pollution level, that is, industrial areas, traffic-congested ways, and so on.
- Study the areas of physical activity and the amount of motorized traffic using the same areas. If possible, the intensity of such traffic should be low. There would be a need in this sense to modify traffic restrictions, speeds on different sections, traffic signals to favor athletes, etc. This links up with the applications already mentioned (Strava and others) that allow city planners to modify the characteristics of routes and tracks [34, 35].

Apart from the semi-public information that would be accessible without restrictions (obviously by local medical and sports experts), it would be offered the possibility, always on a voluntary basis, of issuing reports or personalized recommendations for certain users who request it. They would be informed about the indices obtained, not from the point of view of achieving sporting challenges, but about the adequacy of their activity (both in time, distance, and type of track) to their age and physical condition according to the parameters recorded by the wearable.

Another point that could be addressed from the city authorities would be the gratification for athletes [36]. Apart from the mild gamification that is already implicit in many of the social networks where physical activity data are shared (comparing one's physical results with that of others in terms of performance, distance, and so on), one approach would be the following: for those athletes who wish to do so voluntarily, they would register in a physical activity incentive programme, which would give points for achieving certain milestones in terms of the physical activity carried out (taking into account the characteristics and age of each individual, as well as their objectives). In this way, this achievement of sport milestones could be rewarded by obtaining a certain award in a virtual local currency (or a customer loyalty card), giving the athlete a number of points that would entitle him/her to enjoy free services, such as public transport or free parking for a certain period, discounts in subsidized shops or other establishments, and so on. Although there exist already some approaches on rewarding physical activity, such as Sweatcoin and others [37], they are not very widespread, since they are only located in a small number of countries, and the way in which the credits obtained by physical activity are exchanged is not very clear.

With an approach that transcends the local level, even with a supranational approach, an interesting proposal would be to use the local physical activity results recorded to create a friendly competition between cities: the most sporting or the healthiest city. In this way, public resources and policies would be dedicated to the improvement of public infrastructures for the promotion of physical activity aimed at improving people's health.

4 Conclusions

In this paper, some issues about the current utilization of wearables have been discussed. Firstly, the definition of wearable has been revised in terms of its orientation towards the health care of users. Subsequently, an application of wearables has been described which, although still related to health care, is finally aimed at helping to develop and improve urban infrastructure using aggregated data from the activities of cyclists and runners. Finally, a proposal has been made for the use of wearables in the scope of smart cities, taking up the initial approach of devices oriented towards health care. In this case, it is proposed to go beyond the individual use of the devices for the mere self-evaluation of each individual. Indeed, it is intended that in each city the data massively collected by the wearables will be used to carry out an exhaustive analysis of the quality and quantity of the physical activities carried out, both in average terms and on an individual basis. In this way, local authorities will have valuable information to carry out policies aimed at improving the welfare of the inhabitants from the point of view of physical activity, such as the improvement of infrastructure, medical control of athletes, or incentives to them.

The approach proposed in this article could be considered as a new Research and Innovation initiative within the topics of e-Health and social networking: the public management, preferably in the scope of smart cities, of information shared through social networks to promote physical activity and monitor the health status of a population. That is to say, the joining of social networking, eHealth policies and smart city management with the aim of improving the well-being of a community with the support of technology.

As a future work, the development of a numerical model should be carried out to evaluate in an integral way the previously commented parameters of physical activity. Since it is assumed that the proposal of a mass recording of local athletes' activities is not easy to achieve in the short term, experimentation with a significant sample of open data from Strava and other sports platforms should be considered as a first step.

References

1. Mercer, K., et al.: Acceptance of commercially available wearable activity trackers among adults aged over 50 and with chronic illness: a mixed-methods evaluation. JMIR mHealth uHealth **4**(1) (2016)
2. Guillen-Garcia, F., Castro-Sanchez, J.J., Guillen-Garcia, M.A.: Calidad de vida, salud y ejercicio fisico: una aproximacion al tema desde una perspectiva psicosocial. Rev. Psicol. deporte **6**(2), 0091–110 (1997)
3. Franco, L., et al.: Effectiveness of a customised, unsupervised 4 month exercise programme on exercise tolerance, perception of fatigue and anthropometric variables in sedentary patients with cardiovascular risk factors. Arch. Med. Dep. **33**(175), 325–330 (2016)
4. Miguel-Soca, P.E.: The metabolic syndrome: a high risk for sedentary persons. Rev. Cub. Inf. Ciencias Salud (ACIMED) **20**(2), 1–8 (2009)
5. Warburton, D.E., Nicol, C.W., Bredin, S.S.: Health benefits of physical activity: the evidence. Can. Med. Assoc. J. **174**(6), 801–809 (2006)
6. Behrendt, F.: Why cycling matters for smart cities. Internet of bicycles for intelligent transport. J. Transp. Geogr. **56**, 157–164 (2016)
7. Rhodes, B.J., Maes, P.: Just-in-time information retrieval agents. IBM Syst. J. **39**(3.4), 685–704 (2000)
8. Lymberis, A.: Smart wearable systems for personalised health management: current R&D and future challenges. In: Proceedings of the 25th Annual International Conference of the IEEE Engineering in Medicine and Biology Society. IEEE, New York, pp. 3716–3719 (2003)
9. Starner, T.: The challenges of wearable computing: Part 1. IEEE Micro **21**(4), 44–52 (2001)
10. Yetisen, A.K., et al. Wearables in medicine. Adv. Mater. 1706910 (2018)
11. Rawassizadeh, R., Price, B.A., Petre, M.: Wearables: has the age of smartwatches finally arrived? Commun. ACM **58**(1), 45–47 (2015)
12. Lupton, D.: Self-tracking, health and medicine. Health Sociol. Rev. **26**(1), 1–5 (2017)
13. Piwek, L., et al.: The use of self-monitoring solutions amongst cyclists: an online survey and empirical study. Transp. Res. Part A Policy Pract. **77**, 126–136 (2015)
14. Tomberg, V., Schulz, T., Kelle, S.: Applying universal design principles to themes for wearables. In: International Conference on Universal Access in Human-Computer Interaction, pp. 550–560. Springer, Berlin (2015)
15. Piwek, L., et al.: The rise of consumer health wearables: promises and barriers. PLoS Med. **13**(2), e1001953 (2016)
16. Kamišalić, A., et al.: Sensors and functionalities of non-invasive wrist-wearable devices: a review. Sensors **18**(6), 1714 (2018)
17. Lupton, D.: Lively Data, social fitness and biovalue: the intersections of health self-tracking and social media (September 27, 2015). In: Burgess, J., Marwick, A., Poell, T. (eds.) The sage handbook of social media. Sage, London (2017)
18. Kranz, M., et al.: The mobile fitness coach: Towards individualized skill assessment using personalized mobile devices. Pervasive Mob. Comput. **9**(2), 203–215 (2013)
19. Lytras, M.D., Visvizi, A., Daniela, L., Sarirete, A., Ordonez De Pablos, P.: Social networks research for sustainable smart education. Sustainability **10**(9), 2974 (2018)
20. Cortes, R., Bonnaire, X., Marin, O., Sens, P.: Sport trackers and big data: Studying user traces to identify opportunities and challenges. Doctoral dissertation, INRIA Paris (2014)
21. Griffin, G.P., Jiao, J.: Where does bicycling for health happen? Analysing volunteered geographic information through place and plexus. J. Transp. Health **2**(2), 238–247 (2015)
22. Sun, Y., et al.: Examining associations of environmental characteristics with recreational cycling behaviour by street-level Strava data. Int. J. Environ. Res. Public Health **14**(6), 644 (2017)
23. Sun, Y.: Exploring potential of crowdsourced geographic information in studies of active travel and health: Strava data and cycling behaviour. In: ISPRS Geospatial Week 2017, Wuhan, China, 18–22 Sept 2017, pp. 1357–1361 (2017)

24. Norman, G., Kesha, N.: Using smartphones for cycle planning. In: IPENZ Transportation Group Conference, pp. 22–24 (2015)
25. Sun, Y., Mobasheri, A.: Utilizing crowdsourced data for studies of cycling and air pollution exposure: a case study using Strava Data. Int. J. Environ. Res. Public Health **14**(3), 274 (2017)
26. Sun, Y., et al.: Investigating impacts of environmental factors on the cycling behavior of bicycle-sharing users. Sustainability **9**(6), 1060 (2017)
27. Musakwa, W., Selala, K.M.: Mapping cycling patterns and trends using Strava Metro data in the city of Johannesburg, South Africa. Data Brief **9**, 898–905 (2016)
28. Lemmer, J.T., et al.: Age and gender responses to strength training and detraining. Med. Sci. Sports Exerc. **32**(8), 1505–1512 (2000)
29. Balderrama, C., et al.: Evaluation of three methodologies to estimate the VO2max in people of different ages. Applied ergonomics **42**(1), 162–168 (2010)
30. Munos, B., et al.: Mobile health: the power of wearables, sensors, and apps to transform clinical trials. Ann. N. Y. Acad. Sci. **1375**(1), 3–18 (2016)
31. Amft, O.: How wearable computing is shaping digital health. IEEE Pervasive Comput. **1**, 92–98 (2018)
32. Cintia, P., Pappalardo, L., Pedreschi, D.: Engine matters: a first large scale data driven study on cyclists' performance. In: 2013 IEEE 13th International Conference on Data Mining Workshops (ICDMW), IEEE, pp. 147–153 (2013)
33. Lytras, M.D., Visvizi, A.: Who uses smart city services and what to make of it: toward interdisciplinary smart cities research. Sustainability **10**(6), 1998 (2018)
34. Mora, H., Perez-DelHoyo, R., Paredes-Perez, J.F., Molla-Sirvent, R.A.: Analysis of social networking service data for smart urban planning. Sustainability **10**, 4732 (2018)
35. Stamatiadis, N., Pappalardo, G., Cafiso, S.: Use of technology to improve bicycle mobility in smart cities. In: 2017 5th IEEE International Conference on Models and Technologies for Intelligent Transportation Systems (MT-ITS). IEEE, Bew York, pp. 86–91 (2017)
36. Spil, T., et al.: The adoption of wearables for a healthy lifestyle: can gamification help? In: Proceedings of the 50th Hawaii International Conference on Systems Sciences, pp. 3617–3625 (2017)
37. Derlyatka, A., et al.: Bright spots, physical activity investments that work: Sweatcoin: a steps generated virtual currency for sustained physical activity behaviour change. Br. J. Sports Med., p. bjsports-2018-099739 (2019)

Smart Cities and Smart Villages

Emotion Recognition to Improve e-Healthcare Systems in Smart Cities

Francisco A. Pujol, Higinio Mora and Ana Martínez

Abstract The ability to detect and control patients' pain is a fundamental objective within any medical service. Nowadays, the evaluation of pain in patients depends mainly on the continuous monitoring of the medical staff and, where applicable, on people from the immediate environment of the patient. However, the detection of pain becomes a priority situation when the patient is unable to express verbally his/her experience of pain, as is the case of patients under sedation or babies, among others. Therefore, it is necessary to provide alternative methods for its evaluation and detection. As a result, the implementation of a system capable of determining whether a person suffers pain at any level would mean an increase in the quality of life of patients, enabling a more personalized adaptation of palliative treatments. Among other elements, it is possible to consider facial expressions as a valid indicator of a person's degree of pain. Consequently, this paper presents the design of a remote patient monitoring system that uses an automatic emotion detection system by means of image analysis. For this purpose, a system based on texture descriptors is used together with Support Vector Machines (SVM) for their classification. The results obtained with different databases provide accuracies around 90%, which proves the validity of our proposal. In this way, the e-health systems of a Smart City will be improved by introducing a system as the one proposed here.

F. A. Pujol (✉) · H. Mora · A. Martínez
Department of Computer Technology, University of Alicante,
Carretera de San Vicente s/n, 03690 San Vicente del Raspeig, Alicante, Spain
e-mail: fpujol@ua.es

H. Mora
e-mail: hmora@ua.es

A. Martínez
e-mail: amlopez.ana@gmail.com

A. Visvizi and M. D. Lytras (eds.), *Research & Innovation Forum 2019*,
Springer Proceedings in Complexity,
https://doi.org/10.1007/978-3-030-30809-4_23

1 Introduction

The International Association for the Study of Pain define pain such as the unpleasant sensorial or emotional experience happened due to an actual or a potential tissue damage or injury [1].

The impossibility of some patients to express their pain experience verbally, has created the necessity of using other media for its evaluation and detection. In this way, pain scales based on vital signals and facial changes have been created to evaluate the pain of neonates, which are the main group which cannot express their pain orally.

This paper arises because of the growth of technology in the health environment, which receives the name of e-health. It is defined as a set of technical tools that are used in prevention, diagnosis, treatment, monitoring and health management, saving costs to the health system and improving the quality of life of patients [2].

The tool proposed in this research would be a great advance in Smart Cities, improving the e-health sector, in particular, in the neonatal care. Besides the reduction of the continuous tracking, this tool would allow a more objective diagnosis about if babies are suffering pain or not, since a study on 2007 [3] proved that for different groups of observers (parents, nurses and paediatricians), the pain diagnosis were different, where parents provide the greatest pain scores and the paediatricians the lowest, being nurses in a middle point between both.

If we talk about the supervision at home by parents, this tool would play a crucial role. According to 'The Statistics Portal' [4], health and personal baby care sales have increased 10 billion U.S. dollars since 2010 until 2016 globally, and from 2018 to 2026 [5] baby care products market size worldwide would increase 35.27 billion U.S. dollars. Moreover, if we focus on baby monitors, in accordance with Google Trends, 'baby monitor' has doubled the search interest since 2010 until January of this year [6]. These data show the growth of unrest by parents, wanted to be informed about the health of their children at all times.

When it comes to computer vision and pain detection, several studies have been carried out, such as studies about automatic detection of pain through facial expression [7] or on automatic detection of pain in video through facial actions [8]. However, if we talk about detection of pain in babies, there have been very few studies done.

As we have commented, the evaluation and pain detection in babies requires a continuous monitoring by parents and medical staff, so that is the problem we want to solve in this paper, developing a tool whose principal function is to detect automatically babies pain using computer vision and supervised learning.

In this paper we can find a first section about basic theory about the resources used, a next section on development, followed by a section on experimentation and finally the conclusion.

2 Background

2.1 Pain Perception in Babies

Traditionally, babies' pain has been undervalued, receiving a limited attention due to the thought of babies suffered less pain than adults because of their supposed 'neurological immaturity' [9]. This has been refuted through several studies along these last years, specially by the one conducted by the John Radcliffe Hospital in Oxford in 2015 [10], which come to the conclusion that baby brains react in a very similar way to adult brains when they are exposed to the same pain stimulus.

As we have introduced, the impossibility of expressing pain in a verbal way has created the need of using other media for evaluating and detecting it. For this reason, pain assessment scales based on behavioural indicators has been created, such as PIPP (Premature Infant Pain Profile), CRIES, NIPS (Neonatal Infant Pain Scale) or NFCS (Neonatal Facial Coding System). This last one, NFCS, is the one in which we have based the paper. It is used mainly in the postoperative phase and it is based on facial changes through face muscles, mainly on: forehead protrusion, contraction of eyelids, nasolabial groove, horizontal stretch of the mouth and tense tongue.

2.2 Feature Extraction: Local Binary Pattern

Feature extraction on facial images consists on extract associated information to the different facial muscles activity. This can be done in a global way, where you analyse the face as a hole, or locally, where some parts of the face are selected depending on the purpose. Moreover, feature extraction methods can be divided depending on their objective.

In this paper we propose a solution using texture descriptors, where descriptors use to be sets of numbers which indicate characteristics of the image.

Local Binary Pattern (LBP) LBP is a simple but effective texture descriptor which label every pixel of the image analysing its neighbourhood, studying if the grey level of every neighbour pixel is over a certain threshold and codifying this comparison by a binary number. This descriptor has become very popular due to its discriminative power and its low computational cost, which allows a real-time image processing. In addition, this descriptor has a great robustness in the present of grey level changes produced by light changes [11].

On its first version, LBP works with 3×3 matrix which goes across the image pixel by pixel, taking the values of the eight neighbour pixel and taking as the threshold the central pixel. So, the binary assignation would be done in the next way: if the neighbour pixels has a lower value than the central one, they will be assigned with a 0, but if they have equal or greater value, they will be assigned with a 1. Finally,

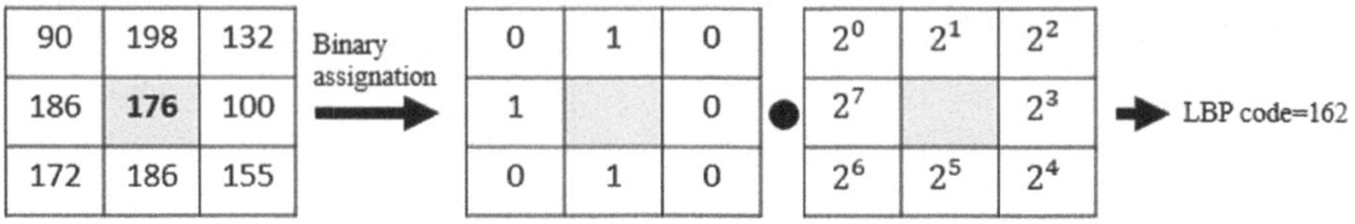

Fig. 1 Graphic example of LBP descriptor

each binary value is weighted by its corresponding power of two and they are added to obtain the LBP code of the pixel (Fig. 1).

This descriptor has evolved over the years, allowing us to work with circular samplings. In this circular samplings, neighbours are equally spaced, allowing the use of any radio and any number of neighbouring pixels.

Once we have obtained the codes of all pixels, a histogram is created with all these values. Moreover, the image can be divided into cells previously, so we would obtain one histogram per cell which would be then concatenated.

Apart from giving the choice of circular samplings, LBP descriptor provide us uniformity, which allows a great reduction of negligible information, and therefore, lower computational cost and invariance to rotations, a very importer issue in our tool [12].

2.3 Classification: Support Vector Machine

Once we have extracted the characteristics from the images, it is necessary its classification, for which we have chosen Support Vector Machine (SVM).

The main idea of SVM is to select a hyperplane that is equidistant to the chosen training examples of every class to be classified to obtain the called maximum margin on each idea of the hyperplane [13]. In this paper, this hyperplane would be the one which separates the characteristics obtained for pain and non-pain images (Fig. 2).

To define this hyperplane, the space of the examples is transformed to a new one called 'the space of characteristics', where a linear separation hyperplane is constructed using kernel functions. There are several types of kernel, but the main ones are: linear, polynomial and Gaussian; and the use of one or the other will depend on the situation, needing to experiment to obtain the best result for each case.

Once the hyperplane is obtained, it will be transformed back into the original space, thus obtaining a non-linear decision boundary.

3 Development

The tool we have developed has been implemented on MATLAB®. For the development of the tool we have used 'Infant COPE database' [14], a database which is

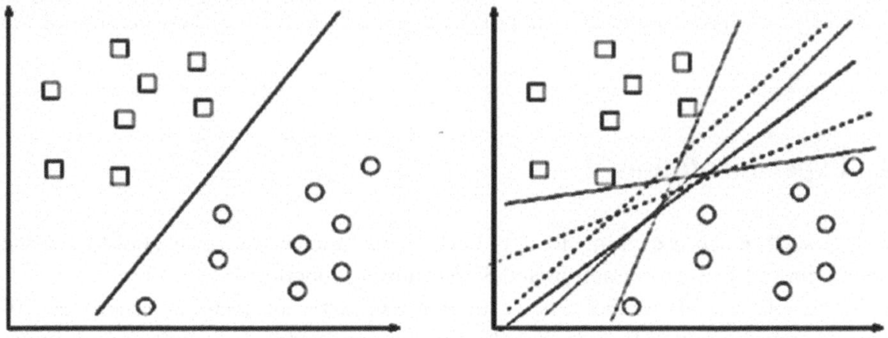

Fig. 2 Representation of hyperplanes: optimal hyperplane and different possible hyperplanes, respectively

composed by 195 colour images of 26 neonates, 13 boys and 13 girls, with an age between 18 h and 3 days. For the images, the neonates have been exposed to the pain of the heel test and to three non-painful stimuli: a corporal disturbance (movement from one cradle to another), air stimulation applied to the nose, and the friction of a wet cotton in the heel. In addition, images of resting infants have been taken. As the detection of pain is a problem of only to states, the database that has been provided has classified all the images with non-painful stimuli and rest as non-pain.

The process of pain detection has been separated in different stages as we can see in the flowchart of Fig. 3.

During the first stage, images has been previously centred on the face of the babies and cropped to a size of 100 × 120 pixels to be more manageable and achieve greater speed when processing the data. Besides this pre-processing, all images have been converted into grey scale. This is because when working with LBP descriptor, it is necessary that images are in grey scale.

For pain detection and following the NFCS scale, the extraction of characteristics would be done only in some parts of the face: right eye, left eye, mouth and brow.

Next step is feature extraction, where we have used some MATLAB's own functions. As images has been cropped, once we have obtained the features of every part, they are concatenated, having in every row the characteristic of eyes, mouth and brow of the image. Finally, we will add at the end of the row a class identifier so that we know which class is every image.

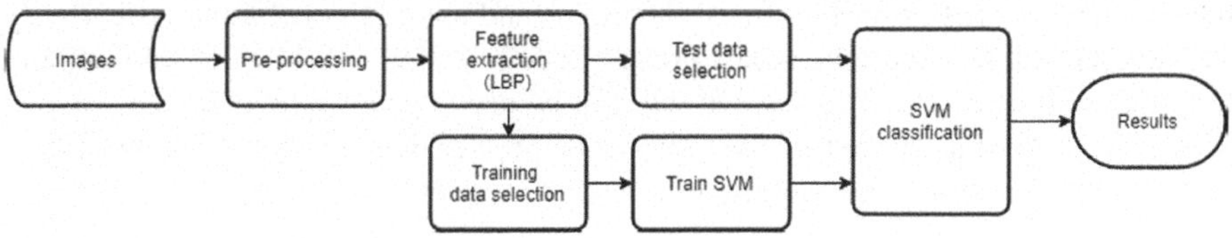

Fig. 3 Flowchart of the different stages

Finally, we have trained the support vector machine for the classification of the two classes.

4 Experimentation

In this section we are going to detail which parameter has been chosen for the functioning of feature extraction and SVM training functions.

To be able to evaluate the tests we are going to carry out, confusion matrices will be used, which allow us to see if the algorithm is confusing classes when classifying. In addition, we can obtain the error rate through the confusion matrix.

On the other hand, we found cross-validation, which is a technique to evaluate if the results are independent of the partition between training and test data.

4.1 LBP Parameters

The parameters that can change on the LBP descriptor are the radius, the number of neighbours and the size of cells.

To evaluate these parameters, one of this parameters will change, leaving fix the others, and finally we would look for the best combination.

Radius and Neighbours To select the radius and the number of neighbours we have made all possible combinations with radius 1, 2 and 3, and neighbours 8, 10, 12, 16, 18, 20 and 24.

As we can see in Fig. 4, the combination with the best recognition rate is the one of radius 2 and 18 neighbours. This combination presents the next confusion matrix:

$$CM = \begin{bmatrix} 27 & 3 \\ 10 & 83 \end{bmatrix} \tag{2}$$

It implies that we have 3 false positives and 10 false negatives, thus having an error rate of 0.1057 and, therefore, a recognition rate of 89.43%.

Cell Size This is another parameter which defines the LBP descriptor. To obtain the one which works better we have used the radius and neighbours obtained previously and vary the cell size with the next values: $5 \times 5, 7 \times 7, 9 \times 9, 11 \times 11$ and without dividing the image. The 3×3 value will not be used because the cell size must be greater than 2*radius, so in this case it must be greater than 4. As we can observe in Fig. 5, the best option is not dividing the image into cells.

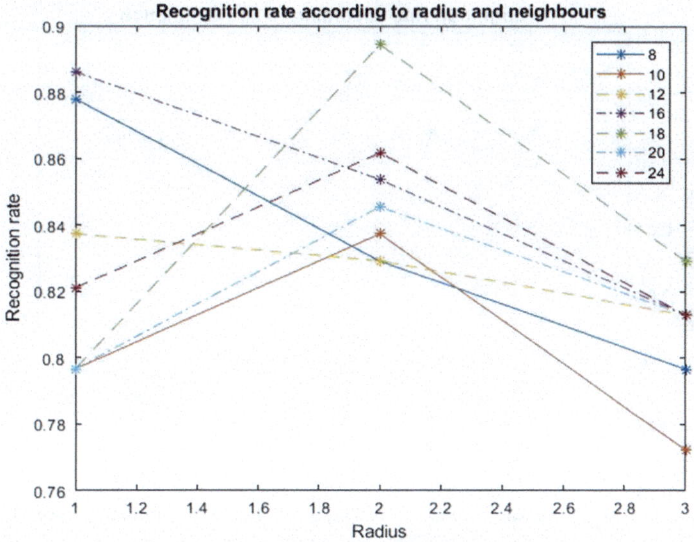

Fig. 4 Recognition rate graphic according to radius and neighbours

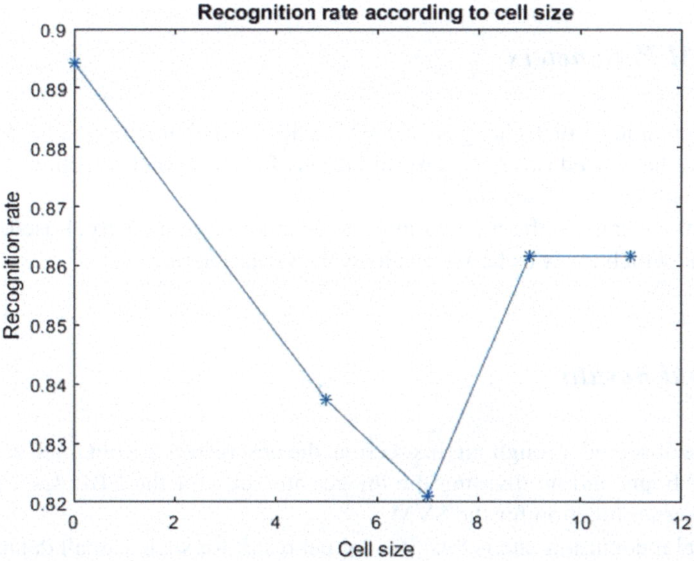

Fig. 5 Recognition rate graphic according to cell size

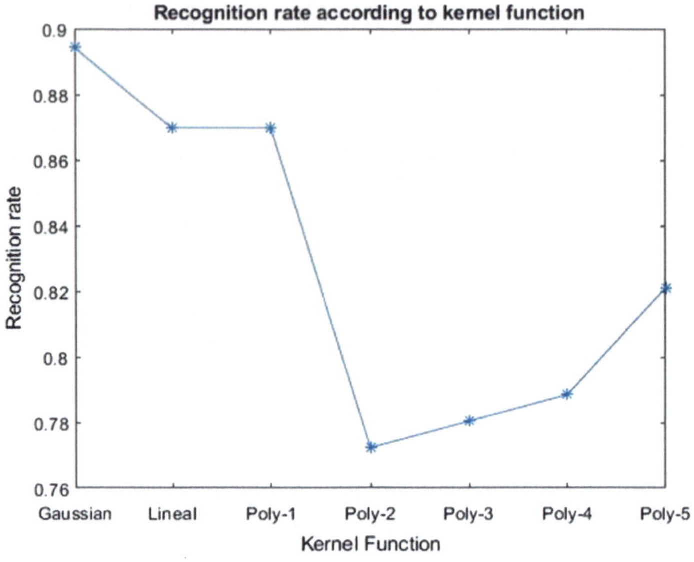

Fig. 6 Recognition rate graphic according to kernel function

4.2 *SVM Parameters*

The only parameter that changes in SVM is the kernel function. The MATLAB function we have used offers us several options for the kernel selection: Gaussian, lineal and polynomic of any order.

In Fig. 6 we can see the recognition rate depending on the kernel used, noticing that the one which gives us better results is the Gaussian one.

4.3 *Final Results*

As we have observed throughout this section, the best results are obtained with radius 2, 18 neighbours and no dividing the image into cells for the LBP descriptor, and Gaussian kernel function for the SVM.

The final recognition rate is 89.43%, a great result for such a small database with so many different stimuli. We also get a cross-validation value of 7.69%, which is the classification error for observations which have not been used for the training with respect to the current training images.

Moreover, our tool recognizes images which can be difficult for a human being to identify, such as the images of Fig. 7.

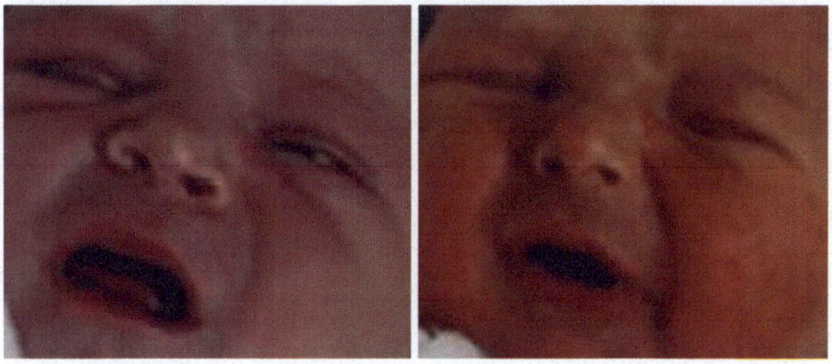

Fig. 7 Doubtful images of baby, pain and no pain respectively

5 Conclusion

In this paper we have implemented a tool which identifies baby's pain through images with a great recognition rate, around 89.43%. Moreover, it can identify doubtful images, a great breakthrough for situations where there may be diversity of opinion about whether the baby is suffering pain or not.

In view of these results, we are working to implement the tool in real time and be able to implement it in a real system, being convenient the collaboration with some hospital for the first tests. Furthermore, we are also working with other databases with babies of other ages to check the functionality and validity of the implemented tool.

Talking about the contribution of the paper to the e-health sector, it would be a great help, mainly in the sectors of neonatology and postoperative in hospitals, reducing the continuous tracking and serve as a tool to assist medical staff.

Acknowledgements This work has been partially supported by the Spanish Research Agency (AEI) and the European Regional Development Fund (FEDER) under project CloudDriver4Industry TIN2017-89266-R.

References

1. IASP: IASP Terminology. https://www.iasp-pain.org/Education/Content.aspx?ItemNumber=1698, last accessed 2019/01/14
2. Ranstad: e-health, cómo la tecnología mejora nuestra salud. https://www.randstad.es/tendencias360/e-health-como-la-tecnologia-mejora-nuestra-salud/, last accessed 2019/01/22
3. Riddell, R.R.P., Craig, K.D.: Judgments of infant pain: the impact of caregiver identity and infant age. J. Pediatr. Psychol. **32**(5), 501–511 (2007)
4. The Statistics Portal: Global retail sales of products specified for babies and children from 2007 to 2016, by category (in billion U.S. dollars), https://www.statista.com/statistics/618389/global-retail-sales-baby-childrens-products/, last accessed 2019/01/18

5. The Statistics Portal, Baby care products market size worldwide from 2018 to 2026 (in billion U.S. dollars), https://www.statista.com/statistics/258435/revenue-of-the-baby-care-products-market-worldwide/, last accessed 2019/01/18
6. Google Trends: Baby Monitor. https://trends.google.es/trends/explore?date=all&q=baby%20monitor, last accessed 2019/01/18
7. Roy, S.D., Bhowmik, M.K., Saha, P., Ghosh, A.K.: An approach for automatic pain detection through facial expression. Procedia Comput. Sci. **84**, 99–106 (2016)
8. Lucey, P., Cohn, J., Matthews, I., Lucey, S., Sridharan, S., Howlett, J., Prkachin, K.: Automatically detecting pain in video through facial action units. IEEE Trans. Syst. Man Cybern. B Cybern. **41**(3), 664–674 (2011)
9. Preciado, A.: Trabajo Fin de Grado: Valoración del dolor en neonatos: Propuesta de un modelo de registro en la unidad de neonatología del complejo hospitalario de Navarra. Universidad Pública de Navarra (2013)
10. Romero, S.: s.f. Muy interesante, Los bebés sienten dolor como los adultos, https://www.muyinteresante.es/salud/articulo/los-bebes-sienten-el-dolor-como-los-adultos-341429608053, last accessed 2019/01/20
11. Pietikäinen, M.: Local binary pattern. Scholarpedia **5**(3), 9775 (2010). http://www.scholarpedia.org/article/Local_Binary_Patterns, last accessed 2019/01/10
12. Lindahl, T.: Master thesis: Study of local binary pattern. Linköping University (2007)
13. Suárez, E.J.C.: Tutorial sobre máquinas de vectores soporte (sVM). ETS de Ingeniería Informática, Universidad Nacional de Educación a Distancia (UNED), Dpto. de Inteligencia Artificial (2014)
14. Nanni, L., Lumini, A., Brahnam, S.: Local binary patterns variants as texture descriptors for medical image analysis. Artif. Intell. Med. **49**(2), 117–125 (2010)

Smart City and Technology Transfer: Towards Sustainable Infrastructure in Bahrain

Rustom Mamlook, Odeh Al-Jayyousi, Samar Alabyooki and Suha Abdullah Ali Attallah

Abstract This paper aims to review the role of smart city innovation for sustainable infrastructure in Bahrain. It studies people's perception and awareness about the smart city. The paper articulates the imperatives for harnessing Information and Communication Technology (ICT) for smart infrastructure. The Gulf Cooperation Council (GCC) countries (Bahrain is a member of GCC) face key challenges in water, energy and food security. This is exacerbated by climate change and energy dependence in spite of the region's potential for solar and wind power and, in some cases, hydropower. Smart city innovations offer new possibilities for innovations to address the challenges in infrastructure. Technological innovations transform natural resources to serve human development. Green energy and ICT are becoming increasingly important in providing energy security and contributing to alleviating climate change risks. Hence, a transition to an innovative economy requires innovation in processes and technologies in infrastructure. The paper recommends that The GCC should be in line with the paradigm of mainstreaming green economy. Document analysis and a structured survey were used to collect data. The survey addressed three dimensions including Technology Enablers (TE), Smart Cities Strategies (SCS), and Social Awareness (SA).

1 Introduction

An innovative city is characterized as a knowledge-based, with sound ICT infrastructure, and technology adoption, and diffusion. The key to an innovative city is human capital that fosters models of innovation. In addition, it embodies models of user innovation, governance, and partnerships.

Since 1999, Bahrain is a signatory to the United Nations Framework Convention on Climate Change (UNFCCC) [1]. However, due to dependence on fossil oil, future

R. Mamlook (✉) · O. Al-Jayyousi · S. Alabyooki · S. A. A. Attallah
Innovation and Technology Management Department, Arabian Gulf University, Manama, Bahrain
e-mail: rustomm@agu.edu.bh

A. Visvizi and M. D. Lytras (eds.), *Research & Innovation Forum 2019*, Springer Proceedings in Complexity,
https://doi.org/10.1007/978-3-030-30809-4_24

energy sector Green House Gas (GHG) emissions through 2030 are expected to rise rapidly. Annual GHG emissions from the energy sector are projected to increase to about 46 million tons of CO_2 by 2030, an increase of over 37 million tons from levels in the year 2000, or a growth rate of about 5.6% per year [2]. There are plans in Bahrain to make a transition to green energy by utilizing solar thermal technologies and advanced natural gas combined cycle technology. Besides, for the industrial sector, GHG mitigation measures include compact fluorescent lamps (CFLs), efficient motors and pumps, waste heat recovery, and combustion efficiency improvements.

The Paris agreement for climate change in 2015 confirms GCC commitment to developing strategies and initiatives to adapt and mitigate climate change by reducing CO_2 emissions. The climate deal includes the transfer of climate technologies for mitigation and adaptation as one of four pillars identified in the Intergovernmental Panel on Climate Change (IPCC). The means of implementation in the global climate arena include three dimensions, namely, finance, technology, and capacity building.

Technology transfer is viewed as a key element in the response to climate change risks. The UNFCC includes the transfer of Environmentally Sound Technologies (ESTs) as part of North-South collaboration. Besides, UNFCC-Annex 2 defines specific measures to support technology transfer between industrial and developing countries. Technology transfer and transitions to green technology play a key role in addressing climate change challenges in Bahrain. This transition requires action across the technology cycle and the socio-technical regime since there are a variety of technological solutions to address climate change risks. For Bahrain, the major sectors with high CO_2 emissions are power generation and industry. The industrial sector is the major energy-intensive economic activity and a major contributor to future growth in GHG emissions. This is mainly due to the presence of Aluminum Bahrain (ALBA), one of the world largest aluminum smelters.

The GCC states can have access to climate funding through the capacity building like national communications and Clean Development Mechanism (CDM). Technology transfer in GCC states is achieved through basic mechanisms as approved by IPCC, which includes setting new eco-standards, technological innovation, and public funding. The IPCC's fourth assessment report presents a set of policy objectives in areas like energy efficiency, fuel switching, renewable energy, and carbon capture and storage. The GCC states can have access to climate funding through the capacity building like national communications and CDM. Technology transfer in GCC states is achieved through basic mechanisms as approved by IPCC, which includes setting new eco-standards, technological innovation, and public funding.

Risks and threats of climate change include the rise of sea levels, floods, and droughts. The trend in population and energy use in Bahrain shows a steady increase in energy use and carbon emissions. This trend is evident since energy is the key to socio-economic development.

To respond to the interlinked risks of climate change, a shift in energy policy to address climate change risks includes the adoption of appropriate technology transfer model and sound policy for technology acquisition and diffusion. Besides, mitigating and adapting to climate change include energy transitions and green solutions including clean energy, energy efficiency, and the reduction of fossil energy use [3].

Small islands represent a unique case in terms of resilience and sustainability. The case of Bahrain illustrates the value of sustainable innovation and adaptation to meet systemic challenges in terms of ecology, economy, and society. Bahrain represents a good example of a transformative society since it was able to make a shift from the natural resources-based economy (pearl and fishing) to the oil-based economy in the early 1930s and to knowledge and innovation economy. To cope with a fluctuation in oil prices, Bahrain is embarking on a strategy to diversify its economy and enhance the role of private sector in development [4]. The energy mix in Bahrain and GCC is mainly based on fossil oil and natural gas [5, 6]. Realizing the changes in the global energy market in Asia, EU, and the USA and the mandate of Paris climate change convention. It is imperative for Bahrain to make a shift to renewable energy and enhance energy efficiency among other options for green growth. In addition, to meet the UN-Sustainable Development Goals (SDGs) and ensure national security [7, 8].

The main islands occur in two groups of unequal size as shown in Fig. 1. The main island of Bahrain accounts for about 85% of the total area and is where Manama, the capital city is located. The next largest island is Hawar, followed by Muharraq Island, Umm Nassan, Jiddah, and Sitra. The many remaining small islands and islets are uninhabited. The paper aims to explore the determinants, enablers, and conditions for a resilient and innovative city using a case study in Bahrain.

Fig. 1 Bahrain map [9]

1.1 Literature Review

Socio-technical landscape, regime, and niche innovations underpin innovative cities lead to a radical and systematic change in current technologies and infrastructure [10, 11]. The interactions among strategy, institutions, and culture contribute to defining the future of smart cities [12, 13]. However, smart cities are likely to follow a path dependency, which implies that each innovative city has a historical trajectory and cultural context [14, 15]. In the new models of creative cities, there is emphasize on measuring the innovation capability [16].

Labs and hubs for innovations contribute to competitive and progress view cities, and to create common knowledge and efficient access for urban populations [12, 17, 18]. A framework for urban innovation included models like industrial districts and clusters that operate in technology parks [19, 20]. Enablers for urban innovations include infrastructure, governance, and leadership. There is evidence for the linkages between urban form, connectivity, and knowledge creation to foster urban innovation [21]. SMEs in cities are primary drivers for economic growth. In addition, crowd-sourcing and participatory innovation platforms constitute key enablers for urban innovation. Creative industries are crucial for shaping a sustainable economic base for cities [22].

The innovative city model attempts to produce conditions for a spatially enabled society through ICT platforms. Indicators for a smart city include smart economy, mobility, environment, living, and governance [23]. The emergence of smart cities is a manifestation of neoliberal urbanization where the state plays the role of an entrepreneur.

Innovation is the process that entails non-linear, dynamic, and multi-actors-sectoral interactions, which include developing and implementing new processes, products, and services to enhance efficiency, effectiveness or quality [23, 24]. Traditionally, research in the field of innovation focused on technological advances and developing new products. Recent research in the field of innovation started to address urban innovation in cities, by employing a diverse and cross-disciplinary body of knowledge including economics, management, administration, and marketing [25, 26].

Institutional inertia is among the barriers that imposed urban innovation due to organization structure, and bureaucracy [27, 28]. Determinants and enablers for innovation include incentives and organizational culture [29]. The frameworks that address urban innovation in the public sector examine relationships between innovation activities, innovation capacity and conditions to measure the performance [30]. Figure 2 illustrates the interactions between innovation activities, which include accessing, selecting, implementing, and diffusing innovative ideas and services. Besides, the innovation capability (which is part of conditions of innovation) entails the management of innovation, leadership, incentives, and culture. The impact on performance includes enhancement in service and efficiency.

Innovation is linked to the abilities of individuals to generate new ideas and put them into practice [22, 31]. Argued that innovation might entail a radical and

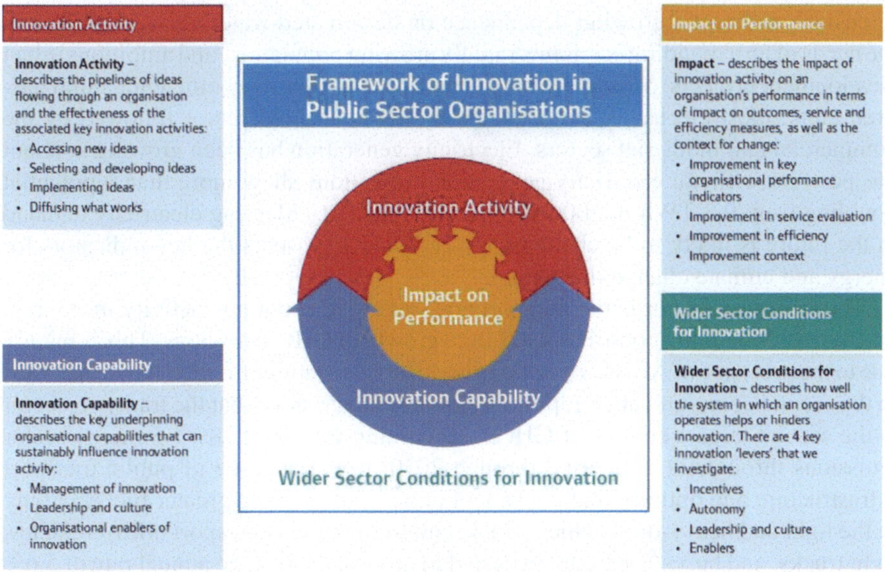

Fig. 2 Framework for Innovation in Public Sector Organizations [27]

structural change in the business model, flow of knowledge and service delivery [32]. Besides, innovation entails changes in operations [22, 33]; promoting public value [11, 34]; and focus on implementation and diffusion new processes [29]; formation of new relationships with users and service providers [19].

In addition, the culture of innovation in the public sector can be enhanced developing the core competencies, avoiding barriers to innovation and developing conditions to foster innovation and performance [35]. There is evidence that innovation is among the drivers for modernizing public administration [36, 37]. Research has shown that innovation in the public sector may play an important role to enhance the performance of the public sectors [30, 38]. Viewing innovation within space and time was manifested in many domains like organizational learning, innovative cities, and municipalities, innovative regions, learning communities, and knowledge hubs [18, 39].

2 Case Study: Bahrain

In 2009, the National Communication for Climate Change in Bahrain highlights key features for socio-economic indicators, which indicate that all natural gas is consumed locally, with 33% used for electricity generation, 27% for aluminum production, 18% re-injected back to oil fields, 8% for the petrochemical production and the remaining 14% is used for assorted industrial applications. Electricity is used intensively to meet a growing dependence on desalinated water, as well as to meet the needs of an expanding economy, rapidly growing population, and ambitious urban development projects. Most electricity is produced in relatively efficient natural gas-fired units; most electricity is consumed by the household sector followed by the commercial and industrial sectors. Electricity generation has been growing at about 9% per year. Official electricity generation projections show more than a potential doubling from 9.3 TWh in 2006 to 20.8 TWh in 2020. Meeting electricity demand in the future is likely to be challenging [2]. Table 1 presents the key indicators for energy and climate change in Bahrain.

The industrial sector is the major energy-intensive economic activity in Bahrain and the overwhelming contributor to future growth in GHG emissions. This is mainly due to the presence of ALBA, one of the world largest aluminum smelters. Projections in the second communication report on climate change show that the transport sector is the second largest emitter of GHGs, accounting for about 18% of energy sector emissions throughout the period through 2030. In the absence of public transport infrastructure and options, and the lack of policies to promote greater fuel economy in the light and heavy-duty vehicle stock, emissions from cars, sport utility vehicles, light trucks, and buses/trucks are expected to grow at an average annual rate of 5.6% per year, roughly double the population growth rate.

Table 1 Bahrain indicators and trends [2, 35]

Domain	Indicators and trends
Population	1,425,171
Population growth	3.88
Oil production (barrel per day)	41,000
Oil refinery capacity	360,000
Electricity production from natural gas sources (% of total)	99.97
Electricity production from oil sources (% of total)	0.03
Average annual energy consumption per capita	12.8 MWh/capita
Electricity generation growth per year	9% per year
Source of Greenhouse Gas (GHG) emissions	85% from the energy sector
Energy intensity (1 million dollars of GDP) during 1990–2007 ranged from about 1325 tons of carbon dioxide equivalent (CO_2e) to 905 tons of CO_2e	An improvement of about 32% during 1990–2007 in total CO_2e emissions.
Energy generation projections (2006–2020)	9.3–20.8 TWh
GHG emissions during 2000–2030	46 million tons of CO_2, an increase of over 37 million tons from levels in the year 2000; a growth rate of about 5.6% per year
Average annual solar radiation	2600 kWh/m^2/year

2.1 *Renewable Energy in Bahrain*

Renewable Energy Technologies (RET) are driven by a rational imperative and shaped by technology push and market pull as shown in Fig. 3. Typically, technological inventions are thought of as moving in a linear fashion from high-technology research laboratories through early stage commercialization and finally into the markets. In the real world, RET innovation is more complex, sometimes its nonlinear pathways, emerging through networks of stakeholders, and frequently involving feedback loops [36].

Bahrain has two sources of renewable energy generation, solar and wind. According to [37], in 2017, the renewable energy generated in Bahrain from solar and wind are 88.8% and 11.1% respectively.

Under the UNFCCC, the facilitation of international collaborative research and development (R&D) is considered vital for achieving climate objectives and enhancing technical capability in technology adoption and localization [38]. It is argued that technical change takes place through a conventional journey of innovation, which includes R&D, demonstration, and deployment with the aim to deliver "development dividends". The paper addressed technology transfer as a key condition to lead to smart cities strategies. Our methodology included three dimensions; i.e., Technology Enablers (TE), Smart Cities Strategies (SCS), and Social Awareness (SA).

3 Methodology

The Kingdom of Bahrain developed its first e-government strategy, covering the period 2007–2010, with the goal of making the government closer to customers. The efforts have been successful and have resulted in more than 200 eServices and 4 channels to deliver these services. In October 2008, Bahrain's Economic Vision

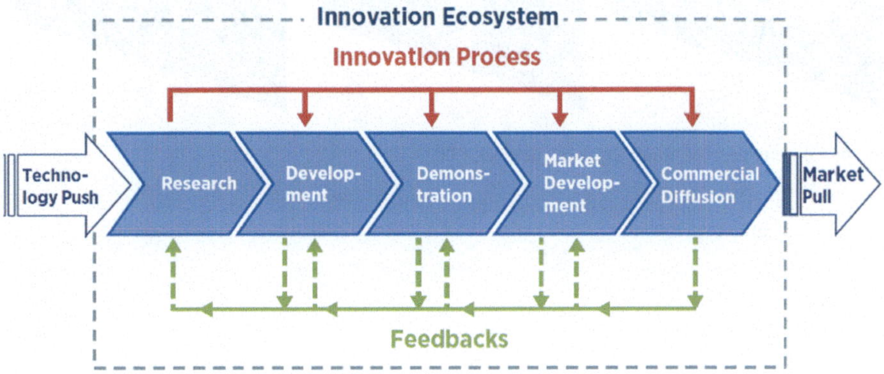

Fig. 3 The ecosystem of innovation [36]

2030, which aims to achieve sustainable development of the national economy, was launched. The Kingdom of Bahrain has developed a methodology for 6 e-government 2016 based on the new eGovernment vision. It was ensured that all dimensions and factors related to the e-Government program were incorporated through the application of an integrated framework. To achieve the eGovernment: Improving the national level in e-dealing and building the capacity of employees in the field of e-government, increased protection of information security and user rights and Create an integrated, cooperative and high-performing government.

In our attempts to frame a methodology for smart city, determinants, enablers, and conditions were outlined. Innovation is a dynamic, non-linear, and inter-sectoral process and hence the research methodology applied in this research is qualitative and exploratory. Data collection was done using document analysis and a structured survey was conducted during December–February in Bahrain. 40 surveys were collected and analyzed based on responses from professionals in the fields, media, education, and public sector. The survey consisted of open-ended questions and multiple choice questions which addressed three dimensions; i. e, Technology Enablers (TE), Smart Cities Strategies (SCS), and Social Awareness (SA) (Fig. 4). Descriptive statistics were conducted to obtain frequencies and correlations.

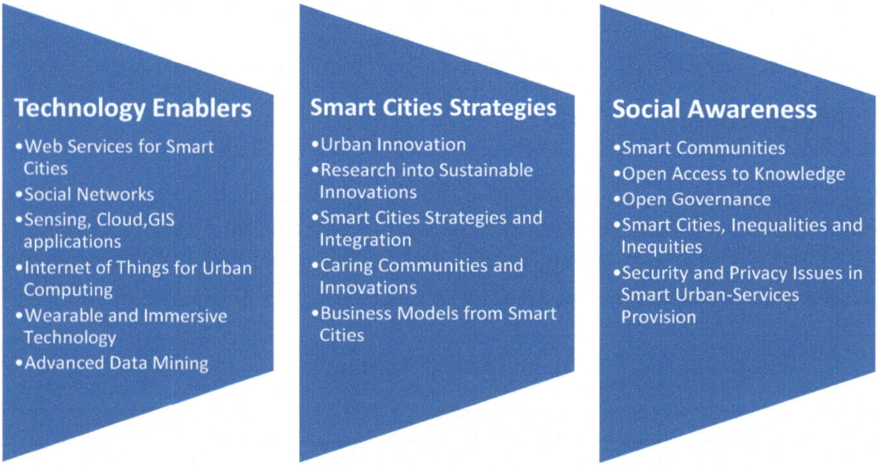

Fig. 4 Dimensions for smart cites framework [38]

4 Results

4.1 Innovation Capability and Government Activities

Bahrain developed a holistic approach to transforming into a smart and innovative city by investing in institutions, infrastructure, technology, and people. Bahrain articulated a vision and a strategic plan to ensure a smart city that is underpinned by efficiency, sustainability, and equity. This transformation was attained by the provision of incentives for investment in e-government, e-business, and e-trade and ICT infrastructure. Bahrain innovation policy was founded on key pillars, which included an orientation to a knowledge economy, smart urban systems, and innovations in logistic, tourism, and services [22, 40]. The intent of these innovative activities was to ensure an enabling environment and the conditions to foster integral innovation capability and performance. These initiatives are within an overarching framework for Bahrain strategy.

Bahrain Vision 2030 articulates a set of innovative urban technologies to adopt smart urban systems, renewable energy, and digital transformations across all sectors. In addition, Bahrain invested in enhancing the enabling environment and conditions for public sector innovation. These include ICT infrastructure, incubators, SMEs support, and accelerators programs. Among the top ranking, five countries in the UN Global Innovation Index (GII) in 2017 are UAE, KSA, Kuwait and Bahrain ranking 35, 55, 56 and 66 respectively. Global cities are correlated with a set of attributes including investing in human capital, leapfrogging new business models, good governance models, and global partnerships. In terms of Global City Index, Bahrain ranked eight and Dubai ranked two, which reflects the quality of business culture, ICT infrastructure, and ease to do business, investment climate, and cultural diversity. Besides, Bahrain succeeded to qualify as one of the most efficient logistic centers and this sector contributes to about 17% of GDP.

4.2 Survey Key Results

The descriptive statistics of the survey reveal the people's awareness and perception with respect to the smart city three dimensions were studied; i. e, Technology Enablers, Smart Cities Strategies, and Social Awareness. Specifically, respondents state that a smart city would include smart buildings, smart parking, and smart utilities as shown in Fig. 5 which refers to technology enablers' dimension.

Which of the following would included in the smart city?	
1	Smart buildings
2	Smart parking

(continued)

(continued)

3	Smart traffic
4	Smart lighting
5	Smart utilities
6	Smart security
7	Others

Moreover, respondents state that the benefits of a smart city include public safety and quality of life in the city as shown in Fig. 6 which represents the Smart Cities Strategies dimension.

What are the most important benefits?	
1	Improve public safety
2	Quick responsiveness
3	Improve efficiency and lower expenses, including saving energy
4	Better meet citizens needs
5	Smoother traffic flow
6	Quality of life in the city
7	Makes it easier run our municipality
8	Good public relations
9	Collect data to facilitate city planning
10	Others

Besides, respondents of the survey stated that the key drawbacks of a smart city include the expense of technology, loss of privacy, and difficulty in implementation as shown in Fig. 7 which refers to Social Awareness dimension.

The major drawbacks related to smart city programs	
1	Expense of technology
2	Implementation difficulty or challenge
3	Maintenance challenge
4	Difficult to manage
5	May be a distraction from more vital projects
6	Loss of privacy
7	Data security
8	Device security
9	System is unreliable
10	No concerns or drawbacks
11	Others

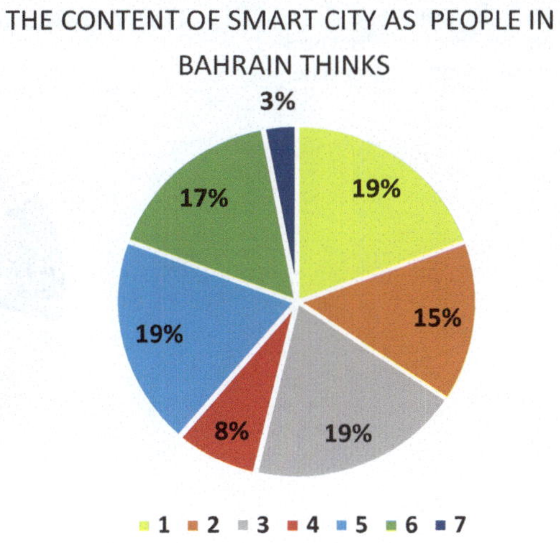

Fig. 5 The respondent's perceptions of what a smart city would include. *Source* The authors

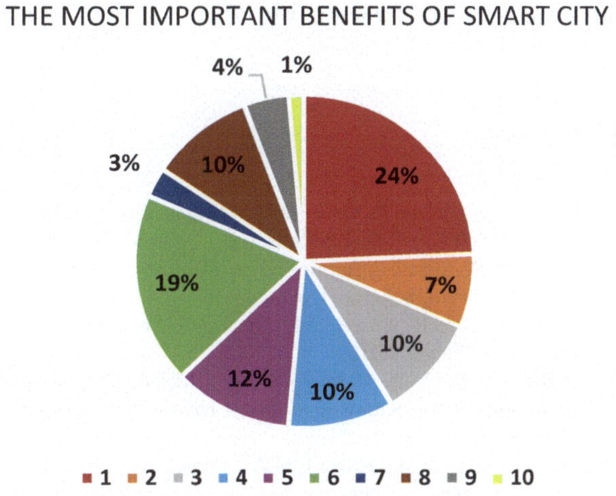

Fig. 6 Respondents' perceptions of the benefits of a smart city. *Source* The authors

5 Conclusion

Adopting an innovative mindset and culture is a key feature of Bahrain journey of innovation. This story is founded on investing in the ecosystem of innovation including human and institutional capital, smart infrastructure and good governance.

The survey reveals the perceptions of respondents of a smart city and the benefits and drawback of the adoption of a smart city. Specifically, respondents state that a smart city would include smart buildings, smart parking, and smart utilities. Moreover, respondents state that the benefits of a smart city include public safety and quality of life in the city. Respondents of the survey stated that the key drawbacks

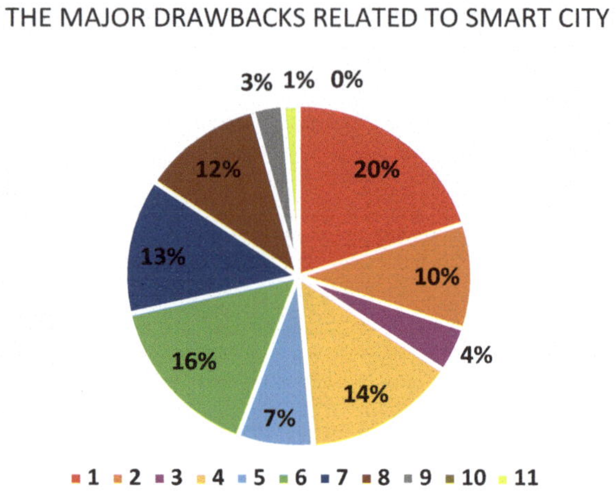

Fig. 7 Respondents' perceptions of the drawbacks of a smart city. *Source* The authors

of a smart city include the expense of technology, loss of privacy, and difficulty in implementation as shown in Fig. 7.

References

1. UNFCCC: Module 4: Measures to Mitigate Climate Change, Bonn (2008)
2. TNC. (2009). TNC. Retrieved March 1, 2018, from https://www.into.ie/NI/Teachers/ConditionsofServiceTermsofEmployment/FlexibleWorkinScheme/TNC2009_6_FlexibleWorkingScheme_Updated.pdf
3. Al-Jayyousi, O.R.: Sustainable development revisited. In: Islam and Sustainable Development, pp. 23–50. Gower. (2016)
4. El-Kholei, A. O., & Al-Jayyousi, O.: Cities of GCC in post-hydrocarbon era: challenges and opportunities. In: Knowledge-Based Urban Development in the Middle East (pp. 248–270). IGI Global (2018)
5. Al-Maamary, H.M., Kazem, H.A., Chaichan, M.T.: The impact of oil price fluctuations on common renewable energies in GCC countries. Renew. Sustain. Energy Rev. **75**, 989–1007 (2017)
6. Mondal, B., Acharyya, K., Howlader, P., Mukherjee, P.S.: Molecular cage impregnated palladium nanoparticles: efficient, additive-free heterogeneous catalysts for cyanation of aryl halides. J. Am. Chem. Soc. **138**(5), 1709–1716 (2016)
7. Crossley, M., Sprague, T.: Education for sustainable development: Implications for small island developing states (SIDS). Int. J. Educ. Dev. **35**, 86–95 (2014)
8. World Health Organization: World Health Statistics 2016: Monitoring Health for the SDGs Sustainable Development Goals. World Health Organization (2016)
9. Bahrain Map: (2018). Retrieved March 12, 2018, from, Google: (www.google.com)
10. Geels, F.W., Schot, J.: Typology of sociotechnical transition pathways. Res. Policy **36**, 399–417 (2007)
11. Noam Bergman, A.H.: Modelling socio-technical transition patterns and pathways. J. Artif. Soc. Soc. Simul. **11**, 3–7 (2008)
12. M. Batty, Axhausen, K.: Smart cities of the future. Eur. Phys. J. Spec. Top., 481–518 (2012)

13. Asheim, B., Isaksen, A., Moodysson, J., Sotarauta, M.: Knowledge bases, modes of innovation and regional innovation policy. Beyond Territ. Dyn. Geogr. Innov. Knowl. Creat., 227–249 (2011)
14. Simmie, J.: Innovation and space: a critical review of the literature. Reg. Stud. **39**, 789–804 (2005)
15. Deloltte. (2015). Smart Cities How rapid advances in technology are reshaping our economy and society. Deloltte, Netherlands
16. Florida, R., Mellander, C., Stolarick, K.: Inside the black box of regional development—human capital, the creative class and tolerance. J. Econ. Geogr. **8**(5), 615–649 (2008)
17. Inkinen, T.: Reflections on the innovative city: examining three innovative locations in a knowledge bases framework. J. Open Innov. Technol. Mark. Complex. **1**, 8 (2015)
18. Kleiman, N., Hillard, T.: Innovation and the city (2016)
19. Anttiroiko, A. V.: City-as-a-platform: the rise of participatory innovation platforms in finnish cities. Sustainability (Switzerland) 8 (2016)
20. Supporting Urban Innovation: The Smart Belfast Framework. Belfast City Council, Belfast (2018)
21. Al-Jayyousi, O.: Knowledge creation in the water sector: towards a learning water organization. Int. J. Water Resour. Dev. **20**, 165–175 (2004)
22. Al-Jayyousi, O.: Integral Innovation. Taylor and Francis, London (2017)
23. Bankins, S., Denness, B., Kriz, A., Molloy, C.: Innovation agents in the public sector: applying champion and promotor theory to explore innovation in the Australian Public Service. Aust. J. Public Adm. **76**, 122–137 (2017)
24. Randhawa, K., Scerri, M.: Service innovation: a review of the literature. In: Agarwal, R., Selen, W., Roos, G., Green, R. (eds.) The Handbook of Service Innovation, pp. 27–51. Springer, Berlin (2015). https://doi.org/10.1007/978-1-4471-6590-3
25. Sanders Jones, J. L., Linderman, K.: Process Management, Innovation, and Efficiency Performance: The Moderating Effect of Competitive Intensity. Bus. Process Manag. Innov. **20**, 335–358 (2014)
26. Hughes, A., Moore, K., Kataria, N.: Innovation in public sector organisations: a pilot survey for measuring innovation across the public sector. NESTA Index Rep., 1–85 (2011)
27. Moussa, M., McMurray, A., Muenjohn, N.A.: Conceptual framework of the factors influencing innovation in public sector organizations. J. Dev. Areas **52**, 231–240 (2018)
28. Allcorn, L.G.: Overcoming organizational inertia: a tripartite model for achieving strategic organizational change. J. Appl. Bus. Econ., 1–14 (2008)
29. Borins, S.: The Challenge of Innovating in Government Innovation. Series Sandford Borins Professor of Public Management University of Toronto (2016)
30. Walker, J.L., et al.: The diffusion of innovations among the American States. Am. Polit. Sci. Rev. **63**, 3, 880–899 (1969 Sept). American Political Science Association Stable
31. Lynn, L.: Innovation and reform in public administration: one subject or two? In: Handbook of Innovation in Public Services, 29–43(2013)
32. Borins, S.: Encouraging innovation in the public sector. J. Intellect. Cap. **2**, 310–319 (2001)
33. Caulier-grice, J., Mulgan, G.: Social innovation the open book of social innovation. Young **30**, 224 (2010)
34. Gow, J.I.: Public sector innovation theory revisited. Innov. J. **19**, 1–23 (2014)
35. Kuhlmann, S., Rip, A.: The challenge of addressing Grand Challenges: a think piece on how innovation can be driven towards the "Grand Challenges" as defined under the prospective European Union Framework Programme Horizon 2020 (2014)
36. Cucciniello, M., Nasi, G.: Evaluation of the impacts of innovation in the Health Care Sector: a comparative analysis. Public Manag. Rev. **16**, 90–116 (2014)
37. Cucciniello, M., Nasi, G.: Transparency for trust in government: how effective is formal transparency? Int. J. Public Adm. **37**, 911–921 (2014)
38. Turnheim, B., Geels, F.W.: The destabilisation of existing regimes: confronting a multi-dimensional framework with a case study of the British coal industry (1913–1967). Res. Policy **42**, 1749–1767 (2013)

39. Yun, J.J., Jeong, E., Yang, J.: Open innovation of knowledge cities. J. Open Innov. Technol. Mark. Complex. **1**, 16 (2015)
40. Bahrain Crude Oil Production. (2018, March 8). Retrieved Maech 8, 2018, from https://tradingeconomics.com/bahrain/crude-oil-production

Smart Cities to Create Opportunities for Young People

Xose Picatoste, Isabel Novo-Corti and Diana Mihaela Ţîrcă

Abstract The urban environments in the Smart Cities create spaces of well-being and coexistence, in a healthy and environmental friendly framework. The technology and respect for the environment are located in the core of these urban areas as a response to the demands of citizens. In the context of the European 2030 Agenda, the main concerns of young Europeans are analysed, particularly those related to technology, balanced development, decent jobs, and the environment. In the basis of a structural equation modeling, the relationship between the concerns about the universal and wide access to technology and those related to the specific problems of young people. This shows the importance of municipal policies in Smart Cities to meet the needs of young people and, as a consequence, achieve more balanced and sustainable development.

X. Picatoste (✉)
Department of Economics, University of A Coruna, Coruna, Spain
e-mail: xose92@gmail.com

Faculty of Economics and Business, Calle Francisco Tomás y Valiente, 5, 28949 Canto Blanco, Spain

I. Novo-Corti
Universidade da Coruña, A Coruña 15071, Spain
e-mail: isabel.novo.corti@udc.es

Faculty of Economy & Business, Campus Elviña s/n, 15071 A Coruna, Spain

D. M. Ţîrcă
Constantin Brancusi University, Targu-Jiu, Târgu Jiu 210007, Romania
e-mail: diana.mihaela.tirca@gmail.com

Faculty of Economics Sciences, 210185 Tineretului Street, no. 4, Targu-Jiu, Romania

A. Visvizi and M. D. Lytras (eds.), *Research & Innovation Forum 2019*,
Springer Proceedings in Complexity,
https://doi.org/10.1007/978-3-030-30809-4_25

1 Introduction

The concept of sustainable development is linked to the future since it is in its own nature [1]. Given that the future relies on the new generations, their own concerns have become a key issue for social and global sustainability.

The industrialization, globalization and urbanization economic process have originated a continuous migration from the countryside to the cities so the number of inhabitants of urban areas has been increasing whilst the rural population has been decreasing. From this perspective, the urban areas have to face the challenge of providing education, health services, housing, employment, etc. for making possible people's lives in a sustainable baseline compatible with technological advancements and environmental preservation. The smart cities try to provide all these requirements.

Knowing the opinion of youth about the importance of technology and environment could help policy makers to drive adequate policies [2] to give a response to their claims.

2 Youth Facing Their Own Future in the Knowledge Society

Smart cities try to make life more human in the context of sustainability. There are two key aspects of this process, one related to the use of technology to facilitate and simplify human life and another related to meeting the proposed goals while preserving the environment. This target is on the line drawn by the Sustainable Development Goal 11 "Make cities and human settlements inclusive, safe, resilient and sustainable" [3].

This paper seeks to analyze the young people's concerns and how they influence their assessment of technology and environmental aspects. In both cases, the urban environment generated in the context of smart cities must be considered paramount. The analysis of these relationships and their evaluation is the main objective of this document, since it could be useful to understand the perceived needs of the younger population about their future, as well as what they expect from society and municipalities.

The research question is: is there any relationship between the concerns of young people and their perception of the importance of technology and the preservation of the environment? And, if so, a second question arises: is it possible to quantify these relationships?

The hypotheses established for this research are:

H1: The proposal of the United Nations in SDG 11 "Make cities and human settlements inclusive, safe, resilient and sustainable" is related to technology

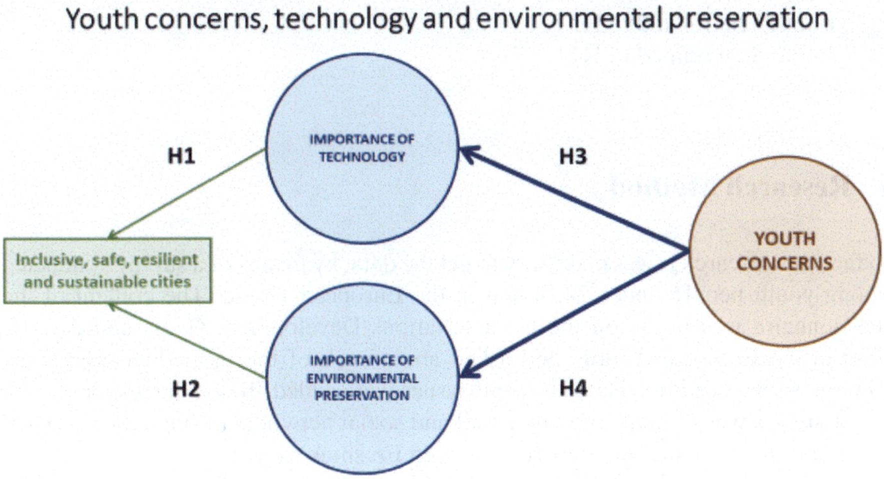

Fig. 1 Youth concerns, technology, and environmental preservation: causal relations

H2: The proposal of the United Nations in SDG 11 "Make cities and human settlements inclusive, safe, resilient and sustainable" is related to environmental preservation
H3: Youth concerns influence their perception about the importance of technology
H4: Youth concerns influence their perception about the importance of environmental preservation.

The explained relation could be summarized in Fig. 1, where the variables are shown as well as their relation, by means of the correspondent arrows.

The framework of smart cities is very important for the youth possibilities of developing in their future lives. This relation could be written as is shown in (1–3).

$$\eta_1 = \beta_1 \xi \tag{1}$$

$$\eta_2 = \beta_2 \xi \tag{2}$$

$$y = \lambda_{11}\eta_1 + \lambda_{21}\eta_2 \tag{3}$$

where

η_1 = Importance of Technology
η_2 = Importance of Environmental Issues
ξ = Youth concerns
y = Importance of Inclusive, Safe, Resilient and Sustainable Cities
λ_{11} = Coefficient related to H1
λ_{21} = Coefficient related to H2

β_1 = Coefficient related to H3
β_2 = Coefficient related to H4.

3 Research Method

To face this research it was necessary to get the data, by means of a survey conducted among youth people under 24, living in the European Union. The content of the questionnaire was based on the United Nations Development Goals and Agenda 2030 in the European Union, and asked about each of the individual objectives. Some specific questions related to youth issues were added. To get the responses, the questionnaire was disseminated by email and social networks through the snowball procedure. It was tested previously among 25 Erasmus students.

One of the most suitable methods for this type of analysis [4, 5] is the Structural Equation Modelling (SEM) [6, 7], which is the method applied for this study. For analyzing the data, the statistical software IBM Statistics 21-SPSS—Statistical Package for Social Sciences- and the AMOS—Advanced Mortar System- were used, in its 21 version.

3.1 The Variables

Some of the variables in this research are not directly observable and some indicators are needed. There are three variables that need to be constructed by means of some adequate indicators, they are the following: The youth concerns, the Importance of Technology and the Importance of Environmental Issues. The data for the variables Importance of Inclusive, Safe, Resilient and Sustainable Cities data can be gotten by means of the direct question (included in the questionnaire) about the importance given to the United Nations Sustainable Development Goal 11 (UN-SDG-11).

The summary of the 23 variables used in this model, distributed between observed and unobserved variables is in Table 1.

4 Results

The Cronbach's Alpha [8] results, as well as composite reliability and the convergent validity are shown acceptable values for accepting the consistence of the measurement model, that is to say, the strength of the constructions of the latent variables. Regarding the Model fit the values of the comparative fit index (CFI > 0.9) shows a good value, according literature [9]. The results are shown in (4–6).

Table 1 Variables in the model

Variables in the model according their nature		
Observed variables	9	• Importance of Inclusive, Safe, Resilient and Sustainable Cities (UN-SDG-11) • Youth specific issues (indicator for ξ) • Youth specific policies (indicator for ξ) • Importance of avoiding digital divide (indicator for η_1) • Importance of Technology access (indicator for η_1) • Importance of Knowledge society and Technology (indicator for η_1) • Importance of "Take urgent action to combat climate change and its impacts" (UN-SDG-13) (indicator for η_2) • Importance of "Conserve and sustainably use the oceans, seas and marine resources for sustainable development" (UN-SDG-14) (indicator for η_2) • Importance of "Ensure access to affordable, reliable, sustainable and modern energy for all" (UN-SDG-7) (indicator for η_2)
Unobserved variables	14	• 9 measurement errors (e_i) • 2 estimation error (z_i) • ξ = Youth concerns • η_1 = Importance of Technology • η_2 = Importance of Environmental Issues

$$\eta_1 = 0.73\,\xi \tag{4}$$

$$\eta_2 = 0.44\,\xi \tag{5}$$

$$y = 0.26\,\eta_1 + 0.56\,\eta_2 \tag{6}$$

All the coefficients are statistically significant ($p < 0.001$) and the values of fitted R-Squared for the Equations above are 0.54, 0.20 and 0.48, respectively.

5 Discussion

The values of the coefficients, their statistical significance and the adjusted R squared indicate that the concerns of young people are related both to the importance given to Technology and to Environmental Preservation, and these concerns can explain 54 and 20% of the changes in those variables, respectively. These results are in accordance with the literature on both areas [10, 11].

In addition, the urban environment (assessed by means of the observable variable "y") is also important for the configuration of both latent endogenous variables (Importance of Technology and Importance of Environmental Preservation).

Table 2 Results for the hypotheses

Hypotheses tested	
H1: The proposal of United Nations in SDG 11 "Make cities and human settlements inclusive, safe, resilient and sustainable" is related to technology	Accepted
H2: The proposal of United Nations in SDG 11 "Make cities and human settlements inclusive, safe, resilient and sustainable" is related to environmental preservation	Accepted
H3: Youth concerns influence their perception about the importance of technology	Accepted
H4: Youth concerns influence their perception about the importance of environmental preservation	Accepted

By analyzing the results as a whole, it is possible to realize the importance of the characteristics of the Smart Cities to respond to the demands of young Europeans.

The obtained results confirmed the four established hypothesis and all of them should be accepted. Those results for the Tested Hypotheses are shown in Table 2.

6 Conclusions

Both Technology and environmental preservation have been shown as very important from the point of view of European young people.

On the other hand, the configuration of cities according to the Sustainable Development Goals of the United Nations Sustainable Development Goals (particularly Goal 11) [12, 13], has been shown as crucial for two pillars of Smart Cities: the aspects technological and environmental.

All these variables are interrelated with youth concerns. Since these concerns were measured in basis to the general "youth issues" and the "specific policies drove to young people", the roadmap detailing municipal policies should take account these variables, not only because their intrinsic importance, but also because the youth needs to believe in their future possibilities to boost sustainable development and the implementation of new technologies for the advancement of the actual and future society.

References

1. WCED Brundtland Commission. Our common future. In: World Commission on Environment and Development, 1 (1987)
2. Bongers, A., Díaz-Roldán, C.: Stabilization policies and technological shocks: towards a sustainable economic growth path. Sustainability **11**(1), 205 (2019)
3. United Nations. Resolution adopted by the General Assembly on 10 September 2015. Assembly, General (2015)

4. Picatoste, J., Pérez-Ortiz, L., Ruesga-Benito, S.M.: A new educational pattern in response to new technologies and sustainable development. Enlightening ICT skills for youth employability in the European Union. Telematics Inf. **35**(4), 1031–1038 (2018)
5. Novo-Corti, I., Badea, L., Tirca, D.M., Aceleanu, M.I.: A pilot study on education for sustainable development in the Romanian economic higher education. Int. J. Sustain. High. Educ. **19**(4), 817–838 (2018)
6. Jöreskog, K.G.: A general method for estimating a linear structural equation system. In: Goldberger, A.S., Duncan, O.D. (eds.) Structural Equation Models in the Social Sciences, pp. 85–112. Academic Press, New York (1973)
7. Teng, J., Mu, X., Wang, W., Xu, C., Liu, W.: Strategies for sustainable development of green buildings. Sustain. Cities Soc. **44**, 215–226 (2019)
8. Cronbach, L.J.: Coefficient alpha and the internal structure of tests. Psychometrika **16**(3), 297–334 (1951)
9. Bentler, P.M.: Comparative fit indexes in structural models. Psychol. Bull. **107**(2), 238 (1990)
10. Fennell, S., Kaur, P., Jhunjhunwala, A., Narayanan, D., Loyola, C., Bedi, J., et al.: Examining linkages between Smart Villages and Smart Cities: learning from rural youth accessing the internet in India. Telecommun Policy **42**(10), 810–823 (2018)
11. European Commission. The Making of a Smart city: Policy Recommendations (2017)
12. United Nations: Convention on Climate Change. Adoption of the Paris Agreement. Draft decision -/CP.21. CCC/CP/2015/L.9/Rev.1 (2015)
13. United Nations: SDG [Internet]. [cited 2019, January, 22th]. Available from: https://sustainabledevelopment.un.org/sdgs

Automatic Speaker Verification, ZigBee and LoRaWAN: Potential Threats and Vulnerabilities in Smart Cities

Adil E. Rajput, Tayeb Brahimi and Akila Sarirete

Abstract While smart cities are becoming continuously connected, various protocol and audio interfaces are widely deployed without the full cognizance of the underlying security vulnerabilities involved. Automatic Speaker Verification (ASV), LoRaWan and Zigbee protocols have been adopted and deployed across the existing infrastructure of smart cities. Such protocols among others form the basis of the Personal Area Networks (PANs) and given the fact that such protocols are embedded in the physical devices along with the high number of such devices makes it extremely difficult to update or replace such devices once certain vulnerabilities are discovered. This paper presents an overview of the technologies and examines the security vulnerability of ASV, ZigBee and LoRaWan. While some of the vulnerabilities might be specific to certain technologies, majority of the vulnerabilities apply to other protocols in the PAN landscape.

1 Introduction

Along the global shift to smart cities and smart villages, many countries have witnessed transformations on an unprecedented scale including capitalization on smart equipment, smart grids, Unmanned Aerial Vehicles (UAVs), and smart urban infrastructure thus promoting creative economy and knowledge-based society [8, 10, 22, 32]. Using connected devices, the Internet of Thing (IoT) makes it possible to boost smart cities to a higher level and to make them sustainable, while improving the quality and safety of the citizens. However, as opposed to traditional networks characterized by devices with high storage, constant power (or strong battery) and high

A. E. Rajput (✉) · T. Brahimi · A. Sarirete
College of Engineering, Effat University, Jeddah, Kingdom of Saudi Arabia
e-mail: aiilahi@effatuniversity.edu.sa

T. Brahimi
e-mail: tbrahimi@effatuniversity.edu.sa

A. Sarirete
e-mail: asarirete@effatuniversity.edu.sa

A. Visvizi and M. D. Lytras (eds.), *Research & Innovation Forum 2019*,
Springer Proceedings in Complexity,
https://doi.org/10.1007/978-3-030-30809-4_26

storage capabilities, the IoT network consists of devices that are limited in power, computing, and storage. The networks formed by such devices are characterized as Personal Area Networks (PANs). Authors in [26] discuss the four underlying protocols that are used heavily in such networks namely (1) Bluetooth, (2) Low Rate WPAN (802.15.4), (3) High Rate WPAN (802.15.3), and (4) Body Area Networks (802.15.6). There are many concerns on the safety of services provided in Smart Cities [5, 22, 28]. In addition, the attackers have become sophisticated constantly finding new vulnerabilities to exploit. Organizations within smart cities are fully connected to third parties and therefore depend on their security measures, as reported in a survey by EY Global Information Security Survey (GISS) [15]. The survey found out that despite investments and improvement in cybersecurity, 77% of organizations still operate in a limited cybersecurity and resilience environment, 15% of organizations have adopted certain safeguards against threats from third parties, while 58% of larger companies and 41% of small organizations have indicated that their incident response program is up-to-date [15].

The concept of IoT is being applied to sectors ranging from manufacturing and utilities to healthcare and consumer electronics and includes different types of network such as Home Area Networks (HAN), Wide Area Networks (WAN), Field Area Networks (FAN) and Wireless Sensor Networks (WSN) [26, 32]. As these smart cities become continuously connected and audio interfaces are widely used, Automatic Speaker Verification (ASV) has become increasingly challenged and security threats and data privacy become a matter of concern [5, 7, 24, 27].

As opposed to traditional networks that saw the evolution and convergence of various protocols over time, PANs are characterized by many nascent protocols. To complicate the matter even further, disparate protocols have been adopted by various manufacturers who in turn have incorporated it as part of the hardware that is being used by many end users. Having a great number of end devices using different protocols not only is a challenge in operational terms but also from a security and privacy perspective.

A typical smart-city environment is composed of smart-homes/smart-offices where the users use various cyber-physical systems to interact with their environment. Typical features of these environments are devices that are controlled by the sound of the user (Alexa etc.) and devices that are connected using a PAN that the user can control via various media. In this paper, we review the security vulnerabilities that are prevalent in ASV, ZigBee and LoRaWan. The rest of the paper is divided as follows: Sect. 2 presents a review of ASV. Section 3 discusses the Low-Rate WPANs and specifically Zigbee and LoRaWan protocols. Section 4 provides conclusion and future research ideas.

2 ASV

2.1 What Is ASV

As explained in [29], Automatic Speaker Verification determines and assures the identity of the speaker. The idea behind it is simple: It classifies whether or not the person who is peaking is indeed the one who is authorized to issue certain commands. In smart environments, the speaker does not necessarily first authenticate to the system but rather simply issue the command. Circumventing such a system can cause serious comprise to the smart ecosystem.

2.2 Replay Attack in ASV

RA is considered the biggest threat in ASV. An RA replays a message sent to a network by an attacker. Specifically, it targets the underlying protocol by recording messages when a transmission is intercepted and then transmitted and replayed from another sender in another context than intended [31]. Replay attacks have been recognized to reduce the efficiency of the ASV system by a significant margin [23]. The use of high quality playback and recording devices in speech processing and machine learning makes it easy to generate these type of attacks [23]. Replay attacks on security protocols have been investigated by many researchers for quite some time. Syverson [31], for example, presented an exhaustive taxonomy of replay attacks and determines the suitability of analytical techniques for representing and/or revealing replays. The replay attack was classified as run-external and run-internal attacks. Liu et al. [21] used an authentication protocol capable of protecting against replay attack and other Denial of service (DoS). However, the schema used was symmetric which may be vulnerable to a network security threat. Soroush et al. [30] developed a scheme using a monotonically increasing counter value in related messages capable of detecting replayed old messages and, ensure their freshness, and eventually reject them. However, the process requires a large amount of time since each node uses a counter to store the timing information. Ghosal et al. [12] proposed a Wireless Sensor Networks processing scheme that is capable of building a security mechanism against replay attacks while ensuring authentication, data integrity, and data freshness. Jurcuts et al. [14] analyzed the vulnerability of security protocols and presented guidelines for the prevention of replay attacks including an empirical study to demonstrate the effectiveness of the new guidelines. Designing new protocols has also been attempted by other researchers to prevent design weaknesses in existing protocols [16, 18] proposed a new scheme for Long Range Wide Area Network (LoRaWAN) replay attack prevention which has been improved by 60–89% compared to current LoRaWAN. Nagarsheth et al. [23] proposed a countermeasure to detect replay attacks by investigating machine learning technique Deep Neural Network (DNN) and using a tandem deep learning architecture. Pfrang and Meier

[25] developed two attack techniques to detect and prevent replay attacks in industrial automation networks. The authors proposed an intrusion detections system (IDS) and investigated the security of the industrial Ethernet protocol Profinet IO. The authors showed how such network segmentation with flow control is able to prevent some of the replay attacks.

3 Low Rate WPAN Standards and Attacks

As pointed out by [26], Low Rate WPAN (LR-WPAN) is a solution that aims at providing a cost-effective way for IoT devices in IoT network to a network that can communicate with other networks. Such devices have limited resources in terms of battery, power, storage, and bandwidth. Consequently, many of the QoS requirements cannot apply to such environments. The details about the LR-WPAN or simply LWPAN are published by IEEE 802.15.4 standard [13]. The simplicity of the protocol lies in the fact that only the Physical and the MAC layer details are defined and various protocols are allowed to operate on top of the MAC layer. On the other hand, the devices in an LR-WPAN network can operate either as a Full-Functional Device (FFD) or a Reduced-Function Device (RFD). An FFD acts as the master node in the PAN network. The slave/RFD devices are tied to the master devices and are assigned specific tasks that involve the use of physical sensors/actuators. It is worth noting that the FFD devices are assigned a higher amount of resources compared to the RFD devices as they are responsible for routing messages and controlling the actions of RFD devices. Lastly, please note that the devices may operate in one of the following two network topologies: (1) Star topology and (2) Mesh topology. In the topology of a star network, devices select a PAN coordinator whose job is to distribute information to other devices. On the other hand, a mesh topology follows a peer-to-peer model where the nodes can function as either a FFD or RFD. Regardless of their status, the FFD devices are not responsible for disseminating the messages in the network.

3.1 Zigbee Protocol

The Zigbee protocol extends the 802.15.4 protocol at the network and the application layer [17]. In addition to the concept described above, the Zigbee protocol adds the following concepts.

1. The Owner: refers to the person who owns the devices and is responsible for the network formed
2. The Coordinator: is responsible for the operation of the network and is also responsible for the security aspect of the network
3. The Router: is responsible for routing the messages to the end devices

4. The End Device is the actual device that contains the physical sensors and actuators.

The security framework of Zigbee depends on two types of keys namely (1) Link key and (2) the Network key [11]. The Link key needs to be preconfigured as this is the key that will be used by all both the coordinators and the routers. The security framework operates in both a centralized and a distributed mode [17].

a. In a distributed mode, the intermediate routers can allow more routers to join the network. Upon discovering a new router or device, a router will issue a network key that will allow the newly found node to join the network. However, the new candidates need to know the link key so they can decrypt the network key and hence use it to pass it to more devices.
b. A centralized mode is a bit more complicated where the network coordinator acts as the Trust Center (TC). The TC establishes a unique TC Link Key for all the devices on the network. Furthermore, the TC also establishes link keys for the devices as requested. All the nodes on this network need to be preconfigured with the link key as once again the link key is used to encrypt the network keys.

We gleaned the following security/privacy issues from the literature [3–5, 11]

1. Denial of Service: The biggest problem with the Zigbee model is the ease with which the Denial of Service can be performed. Given the fact that the devices are characterized as low power devices with limited resources, there are many ways such devices can be disrupted. Specifically, noise can be sent on the communication channel to distort the communication or repeated messages can be sent to the devices themselves who would consume energy to authenticate the message and eventually run out of power.
2. A typical device affected by the DOS attack might require to send a repairing request which would require for it to transmit the network key. An attacker can sniff the network key that can be used in a replay attack scenario allowing the malicious node to join the network.
3. Insecure key storage: Given that the devices are limited in resources, the keys stored on the devices (including the Trust Center) are not stored in a secure fashion. A hacker can compromise the device or sniff the network key and use various techniques described in [2] to compromise the entire network.
4. The Factory reset on the devices allows an owner to remove a device from a current network and move it to another network. Therefore, physical access to such devices allows a malicious user to move a device into another network by installing a link key that has been spoofed by other methods.

3.2 LoRaWan

LoRaWan is built upon a proprietary standard Long Range (LoRa) that is prevalent in the field of Low Rate WPAN [4, 6, 19]. The LoRaWan also operated on the Physical and MAC layer level in line with the 802.15.4 protocol. However, the LoRaWan does not follow the standards whereas the Zigbee protocol is built upon the Physical and MAC layer.

The following components are parts of LoRaWan [35]:

1. End Devices: These are the devices that contain the physical sensors and actuators
2. Gateway: Gateways are responsible to relay messages between end devices and the Server
3. Server: The server is divided into two parts.
   a. Network Server: The network server is responsible for gathering the information from the network devices
   b. Application Server: The application server is responsible for transmitting the information from

The Security model in LoRaWan operates in three modes as follows [35]:

1. **Class A**: In class A, the end devices open the communication links only when needed. Specifically, they will open 2 downlink receive windows after performing uplink message transmission. These two links are used to send and receive messages. This mode preserves the power for longer life.
2. **Class B**: In this mode, the end devices will open extra channels in addition to the two uplink windows described in class A. The gateway sends beacons to establish time references for the end devices. This is important for security purposes.
3. **Class C**: In this mode, the end devices receive constant messages and obviously consumes the most power.

The LoRaWan model allows the end devices to join the network using either Over the Air Activation (OTAA) mode or Activation by Personalization (ABP) mode. Simply put, the OTAA mode allows an end device to send a join request by sending its own device key, the application key for the application to join and a random counter that would ensure that a malicious node cannot sniff the device key and the application key and send a fake request. In the ABP model, the device and the application key are preprogrammed into the end devices and the servers. In such cases, the devices cannot send a join request. However, the keys are preprogrammed once in both the end devices and the server which makes them vulnerable to compromise as we will discuss next.

3.3 Security Vulnerabilities

The following three vulnerability attacks have been discussed in the literature in detail ([1, 4, 33–35]).

i. **Replay Attack**

As discussed throughout this paper, the replay attack simply allows a malicious device to sniff certain information to be able to fake its identity with the Network and App Server. Given the fact that the end devices also issue a unique counter makes it difficult for the malicious node to resend the same information. The App and Network servers would note the last counter that came from the end device and hence ignores the message as a fake one. However, given the limited computing power of the end devices, each device is bound to reuse the counter again as the counter will be reset at some point. It is at this stage the malicious node will be able to send a fake message that would get through without detection.

ii. **Eavesdropping**

A wide range of applications, including tcpdump and ethereal, can be used for eavesdropping [9]. Eavesdropping aims at compromising the key encryption between the end devices and the Network/Application server. Described in Ahmed et al. [2], password/key cracking algorithms are used extensively by hackers. The malicious node will sniff the packets between the end devices and the server to glean the key. The underlying factor here is that the counters used by the end devices are not secure. Furthermore, the vulnerability in the ABP mechanism makes it easier for the malicious node to perform this task.

iii. **Bit Flipping Attack**

A "bit-flipping attack" is based on the protocol's vulnerability [20]. It consists of a "man-in-the-middle attack" in which the malicious node tries to compromise the security between the Application and the Network server by flipping a bit of the message payload and testing whether or not the Application server accepts the message. While the attack cannot change the plaintext significantly, such attacks can reroute the messages to a different encryption server or change the content of the message to a degree where the message can be changed significantly.

4 Conclusion and Future Work

Securing traditional networks had posed many challenges to experts. However, IoT and in turn Personal Area Networks have complicated the security and privacy matter exponentially. In this paper, we have focused on three techniques popular in the IOT/PAN environment namely ASV, ZigBee and LoRaWan. The rationale for choice was driven by their widespread use in the US, Europe and the rest of the world. While we have provided a brief overview of such protocols, we plan to expand the work to provide a taxonomy of the protocols used in PAN and the associated vulnerabilities.

References

1. Abomhara, M.: Cybersecurity and the internet of things: vulnerabilities, threats, intruders and attacks. J. Cyber Secur. Mobil. **4**(1), 65–88 (2015)
2. Ahmed, S., Rajput, A.: Threats to patients' privacy in smart healthcare environment. In: Lytras, M., et al. (eds.) Innovation in health informatics: a smart healthcare primer. Elsevier, Amsterdam, Netherlands (2019)
3. Ahmed, S.: BYOD, Personal Area Networks (PANs) and IOT: Threats to Patients Privacy (2019). arXiv:1902.06462
4. Alliance, L.: LoRaWAN specification. LoRa Alliance (2015)
5. Alromaihi, S., Elmedany, W., Balakrishna, C.: Cyber security challenges of deploying IoT in smart cities for healthcare applications. In: 2018 6th International Conference on Future Internet of Things and Cloud Workshops (FiCloudW) (2018). https://doi.org/10.1109/w-ficloud.2018.00028
6. Blenn, N., Kuipers, F.: LoRaWAN in the wild: measurements from the things network (2017). arXiv preprint arXiv:1706.03086
7. Botta, A., de Donato, W., Persico, V., Pescapé, A.: Integration of cloud computing and internet of things: a survey. Future Gener. Comput. Syst. **56**, 684–700 (2016)
8. Boyd, C.: The Top 10 Smart Cities on The Planet (2018, Dec 12). Retrieved from http://www.fastcoexist.com/1679127/the-top-10-smart-cities-on-the-planet
9. Brustoloni, X., Brustoloni, J.C.: Hardening Web browsers against man-in-the-middle and eavesdropping attacks. In: Proceedings of the 14th International Conference on World Wide Web—WWW 05 (2005). https://doi.org/10.1145/1060745.1060817
10. Das, A., Sharma, S.C., Ratha, B.K.: The new era of smart cities, from the perspective of the internet of things. Smart Cities Cybersecur. Priv., 1–9 (2019). https://doi.org/10.1016/b978-0-12-815032-0.00001-9
11. Fan, B.: Analysis on the security architecture of zigbee based on IEEE 802.15. 4. In: 2017 IEEE 13th International Symposium on Autonomous Decentralized System (ISADS), pp. 241–246. IEEE, New York (2017 Mar)
12. Ghosal, A., Halder, S., Sur, S., Dan, A., Dasbit, S.: Ensuring basic security and preventing replay attack in a query processing application domain in WSN. In: Computational Science and Its Applications—ICCSA 2010 Lecture Notes in Computer Science, pp. 321–335 (2010). https://doi.org/10.1007/978-3-642-12179-1_28
13. IEEE 802.15 WPAN™ Task Group 4 (TG4). (n.d.). Retrieved from http://www.ieee802.org/15/pub/TG4.html
14. Jurcut, A.D., Coffey, T., Dojen, R.: On the prevention and detection of replay attacks using a logic-based verification Tool (2014 June 23). Retrieved from https://link.springer.com/chapter/10.1007/978-3-319-07941-7_13
15. Kessel, P.V.: Is cybersecurity about more than protection? (2018 Oct 10). Retrieved from https://www.ey.com/en_gl/advisory/global-information-security-survey-2018-2019
16. Kim, J., Song, J.: A simple and efficient replay attack prevention scheme for LoRaWAN. In: Proceedings of the 2017 the 7th International Conference on Communication and Network Security (2017)
17. Kinney, P.: Zigbee technology: wireless control that simply works. In: Communications Design Conference, vol. 2, pp. 1–7 (2013 Oct)
18. Kumar, M., Gupa, M.K., Kumari, S.: An improved smart card based remote user authentication scheme with session key agreement during the verification phase. J. Appl. Comput. Sci. Math. **5**(11) (2011)
19. Lavric, A., Petrariu, A.I.: LoRaWAN communication protocol: the new era of IoT. In: 2018 International Conference on Development and Application Systems (DAS), Suceava, pp. 74–77 (2018). https://doi.org/10.1109/daas.2018.8396074
20. Lee, J., Hwang, D., Park, J., Kim, K.-H.: Risk analysis and countermeasure for bit-flipping attack in LoRaWAN. In: 2017 International Conference on Information Networking (ICOIN). IEEE, New York, pp. 549–551 (2017)

21. Liu, D., Ning, P.: Efficient distribution of key chain commitments for broadcast authentication in distributed sensor networks. In: Proceedings of 10th annual Network and Distributed System Security Symposium, pp. 263–276 (2003)
22. Lytras, M.D., Visvizi, A.: Who uses smart city services and what to make of it: toward interdisciplinary smart cities research. Sustainability **2018**(10), 1998 (2018). https://doi.org/10.3390/su10061998
23. Nagarsheth, P., Khoury, E., Garland, M.: Replay attack detection using DNN for channel discrimination (2017). Retrieved from https://www.isca-speech.org/archive/In-terspeech_2017/pdfs/1377.PDF
24. Pathan, A.-S.K., Lee, H.-W., Hong, C.S.: Security in wireless sensor networks: issues and challenges. In: The 8th International Conference on Advanced Communication Technology, 2006. ICACT 2006, vol. 2, pp. 1043–1048. IEEE, New York (2016)
25. Pfrang, S., Meier, D.: Detecting and preventing replay attacks in industrial automation networks operated with profinet IO. J. Comput. Virol. Hacking Tech. **14**(4), 253–268 (2018). https://doi.org/10.1007/s11416-018-0315-0
26. Rajput, A., Brahimi, T.: Characterizing IOMT/personal area networks landscape. In: M. Lytras, et al. (Eds.) Innovation in Health Informatics: A Smart Healthcare Primer. Elsevier Amsterdam, Netherlands (2019). (Earlier version available as arXiv preprint arXiv:1902.00675)
27. Rathore, M.M., Ahmad, A., Paul, A., Rho, S.: Urban planning and building smart cities based on the Internet of Things using Big Data analytics. Comput. Netw. **101**, 63–80s (2016)
28. Rawat, D.B., Ghafoor, K.Z.: Smart Cities Cybersecurity and PriVacy. Elsevier, Amsterdam (2019)
29. Reynolds, D.A.: An overview of automatic speaker recognition technology. In: 2002 IEEE International Conference on Acoustics, Speech, and Signal Processing, Vol. 4, pp. IV-4072. IEEE, New York (2002 May)
30. Soroush, H., Salajegheh, M., Dimitriou, T.: Providing transparent security services to sensor networks. In: ICC, pp. 3431–3436 (2007)
31. Syverson, P.: A taxonomy of replay attacks. In: Proceedings of the 7th IEEE Computer Security Foundations Workshop, pp. 131–136. IEEE Computer Society Press (1994)
32. Talari, S., Shafie-Khah, M., Siano, P., Loia, V., Tommasetti, A., Catalao, P.S.: Review of smart cities based on the internet of things concept. Energies **10**, 421 (2017). https://doi.org/10.3390/en10040421
33. Toussaint, J., El Rachkidy, N., Guitton, A.: Performance analysis of the on-the-air activation in LoRaWAN. In: 2016 IEEE 7th Annual Information Technology, Electronics and Mobile Communication Conference (IEMCON), pp. 1–7. IEEE, New York (2016 Oct)
34. Wixted, A.J., Kinnaird, P., Larijani, H., Tait, A., Ahmadinia, A., Strachan, N.: Evaluation of LoRa and LoRaWAN for wireless sensor networks. In 2016 IEEE SENSORS, pp. 1-3. IEEE (2016, October)
35. Yang, X.: LoRaWan: vulnerability analysis and practical exploitation. Delft University of Technology. Master of Science (2017)

Mexico City Traffic Analysis Based on Social Computing and Machine Learning

Magdalena Saldaña Pérez, Miguel Torres Ruiz and Marco Moreno Ibarra

Abstract Nowadays artificial intelligence is immersed in all the people's activities. Internet and mobile devices let us produce and consult information related to social and urban aspects. The crowd sourcing information and the social computing analyze and implement solutions to real world problems using the web content generated by social media and internet users. One of the urban factors that affect people's activities is the vehicular traffic, every day traffic produces high stress levels and time delays when people are trying to move from one place to another using their cities highway. Vehicular traffic problems impact directly over the human's health and over the financial dynamics of the affected cities. In the present approach, social computing is implemented by analyzing crowd sourcing information related to vehicular traffic, and computing regressions over the identified traffic events, to determine how traffic would affect an urban area at different hours. The consulted crowd sourcing information is obtained from Twitter. The traffic events forecast is implemented using a machine learning regression algorithm; the retrieved data from the social network and the regression progress results are visualized in the study area's map, using a geographic information system. The goal of the geospatial visualization is show to the citizens the places where traffic events probably would occur, giving them the opportunity to change their routes avoiding traffic problems. One of the main characteristics of this approach is its use of volunteered geographic information.

M. S. Pérez (✉) · M. T. Ruiz · M. M. Ibarra
Instituto Politécnico Nacional, Centro de Investigación en Computación,
Nueva Industrial Vallejo, Gustavo a. Madero 07738, Mexico
e-mail: amagdasaldana@gmail.com

M. T. Ruiz
e-mail: mtorres@cic.ipn.mx

M. M. Ibarra
e-mail: marcomoreno@cic.ipn.mx

A. Visvizi and M. D. Lytras (eds.), *Research & Innovation Forum 2019*,
Springer Proceedings in Complexity,
https://doi.org/10.1007/978-3-030-30809-4_27

1 Introduction

Every day people have to displace from one place to other another in order to study, work, or make different outdoor activities; car accidents, closures, vehicle crashes, accidents, protest marches, among other events, produce delays and displacement irruptions. Thanks to the technological advances on mobile devices and computers, people have the opportunity to know what is happening at their surroundings and report their perceptions [25].

Social media such as Facebook and Twitter let people share information about the traffic events they witness creating crowd information. At social media, users act as intelligent sensors, the data they produce provide a source of information about what happens in real world.

In the present approach, crowd information related to traffic is analyzed in order to compute regressions of possible traffic events that would occur in a study area, affecting the displacement of the people. One of the strengths of the approach is the managing of users-generated content using machine learning algorithms and social computing.

Social computing refers to the systems that process and analyze information generated by groups, that has been disseminated across communities and organizations [28]. Social computing and the actual technological advances improve people's life and boost the development of smart cities.

Smart cities, according to Gartner [8], are urban areas where multiple entities converge to formulate sustainable solutions for human problems, including contextual and real-time data analysis. Smart cities use technology at their social and urban aspects. For 2020 it is estimated that the 60% of smart cities initiatives would use the Internet of Things (IoT).

Urban areas are severely affected by vehicular traffic, the problem is even worst at crowded cities such as New York, London, Beijing, Tokio, and Mexico [20]. According to the *Work and Health Minister of Japan*, the survival probability at a car accident gets decreased 50%, traffic delays the displacement of the first aid services [20]. Some other problems derived from traffic are the air pollution, contingences, and displacement delays; some governments are really interested in mitigating these factors. In London city, for example, the car accidents rate has decreased thanks to a system that manages the flux of vehicles driving in a road at the same time; in China, sensors in the streets help to model the traffic conditions such as bottlenecks [14].

In this approach Twitter is used to extract tweets describing the traffic at a study area; the tweets are analyzed and classified with a machine learning model implemented to calculate regressions over the processed tweets. The resultant regressions let forecast traffic events; such events are geo-visualized in a map of the studied city to identify the places where traffic events would occur and the congested roads, this could help at the designing of routes that avoid road problems and traffic jams.

The approach is organized as follows, in Sect. 2 the background of the research is presented. In Sect. 3, the research method is presented, on it the data collection, data preprocessing, data classification, and the regressions computing are treated. In Sect. 4 the findings are described. Section 5 presents the general discussion of the approach.

2 Background

People are connected everywhere, and all time they are sharing their perspectives and points of view about what happens at their surroundings and about socio politics events.

The data they generate helps technology and science to have information about what happens in real world. One characteristic of the user generated content is the geo information included on it, and the way in which geographic information can be inferred from it [13].

The volunteered geographic information (VGI) lets everybody to produce information even if the user has or not specialized knowledge about the geospatial data treatment [9]. VGI sources provide useful information about what happens in real world, this can be used in social computing to forecast different situations that modify the humans' life. VGI data can be used with Artificial Intelligence and computing science in order to solve specific urban problems such as air pollution, traffic, overpopulation, economy inequality, dispersion of diseases, and to facilitate the smart life development [10].

The microblogging service Twitter is a social media where people express their opinions and subjective impressions about general subjects. Multiple analysis have been done over the tweets content with a view to have information for different studies about the peoples interest and how they make use of the social media to interact with others.

Despite the facilities to include coordinates in the tweets metadata, just 3% of the microblogs tweeted a day have coordinates, and just 17% contain information to be georeferenced [2].

When tweets are missing of coordinates the geolocation can be used, this consist of searching a specific place in a digital map using a geospatial information system (GIS), or a specialized web service such as Google Earth. The web services let make dynamic searches of addresses in their maps [23].

Even if the geolocation of tweets is possible, the process is a challenge due to the short longitude of their texts and the missing of context to infer information. Previous works have treated the tweets geolocation correlating the tweets content and the area where they were posted [3, 29].

For example, Hahmann et al. [11] analyze the relation between tweets and the geographic area where they are posted; tweets are manually classified using supervised and semi-supervised classification methods. The Hahmann et al. project [11] lets to evaluate the relation degree between the tweets text and their geographic location, which could be useful for some other researches regarding critical situations control. The authors conclude that when a critical event occurs a lot of information is produced in social networks; since many of the users are not in the place of the reported events, the publications are redundant hindering the location of the event.

Most of the times data need to be processed and classified in order to be used on regression machine learning algorithms [1]. When working with machine learning algorithms it is important pay attention to the data features, the task where the data will be used (commonly it is necessary to process the data to be used as valid inputs), and the *bias* that can be tolerated at the learning process. Many times a good moment to choose the machine learning algorithm to be used is during or after the data text mining process [30].

The *No-free-lunch* theorem stablishes that, if an algorithm has a good development at certain process, probably it will have a bad development at some other tasks [16]; this means that any machine-learning algorithm could be better for any circumstances.

Two algorithms implemented in the present approach are K Nearest Neighbors, and Support Vector Machine.

K Nearest Neighbors (KNN) is a classification learner, its main characteristic is that it stores data and classifies the unlabeled data inputs by searching similar data in its classes [18]. First it looks the k nearest neighbors to the sample, then considering the neighbor's classes determines the class for the input [15, 27].

The main purpose of the SVM regression algorithm is create a hyperplane to best divide a dataset into two classes; the points represent data in a multi-dimensional space with many dimensions as the data features considered [6]. The support vectors are data points that represent the critical elements of the data set and are the ones located nearest to the hyperplane [24].

Regression models predict data regarding the possible relations between a target and a predictor, usually such models find the causal relationship between variables [26]. Data regression provides two main results, the first can be the relation between a dependent and an independent variable, the second is the impact factor of multiple independent variables over a dependent one.

Most of the times regression algorithms take into account the current and the past status of a data when forecasting [26]. The logistic regression is a kind of regression used when the dependent variable is binary; binomial distribution and logit functions are applied. Equation (1) represents the odds of an event occurrence, since the variable is binary, it is necessary to consider its presence and absence probabilities. Equation (2) represents the link between the odds and the logit function.

$$odds = \frac{p}{(1-p)} \tag{1}$$

$$\ln(odds) = \ln\left(\frac{p}{(1-p)}\right) \tag{2}$$

where: p is the occurrence probability of the event, and $(1-p)$ the probability of not occurrence.

Logistic regression makes use of non-linear log transformation over the data and considers the maximum likelihood, there must not be collinearity between data. When the dependent variable is ordinal, the regression is called ordinal logistic, if the dependent variable is multiclass, the regression is called multinomial logistic [17].

When data has not a linear behavior the use of Support Vector Machine (SVM) with a no linear kernel such as Radial Basis Function (RBF) is a good option. SVM and the logistic regression have a similar performance in practice [4].

In the present approach a SVM regression model is implemented to forecast the coordinates of possible traffic events in the study area. The SVM regression output is a real number; a margin of tolerance called *epsilon* must be considered [7]. In the approach a SVM with RBF kernel regression model is implemented. Equation (3) describes de RBF kernel.

$$k(x_i, x_j) = \exp\left[-\frac{|y_i - x_j|^2}{2\sigma^2}\right] \quad (3)$$

where $k(y_i, x_j)$ describes the Gaussian radial basis function of the kernel, x_i is the known feature, x_j is the forecasted feature, and σ is the proposed margin of tolerance.

The next section describes the methodology proposed to forecast traffic events from tweets using a SVM regression model, and the geospatial data treatment applied over traffic related tweets.

3 Research Method

In this approach information about traffic is extracted from tweets published in a specific study area, and classified in order to identify the most common ones. The classification is done using machine learning; after that, tweets are geo referenced in the study area map. A regression model is implemented to forecast the coordinates where traffic events possibly occur, considering the information from the analyzed tweets.

The methodology to complete the purposes defined above is described in this section. The methodology considers five stages: *data collection and storage, data preprocessing, data classification, regression model, and the geovisualization* as shown in Fig. 1.

3.1 Data Collection and Storage

This stage consists of extracting tweets related to traffic conditions and traffic events presented in the study area. It has been created a database called *Events* to store

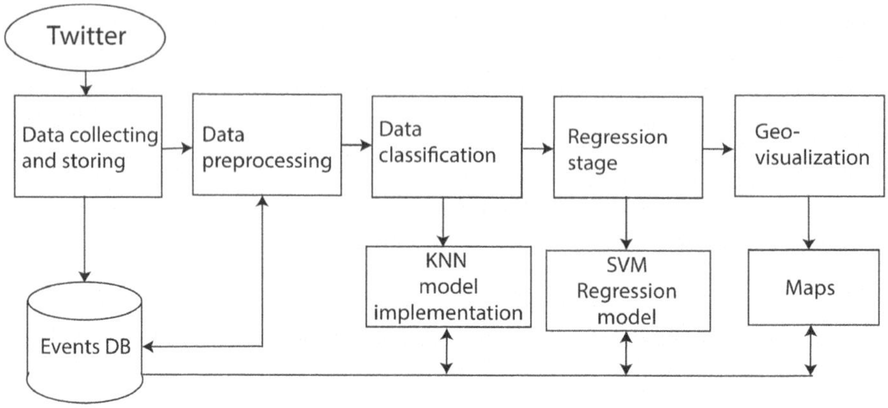

Fig. 1 Methodology. The diagram shows the five stages and the relation between them

all the extracted tweets. The *Events* database has five tables: *collected_tweets, cleaned_tweets, classified_tweets, events_regression, and geographic_data.*

In *collected_tweets* table, tweets are stored the same as they were originally posted. Tweets are processed through a text mining process and after that, stored at the *cleaned_tweets* table, such table compounds a data corpus of traffic events descriptions.

After the machine learning classification, tweets are stored at the *classified_tweets* table. The regressions calculated to forecast traffic events are stored at the *events_regression* table. Such results are located in the study area's map.

3.2 Data Preprocessing

For this approach a Python script to extract tweets from traffic reporting accounts has been programmed. Every time a followed account posts a tweet, it is stored in the *collected_tweets* table. The number of useful tweets considered in this approach to make probes is 44,000.

Tweets are text mining processed with a view to identify key words related to the traffic, after the process, the tweets are stored in the *cleaned_tweets* table. The text mining process steps are listed below.

1. The text is divided in words, and transformed into a strings sequence.
2. Special characters and stop words are removed.
3. Uppercase letters are replaced by lowercase ones.
4. Key words related to traffic are identified. Road names and points of interest are identified to geo-locate to the registry if needed.

5. All the words identified in the messages are stored in a dictionary structure where the key is the word, and its value is the number of times such word appeared on the entire corpus.
6. Vectorization, each chain of strings is represented as a vector, its elements are the words of the entirely corpus, the value at each position is the number of times that the word appears in the tweet text.

After this process, tweets are ready to be used at the classification stage.

3.3 Data Classification

For the classification the implemented algorithm was KNN (it throw up the highest accuracy when was compared against the Naïve Bayes classification algorithm results obtained for the approach's dataset, Naïve Bayes is one of the most commonly used algorithms when classifying short texts) [31]. Below, the KNN methodology for data classification is described.

Given a set W of training vectors of text, the classification of the instance x is done as follows.

a. The algorithm calculates the Euclidean distance of each sample w_{ni}, to the instance x.
   The algorithm assigns a specific weight to each sample. If the sample w_{ni} is near to x then $weigth(w_{ni}) = 1/K$. If the sample w_{ni} is far from x then $weigth(w_{ni}) = 0$. The assigned weights addition must be one, $\sum_i^n w_{ni} = 1$.
b. The algorithm adds the weights of the samples of the same class nearer to x, assigns the instance to the class with the highest weight.
c. After the model implementation, it is applied over the text vectors, and the resultant precision metrics are calculated.

The algorithm requires of being trained and tested. The algorithm training was done using a previous manual classification of the samples, this classification results were stored in the *classification* feature of the text vector. Table 1 describes the classes considered in this approach.

The classified vectors were considered as the *training corpus* (70% of the total samples), the remaining 30% was considered as the *test corpus*. To prove the precision of the classification model, the classification assigned by the KNN algorithm was compared to the manual classification of the samples. Finally, the algorithm is used to classify the remaining 30% of the samples without classification.

Once the classification has been applied over the test samples, the classification label of the registries is saved in the *classification* field of the registries, in the table *classified_tweets*, as well as the text vector that provided the data to classify the instance, and the coordinates of the tweet, in case that they exist.

The KNN classification algorithm implemented obtained a precision score of 94% when proved with the researching samples. The algorithm was coded in Python

Table 1 Classification proposed for the approach

Class	Description
Accident	The tweet describes an accident that affects the vehicles displacement
Closure	Closures in roads that prohibit the transit of vehicles through them
Traffic	Congested roads and traffic jams reported
Good	Good displacement conditions
Rule	The tweet provides information about vehicular regulations of the area
Un-classified	There is not enough information in the tweet's text to classify it as a traffic event

The classes have been defined considering different traffic problems reported in the tweets published in the study area

using the Sci-kit learn library [21], specialized in machine learning and statistical processes.

The coordinates of the treated tweets are stored in the fields *latitude* and *longitude* in the table *tweets-classified*. In this table, the *latitude* and *longitude* coordinates are merged into a geometric point that represents the location of the tweet in a map of the study area. The coordinates merging is done using the *PostGIS* functions. *PostGIS* is a geospatial tool for *PostgreSQL* [22]. The geolocation of the tweets lets represent them in the study areas map with a view to identify the areas with more traffic problems.

3.4 Regression Stage

In this stage a regression model is implemented to calculate the possible coordinates where traffic events probably will occur. After its implementation, the model is applied over the classified tweets to calculate traffic events in the study area.

A regression analysis searches the relation between the target and the predictor features; it also indicates the impact of multiple independent variables over a dependent variable [27].

In the approach, the implemented SVM algorithm uses the RBF kernel to calculate the regression of coordinates; Equation (4) describes the RBF kernel.

$$k(x_i, x_j) = \exp\left[-\frac{|y_i - x_j|^2}{2\sigma^2}\right] \quad (4)$$

where $k(y_i, x_j)$ is the Gaussian radial basis function of the kernel, x_i is the known feature, x_j is the forecasted feature, and σ is the proposed margin of tolerance.

The SVM regression algorithm has been implemented using the sci-kit library. The regression process consists of two main stages. In the first one, the algorithm is programmed and trained, in the second one, the algorithm is tested in order to produce pairs of coordinates where possible traffic events would appear.

The registries from the table *classified_tweets* are divided into two sets, the 70% of the registries compound the *training set*, meanwhile the remaining 30% of samples are the *test set*. The data features are the *id* of the registry, the *day* and *time* when the tweet was posted, and its coordinates of *latitude* and *longitude*.

To process data, the feature *day* of the registry is considered as two elements, the *day_of_week* and the *day_of_month*. The *day_of_month* is the day pointed in the original tweet, the *day_of_week* is the number of the day that corresponds to the tweets day (0: Monday, 1: Tuesday, 2: Wednesday, 3: Thursday, 4: Friday, 5: Saturday and 6: Sunday). These considerations over the day of the tweet are useful to calculate regressions for specific days of the week.

The *time* of the tweet is also separated into the features *hour* and *minutes*, with a view to increase the number of vectors considered at each regression; two vectors posted at the same hour but at different minutes can be considered in the same regression calculation.

The *latitude* and *longitude* coordinates are separated into two vectors, one that contains the *latitude* coordinate for each sample of the training set, and a vector that stores the *longitude* coordinate for each training sample. The coordinates are separated since the SVM algorithm calculates regressions for one data at a time, thus a model calculates latitude regressions and another model computes longitude regressions. The functioning of both models is the same; they only differ in the feature to calculate.

The SVM regression algorithm purpose is to find a function that deviates from x_j a value less than σ for each y_i, being as lineal as possible considering the norm value (β, β'). The problem then is a minimization problem for (5) [5].

$$J(\beta) = \frac{1}{2}\beta'\beta \tag{5}$$

Since it is possible that the proposed function does not satisfy the condition shown in (7) for all the samples, slack variables are used as shown in (6).

$$\forall n : \left| y_n - \left(x_n'\beta + b \right) \right| \leq \sigma \tag{6}$$

The value C represents the offset between $J(\beta)$ and the number of deviations bigger than σ tolerated, as shown in (7).

$$J(\beta) = \frac{1}{2}\beta'\beta + C\sum_{n=1}^{N}(\sigma_n + \sigma_n^*) \tag{7}$$

Some queries are applied over the training data; this creates a set of vectors with similar features to facilitate the SVM learning process. The queries obtain the registries with the same *day_of_week* and their *hour* of publication is found at a defined period of time. Such samples are considered as support vectors for both regression models.

As result from the regression process, each model generates a vector of coordinates (latitude or longitude) for each sample (*vector_latitude* and *vector_longitude* for each case). The sample and the computed coordinates are related with the sample id.

To compute the regressions accuracy, the coordinates of the training samples are compared with the regression coordinates. The achieved precision was 96%. The models are applied over the test samples, the obtained vectors of coordinates are joint into a geometric point, in order to visualize the geographic place where a traffic event has been forecasted. The regression coordinates and the geometric point are stored in the *events_regression* table.

In the geovisualization the classified tweets and the regression results are mapped in the study area map. The maps used belongs to a volunteer geographic information source, Open Street Maps® [19], a cartography service where users can create and edit web maps, providing updated cartographic information; its main purpose is motivating people to produce and share data with a view to optimize computing processes [12].

In the Findings section the maps designed to show the classification and regressions results are shown.

4 Findings

In this approach stage, volunteer geographic information and social media data intertwine to analyze traffic events.

Regarding the results, the classification process relevance is the number of traffic events identified at each class, and the algorithm achieved precision. Figure 2 shows

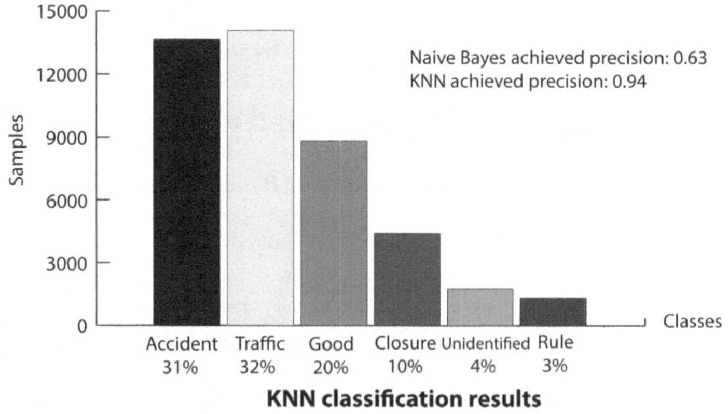

Fig. 2 Classification results obtained from the classification of the test samples using the KNN algorithm

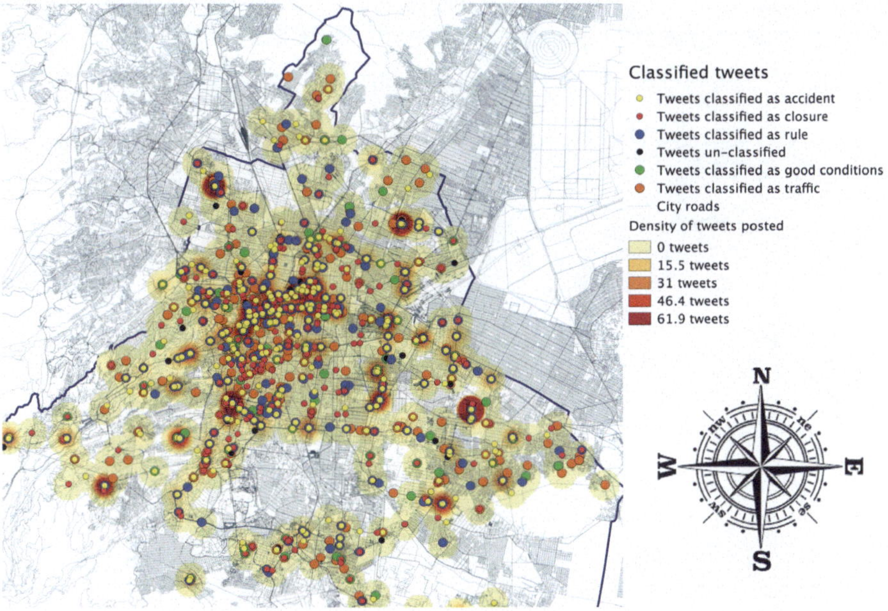

Fig. 3 The Mexico City maps shows the roads of the city. The points represent tweets related to the traffic reported in the area. Each class has been represented using a different color

the classification precision obtained from KNN and Naïve Bayes algorithms, and the percentage of test samples classified at each class.

In Fig. 3, the map shows the tweets classified using the KNN represented as circles, each proposed class has a specific color. In the map the color shadows represent a heat map (gradually faded colors that increase their intensity according to the number of samples over them). As it is visible in the map, most of the traffic events are reported in the center of the city, the class with more reports is *closure*. The plotted points represent the entire corpus used in the approach (44,000 samples).

The concentration of events at the center has a direct relation with the commercial and industrial activities developed in the city.

The classified tweets are used in the regression model, in order to generate predictions of traffic in certain periods. When a prediction of traffic is needed, a request to the database is done to acquire the registries to generate the support vectors; such registries correspond to the *day_of_week* and the *hour* requested, this day and hour filter let compute more accurate predictions.

Figures 4, 5 and 6 show the traffic events at three different time lapses in the same day. The points represent traffic events that would occur in the study area meanwhile the heat map represents the density of samples considered in the computing, in other words the heat map represents different traffic events reported in that area that helped to compute the traffic regressions.

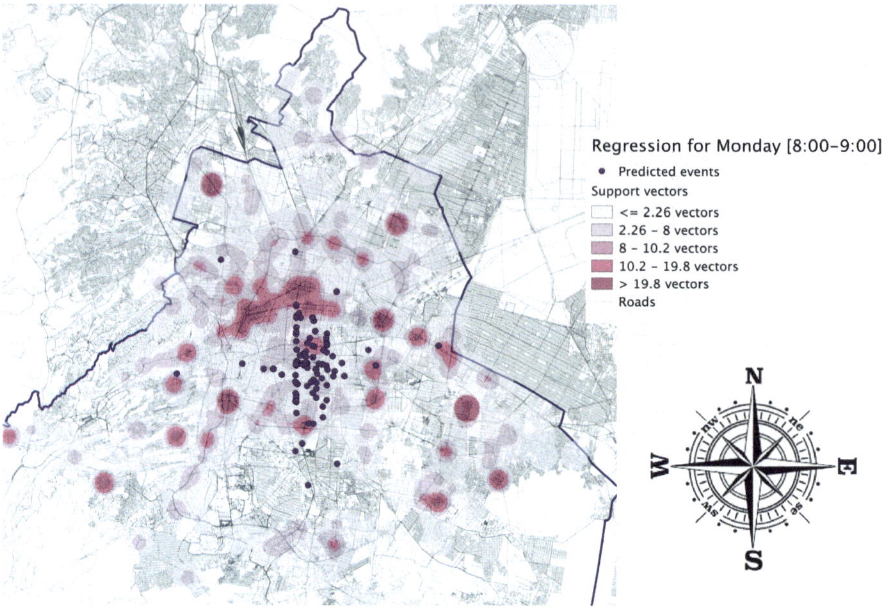

Fig. 4 Regression results obtained for Monday 12th March between 8:00 and 9:00 h. The points represent the possible traffic events in the study area

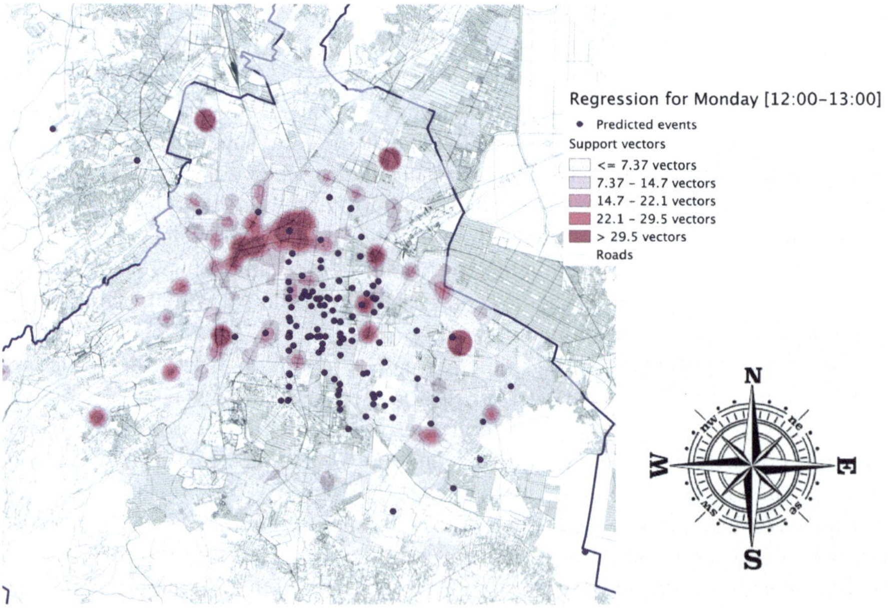

Fig. 5 Regression results obtained from Monday 12th March between 12:00 and 13:00 h. The points represent the possible traffic events in the study area. The heat map represents the support vectors density

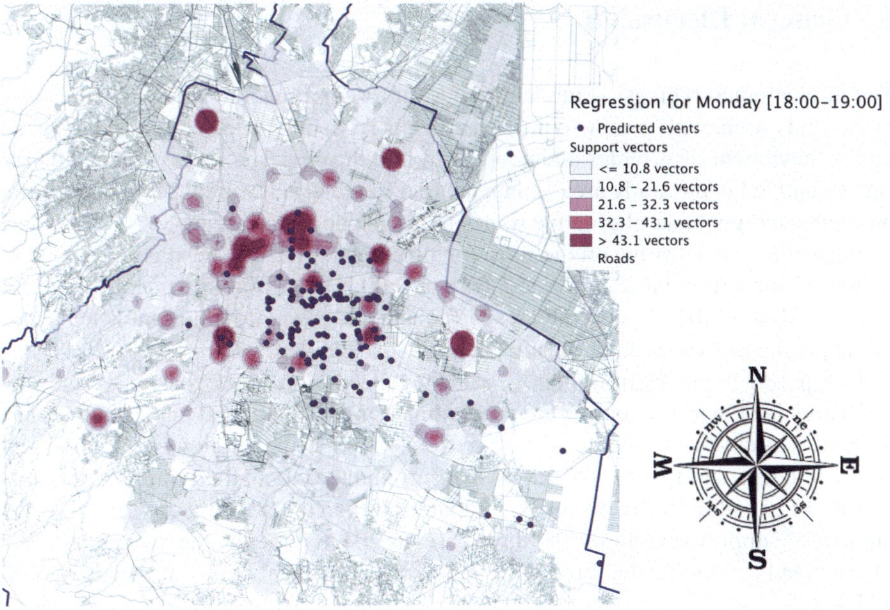

Fig. 6 Regression results obtained from Monday 12th March between 18:00 and 19:00 h. The points represent the possible traffic events in the study area

The three maps (Figs. 4, 5 and 6) correspond to a Monday. From Fig. 4 we can observe that between the 8:00 and 9:00 h, most of the support vectors considered for the regression are located in the center and around the center, at this hour more people move from their homes to their activities areas near downtown. Figure 5 shows that at 12:00 h people are already in the city center, the vectors are less since the activity at social media decreases meanwhile people develop their personal activities such as study, work, among others. Is at this hour when less traffic reports are posted.

In Fig. 6 more activity is observable, the support vectors distributed in all the area reflect the moving of people returning to their homes, or moving to other places, not exactly to the areas where they work. The regressions are computed in the center of the city since it represents the starting point of the new routes that people would take in order to go to another place. It is important to mention that the time lapse between 18:00 and 20:00 h is the time at which most of the offices and schools finish their daily activities.

The support vectors are structured from the processed data of tweets related to traffic reported in the study area; the mapping of the regression outcomes show the congested areas in the city.

The numerous movements of people from one area to another become traffic problems; also, the overcrowding produces some other urban problems such as air pollution, economic losses, social problems, healthy problems such as citizens' stress, and time delays.

5 General Discussion

The implemented regression model provides information about future traffic events in the study area, with a view to prevent citizens from congested roads, and to speed up the movement of vehicles and people through the city. The regression results are geo-visualized in the study area's map, in order to identify the areas with more traffic problems and generate alternative routes to avoid congested areas.

Figure 7, show traffic predictions for different days, all the predictions were requested for a time lapse between 8:00 and 9:00 h for different days of the fourth week of March 2018. Such time lapse is considered since it is the time when most of the people moves through the city, and the roads get crowded.

Comparing the maps from Figs. 7, 8 and 9, it is possible to see that on Wednesday and Thursday the traffic reports are concentrated in the center and in the southeast of the city. Moreover, the predicted traffic events are distributed along the mentioned areas, which probes that people moves from around the city to the center. The biggest density of possible traffic events is computed in the center of the city; the average number of support vectors for the three cases is 2017 elements, the average number of predicted events for the three days is 597 possible events to occur between 8:00 and 9:00 h.

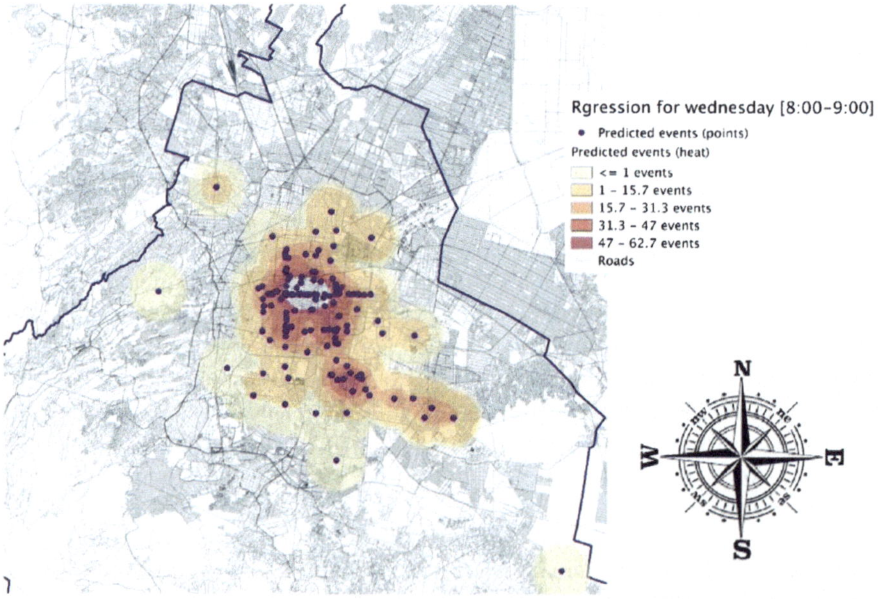

Fig. 7 The map shows the regression results obtained from Wednesday 21st March between 8:00 and 9:00 h

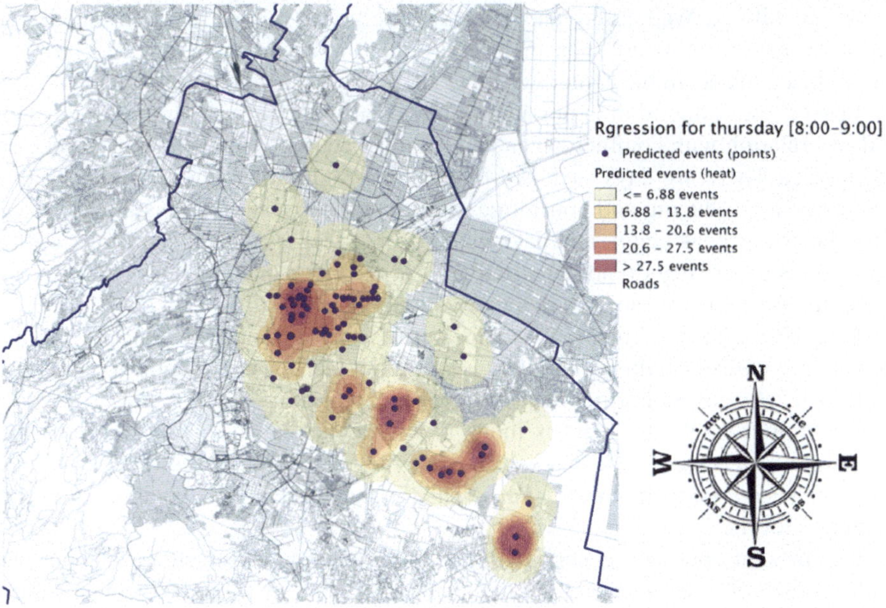

Fig. 8 The map shows the traffic events predicted from Thursday 22th March between 8:00 and 9:00 h

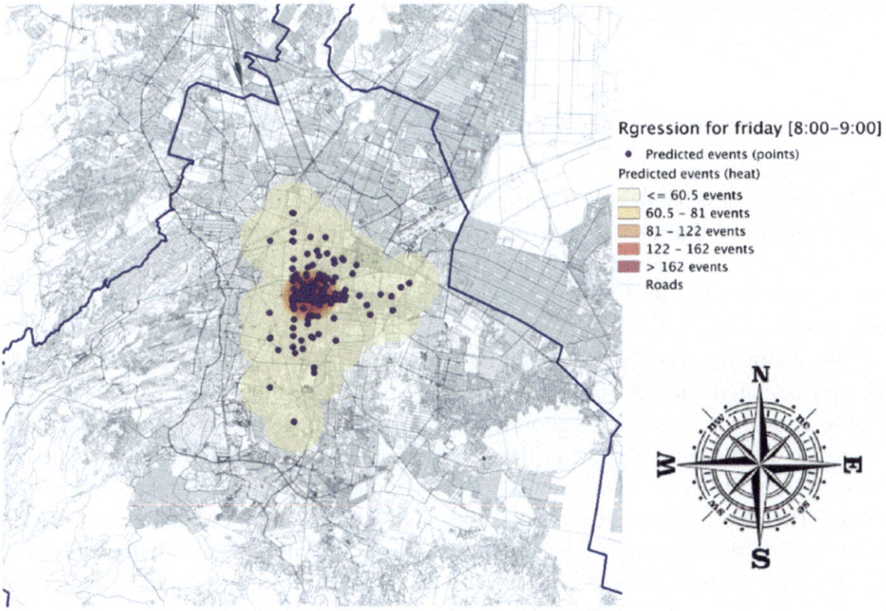

Fig. 9 The map shows the predictions computed from Friday 23rd March between 8:00 and 9:00 h

Meanwhile on Wednesday and Thursday the traffic reports are distributed in a similar way generating hot spots in different areas, on Friday there is a concentration of predicted events in the center, which generates just one hot spot in the map.

From Figs. 7, 8, and 9, is possible to infer that the number of traffic events has a direct relation with commercial and educational activities developed in the center of the city. There are days such as Monday and Wednesday when people is more interested in producing and consulting information about the traffic situation since their labor journeys begin, and days when people feel less interested in the traffic situation because they do not have activities in the city center, or they perceive less concurrence in the city such as in the weekend.

One of the main problems in urban areas is the pollution, specifically the air pollution. In Mexico City this problem is a constant increased by weather conditions. Since 2015 the city authorities have implemented precautionary measures to dismiss the air pollution. When vehicles move at low velocities (less than 40 km/h) the pollution is increased, as it is represented in the heat maps, in the hot spots areas the vehicles are trapped in traffic jams, which reduces their speed and increases the pollution levels.

The present approach generates the opportunity to have information about the traffic situation in urban areas and to calculate predictions to know the near future behavior of traffic. The main characteristic of the approach is that joints social computing, machine learning algorithms, and data from a social network. Another computing field used in the approach is the geo spatial processing, applied during the geo-referencing of the messages and in the geo-visualization of the data.

6 Conclusions

In the present approach, social computing is implemented by analyzing posts from a social network that describe traffic events in an urban area. The posts reflect the citizens interest on providing information about what happens in real world, such information can be used with different computing technologies such as Machine learning.

The obtained posts are processed by a text mining procedure and classified into six traffic related classes using a machine learning algorithm.

The classified messages are georeferenced in order to identify the geographic place where the events occurred. After the georeferencing of the messages, their features are treated to structure data vectors used to implement, train and test a regression model. The implemented regression model is based on the SVM machine learning algorithm, its structure computes predictions of future traffic events in the study area for specific days of the week at different time lapses. The presented methodology can be adapted to be used at different urban areas, it can be also separated to just classify tweets related to traffic, or to compute regressions of traffic events, both procedures are independent.

As future work, new data sources to feed the classification and regression models could be used, such as different social networks, or data obtained from mobile applications designed to monitor citizen's itineraries and authoritative data. Another improvement to the methodology could be the implementation of different regression algorithms in order to achieve results to be compared against the presents approach methodology.

The motivation of the approach is to use the social computing and data created by citizens to ameliorate the urban transporting conditions and the people's life.

References

1. Bishop, C.M.: Pattern Recognition and Machine Learning. Springer, Berlin (2006)
2. Burton, S.H., Tanner, K.W., Giraud-Carrier, C.G., West, J.H., Barnes, M.D.: Right time, right place health communication on Twitter: value and accuracy of location information. J. Med. Internet Res. **14**(6), e156 (2012)
3. Cheng, Z., Caverlee, J., Lee, K.: You are where you Tweet: a content-based approach to geo-locating twitter users. In: Huang, J. (ed.) Proceedings of 19th ACM International Conference on Information and Knowledge Management (CIKM) (2010), pp. 759–768. ACM, New York. https://doi.org/10.1145/1871437.1871535
4. Cherkassky, V., Ma, Y.: Practical selection of SVM parameters and noise estimation for SVM regression. Neural Netw. **17**(1), 113–126 (2004)
5. Cortes, C., Vapnik, V.: Support-vector networks. Mach. Learn. **20**(3), 273–297 (1995)
6. Cui, R., Gallino, S., Moreno, A., Zhang, D.J.: The operational value of social media information. Product. Oper. Manag. **27**(10), 1749–1769 (2018)
7. Flake, G.W., Lawrence, S.: Efficient SVM regression training with SMO. Mach. Learn. **46**(1–3), 271–290 (2002)
8. Gartner, G.: Emotional response to space as an additional concept of supporting wayfinding in ubiquitous cartography. In: Mapping Different Geographies, pp. 67–73. Springer, Berlin (2011)
9. Goodchild, M.F.: Citizens as sensors: the world of volunteered geography. GeoJournal **69**(4), 211–221 (2007)
10. Goodchild, M.F., Glennon, J.A.: Crowdsourcing geographic information for disaster response: a research frontier. Int. J. Digit. Earth **3**(3), 231–241 (2010)
11. Hahmann, S., Purves, R.S., Burghardt, D.: Twitter location (sometimes) matters: Exploring the relationship between georeferenced tweet content and nearby feature classes. J. Spat. Inf. Sci. **2014**(9), 1–36 (2014)
12. Haklay, M., Weber, P.: Openstreetmap: User-generated street maps. IEEE Pervasive Comput. **7**(4), 12–18 (2008)
13. Han, B., Baldwin, T.: Lexical normalisation of short text messages: Makn sensa #twitter. In: Proceedings of 49th Annual Meeting of the Association for Computational Linguistics: Human Language Technologies (HLT) (2011), vol. 1, ACL, pp. 368–378
14. Isaka, H., Nagayoshi, H., Yoshikawa, H., Yamada, T., Kakeno, N.: Next generation of global production management using sensing and analysis technology. Hitachi Rev. **65**(5), 47–52 (2016)
15. Kibanov, M., Becker, M., Mueller, J., Atzmueller, M., Hotho, A., Stumme, G.: Adaptive kNN using expected accuracy for classification of geo-spatial data. In: Proceedings of the 33rd Annual ACM Symposium on Applied Computing, pp. 857–865. ACM, New York (2018 Apr)
16. Koppen, M., Wolpert, D.H., Macready, W.G.: Remarks on a recent paper on the "no free lunch" theorems. IEEE Trans. Evol. Comput. **5**(3), 295–296 (2001)

17. Menard, S.: Applied Logistic Regression Analysis, vol. 106. SAGE Publications (2018)
18. Mu, X., Zhu, F., Liu, Y., Lim, E.P., Zhou, Z.H.: Social stream classification with emerging new labels. In: Pacific-Asia Conference on Knowledge Discovery and Data Mining, pp. 16–28. Springer, Cham (2018 June)
19. OSM. ©OpenStreetMaps Contributors. https://www.openstreetmap.org. Consulted on Mar 2018
20. Oatley, G., Crick, T., & Howell, R. (2015). Data exploration with GIS viewsheds and social network analysis
21. Pedregosa, F., Varoquaux, G., Gramfort, A., Michel, V., Thirion, B., Grisel, O., … Vanderplas, J.: Scikit-learn: Machine learning in Python. J. Mach. Learn. Res. **12**, 2825–2830 (2011 Oct)
22. Postgis. https://postgis.net/. Consulted on Mar 2018
23. Rahimi, A., Vu, D., Cohn, T., Baldwin, T.: Exploiting text and network context for geolocation of social media users (2015). arXiv preprint arXiv:1506.04803
24. Raiyani, K., Goncalves, P.Q.T., Beires-Nogueira, V.: Multi-language neural network model with advance preprocessor for gender classification over social media. In: Proceedings of the Ninth International Conference of the CLEF Association (CLEF 2018) (2018 Sept)
25. Saldana-Perez, A.M.M., Moreno-Ibarra, M., Tores-Ruiz, M.: Classification of Traffic Related Short Texts to Analyse Road Problems in Urban Areas. Int. Arch. Photogramm. Remote Sens. Spat. Inf. Sci. **42**, 91 (2017)
26. Seber, G.A., Lee, A.J.: Linear Regression Analysis, vol. 329. Wiley (2012)
27. Tokala, S., Gambhir, V., Mukherjee, A.: Deep Learning for social media health text classification. In: Proceedings of the 2018 EMNLP Workshop SMM4H: The 3rd Social Media Mining for Health Applications Workshop and Shared Task, pp. 61–64 (2018)
28. Wang, F.Y., Carley, K.M., Zeng, D., Mao, W.: Social computing: From social informatics to social intelligence. IEEE Intell. Syst. **22**(2) (2007)
29. Wing, B.P., Baldridge, J.: Simple supervised document geolocation with geodesic grids. In: Lin, D. (ed.) Proceedings of 49th Annual Meeting of the Association for Computational Linguistics: Human Language Technologies (HLT), New York, NY and USA, vol. 1, ACL, pp. 955–964 (2011)
30. Wolpert, D.H., Macready, W.G.: No Free Lunch Theorems for Search, vol. 10. Technical Report SFI-TR-95-02-010, Santa Fe Institute (1995)
31. Yong, Z., Youwen, L., Shixiong, X.: An improved KNN text classification algorithm based on clustering. J. Comput. **4**(3), 230–237 (2009)

Management City Model Based on Blockchain and Smart Contracts Technology

Higinio Mora, Raquel Pérez-delHoyo, Rafael Mollá Sirvent and Virgilio Gilart-Iglesias

Abstract Recent developments of Information and Communication Technologies (ICT) have led to new and innovative ways to face the challenges of modern societies. We could even say that a revolution is taking place the world over as a consequence of the new capabilities of provided by technology. Blockchain are one of these new disruptive technologies emerging in modern world. This technology provides a decentralized, efficient and secure management that enables new asset administration models. The main contribution of this work is the development of a management model based on blockchain technology and smart contracts. This research is within the research track smart cities and smart villages development and aims to facilitate the city management, the development of e-government and improve the quality of life of citizens when they interact with the administration.

1 Introduction

The procedures that have to be done to get any kind of license are usually tedious and slow. They oblige the applicant to travel to different places to request a multitude of forms, with the loss of time that this implies. Recent developments of Information

H. Mora (✉) · R. M. Sirvent · V. Gilart-Iglesias
Department of Computer Technology and Computation, University of Alicante,
Campus de San Vicente del Raspeig, 03690 Alicante, Spain
e-mail: hmora@dtic.ua.es

R. M. Sirvent
e-mail: rmolla@dtic.ua.es

V. Gilart-Iglesias
e-mail: vgilart@dtic.ua.es

R. Pérez-delHoyo
Department of Building Sciences and Urbanism, University of Alicante,
Campus de San Vicente del Raspeig, 03690 Alicante, Spain
e-mail: perezdelhoyo@ua.es

A. Visvizi and M. D. Lytras (eds.), *Research & Innovation Forum 2019*,
Springer Proceedings in Complexity,
https://doi.org/10.1007/978-3-030-30809-4_28

and Communication Technologies (ICT) have led to new and innovative ways to face the challenges of modern societies, as in this case obtaining licenses.

Virtual currencies and the technology that support them, such as blockchain are one of these new disruptive technologies emerging in modern society [1–3]. The propose a disruptive way of doing things differently that can change the world [4]. Thus, Blockchain technology provides a decentralized, efficient and secure management that enables new asset administration models [5]. This approach can be used in new and traditional areas, for example, for the development of smart city concept [6].

The use of blockchain and smart contracts technologies allow us to automate administrations processes without the need to have a centralized IT infrastructure, which would be costly and complex. This also will help to generate trust between administrations and citizens, reducing the time of concessions, eliminating the possibility of corruption and improving the image of administrations [7].

Given this scenario, we describe in this paper an idea for implementing a secure and agile management system based on new disruptive technologies such as blockchain and smart contracts. This system aims to facilitate the city management and improve the quality of life of citizens when they interact with the administration to make arrangements with the government. An analysis of the potential of the technology is performed and a quantitative study is conducted to find out the advantages over traditional models for both the citizens and public administrators.

This is just a preliminary work of this research to describe the main idea and the role played by new technologies, but further in-depth work is conducted to bring out the main benefits of this proposal and drive its implementation in real public administrations. Additionally, a further understanding of new technologies and their impact of society will help us to integrate these ICT advances in the domain of education in order to improve the training of future professionals [8, 9].

The remainder of this work is organized as follows: Sect. 2 describes the technologies on which or model is based; Sect. 3 describes the traditional models of citizen-administration interaction in urban development process; Sect. 4 presents our proposal; and finally, the conclusions are presented in Sect. 5.

2 Technologies

There are many technologies involved on a smart city development [10]. In this section, we will now describe the technologies on which it is based.

2.1 Blockchain

In recent years, the development of distributed computing and advances in cryptography have made possible the creation of blockchain technology. This technology was presented by Nakamoto in 2008. Basically, it is a network of nodes that collaborate with the objective of maintaining a distributed and secure database. It is the technology on which Bitcoin and other cryptocurrencies are supported [11].

A Blockchain is a distributed database (ledger) through a peer-to-peer network. Each node in the network maintains a copy of the ledger to prevent counterfeiting. All copies are updated and validated simultaneously.

The blockchain consists of a set of protocols and cryptographic methods applied to a network of nodes that collaborate to achieve secure data registration within a distributed database consisting of encrypted blocks that encapsulate the data.

The trust factor is one of the most important elements in blockchain technology [12]. In the blockchain, trust is ensured by the use of open-source code backed by cryptography.

Thanks to the encryption methods, each data block is wrapped in a protective layer in a secure way. The content of the blocks is delegated to the miners, who validate them by solving mathematical puzzles and consensus.

Blockchain can be seen as both a technical innovation and an economic innovation. This technology can be used to solve any problem where a transaction log is needed in a decentralized environment where all the parties involved are not trusted.

2.2 *Smart Contracts*

Smart contracts are pieces of code that code for real-world contractual agreements.

The contracts represent a binding agreement between two or more parties, where each entity must comply with its obligations presented in it. In this way, transactions can be processed by code without the requirement of intermediaries nor office clerks interventions. This can be carried out thanks to the automatic execution of code that distribute and verify the nodes of a blockchain network [13]. Smart contracts also offer the possibility of making transactions between unreliable parties without intermediate commissions, dependence on third parties and direct interaction between counterparts. It promotes efficiency and ensures transparency [14].

Another important part of a contract is that it can be enforceable by law, generally through an organization (centralized legal entity). The potential applications for e-government seem limitless.

3 Traditional Models of Citizen-Administration Interaction in Urban Development Process

The regulation and study of urban planning has traditionally been differentiated in three areas: planning, management and discipline.

The urban planning discipline includes all activities of the administration competent in urbanism that aims to verify and control that the ultimate action in the development process, consisting of construction, building and land use, complies with the law and planning.

This control is carried out in two stages:

(a) Ex ante control (before the action is carried out), i.e. the administrative activity of controlling urban legality. The construction, building or use of the land needs to obtain an enabling title beforehand. It consists of the traditional regime of planning permissions, which authorizes the individual to carry out the construction after verifying that it conforms to the planning; although today this system has been integrated with the concepts of the responsible declarations and previous communications, as we shall see.
(b) Ex post control (after the implementation of the action), i.e. protection of the urban legality or regime of urban infractions, which is repressive against the commission of an urban wrongful act that does not conform to the regulation or planning. This second control gives rise to two types of measures:

- Restorative measures: the aim is to ensure that the action complies with the law, either by obtaining the appropriate permission or by demolishing what has been illegally built.
- Sanctions measures: these are punitive in nature and seek to punish the person or persons responsible for the illicit act.

3.1 Planning Permissions

Cities or municipalities and regions are usually responsible for planning permissions and the proper application of zoning and building regulations.

Concept and legal nature. The urban planning confers only an initial aptitude for the construction of a land, but the right to build, is reached after the compliance of some burdens and obligations. The final verification of compliance with legislation and urban planning is embodied in the municipal license.

It is interesting to highlight two notes of the planning permission:

(a) It has constitutive effects: the license grants the right to build or to carry out urban planning actions. It is necessary the previous license to develop these activities. The building permit is also essential to authorize and register deeds of declaration of new building.
(b) It is a regulated administrative act: the administration is obliged to grant it if the request is in accordance with the applicable urban planning regulations, and legal duties and obligations have been complied with.

License, responsible declaration and previous communication. The traditional system for authorizing constructions, buildings and land uses is that of urban planning permissions or licenses. The interested party, in order to be authorized to carry out the construction, building and use of the land, requires a prior and favorable pronouncement by the administration.

This traditional system has been deeply altered after the approval of the Directive 2006/123/EC on services in the internal market [15], which has meant a transformation of the administrative intervention of citizens. This Directive states that the system of licenses should be replaced by less restrictive alternatives, provided that certain public interests are not affected.

In its development, the common administrative procedure legislation has provided, as alternatives to licenses, for the system of responsible declaration and previous communication.

The responsible declaration is the document that has to be subscribed by the interested parties in which they state, under their responsibility, that they comply with the legal requirements established in current legislation to access the recognition of a right, in our case that it meets all the requirements to execute some building works.

Previous communication is the document by means of which interested parties inform the administration of their identification data and other requirements for exercising a right or starting an activity.

The submission to the administration of the responsible declaration and previous communication enables the interested party to execute the building works without the need for any statement on the part of the administration, without prejudice to the faculties of control and inspection that may be exercised to verify that it complies with the law.

Acts subject to license. Despite the existence of the system of responsible declaration and prior communication, many actions in the process of urban development are still subject to license. A summary of the most common acts subject to license is shown in Table 1.

Conceptual description of the licensing procedure. As explained above, the general time limit for obtaining a building permit is two months. However, this time limit is increased to six months if an environmental permit is also required, as the environmental permit must first be obtained and then the building permit. In order to reduce these time limits, there is the possibility of the prior concession of the building permit but, to this end, the holder must renounce any type of compensation claim against the administration in the event that the environmental license is not subsequently granted.

The following Fig. 1 shows the fundamental stages of a licence application procedure.

In order to apply for a building permit and an environmental permit, it is necessary to present the basic documentation explaining the project to be carried out. The municipality performs a first formal check of the documentation submitted by the applicant. If the documentation is correct, the municipality informs all those directly affected about the project to be carried out and requests the corresponding reports from all the affected administrations. In addition, there is also a period of public information aimed at the general public. Within six months from the date of the application, taking into account all the allegations received in the period of public

Table 1 Regulation of urban intervention instruments: common acts subject to license

Acts requiring planning permission
The works of new plant of construction, building and implementation of facilities
The works of extension of existing constructions, buildings and facilities
The works of modification or refurbishment affecting the structure of constructions, buildings and facilities, whatever their use
The modification of the use of constructions, buildings and facilities
Works and uses to be carried out provisionally
Acts of intervention on heritage protected or catalogued buildings
Demolition of buildings
Placement of posters and billboards
The acts of parceling land
The execution of urbanization works
Earth moving that goes beyond the ordinary practice of agricultural work
The accumulation of discharges and the deposit of materials alien to the characteristics of the natural landscape
The construction of walls and fences
The opening of roads, as well as their modification or paving and, in general, any type of works or uses that affect the configuration of the territory
The felling of trees which are forests, groves or parks

exhibition, the municipality resolves favorably or negatively the licence application and informs the interested parties.

New detailed technical documentation is required in order to start the building work. Once the works have been completed, the municipality must be informed that it will start using the building. It is also necessary for the developer to declare that the completed building meets the conditions required by the regulations. One month after the presentation of the 'communication of the start of the activity' and the 'responsible declaration of first occupation', the activity and occupation of the building can start.

4 Management City Model

Given the complex process of obtaining licenses for the specific case of urban planning, we propose a model with which the processes are simplified and the time for obtaining permits is reduced considerably, making use of blockchain technologies and smart contracts *scripts*. In such a way, this management model aims to simplify and automatize the licensing process through these new disruptive technologies.

The smart contracts will be created and codified by the administration depending on the type of license and requirements needed. All features and requirements should

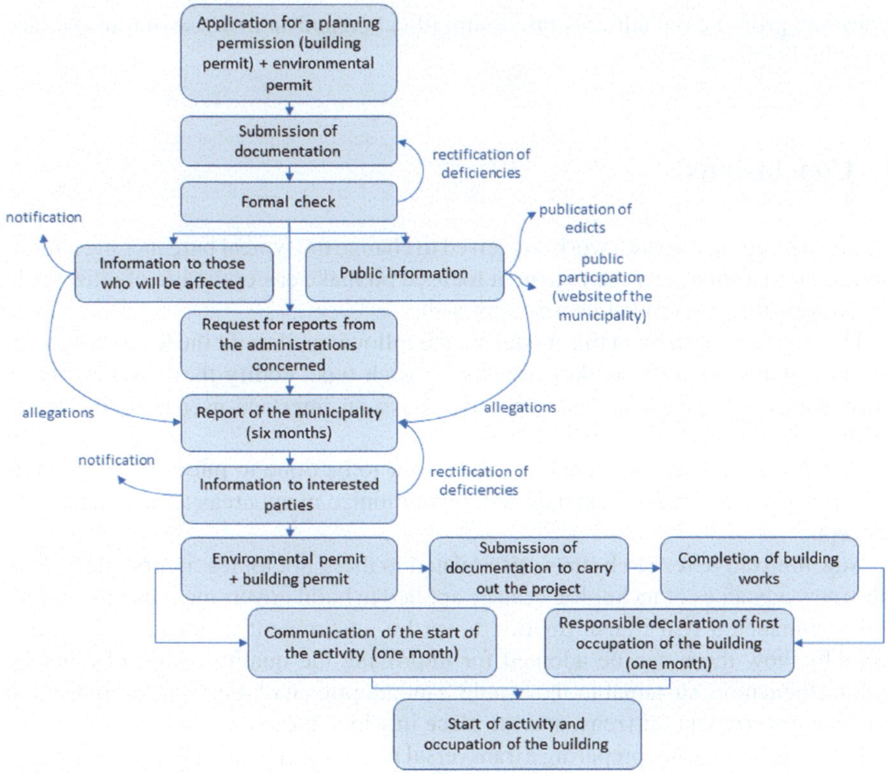

Fig. 1 Urban planning discipline: fundamental stages of a license application procedure

be specified and coded for automatically check. It is a complex task at first glance, but it can be carried out starting with obtaining licenses that require a minimum of formalities, and building up to more complex administrative proceedings.

In the first place, those types of licenses must be identified and published in order to be able to codify the requirements and data that the applicant must provide. Once the contract is codified, the citizens can fulfill the fields of the contract with the key features of the specific demanding license through a web interface or a mobile App. Next, a smart contract validates that the data provided by the applicant meets the requirements for granting the license.

The licenses granted will be added to the blockchain so that they are publicly accessible. In this way, the smart contract provides validity, without depending on authorities, due to its nature: it is a code visible by all community and that cannot be changed because it exists on the blockchain technology, which gives it decentralized, immutable and transparent features.

Consulted city managers believe that this technology facilitates the management, increase the security and dramatically reduces the delay for obtaining a license. This procedure promotes an administrative simplification and an increase of transparency in public licensing process.

5 Conclusions

This is an ongoing research work conceived to change the typical bureaucratic administration into a smart city management focused on make easier the life of citizens. In such a way, this is a citizen-centric approach.

The steps for implement this model are the following: classify the licenses according its features, identify the key aspects for each type, codify these aspects into a smart contract for each license type, and make an interface user-friendly for the citizen.

Nevertheless, there still remains much work to be done to put into practice this idea. It is expected that blockchain will revolutionize many areas around smart city concept.

New innovative research lines arise related to these hot topics: in first place, how these new advances of technology can be applied to build government infrastructures and to enhance a real transformation toward e-government in modern cities, and secondly, how these can be adopted for improving the quality of life of cities by making them more sustainable and livable. This adoption will speed up the connection between government, citizens, and the place in which they live.

In this regard, we are preparing a transversal new project on how ICTs can enhance the achievement of Sustainable Development Goals in smart cities. In this project a multidisciplinary team consisting of urban planners, public authorities of the city and computer engineers are working together to go ahead with these innovative ideas.

References

1. Phillip, A., Chan, J.S.K., Peiris, S.: A new look at cryptocurrencies. Econ. Lett. **163**, 6–9 (2018)
2. Mendoza-Tello, J.C., Mora, H., Pujol-López, F.A., Lytras, M.D.: Social commerce as a driver to enhance trust and intention to use cryptocurrencies for electronic payments. IEEE Access **6**, 50737–50751 (2018)
3. Mendoza-Tello, J.C., Mora, H., Pujol-López, F.A., Lytras, M.D.: Disruptive innovation of cryptocurrencies in consumer acceptance and trust. Inf. Syst. e-Bus. Manage. (2019)
4. Andoni, M.: Blockchain technology in the energy sector: a systematic review of challenges and opportunities. Renew. Sustain. Energy Rev. **100**, 143–174 (2019)
5. Sicilia, M.A., Visvizi, A.: Blockchain and OECD data repositories: opportunities and policy-making implications. Library Hi Tech. **37**(1), 30–42 (2019)
6. Lytras, M.D., Visvizi, A.: Who uses smart city services and what to make of it: toward interdisciplinary smart cities research. Sustainability **10**, 1998 (2018)
7. Pradip, K.S., Jong, H.P.: Blockchain based hybrid network architecture for the smart city. Futur. Gener. Comput. Syst. **86**, 650–655 (2018)
8. Lytras, M.D., Visvizi, A., Daniela, L., Sarirete, A., Ordonez De Pablos, P.: Social networks research for sustainable smart education. Sustainability **10**, 2974 (2018)

9. Mora, H., Pujol-López, F.A., Mendoza-Tello, J.C., Morales-Morales, M.R.: An education-based approach for enabling the sustainable development gear. Comput. Hum. Behav. (2018)
10. Lu, H.-P., Chen, C.-S., Yu, H.: Technology roadmap for building a smart city: an exploring study on methodology. Futur. Gener. Comput. Syst. (2019)
11. Daniel, M., Cristian, C., Shang, G.: Smart contract applications within blockchain technology: a systematic mapping study. Telemat. Inform. **35**(8), 2337–2354 (2018)
12. Axel, M., Benoît, D., Jean-Luc, B.: Blockchain based trust & authentication for decentralized sensor networks. CoRR abs/1706.01730 (2017)
13. Jonathan, L., Silvia, E.C.: Blockchain innovation beyond bitcoin and banking (2016)
14. Hughes, A., Park, A., Kietzmann, J., Archer-Brown, C.: Beyond bitcoin: what blockchain and distributed ledger technologies mean for firms. Bus. Horizons (2019)
15. EU: Directive 2006/123/EC of the European Parliament and of the Council of 12 December 2006 on services in the internal market (2006)

Emerging Computer Vision Based Machine Learning Issues for Smart Cities

Muhammad Murtaza Khan, Muhammad U. Ilyas, Sajid Saleem, Jalal S. Alowibdi and Mohammed Saeed Alkatheiri

Abstract Machine learning algorithms boast remarkable predictive capabilities and deep learning, a branch of machine learning, has already provided the much required breakthroughs for recognition and authentication. This has enabled the deployment of a face recognition based biometric identification system by the Department of Homeland Security at U.S. airports. Deep learning algorithms require huge amounts of data for training. However, this shall not be an issue in smart cities. Instead, the ability to use the same deep learning technology for ulterior motives raise some issues. Machine learning algorithms have already been developed which can generate fake images and videos and rendering humans incapable of differentiating between real and generated content. Such content can be used for spreading disinformation regarding individuals and may lead to legal issues. Any system, relying on face recognition may be mislead using such technology. Similarly, researchers were able to create master finger prints, i.e., a set of finger prints that may be used instead of an original finger print to defeat authentication biometric systems. Alongside data pollution, these image and video processing issues may present significant

M. M. Khan (✉) · M. U. Ilyas · S. Saleem · J. S. Alowibdi · M. S. Alkatheiri
College of Computer Science and Engineering, University of Jeddah,
Jeddah 23890, Saudi Arabia
e-mail: mkhan@uj.edu.sa; muhammad.murtaza@seecs.edu.pk
URL: http://www.uj.edu.sa/

M. U. Ilyas
e-mail: milyas@uj.edu.sa; usman.ilyas@seecs.edu.pk
URL: http://seecs.nust.edu.pk/

S. Saleem
e-mail: sssaleem@uj.edu.sa; sajid.saleem@seecs.edu.pk

J. S. Alowibdi
e-mail: jalowibdi@uj.edu.sa

M. S. Alkatheiri
e-mail: msalkatheri@uj.edu.sa

M. M. Khan · M. U. Ilyas · S. Saleem
School of Electrical Engineering and Computer Science, National University of Sciences and Technology, Islamabad 44000, Pakistan

A. Visvizi and M. D. Lytras (eds.), *Research & Innovation Forum 2019*,
Springer Proceedings in Complexity,
https://doi.org/10.1007/978-3-030-30809-4_29

challenges to governance of smart cities that shall rely on automated processing of such data. This paper presents an introduction to interpretability, attempts made to improve understanding of ML algorithms and their results. Area where we believe the definition of interpretability is lacking are also highlighted and a few example scenarios are presented. Finally, some possible directions for addressing the raised issues are introduced.

1 Introduction

1.1 Background

Smart cities shall rely on data collection not only from their citizens but also from devices and infrastructure capable of communicating with each other. This Internet of things (IoT) based environment shall thrive on availability of data and its processing for automation of daily activities. In this regard, machine learning (ML) shall play a critical role alongside availability of 5G communication network and development of smart infrastructure. The focus of machine learning algorithms will be automation and facilitation of daily tasks encompassing, but not limited to, driving, shopping, entertainment, seamless authentication, security and surveillance. Thus, not only shall machine learning assist residents in their daily lives but it shall be of greater importance to manage these smart cities and provide security to its residents. Policing, border control, security assurance, crime identification and punishment shall all be implemented with the aid of this technology. Although all of this may sound like science fiction, it is not, as certain machine learning systems are already being widely used to assist in decision making. However, these advancements may lead to issues and scenarios in which the advantages of automation may be questioned, especially relating to authentication.

1.2 Bias in Machine Learning Algorithms

Machine learning algorithms are already finding applications in decision making in the context of smart cities. They are being employed in areas such as recidivism prediction, i.e., helping estimate the likelihood that a criminal defendant will commit another crime in future. An example of such a system is Correctional Offender Management Profiling for Alternate Sanctions (COMPAS), which is being used for pre-trial decision making pertaining to parole [1]. However, this system came under criticism because of identification of certain inherent biases that have been exposed by researchers [2]. It has been observed that COMPAS assigns a black non-recidivating defendant twice the probability relative to a white defendant with similar profile for being high risk. Also, the probability to assign a black defendant low-risk is half that of a white defendant. One reason for this bias may be that many criminal offenders are never identified and hence there is a difference in the rates at which offenders are caught [3]. The machine learning community is aware of the problem and is trying to identify how decision making can be made more reliable and acceptable. In this regard an effort is being made to make the results and algorithms related to machine learning more interpretable.

1.3 Paper Organization

Section 2 covers some of the efforts being made to address the issues related to machine learning algorithms interpretability. Section 3 presents some issues related to use of machine learning algorithms in the context of image and video processing and their categorization in the categories presented in Sect. 2. Section 4 presents a discussion on addressing these issues and concludes the paper.

2 Interpretability of Machine Learning Algorithms

Predictive capabilities of machine learning algorithms have increased remarkably in the past few years. Faith in their ability to classify and predict soared after the success of AlexNet [4], which popularized deep learning and led to object recognition rates comparable to humans. The aftermath was development of bigger and deeper machine learning networks with focus on achieving better results as compared to humans without much focus on how they are being achieved. However, as the deployment of these systems started, questions arose about their reliability, i.e., can the model developed for prediction be trusted? Should algorithms that are unfamiliar with the concept of loan decide who qualifies for one? [5]. In this regard some efforts have been made to define interpretability of results [6]. It may be misconceived that there is a unanimous definition of interpretability, on which the machine learning community agrees. However, a quick search reveals that this is not the case. There are two major divisions when discussing interpretability of machine learning algorithms. One focuses on algorithms with accurate results, that are explainable or their work process can be understood [7]. While the second approach attempts to highlights desiderata of interpretability research and then the properties of interpretable models [6].

In [7], Bibal et al. highlight that as early as 1994 there was awareness about the interpretability of machine learning results, as highlighted by Feng and Michie [8], and that it has been addressed throughout the evolution of machine learning processes [9]. However, the focus was divided into coming up with what was termed as '*mental fit*' and '*data fit*'. Mental fit referred to the ability of a human to grasp and evaluate the model, while data fit referred to predictive accuracy.

Recently, the focus has shifted more towards the mental fit part of the problem. Current machine learning algorithms have reached super human accuracy in certain tasks whereas the focus is to achieve similar results for open challenges. However, the methodology or functionality by which the algorithms have reached the accuracy is still not clear. In this regard Lipton [6] has proposed to consider the following desiderata for interpretability.

Trust may be considered as the first desiderata, focusing on the ability of the model to be accurate. Accuracy may be defined as the ability of the algorithm to perform accurately when a human performs accurately and make mistakes when humans make mistakes, i.e., the ability of the algorithm to match human like capability [6]. It may be noticed that machine learning algorithms have achieved or even bettered human performance in specific tasks like object detection in images and playing chess.

Secondly, ***causality*** is considered important as it is considered that machine learning models should help in making inferences based on training data. However, it should not be overlooked that there may be unobserved causes that may be responsible for establishing some relationship. Humans have a great ability to generalize in unfamiliar situations so the same ***transferability*** may be desired from machine learning algorithms. It may be considered that this is exactly what machine learning algorithms do, i.e., training, validation and testing data sets are non-overlapping, yet the algorithms perform well [6]. However, it has been demonstrated that in certain adversarial environments the algorithms can be made to falter; algorithms inability to recognize objects that humans were able to recognize in case of addition of noise.

Informativeness is another desiderata as sometimes the machine learning model may be used to provide information to decision makers. An example can be identification of cases which are similar to the one for which the data has been presented to the algorithm. Last but not least, the general public is most interested to determine if the decisions being provided by machine learning algorithms can be validated as ***fair and ethical***.

In the context of machine learning algorithms for image and video processing, efforts have been made to understand the underlying process that an algorithm is utilizing to reach the desired accuracy and results [10–13]. Zeileret al. [10] presented a visual representation of reconstructed features at five different layers of a convolutional neural network to identify increasing invariance and class discrimination passing through each layer. Yosinski et al. [11] improved upon the idea by using regularization based optimization to produce better representative images and also presented a visual representation of activations at each level. In [12], authors focused on representation of long short term memory (LSTM) networks. Zhang et al. have proposed a filter based approach in which each filter corresponds to a particular object/part in each image used for recognition. Hence, ensuring that the algorithms functionality is understood.

These efforts address both 'mental fit' and 'data fit' properties desired in machine learning. They also address 'trust', 'causality', 'transferability'. Thus, addressing the basic concerns of the machine learning community. However, the focus on identifying the fairness of a result or if the result is ethical or not has not been addressed. Recently, deep fakes [14] have surfaced. This technology allows replacement of face of a person in a video with the face of another person. This machine learning based technology resulted in generation of video content which may be considered immoral and placed some people in compromising situations. Similarly, until recently, finger print based authentication was being used across multiple platforms and applications. However, development of master finger prints have resulted in questioning the efficacy of this method. In the following section we shall present a brief discussion on the above mentioned issues.

3 Image Processing Related Issues for Smart Cities

Authentication, access, surveillance, security shall all be seamless in cities of the future. Audio and video captured via IoT devices shall be used for identifying ongoing malicious activities. At present, the systems are not sufficiently integrated to autonomously identify crimes or process malicious events. However, in sporadic events machine learning and technology is being used for policing and criminal activity identification. The first example is the car speed surveillance systems deployed to identify an overspeeding vehicle. The machine learning based system detects overspeeding, takes a picture and archives it, processes the number plate from the image and generates a ticket. Wearable and personal technology is also being used for policing. In 2017 a man was charged with murder when his account of events, associated to the murder of his wife, did not adhere to the information obtained from his wife's Fitbit device [15]. Video evidence obtained from the mobile phone of the accused was introduced in a 2015 murder trial in which the prosecutors claimed to have found video footage of other victims [16]. DNA technology and other forensic science measures have improved law enforcement. However, consider the impact that technologies like deep fakes may have on video evidence. Fake video evidence against any one can be generated and this may result in false prosecution or conviction of innocent people. The same tactics of generating false evidence may be used to malign a criminal process by creating doubts about evidence in the case. In such scenarios the source of information will become increasingly important. This does not mean the individual but the physical source, i.e., device capturing the video. Currently, EXIF tags are added into images and XMP tags are added in videos, however, they may be edited and hence can be falsified.

Consider the case in which an actor performs some actions which are then reenacted by a target actor. This is exactly what Face2Face a real time face capture and reenactment tool does [17]. Using this tool the authors generated videos of celebrities performing strange facial expressions. However, this tool may be used for generating fake news. Consider another recent development in which Nvidia generated fake human faces [18]. Contrary to their previous efforts this time around it is very difficult to determine if the face is a real person or has been generated using Artificial Intelligence. To demonstrate the effectiveness of the proposed methodology they even demonstrated how the technology may be used to generate fake car images. Another recent application of machine learning is development of a virtual holocaust survivor [19]. To preserve history, Pinchas Gutter, a Nazi death camp survior, has been immortalized as a hologram. The idea called new dimensions in testimony, focuses on preserving history interactively where a user can ask questions and a hologram can answer them. This is an interesting concept that would make children interested in history, however, its malevolent use can create confusion a few hundred years from the time of origin. It may be impossible to identify the correct history if multiple such testimonies are available.

Facial impersonation is just one part of the problem. The current security systems comprise of authentication methods including facial recognition and finger print recognition. This is specially true when it comes to accessing information on personal devices. So far both methods were considered safe and one acted as a backup if the other failed. However recently Roy et al. [20] have been able to generate partial finger prints that may be used as master fingerprint. Thus, both facial and finger print based authentication measures are suspect.

Looking at the above examples holistically it can be said that machine learning algorithms have become powerful enough to deceive humans and generate data to impersonate them. This means that the algorithms can be *Trusted* as they produces accurate results and hence the *datafit* quality is good. The fact that the algorithms are becoming capable of generating new content indicates that *transferability* is being achieved. Since efforts are already being made to understand the output of each stage of computer vision based machine learning algorithms the area of *informativeness* is being addressed as well. The area that still requires attention is *fair and ethical* use. All of the above mentioned scenarios fail to ensure fair and ethical use of the algorithms Table 1. It may be argued this desiderata was conceived to identify bias in the output of a machine learning algorithm but alternatively it may be used as a classifier related to the capacity of an algorithm to cause harm. This is true for the COMPAS study [1] where an innocent person may be convicted because of the bias in the training of the algorithm or because of unethical capabilities of the algorithm.

Table 1 Machine learning applications and interpretability

Application	Data fit, truth	Data fit, causality	Data fit, transferability	Mental fit, informative-ness	Fair and ethical use
Face2Face	Yes	Yes	Yes	Yes	No
Face Gen.	Yes	Yes	Yes	Yes	No
Car Gen.	Yes	Yes	Yes	Yes	No
Fingerprint Gen.	Yes	Yes	Yes	Yes	No
Virtual history	Yes	Yes	Yes	Yes	No

4 Discussion and Conclusions

It can be argued that it shall be difficult to change information being obtained from smart surveillance and authentication infrastructure since this information shall be available only to competent authorities and hence the data shall be tamper proof. This may have been correct a decade ago, however, in today's world authorities cannot ignore the use of *user generated evidence* [21]. In 2017, the International Criminal Court issued arrest warrants for Libyan nationalist Mahmoud Mustafa Busayf Al-Werfalli for war crime of murder based on video footage posted by a Facebook user [22]. Considering the verdict, a scenario can be envisioned where a person under suspicion may be falsely accused if video or finger print evidence against them were generated using readily available machine learning tools. This becomes more problematic in a society which relies extensively on automation of surveillance and authentication for its functioning and that may employ smart technologies for identifying crimes and determining punishment.

With the passage of time, machine learning algorithms will become more sophisticated and hence will be able to produce results which humans will find difficult to identify as real or otherwise. Machine learning algorithms will be developed to differentiate between captured and modified (generated) images. However, the issue of fair and ethical use will remain. Therefore, it is imperative that guidelines be developed regarding use of facial recognition and finger print recognition for authentication and specially for the purpose of judicial proceedings. It is proposed that all devices capable of capturing audio, images or videos should add identification information in their streams, i.e., the identity of the device. This information could be in the form of a watermark of the device. This is already being done by printer manufacturers and a similar procedure may be proposed for audio, video capturing devices. Unlike EXIF tags or XMP tags, it should be illegal to edit watermark information in captured data. Any processing of such multimedia content will result in modification of the watermark and hence will indicate that the content has been tampered with. Since smart city infrastructure shall revolve around IoT time stamp information from the network along with location may be added to multimedia data that can later be retrieved for authentication. The above mentioned measures may be just a first step and a more comprehensive strategy must be developed to prevent unethical use of machine learning algorithms.

References

1. California Department of Corrections and Rehabilitation. https://www.cdcr.ca.gov/Rehabilitation/docs/FS_COMPAS_Final_4-15-09.pdf. Last accessed 25 Jan 2019
2. Blomberg, T., Bales, W., Mann, K., Meldrum, R., Nedelec, J.: Validation of the COMPAS risk assessment classification instrument. Florida State University, Tallahassee, FL, College of Criminology and Criminal Justice (2010)
3. Chouldechova, A.: Fair prediction with disparate impact: a study of bias in recidivism prediction instruments. Big Data **5**(2), 153–163 (2017)

4. Krizhevsky, A., Sutskever, I., Hinton, G.: Imagenet classification with deep convolutional neural networks. Adv. Neural Inf. Process. Syst. 1097–1105 (2012)
5. Lipton, Z.: The doctor just won't accept that! arXiv:1711.08037v2 (2017)
6. Lipton, Z.: The Mythos of Model Interpretability. arXiv:1606.03490v3 (2017)
7. Bibal, A., Frenay, B.: Interpretability of machine learning models and representations: an introduction. In: European Symposium on Artificial Neural Networks, Computational Intelligence and Machine Learning. Bruges, Belgium, 27–29 April 2016
8. Feng, C., Michie, D.: Machine learning of rules and trees. Machine Learning, Neural and Statistical Classification. Ellis Horwood, Hemel Hempstead (1994)
9. Weihs, C., Sondhauss, U.: Combining mental t and data t for classification rule selection. In: Exploratory Data Analysis in Empirical Research, pp. 188–203. Springer, Berlin (2003)
10. Zeiler, M., Fergus, R.: Visualizing and Understanding convolutional networks. In: European Conference on Computer Vision (ECCV), pp. 818–833 (2014)
11. Yosinski, J., Clune, J., Nguyen, A. M., Fuchs, T. J., Lipson, H.: Understanding Neural Networks through Deep Visualization. arxiv:1506.06579
12. Li, J., Chen, X., Hovy, E., Jurafsky, D.: Visualizing and Understanding Neural Models in NLP. arxiv:1506.01066
13. Zhang, Q., Wu, Y. N., Zhu, S-C.: Interpretable convolutional neural networks. In: IEEE Computer Vision and Pattern Recognition, pp. 8827–8836 (2018)
14. Deepfakes online application. https://deepfakesapp.online/. Last accessed 25 Jan 2019
15. Fitbit fitness tracker cracks Connecticut murder case. https://www.nydailynews.com/news/crime/police-solve-connecticut-murder-clues-fitbit-activity-article-1.3094802. Last accessed 25 Jan 2019
16. Detectives find new evidence in a high profile murder case. https://www.youtube.com/watch?v=JfBAg5t5Zoc. Last accessed 25 Jan 2019
17. Thies, J., Zollhofer, M., Stamminger, M., Theobalt, C., Niessner, M.: Face2Face: real-time face capture and reenactment of RGB videos. In: IEEE Conference on Computer Vision and Pattern Recognition (CVPR), pp. 2387–2395 (2016)
18. Karras, T., Laine, S., Aila, T.: A Style-Based Generator Architecture for Generative Adversarial Networks. arxiv:1812.04948
19. Virtual Holocaust Survivor. https://www.theguardian.com/technology/2016/jun/18/holocaust-survivor-hologram-pinchas-gutter-new-dimensions-history. Last accessed 25 Jan 2019
20. Roy, A., Memon, N., Ross, A.: MasterPrint: exploring the vulnerability of partial fingerprint-based authentication systems. IEEE Trans. Inf. Secur. **12**(9), 2013–2025 (2017)
21. Hamilton, R.J.: User-generated evidence. Columbia J. Trans. Law **1**(57) (2018) (American University, WCL Research Paper No. 2018-11)
22. ICC issues arrest warrant for Mahmoud Mustafa Busayf Al-Werfalli. https://www.icc-cpi.int/Pages/record.aspx?docNo=ICC-01/11-01/17-2. Last accessed 25 Jan 2019

Information Systems

The CRISP-DCW Method for Distributed Computing Workflows

Marco Spruit and Stijn Meijers

Abstract Big data analysis is increasingly becoming a crucial part of many organizations, popularizing the distributed computing paradigm. Within the emerging research field of Applied Data Science, multiple notable methods are available that help analysists and scientists to create their analytical processes. However, for distributed computing problems such methods are not available yet. Therefore, to support data analysts, scientists and software engineers in the creation of distributed computing processes, we present the CRoss-Industry Standard Process for Distributed Computing Workflows (CRISP-DCW) method. The CRISP-DCW method lets users create distributed computing workflows through following a predefined cycle and using reference manuals, where the critical elements of such a workflow are developed for the context at hand. Using our method's reference manuals and predefined steps, data scientists can spend less time on developing big data processing workflows, thus increasing efficiency. Results were evaluated with experts and found to be satisfactory. Therefore, we argue that the CRISP-DCW method provides a good starting point for applied data scientists to develop and document their distributed computing workflow, making their processes both more efficient and effective.

1 Introduction: Distributed Computing Workflows

The production of data is growing, and with it is the need to analyze data. Organizations are becoming more dependent on data when it comes to decision making, and knowledge discovery and data mining are well-researched topics [1]. Big data is a phenomenon that is written about a lot, and projections indicate that this trend is not stopping anytime soon [2]. A crucial part of big data analytics is the need for a scalable architecture to perform the analysis [3]. While data mining and data analytics

M. Spruit (✉) · S. Meijers
Department of Information and Computing Sciences, Utrecht University, Princetonplein 5, 3584 CC Utrecht, The Netherlands
e-mail: m.r.spruit@uu.nl

A. Visvizi and M. D. Lytras (eds.), *Research & Innovation Forum 2019*, Springer Proceedings in Complexity,
https://doi.org/10.1007/978-3-030-30809-4_30

methods are well-researched, and methods like CRISP-DM and SEMMA are used in practice [4], methods for creating and documenting these scalable architectures are not.

The most significant trend in the field of distributed computing is the emergence of big data processing platforms such as Hadoop and Spark. These big data processing platforms leverage distributed systems to perform complex and parallel computations over massive amounts of data. Platforms like these grow in popularity due to another phenomenon in the field: cloud computing. Cloud computing providers, such as Amazon Web Services and Microsoft Azure, make leveraging big data processing platforms such as Hadoop and Spark more accessible and make it a financially attractive option for organizations due to its elasticity [5]. Some researchers even argue that this particular type of computing will be seen as a utility, just like gas, water and electricity somewhere in the future [6].

Big data problems that occur in practice, often are repeated and not only designed to run once. The automation and modeling of these big data analytics processes can be done using workflows. Scientific workflows and workflow management systems are used to schedule analytical processes, make them more understandable and make them easily adjustable [7].

This research pursues an applied data science approach [8], with a particular focus on the information infrastructure dimension, by proposing a knowledge discovery method that helps data scientists set up big data processing platforms and workflows to make their analytical processes more effective and efficient. The following research question is answered:

> How can we design a distributed computing workflow for knowledge discovery, such that data scientists can improve their analytical processes' efficiency and effectiveness in a scalable manner?

The contribution that this paper delivers is the creation of the CRoss-Industry Standard Process for Distributed Computing Workflows (CRISP-DCW) method that makes big data processing workflows more accessible for data scientists and lets data scientists document their process and results in a structured manner, making the processes they create improvable. The method is evaluated with experts.

The paper is structured as follows: the research method and design are described in section two, the designed method is described in section three. Section four describes the conclusions and discussion.

2 Research Method: Applied Data Science

The primary purpose of this research is to create a method that aids data scientists in the development of well-documented and deployable distributed computing workflows. We use an Applied Data Science approach which embeds Design Science research as described by Wieringa [9] as our research method [10]. Problem investigation is the first step in Design Science. This is done through literature, and market research and interviews with experts. The second step in the Design Science cycle is the design of the artifact that will influence the context to solve the problem. The third step in the Design Cycle is the validation of the artifact, which will be explained in future work to fully conform to the applied data science approach.

The problem investigation is carried out through thorough literature research. A snowballing technique is applied to find and select papers for research and investigating the problem context. Interviews with experts are performed to get a better grip on the state of the market.

The artifact design is carried out by researching relevant literature and adjusting existing methods such as the Three Phases (3PM) Method [11, 12] and the CRISP-DM method [13] with the help of the findings from literature. A new method is created, to fit our current context of distributed computing. We create a method description and provide Process Deliverable Diagrams, which help create insights into the methods deliverables, actions to be carried out, and their interdependencies. Afterwards, we evaluate our model and method with experts through interviews to further discuss the created artifact.

3 Method Design

This research proposes the Cross-Industry Standard Process for creating Distributed Computing Workflows (CRISP-DCW) method to enable data scientists to create and document distributed computing workflows. The CRISP-DCW method has an iterative cycle consisting of three phases, each containing activities. The first phase, Problem Context, contains one activity (1) Define Context and Goals. The second phase, Design, consists of three activities: (2) Determine Input, (3) Estimate Output and (4) Determine Processing. The third phase, Implementation, consists of two activities: (5) Evaluate Results, and (6) Deploy Outcomes. The full method cycle is displayed in Fig. 1.

Each activity in the method is supported by tasks, to be executed by the user. Each task is associated with deliverables, which make up the documentation of the distributed computing workflow.

3.1 Define Context and Goals

This activity in the cycle describes the context in which the big data problem occurs. Tasks and sub-tasks are defined to be performed in this activity, helping the data scientist to structure this activity. These can be found in Table 1.

The focus of this activity is finding the goals and constraints, which will be used to develop the workflow. Time constraints, cost constraints, and available resources are listed to find out what resources are already available in the organization, as this

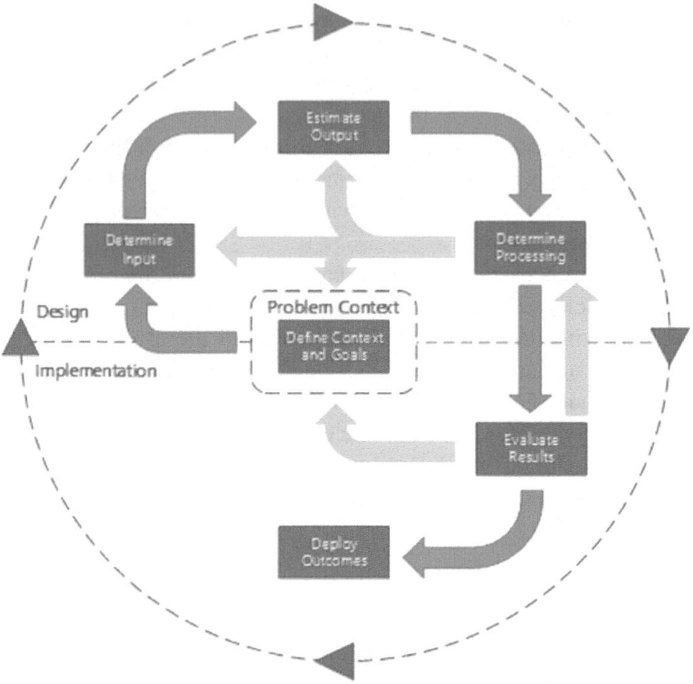

Fig. 1 Overview of the CRISP-DCW method

influences decisions on for example; cluster size and storage solutions [5]. Knowledge discovery goals are described to help pick the right tools and output format, as well as create a more clear view of the future use of results [14]. When there is a model available to be deployed, this is extensively described as this naturally influences among others; our choice of tool, cluster size, and storage.

3.2 Determine Input

The Determine Input activity describes the data which will be processed by the workflow. We use the four V's of big data as defined by the NIST Big Data Public Working Group [3] to describe our input data, as well as a fifth V: Volume, Velocity, Variety, Variability, and Value. Tasks and sub-tasks for this activity can be found in Table 2. As the choice in big data processing storage, engines and tools is heavily dependent on the type of data to be handled; it is essential that we cover all aspects of the input data.

Table 1 Tasks and sub-tasks in define context and goals activity

Task	Sub-task	Description
Determine data processing objectives	Describe time constraints	What are limitations to the processing time of this project?
	Describe cost constraints	Describe whether there are constraints on costs to be made in this project
	Describe knowledge discovery goals	Describe what kind of insights we aim to gain from the data processing
	Describe interpretability goals	Describe whether the process needs to be understandable from a computing perspective
Determine available resources	Describe hardware availability	Describe what hardware is available to use
	Describe software availability	Describe what software is available to use
	Describe knowledge availability	Describe what knowledge is available in the organization
Determine model specifications	Describe model or processing type	Describe the model or processing that will be performed in this project
	Describe language/platform of origin	Describe where the model originated
Create report	Summarize objectives	Summarize objectives
	Summarize resources	Summarize resources
	Summarize model	Summarize model
	List goals and constraints in a central table	Make all information easily available via a central table to be found by stakeholders

3.3 Estimate Output

We chose to first describe the envisioned output of the big data processing workflow, as this influences the decisions that are made in the data transformation process. The four Vs; Volume, Velocity, Variety, and Variability are described once more, now from the perspective of the estimated output data. Results and Visualization are also described. The Tasks and Sub-tasks that are described in this activity are found in Table 3.

Table 2 Tasks and sub-tasks in the determine input activity

Task	Sub-task	Description
Determine volume	Describe data size	Describe the size of the dataset(s), us gigabytes
Determine velocity	Describe batch versus stream	Describe at what speed the input is created, are we dealing with batch or stream data
	Describe timing	When does the data arrive?
Determine variety	Describe data format	Describe the data files' format. For example: .txt, .JSON, etc.
	Describe data structure	Is the data structured, semi-structured, or unstructured?
	Describe data type	Describe the data type to create an unobstructed view of the data
Determine variability	Describe data variability in volume	Describe in what way input data is expected to change regarding volume
	Describe data variability in velocity	Describe in what way input data is expected to change regarding velocity
	Describe data variability in variety	Describe in what way input data is expected to change regarding variety
Determine value	Describe origin of data	Describe where the data originated, this gives us insight into how to handle data, privacy
	Describe value insights	Describe what initial goals are set for getting value from this data
Assemble input report		Assemble all outputs in a central input report

3.4 Determine Processing

The Determine Processing activity is the most practical and hands-on activity in the cycle. This is where the choices are made for tools and where they are put into practice, using the previous activities and their reports. The CRISP-DCW method provides several reference manuals, which can be used to choose between big data processing tools, storage, and workflow engines. These reference manuals let the user make decisions based on the constraints and goals previously described.

Figure 2 shows the reference manual storage solutions. The manual shows a choice between three providers: Azure HDInsight [15], Amazon EMR [16], and On-Premise servers. These three are provided, as they are the most prominent in the field, and when no other resources are available, these are most reasonable [17]. The

Table 3 Tasks and sub-tasks in activity estimate output

Task	Sub-task	Description
Determine volume	Describe data size	Describe the size of the dataset(s), us gigabytes
Determine velocity	Describe batch versus stream	Describe at what speed the output is created, are we dealing with batch or stream data
	Describe timing	When does the data arrive?
Determine variety	Describe data format	Describe the data files' format. For example: .txt,. JSON, etc.
	Describe data structure	Is the data structured, semi-structured, or unstructured?
Determine variability	Describe data variability in volume	Describe in what way output data is expected to change regarding volume
	Describe data variability in velocity	Describe in what way output data is expected to change regarding velocity
	Describe data variability in variety	Describe in what way output data is expected to change regarding variety
Determine results and visualization	Describe results and visualization	Describe the envisioned results and visualizations, what would the ideal end-product look like?

four characteristics on which the choice of storage solutions is made are identified by Van Steen and Tanenbaum [18] and Voorsluys et al. [19].

The second reference manual provided with the CRISP-DCW method is the reference manual for Input and Output Processing. The reference manual is shown in Fig. 3. These tools can be used for the transfer of data from storage to processing engines or are used for moving the data from processing to storage. The user is given a choice between tools that can handle Structured, Semi-Structured and Unstructured data, and these tools are split up between Stream and Batch processing as well. Finally, the user chooses as to what data format is to be processed. The proposed tools and techniques are not exhaustive, but function as aid when users are inexperienced with tooling like this [20–24].

The third reference manual that is available in the CRISP-DCW method is shown in Fig. 4. This reference manual helps the user pick the right tools and processing engines to transform the data n the workflow. The method only focuses on the Hadoop and Spark engines as of now; this is due to their adoption in industry and high availability of functionalities on the primary cloud providers [25].

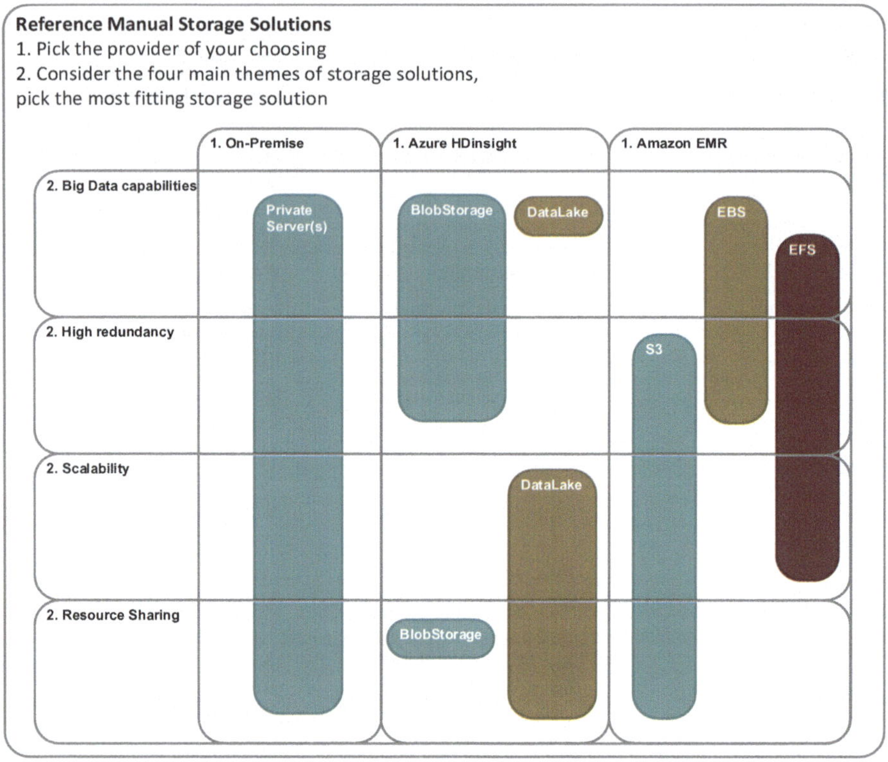

Fig. 2 Reference manual storage solutions

The user is guided in picking the right tools and engines by specifying the type of processing that the user wants to perform, the type of data that is to be transformed, and the language that has a preference. These are all described in the previous activities. The tools in this reference manual are not exhaustive, but are the most common on these platforms [26–30].

The fourth and last reference manual that is provided in this method is shown in Fig. 5. This reference manual is used to select the right workflow engine to be used in the project. Bchoosingng the processing engine of choice, the provider and the language of choice, the user is presented with the most suitable workflow engine. The reference manual is inspired by the documentation on these workflow engines [31, 32, 22].

In order to structure the Determine Processing activity, tasks and sub-tasks are defined. Table 4 shows all tasks, and sub-tasks in the processing activity, as well as a short description of what the tasks entail.

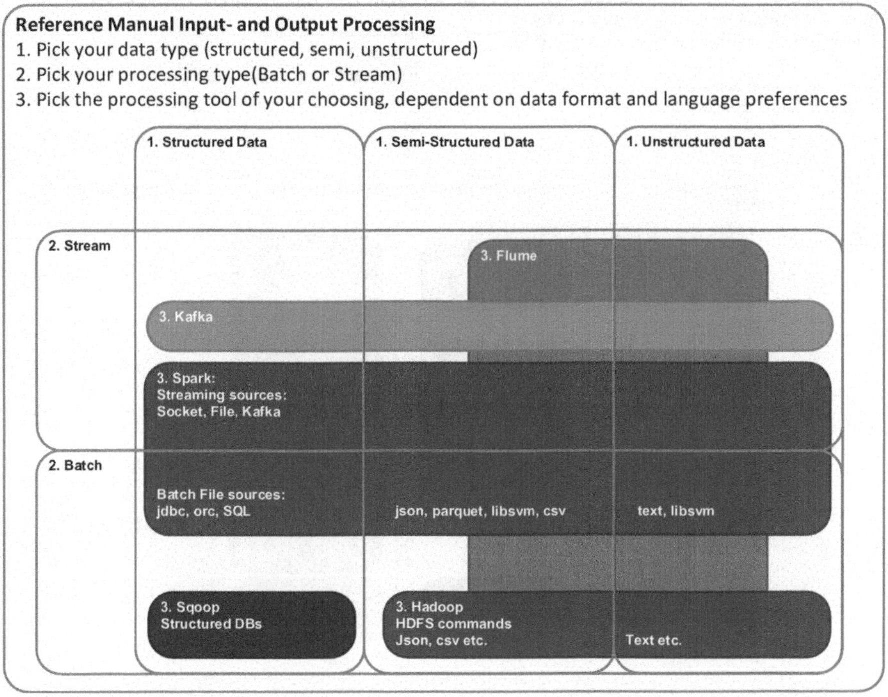

Fig. 3 Reference manual input- and output processing

3.5 Evaluate Results

The activity Evaluate Results described the results that are achieved when putting the workflow in practice. The workflow is evaluated using Performance metrics. Characteristics found by Van Steen and Tanenbaum [18] are used to evaluate the Distributed System that the project created. Complete tasks and sub-tasks performed in this activity can be found in Tables 5 and 6.

The phases, activities, and tasks in the method, their dependencies and deliverables are elaborated through a Process Deliverable Diagram (PDD). The PDD is found in Fig. 6. This modeling technique by Van de Weerd and Brinkkemper [33] is chosen because it makes the steps that the user has to take more explicit and shows what deliverables are expected after completing each activity.

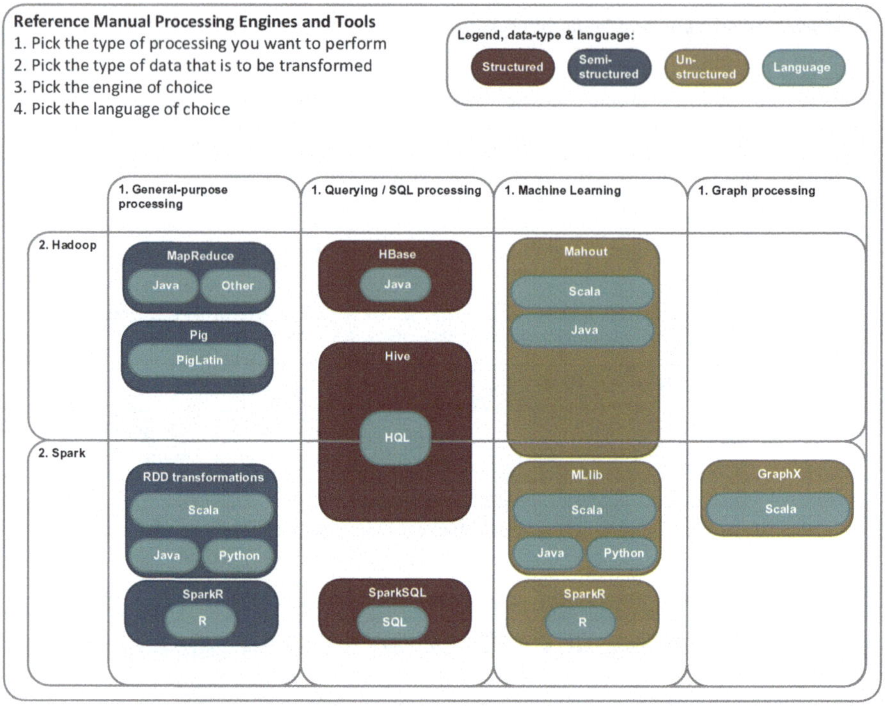

Fig. 4 Reference manual processing engines and tools

4 Evaluation and Discussion

The applied CRISP-DCW method and its results have been evaluated with experts. A data scientist and a software engineer, who are potential users of the method are interviewed on our results. The tool is found to be usable and understandable. Experts indicate that they would be able to reproduce the results using the documentation provided by the method. A data scientist remarked: "[Regarding the results] A focus on costs is not necessarily relevant for data scientists, but I can see the value of that for management", this is valuable information to consider, as each user will look at this method differently. A software engineer remarked: "It's a bit of a downside that there is an XML-like tool in play (XML feels a bit verbose), and it is not compatible with the latest version of Spark.". In further research regarding this method, it needs to be considered that the tools used are completely up-to-date, as some users might need these latest installments of the tools.

The method is considered to increase effectiveness in comparison to current methods, like CRISP-DM and SEMMA, as these offer no clear guidelines on how to leverage big data processing technologies. New insights can be created from data that were previously inaccessible due not being able to process these data on a single

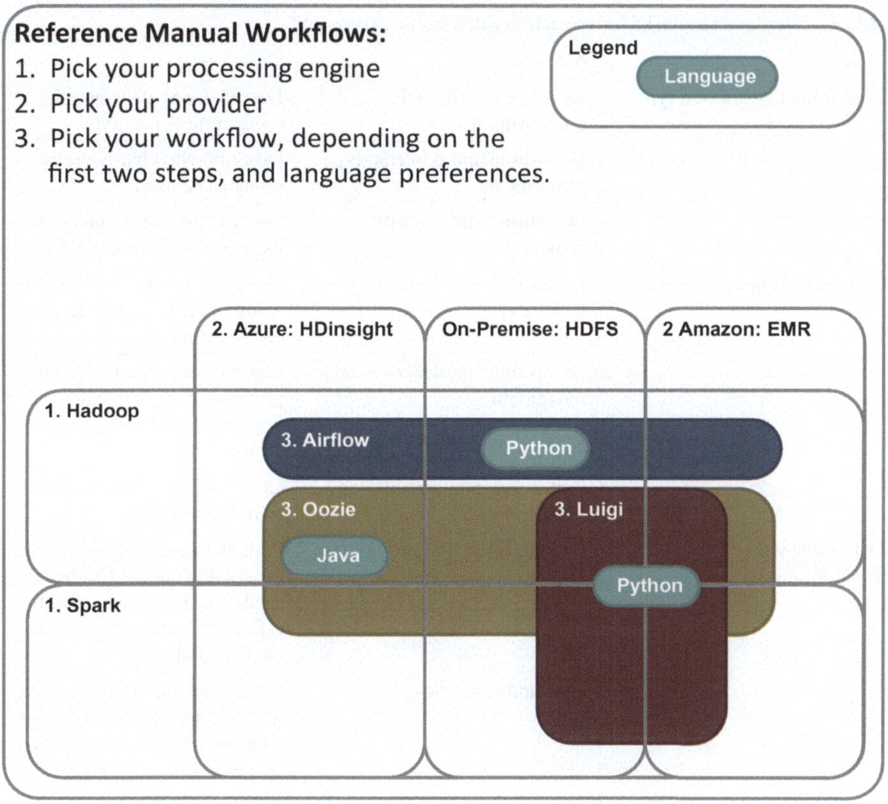

Fig. 5 Reference manual workflow engines

machine, and data scientists not having clear guidelines how to leverage distributed computing.

Efficiency of analytical processes should increase due to the guidelines and reference manuals provided with this method, whereas data scientists would previously need to take time to study what tools and techniques are available for big data processing and how these interact with each other, the CRISP-DCW method provides structure and reference manuals to easily find a fitting solution.

The rigor to the research was found to be adequate but needs to be improved by validating the method by exposing stakeholders and contexts to the method. This fell out of the scope of this research due to page constraints, but will be addressed in future research.

Table 4 Tasks and sub-tasks for the activity determine processing

Task	Sub-task	Description
Determine distribution type	Describe distributed computing need	Describe why distributed computing is needed
	Determine and describe type of hardware	Describe what hardware is going to be used
	Determine and describe provider(s)	Describe what providers are used, Azure versus AWS
Determine input processing	Determine and describe storage type	Use reference manual storage solutions to pick and describe chosen storage
	Determine and describe data movement	Use reference manual input processing to pick and describe chosen processing tool(s)
	Report on input processing	Create report of all results found in this task
Determine data transformation	Pick and describe processing engine	Pick and describe the processing engine for the models to be run on. Use the reference manual processing engines and tools
	Pick and describe type of job(s)	Determine and describe what the right jobs to run are when wanting to achieve the in- and output goals
	Report on set-up engine and job(s)	Write a report on the processing, provide scripts where needed
Determine output processing	Determine and describe output storage	Use reference manual storage solutions to pick and describe chosen storage
	Determine and describe data movement	Use reference manual input processing to pick and describe chosen processing tool(s)
	Report on output processing	Create report of all results found in this task
Set up workflow	Describe input, output, and job succession	Describe right succession for the developed input, processing, and output jobs
	Determine and describe workflow engine	Pick the right workflow engine for the goals and constraints. Use the reference manual workflow engines

(continued)

Table 4 (continued)

Task	Sub-task	Description
	Determine and describe processing timing	Describe how the workflow engine is set up and how the timing of running the workflow is set up. Use the constraints described in the first activity
	Embed processing in workflow	Embed the jobs that are developed in the workflow, use documentation of the provider, provide documentation and scripts where needed
	Provide report on workflow	Create a report of all results found in this task
Assemble processing report		Assemble all reports in this activity in a central document, which can be communicated to stakeholders

Table 5 Tasks and sub-tasks for the activity evaluate results

Task	Sub-task	Description
Evaluate performance	Describe results	Describe whether the results are effective regarding the analytical goals
	Describe timing and latency	Describe whether the timing of the workflow is acceptable and within constraints
	Describe costs	Describe what costs are associated with the newly developed workflow
Evaluate distributed system	Describe scalability	Describe the scalability of the distributed system, is it scalable enough for future growth, of shrinking of the needed capacity?
	Describe distribution transparency	Describe whether the distributed system is transparent enough for its users
	Describe openness	Describe the openness of the distributed system. Is it easily transformed and adaptable?
	Describe resource sharing	Describe whether the distributed system can share resources and results with other systems
Assemble evaluation report		Use all outputs in this activity and use them to make a report on the evaluation metrics

Table 6 Tasks and sub-tasks for the activity deploy outcomes

Task	Sub-task	Description
Evaluate performance	Describe results	Describe whether the results are effective regarding the analytical goals
	Describe timing and latency	Describe whether the timing of the workflow is acceptable and within constraints
	Describe costs	Describe what costs are associated with the newly developed workflow
Evaluate distributed system	Describe scalability	Describe the scalability of the distributed system, is it scalable enough for future growth, of shrinking of the needed capacity?
	Describe distribution transparency	Describe whether the distributed system is transparent enough for its users
	Describe openness	Describe the openness of the distributed system. Is it easily transformed and adaptable?
	Describe resource sharing	Describe whether the distributed system can share resources and results with other systems
Assemble evaluation report		Use all outputs in this activity and use them to make a report on the evaluation metrics

5 Conclusion

We introduce the CRoss-Industry Standard Process for Distributed Computing Workflows (CRISP-DCW) method to facilitate a more structured design and more insightful documentation of big data processing workflows. The proposed method serves two purposes. Firstly, it helps data scientists and software engineers to construct big data processing workflows, by guiding them through the process and determining the deliverables that are needed to create such workflows. The method offers several reference manuals to leverage big data processing tools and techniques. Secondly, the method supports the documentation and improvement of created workflows. By setting the standard for documenting workflows, the workflows are iteratively improvable, even when the data scientist currently working on the workflow has not originally created the workflow. Before the CRISP-DCW method, there was no standard process in place to do this kind of documentation. When organizations implement the CRISP-DCW method as a standard practice, time and resources will

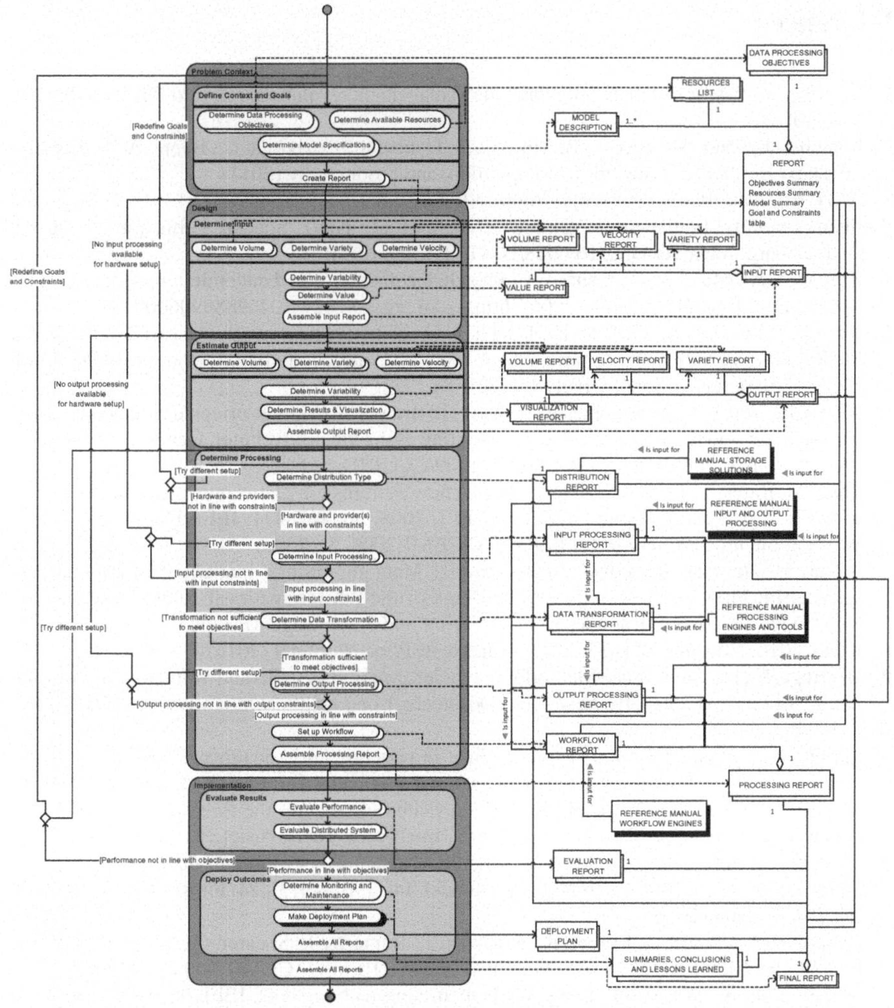

Fig. 6 The CRoss-Industry Standard Process for Distributed Computing Workflows (CRISP-DCW) as a Process Deliverable Diagram (PDD)

be saved as the method provides clear and concise instructions and documentation on how to develop big data processing workflows.

References

1. McAfee, A., Brynjolfsson, E.: Big data: the management revolution. Harvard Bus. Rev. **90**(10), 3–9 (2012)
2. Manyika, J., Chui, M., Brown, B., Bughin, J., Dobbs, R., Roxburgh, C., Byers, A.H.: Big data : the next frontier for innovation, competition, and productivity (2011)
3. NIST Big Data Public Working Group: NIST Special Publication 1500-1—NIST Big Data Interoperability Framework: Volume 1, Definitions. NIST Special Publication (Vol. 1). Gaithersburg. https://doi.org/10.6028/NIST.SP.1500-1 (2015)
4. Kurgan, L.A., Musilek, P.: A survey of knowledge discovery and data mining process models. Knowl. Eng. Rev. **21**(1), 1–24 (2006). https://doi.org/10.1017/S0269888906000737
5. Armbrust, M., Fox, A., Griffith, R., Joseph, A.D., Katz, R., Konwinski, A., Lee, G., Patterson, D., Rabkin, A., Stoica, I., Zaharia, M.: A view of cloud computing. Commun. ACM **53**(4), 50–58 (2010). https://doi.org/10.1145/1721654.1721672
6. Buyya, R., Yeo, C.S., Venugopal, S., Broberg, J., Brandic, I.: Market-oriented cloud computing: Vision, hype, and reality of delivering computing as the 5th utility. Futur. Gener. Comput. Syst. **25**(6), 599–616 (2009). https://doi.org/10.1109/CCGRID.2009.97
7. Zhao, Y., Raicu, I., Foster, I.: Scientific workflow systems for 21st century, new bottle or new wine? In: IEEE Congress on Services—Part I, 2008, pp. 467–471. IEEE Computer Society, Washington. https://doi.org/10.1109/SERVICES-1.2008.79 (2008)
8. Spruit, M., Jagesar, R.: Power to the people! Meta-algorithmic modelling in applied data science. In: Fred, A. et al. (eds.) Proceedings of the 8th International Joint Conference on Knowledge Discovery, Knowledge Engineering and Knowledge Management, pp. 400–406. KDIR 2016, November 11–13, 2016. ScitePress, Porto, Portugal (2016)
9. Wieringa, R.: Design Science Methodology for information Systems and Software Engineering, vol. 2. Springer, Heidelberg, New York, Dordrecht, London. https://doi.org/10.1145/1810295.1810446 (2010)
10. Spruit, M., Lytras, M.: Applied data science in patient-centric healthcare: adaptive analytic systems for empowering physicians and patients. Telematics Inform. **35**(4), 643–653 (2018)
11. Ooms, R., Spruit, M., Overbeek, S.: 3PM revisited: dissecting the three phases method for outsourcing knowledge discovery. Int. J. Bus. Intell. Res. **10**(1), Article 5 (2019)
12. Vleugel, A., Spruit, M., Van Daal, A.: Historical data analysis through data mining from an outsourcing perspective: the three-phases model. Int. J. Bus. **1**(3), 24. https://doi.org/10.4018/jbir.2010070104 (2010)
13. Chapman, P., Clinton, J., Kerber, R., Khabaza, T., Reinartz, T., Shearer, C., Wirth, R.: CRISP-DM 1.0. In: CRISP-DM Consortium. https://doi.org/10.1109/ICETET.2008.239 (2000)
14. Wu, X., Zhu, X., Wu, G.-Q., Ding, W.: Data mining with big data. IEEE Trans. Knowl. Data Eng. **26**(1), 97–107 (2014). https://doi.org/10.1109/TKDE.2013.109
15. Microsoft: Microsoft Azure. Retrieved from https://azure.microsoft.com/ (2017 November 24)
16. Amazon Web Services Inc.: Amazon Web Services. Retrieved from Amazon Web Services: https://www.aws.amazon.com/ (2017 December 13)
17. Leong, L., Bala, R., Lowery, C., Smith, D.: Magic Quadrant for Cloud Infrastructure as a Service, Worldwide (2017)
18. Van Steen, M., Tanenbaum, A.S.: Distributed Systems, Third, vol. 1. Maarten van Steen (2017)
19. Voorsluys, W., Broberg, J., Buyya, R.: Introduction to cloud computing. In: Buyya, R., Broberg, J., Goscinski, A. (eds.) Cloud Computing: Principles and Paradigms, 1st ed., pp. 3–41. Wiley (2011)
20. Apache Spark: Spark Overview. Retrieved from Apache Spark. https://spark.apache.org/docs/2.3.0/index.html(2018 April 17)
21. The Apache Software Foundation: Apache Hadoop. Retrieved from Apache Hadoop: http://hadoop.apache.org/ (2017 November 28)
22. The Apache Software Foundation: Documentation. Retrieved from Apache Kafka a distributed streaming platform: https://kafka.apache.org/documentation/#uses (2017 April 17)

23. The Apache Software Foundation: Flume 1.8.0 User Guide. Retrieved from Apache Flume: https://flume.apache.org/FlumeUserGuide.html (2018 April 17)
24. The Apache Software Foundation: User Guide. Retrieved from Apache Sqoop: http://sqoop.apache.org/docs/1.99.7/user.html (2018 April 17)
25. Allen, R., Li, M.: Ranking Popular Distributed Computing Packages for Data Science. Retrieved from KDnuggets. https://www.kdnuggets.com/2018/03/top-distributed-computing-packages-data-science.html (2018, March 29)
26. Apache Spark: SparkR (R on Spark). Retrieved from Apache Spark. https://spark.apache.org/docs/latest/sparkr.html (2018 April 17)
27. Armbrust, M., Ghodsi, A., Zaharia, M., Xin, R.S., Lian, C., Huai, Y., Liu, D., Bradley, J.K., Meng, X., Kaftan, T., Franklin, M.J.: Spark SQL: relational data processing in spark. In: Proceedings of the 2015 ACM SIGMOD International Conference on Management of Data—SIGMOD '15, pp. 1383–1394. https://doi.org/10.1145/2723372.2742797 (2015)
28. Meng, X., Bradley, J., Yavuz, B., Sparks, E., Venkataraman, S., Liu, D., Freeman, J., Tsai, D.B., Amde, M., Owen, S., Xin, D., Xin, R., Franklin, M.J., Zadeh, R., Zaharia, M., Talwalkar, A.: MLlib: machine learning in Apache Spark. J. Mach. Learn. Res. **17**, 1–7 (2016). 10.1145/2882903.2912565
29. White, T.: Hadoop: The Definitive Guide (Third). O'Reilly, Beijing, Cambridge, Farnham, Koln, Tokyo (2015)
30. Xin, R.S., Gonzalez, J.E., Franklin, M.J., Stoica, I.: GraphX: a resilient distributed graph system on spark. In: First International Workshop on Graph Data Management Experiences and Systems (GRADES 2013), p. 6. https://doi.org/10.1145/2484425.2484427 (2013)
31. Islam, M., Huang, A.K., Battisha, M., Chiang, M., Srinivasan, S., Peters, C., Srinivasan, S., Peters, C., Neumann, A., Abdelnur, A.: Oozie: towards a scalable workflow management system for Hadoop. In: Proceedings of the 1st ACM SIGMOD Workshop on Scalable Workflow Execution Engines and Technologies—SWEET '12 (pp. 1–10). Scottsdale: ACM. https://doi.org/10.1145/2443416.2443420 (2012)
32. Spotify AB: Luigi is now open source: build complex pipelines of tasks. Retrieved from Spotify Developer: https://developer.spotify.com/news-stories/2012/09/24/hello (2012 September 24)
33. Van de Weerd, I., Brinkkemper, S.: Meta-modeling for situational analysis and design methods. In: Syed, M.R., Syed, S.N. (eds.) Handbook of Research on Modern Systems Analysis and Design Technologies and Applications, pp. 38–58. Information Science Reference, Hershey. https://doi.org/10.4018/978-1-59904-887-1.ch003 (2009)

Towards a Software Engineering Framework for the Design, Construction and Deployment of Machine Learning-Based Solutions in Digitalization Processes

Ricardo Colomo-Palacios

Abstract There is an increasing demand of digitalization technologies in almost all aspects in modern life. A swelling part of these technologies and solutions are based on Machine Learning technologies. As a consequence of this, there is a need to develop these solutions in a sound and solid way to increase software quality in its eight characteristics: functional suitability, reliability, performance efficiency, usability, security, compatibility, maintainability and portability. To do so, it is needed to adopt software engineering and information systems standards to support the process. This paper aims to draw the path towards a framework to support digitalization processes based on machine-learning solutions.

1 Introduction

The world is now facing a deep transformation powered by digitization [1] towards the, so called, fourth industrial revolution [2]. According to a recent study on Maturity Models for Digitalization [3], the five main aspects in the topic are Strategy, Processes, Technologies, Products & Services and People. Focusing on technology, artificial intelligence technologies, help to implement the solutions towards digitalization [4] by means of the capabilities to analyse process data. This enables the shift in the fourth industrial revolution in the area of decision-making [1].

Given the importance of Digitalization and Digitalization Strategy, main consultancy companies in the world presented frameworks to help organizations in their strategies. This includes Boston Consultancy Group [5], Cognizant [6], IDG [7] or McKinsey [8], citing just a few of the most important frameworks. Most of the frameworks include a step to develop the platform to support new business strategies, including analytics and machine learning. However, these frameworks are not analysing in deep how these systems must be developed to align them to the new business strategy rooted in Digitalization. Apart from pure software development

R. Colomo-Palacios (✉)
Østfold University College, 1757 Halden, Norway
e-mail: ricardo.colomo-palacios@hiof.no

A. Visvizi and M. D. Lytras (eds.), *Research & Innovation Forum 2019*,
Springer Proceedings in Complexity,
https://doi.org/10.1007/978-3-030-30809-4_31

aspects, alignment aspects must be ensured by means of Information Technology (IT) Governance frameworks e.g. ISO 38500 [9] or Control Objectives for Information and Related Technologies (COBIT) [10, 11].

Literature reported Machine Learning, as a field inside artificial intelligence that uses statistical and novel techniques such as, regression analysis and neural network respectively, to give computer systems the ability to learn, as one of the most promising fields of research in the area [12–15]. However, software quality, remote software updates, in-house software development, and security will become critical success factors in Machine Learning solutions [16]. On the other hand, initial doubts on reliable engineering appear in machine learning-based systems as they become more prevalent.

Traditionally, Machine Learning is a tool for software engineering and its developments to support software engineering at large have been reported vastly in the literature e.g. [17–21]. Conversely, recently machine learning and software engineering communities launched a call for an integration between Machine Learning and Software Engineering disciplines towards cross fertilization and not just a unidirectional influence. As a consequence of this nascent importance, the first International Workshop on Machine Learning and Software Engineering in Symbiosis took place in Montpellier in September 2018 [22]. In this conference, organizers called for more research on the topic. Another interesting venue is the First Symposium on Software Engineering for Machine Learning Applications at Polytechnique Montréal (12–13 June 2018). In this event, reported in [23], organizers encouraged Software Engineering and Machine Learning communities to collaborate to solve the critical aspects of assuring the quality of artificial intelligence and software systems.

Taking this into account, to the best of authors' knowledge, there is not a framework to guide the analysis, design, construction and deployment of machine-learning oriented solutions to support smart digitalization processes, aligned with new business processes designed in the digitalization and presenting a rigorous observance of software engineering methods, standards and practices. This paper is aimed to trace the path to bridge this research and practice gap.

2 Relevance

Technology is a key aspect for the Digital Society. In digitalization settings, the software aspect is together with hardware, the backbone of the concept and its instantiation in organizational settings. However, the development of technology in general and software in particular presents several challenges. Some of these concerns are aspects revealed fifty years ago when the Software Engineering term was coined: low maintainability, increasing costs and time to market and technical debt, naming just some of the main issues. Focusing on the Machine Learning aspects of the software developed inside digitalization processes, software development process in such settings can present specificities that must be studied in order to create a path to guide practitioners in the development of such solutions. The framework developed in this framework must be rooted in four different aspects:

1. Digitalization frameworks. The study of modern practices in digitalization will ensure a proper digitalization strategy including the most important aspects of it that deals with technology.
2. IT and Business alignment. Main alignment and IT Governance tools and frameworks will be studied to support the development and deployment of new systems.
3. Software Development standards and practices. Following previous efforts to support software process in specific functional areas e.g. automotive or health, a set of recommendations will be taken from main standards and practices to support the development of solutions in the settings of this project.
4. Machine Learning Specifics. Solutions designed must include best of breed algorithms and as a part of the framework, a guiding tool to select most convenient algorithms as well as a set of practices will be included in the framework.

As a consequence of its cross-disciplinary nature, this topic is, from the author's perspective, central to the development of research and innovation, since it is aimed to enhance the technology sustaining the Digital Society as a whole. Moreover, the particularization of the framework to be developed inside digitalization projects, makes the proposal even more attractive for the local industry, aimed to embrace the concept and evolve in their digitalization process. Given that it is aimed to test the framework in a set of local industries, the benefit for these organizations will be twofold. Firstly, they will benefit from a specific software developed as a product for them and secondly, they will benefit from the knowledge gained as best-practices in the process of development of the framework.

End users of the framework will be software practitioners aimed to develop machine learning-based solutions including colleagues in academia, industry and practitioners worldwide.

Stakeholders in the development will be all organizations and practitioners that will help in the construction of the framework but also those companies that will adopt the framework to guide part of their digitalization process including companies in the region that will validate the framework.

3 Objectives and Hypothesis

This project focuses on the design of a comprehensive framework based on the approach presented herein, and with the capabilities of addressing the proposed research challenges. Thus, the research objectives can be broken down into the following sub-objectives:

Objective 1. Investigate and gather existing models, constructs and approaches within the industry and the research community related to the aims of this work.

Objective 2. Collect, unify and improve existing approaches if any, and propose new techniques and standards if required to solve the described problem.

Objective 3. Devise and design an approach, based on study previously performed, and with the capabilities for meeting the research challenges pointed out in this document.

Objective 4. Develop a framework as an artefact that permits its evaluation in terms of applicability, quality efficiency, and efficacy aimed to demonstrate its feasibility to solve the business problem.

Objective 5. Validate the framework in real-world scenarios.

Objective 6. Evaluate the proposed framework and compare it with related research contributions in the area and other existing approaches in the industry, if any.

Taking into consideration what is stated above, it is possible to formulate the hypothesis aimed to be validated in this doctoral thesis, as follows:

> If there exists a framework that allows corporations to guide them in the construction and deployment of machine learning-based solutions in digitalization processes in a timely manner, and regardless of their nature, then such framework can be adopted by organizations to manage and optimize their development and adoption of machine learning-based solutions in digitalization processes.

4 Method

The proposed research method for this project will follow the design-science paradigm for Information Systems research. More specifically, the Information Systems Research Framework described in [24] will be extended and adapted for the purpose of this work. According to its authors, the fundamental principle of design-science research is that "knowledge and understanding of a design problem and its solution are acquired in the building and application of an artifact".

The seven foundational guidelines of the proposed framework are as follows [24]:

Guideline 1: Design as an artifact. "Design-science research must produce a viable artifact in the form of a construct, a model, a method, or an instantiation". It is proposed to build a framework that ensures the proper alignment, construction and testing of machine learning solutions inside digitalization efforts.

Guideline 2: Problem relevance. "The objective of design-science research is to develop technology-based solutions to important and relevant business problems". The construction of the proposed framework will provide a guidance to practitioners and organizations alike. This will allow the organizations to ensure a proper solution aligned to their business objectives and the practitioners a way to develop software in an integrated and secure way.

Guideline 3: Design evaluation. "The utility, quality, and efficacy of a design artifact must be rigorously demonstrated via well-executed evaluation methods". The design evaluation approach used in this project encompasses an empirical research methodology. More specifically, key aspects of the proposed framework will first be verified by expert interviews, then by experimental settings in specific organizations.

Results will be evaluated qualitatively and also quantitatively (where the data basis allows it). Examples of quantitative assessments can be the changes in specific key performance indicators (KPIs) in the organization.

Guideline 4: Research contributions. "Effective design-science research must provide clear and verifiable contributions in the areas of the design artifact, design foundations, and/or design methodologies". The proposed compliance framework is novel and it will address the specific intersection of (a) Machine Learning (b) Software Engineering (c) Digitalization and (d) IT Governance. The main research contributions are focused on filling the gaps between these areas and thereby providing a generally applicable framework that is domain-independent and technology-agnostic.

Guideline 5: Research rigor. "Design-science research relies upon the application of rigorous methods in both the construction and evaluation of the design artifact". The proposed framework will be developed after a structured and in-depth assessment of related works in the research landscape. Each important construct of the framework will be rigorously evaluated with respect to its contribution to the overall framework aims while both framework and individual constructs will be considered through the prism of whether they verify or falsify to posited hypotheses.

Guideline 6: Design as a search process. "The search for an effective artifact requires utilizing available means to reach desired ends while satisfying laws in the problem environment". The artifact design will be undertaken incrementally and iteratively by refining the framework and the underlying assumptions against new research challenges, alternatives or issues.

Guideline 7: Communication of research. "Design-science research must be presented effectively both to technology-oriented as well as management-oriented audiences".

Apart from the overall approach, several techniques will be adopted in the project:

Systematic Literature Reviews. Based on the guidelines by [25], these tools will be used to investigate literature in a rigorous way.

Mappings among standards and practices will be used to support the development of the framework based on the guidelines published in [26].

Qualitative Techniques. These techniques will be used for gathering information in the construction of the framework but also in the assessment of it, and include mostly Grounded Theory approaches [27], but also other approaches like focus groups, Delphi studies and interviews will be used.

Quantitative Techniques. Although the approach on the thesis in the assessment will be mostly qualitative, several quantitative measures and analysis will be employed also to support qualitative studies.

Systematic Literature Reviews and Mappings will be relevant for Objectives 1 and 2 while Qualitative techniques are key in Objectives 4, 5 and 6 and to a lesser extent to Objective 3. Finally, Quantitative Techniques are meant to be used to fulfill Objectives 5 and 6.

5 Conclusions

In this paper, a proposal for the development of a framework to guide the design, construction and deployment of machine learning-based solutions in digitalization processes based on software engineering is presented. Given the importance of the topic results of the proposed research could benefit society as a whole.

As future work, author propose the development of the proposed framework, its validation and improvement.

References

1. Syam, N., Sharma, A.: Waiting for a sales renaissance in the fourth industrial revolution: machine learning and artificial intelligence in sales research and practice. Ind. Mark. Manag. **69**, 135–146 (2018). https://doi.org/10.1016/j.indmarman.2017.12.019
2. Marr, B.: Why Everyone Must Get Ready For The 4th Industrial Revolution. https://www.forbes.com/sites/bernardmarr/2016/04/05/why-everyone-must-get-ready-for-4th-industrial-revolution/ (2016)
3. Canetta, L., Barni, A., Montini, E.: Development of a digitalization maturity model for the manufacturing sector. In: 2018 IEEE International Conference on Engineering, Technology and Innovation (ICE/ITMC), pp. 1–7 (2018). https://doi.org/10.1109/ICE.2018.8436292
4. Wang, S., Wan, J., Zhang, D., Li, D., Zhang, C.: Towards smart factory for industry 4.0: a self-organized multi-agent system with big data based feedback and coordination. Comput. Netw. **101**, 158–168 (2016). https://doi.org/10.1016/j.comnet.2015.12.017
5. BCG: Digitalization Strategy Framework. https://www.bcg.com/capabilities/technology-digital/digitalization-strategy-framework.aspx
6. Cognizant: A Framework for Digital Business Transformation, https://www.cognizant.com/InsightsWhitepapers/a-framework-for-digital-business-transformation-codex-1048.pdf
7. Schmarzo, B.: Digital Business Transformation Framework, https://www.cio.com/article/3130103/analytics/digital-business-transformation-framework.html
8. Catlin, T., Lorenz, J.-T., Sternfels, B., Willmott, P.: A roadmap for a digital transformation| McKinsey. https://www.mckinsey.com/industries/financial-services/our-insights/a-roadmap-for-a-digital-transformation
9. Bin-Abbas, H., Bakry, S.H.: Assessment of IT governance in organizations: a simple integrated approach. Comput. Hum. Behav. **32**, 261–267 (2014). https://doi.org/10.1016/j.chb.2013.12.019
10. Oliver, D., Lainhart, J.: COBIT 5: adding value through effective geit. EDPACS **46**, 1–12 (2012). https://doi.org/10.1080/07366981.2012.706472
11. COBIT 5: A Business Framework for the Governance and Management of Enterprise IT. http://www.isaca.org/cobit/
12. Lu, Y.: Industry 4.0: A survey on technologies, applications and open research issues. J. Ind. Inf. Integr. **6**, 1–10 (2017). https://doi.org/10.1016/j.jii.2017.04.005
13. Oliff, H., Liu, Y.: Towards industry 4.0 utilizing data-mining techniques: a case study on quality improvement. Proc. CIRP. **63**, 167–172 (2017). https://doi.org/10.1016/j.procir.2017.03.311
14. Miškuf, M., Zolotová, I.: Comparison between multi-class classifiers and deep learning with focus on industry 4.0. In: 2016 Cybernetics Informatics (K I), pp. 1–5 (2016). https://doi.org/10.1109/CYBERI.2016.7438633
15. Lee, J., Davari, H., Singh, J., Pandhare, V.: Industrial artificial intelligence for industry 4.0-based manufacturing systems. Manuf. Lett. **18**, 20–23 (2018). https://doi.org/10.1016/j.mfglet.2018.09.002

16. Traub, M., Vögel, H., Sax, E., Streichert, T., Härri, J.: Digitalization in automotive and industrial systems. In: 2018 Design, Automation Test in Europe Conference Exhibition (DATE), pp. 1203–1204 (2018). https://doi.org/10.23919/DATE.2018.8342198
17. Zhang, D., Tsai, J.J.P.: Machine learning and software engineering. Softw. Qual. J. **11**, 87–119 (2003). https://doi.org/10.1023/A:1023760326768
18. Meinke, K., Bennaceur, A.: Machine learning for software engineering: models, methods, and applications. In: Proceedings of the 40th International Conference on Software Engineering: Companion Proceedings, pp. 548–549. ACM, New York, NY, USA (2018). https://doi.org/10.1145/3183440.3183461
19. Garcia-Crespo, A., Colomo-Palacios, R., Gomez-Berbis, J.M., Mencke, M.: BMR: benchmarking metrics recommender for personnel issues in software development projects. Int. J. Comput. Intell. Syst. **2**, 257–267 (2009)
20. Colomo-Palacios, R., González-Carrasco, I., López-Cuadrado, J.L., Trigo, A., Varajao, J.E.: I-Competere: Using applied intelligence in search of competency gaps in software project managers. Inf. Syst. Front. **16**, 607–625 (2014). https://doi.org/10.1007/s10796-012-9369-6
21. Colomo-Palacios, R., González-Carrasco, I., López-Cuadrado, J.L., García-Crespo, Á.: ReSySTER: a hybrid recommender system for Scrum team roles based on fuzzy and rough sets. Int. J. Appl. Math. Comput. Sci. **22**, 801–816 (2012). https://doi.org/10.2478/v10006-012-0059-9
22. The 1st International Workshop on Machine Learning and Software Engineering in Symbiosis (MASES 2018), https://mases18.github.io
23. Khomh, F., Adams, B., Cheng, J., Fokaefs, M., Antoniol, G.: Software engineering for machine-learning applications: the road ahead. IEEE Softw. **35**, 81–84 (2018). https://doi.org/10.1109/MS.2018.3571224
24. Hevner, A., Chatterjee, S.: Design Research in Information Systems—Theory and Practice (2010)
25. Kitchenham, B.: Procedures for Performing Systematic Reviews. Keele UK Keele University, vol. 33, 2004 (2004)
26. Baldassarre, M.T., Caivano, D., Pino, F.J., Piattini, M., Visaggio, G.: Harmonization of ISO/IEC 9001:2000 and CMMI-DEV: from a theoretical comparison to a real case application. Softw. Qual. J. **20**, 309–335 (2011). https://doi.org/10.1007/s11219-011-9154-7
27. Glaser, B.G.: Theoretical Sensitivity: Advances in the Methodology of Grounded Theory. Sociology Press (1978)

Applied Data Science in Financial Industry

Natural Language Processing Techniques for Bank Policies

Marco Spruit and Drilon Ferati

Abstract In a time when the employment of Natural Language Processing techniques in domains such as biomedicine, national security, finance and law, is flourishing, this study takes a deep look in its application in policy documents. Besides providing an overview of the current state of the literature that treats these concepts, the study at hand implements a set of unprecedented Natural Language Processing techniques on internal bank policies. The implementation of these techniques, together with the results that derive from the experiment and the experts' evaluation, introduce a Meta-Algorithmic Modelling framework for processing internal business policies. This framework relies on three Natural Language Processing techniques, namely information extraction, automatic summarization and automatic keyword extraction. For the reference extraction and keyword extraction tasks we calculated Precision, Recall and F-scores. For the former we obtained 0.99, 0.84, and 0.89; for the latter we obtained 0.79, 0.87 and 0.83, respectively. Finally, our summary extraction approach was positively evaluated using a qualitative assessment.

1 Introduction

Data are the pollution of information age, since they are created and they are here to stay [1]. This increase in the flow of data that organizations create and collect, necessitates the need to leverage from these resources and extract information and subsequent knowledge. The large stream of data unveiled two data formats, namely structured and unstructured data, with each of them requesting different treatment methodologies to derive knowledge. Although many argue that this process is less-time consuming when the data have a consistent representation and a predefined structure, only 20% of the data that organizations have, are actually found in this manner. The rest of the data are found in an unstructured format [2]. These data

M. Spruit (✉) · D. Ferati
Department of Information and Computing Sciences,
Utrecht University, Princetonplein 5, 3584 CC Utrecht, The Netherlands
e-mail: m.r.spruit@uu.nl

A. Visvizi and M. D. Lytras (eds.), *Research & Innovation Forum 2019*,
Springer Proceedings in Complexity,
https://doi.org/10.1007/978-3-030-30809-4_32

have no consistency in appearance and are usually text-heavy, making it a challenge to extract patterns and relationships from and among them. This has called for the introduction of Text Mining (TM) as a discipline that "analyses text to extract information that is useful for particular purpose" [3]. The wide range of techniques that it fosters, together with its applicability in domains such as biomedicine, national security, finance, social studies, law and so on, show the prominence of such data analyzing techniques [4–8]. Nevertheless, not all disciplines have been able to taste the same riches. Policies are industry-wide documents that represent written guidelines of acceptable actions to which organizations must adhere. Financial institutions, especially banks, can be thought of industries that have a high number of policies in place. These organizations continuously introduce policies, in order to be fully complaint with regulations that governing bodies impose. Nevertheless, even though such documents are found across industries, they still lack in standardization [9] and their domain-specific language often makes them incomprehensible [10]. Considering the importance of these documents for the business, but at the same time their inconsistent and exhausting representation, a perception was created that TM and its techniques can be used to bring order and understanding into them. This is what motivated the compilation of the following three related research questions (RQ):

1. To what extent has TM been applied on policy documents?
2. Which TM techniques or frameworks have been applied on policy documents?
3. Which TM techniques can be used to obtain information that would enable an easier navigation through the policies?

Nevertheless, in our attempt to validate this perception, the scientific body of literature showed a landscape different from what was anticipated. Thus, next to providing an overview of the current state of literature that treats the concept of using TM on internal bank policies, this paper also introduces a novel TM framework for processing internal business policies. The framework represents a meta-algorithmic model (MAM) of the approach that was validated in a case study at one of the biggest banks in Netherlands [11]. The rest of the paper is structured as follows. In Sect. 2 we present the research approach, which is followed by the related literature in Sect. 3. Section 4 presents the case study methodology, whereas the results of the case study and implications are presented in Sect. 5. Conclusions are formulated in Sect. 6.

2 Research Approach

The answer to the research questions calls for an extensive review of the literature that explores the use of TM on policy documents. This examination employed the snowballing approach, as defined in [12] to determine to what extent these concepts have been addressed before. Such a review, besides providing empirical evidence of the state of literature, also provided the necessary knowledge foundations to base the development of the artefact on. The method development is in line with the Applied Data Science approach outlined in [13] which integrates the Design Science (DS) approach as introduced by [14] and its guidelines for acquiring a better understanding of the requirements which subsequently leads to an effective execution of the research. In short, when these guidelines are applied to the study at hand, they are as follows:

1. *Problem Identification*: The extensive amount of internal bank policies, together with their text-heavy format, makes it challenging to read and extract information from them.
2. *Definition of the objectives of a solution*: We devise a new method that will enable the automatic analysis of such documents and extract valuable information from them.
3. *Design and Development*: The method will rely on the combined use of TM and Natural Language Processing (NLP) techniques, that will be derived from the literature.
4. *Evaluation*: The use of statistical methods together with expert opinions will be employed to evaluate the framework.
5. *Communication*: The validity and generalizability of the devised framework will be communicated based on the results.

The validation of the framework relied on a case study at one of the most prominent international banks in The Netherlands. Considering that this experiment utilizes domain-specific documents, a particular set of domain knowledge is needed to evaluate the outcome. Since such a knowledge is hardly found on engineers, the evaluation of the algorithmic outcome was done in a form of acceptance testing, by experts that reside within the organization that facilitated the study. Acceptance testing is an entity of Black-Box testing [15] where the evaluators provide comments regarding the results that have derived from the artefact. These comments indicates how relevant a result is with respect to a policy document.

3 Case Study Methodology

The following three TM techniques were examined in this study: information extraction, automatic summarization and keyword extraction. Given that the literature did not give much evidence on the use of TM techniques on internal business policies, the use of these techniques on such documents is considered a novelty. Moreover, the selection of the techniques was also in accordance with the requirements that the facilitator of the study expressed. Within their content, policies often refer to other internal policies, for various purposes. These references can be found dispersed throughout the whole content of policies which can be of numerous pages. Stakeholders find these information important, indicating that the automatic extraction of such information would be beneficial. Additionally, for each document they requested a list of most descriptive words, which can be later used to tag and index these documents in a centralized repository. Next to this, having a concise summary of each policy would enable the stakeholder to get a timelier understanding of the context of the document. To endorse this study, the stakeholders provided a corpus of 25 internal policies, from which two policies were unusable, since one of them was written in the Dutch language and the other was inexistent. This resulted in a final corpus of 23 internal bank policies. Upon investigating the corpus of documents, some data preparation steps took place. Initially all the policies were converted from PDF format to a plain text format (.txt). This transformation was done with the help of a script written in Python. Here, Python was selected, given that it is a high-level general-purpose programming language, with a large bag of standard libraries that enable it to be used efficiently in scientific computations [16]. Next to this, some data cleansing took place, by removing the cover page, table of content, tables and appendices. Since the investigation concerned only the main content of the document, these steps were taken as a preventive measure so that they will not skew the data. Furthermore, Python was also used in constructing the algorithm for processing the textual data. This was based on the decision to use the Natural Language Toolkit (NLTK) package, which is one the most popular Python libraries that is specialized in processing textual data [17].

3.1 Reference Extraction

When dealing with IE, one can chose to rely on linguistic processing of the text or on keyword matching approach, to extract relevant information from the document. In this research we rely on the linguistic processing of text approach, more precisely we rely on the IE architecture that NLTK supports (Fig. 1). This process was initiated with the splitting of the raw text into sentences. This was made possible by sentence segmenting module that NLTK provides. Furthermore, the split sentences were further tokenized into respective words, which was done by using the NLTK word tokenizer module. The next step in the pipeline was to tag the words based on their semantic and syntactic structure. Such a step assigned a Part-of-Speech (POS) tag to the words based on their role in the textual content [18]. Suffering from the lack of a domain-specific tagger, here we used the default NLTK POS tagger, which utilizes the Penn Treebank [19] annotation corpus. This step transformed the token representation of the words into a (word, tag) tuple representation. Having such a representation enables us to investigate the tagged corpus and identify the representation of the relevant entities. The significance of this step is that it enabled to distinguish the signal from the noise. Thus by investigating the corpus, we distinguished the relevant entities and constructed a chunking grammar based on their POS representation. We constructed this chunking grammar with the help of Regular Expressions, which was subsequently parsed on the annotated text. When parsed on the textual content, the chunker identified entities that we deemed as relevant, and grouped them together in a single tree representation.

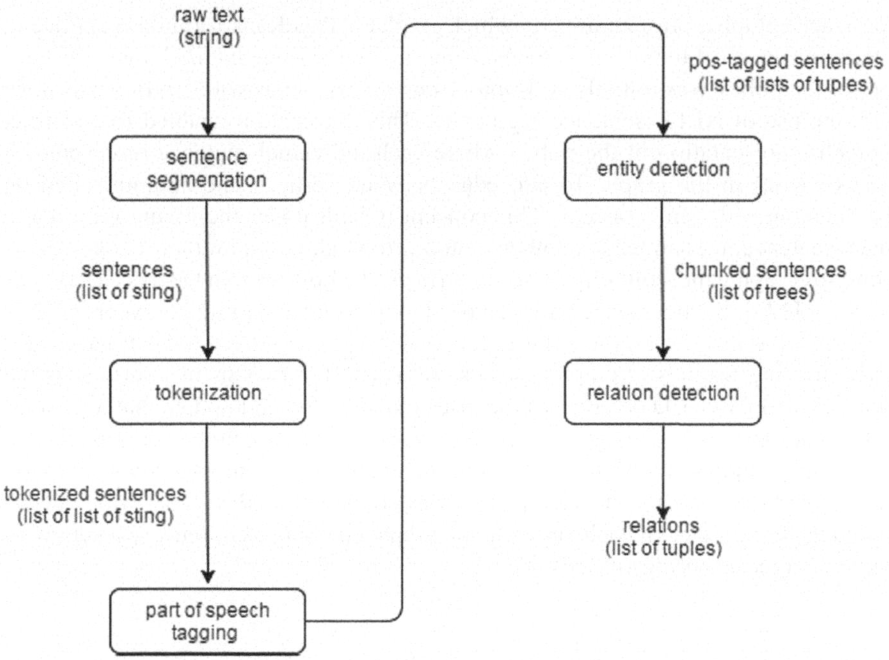

Fig. 1 Information extraction architecture

3.2 Automatic Summarization

When dealing with automatic summarization, [20] defines two groups of summarizing algorithms, namely extractive and abstractive algorithms. These algorithms differ in the approach that they employ for constructing summary representations. Extractive algorithms construct the summaries by using the most important sentences of the textual document and concatenating them into a consisted summary. Contrarily, the abstractive summaries may not always draw on the same concepts as the ones that the original text contains. It usually reuses the main phrases of the document and constructs them in a manner that would convey the message. In addition to this, [20] further categorizes these two algorithms based on their appliance, which can be either a multi-document or single-document summarization algorithm. As the name implies, a multi-document summarization algorithm generates a single summary from the entire corpus of documents, whereas a single-document summarization algorithm generates an individual summary for each document. These factors scoped our algorithm selection options to an extractive, single-document summarization algorithm. This decision, in turn, led us to use the TextRank approach of [21].

TextRank has been introduced more than a decade ago and it has its roots embedded in the PageRank algorithm of [22], utilizing the same logic. As such, TextRank is a graph-based approach, that uses the knowledge drawn from the entire text to construct a graph representation, on which graph the PageRank formula is applied to determine the most important vertices. Adhering to the same methodology, the content of the policy was initially split into sentences. Same as with IE, this was done with the use of NLTK sentence segmenter. This segregation enabled to construct a graph representation of the policy where each individual sentence represented a vertex (node) in the graph. To add edge between vertices, the authors relied on the "recommendation" concept. This concept is built under the assumption that a given sentence, recommends another sentence to read, based on their resemblance. Thus, to compute the similarity between vertices, the Levenshtein Distance (LD) was employed [23]. Such a metric has been used to measure error rate in text entry [24], measure the syntactic variation of different dialects [25, 26], keyword extraction [27] and extracting features of graph representation [28]. For two distinct vertices in the graph (V_1 and V_2), LD determines the insertion, deletion and update that V_1 needs, to become same as V_2, assigning this score as an edge between the vertices. Such a score was computed for all the vertices in the graph, adding an edge between them. With this in place the PageRank equation was computed on the graph, which determined the importance of each vertex in the graph, enabling us to retrieve a summary with the highest scoring vertices.

3.3 Keyword Extraction

Another reason that made TextRank appropriate for this study, was its ability to extract keywords as well. When extracting keywords, the analysis took place on a lower level, more precisely on a word level. Thus following the same logic as in automatic summarization, we further split the sentences into individual words with the use of the NLTK word tokenizer. These tokenized words represented the vertices of the graph. Nevertheless, these changes had an impact on the size of the graph which increased exponentially given the large number of vertices. Thus, in order to reduce the density of the graph, three word filtering methods were applied. Initially, all the "stop words" were removed. These are frequently appearing words that assist in creating an idea, when used in a sentence, but do not represent a significant meaning in themselves [29] (i.e. the, and, or, that, this). Next, a syntactic filter was applied. What this filter did, is that it determined valuable and invaluable words based on their POS representation. Similar as with IE, initially the words were tokenized and annotated with their POS tag. From here, all the tokens with an invaluable syntactic representation were filtered out. The third filter was designed to remove duplicate values from the graph. These filtering steps are also advocated by [21]. Then, the similarity between vertices was computed using LD, thus creating the edges between them. Subsequently, the PageRank formula was applied to the graph to determine the most important vertices. When extracting a list of references, the authors advocate that 1/3 of the content should be retrieved as potential keywords. Nevertheless, even with all the filtering, the graph still contained several thousands vertices. Thus adhering to such a ratio may have resulted in an exhausting list of potential keywords. A better solution was achieved when only 1% of the highest scoring vertices were considered as potential keywords. This gave us a number of keywords ranging from 8 up to 51, per document. Additionally, since up to this point the list of potential keywords only consisted of single word entities, the authors advise that key-phrases can also be generated from this list. This was done by combining two unique keywords with one-another. To validate such key-phrases, the mutated entities were checked against the entire content of the document. This determined whether such a phrase indeed existed in the text.

4 Results

Thus upon selecting, constructing and implementing the necessary artefacts, a collection of results were retrieved from each executed technique. These results were a direct output of all statistical and linguistic computations that the textual content underwent. This made the generated output available for evaluation. Given that the study deals with domain-specific documents, domain knowledge was needed to also evaluate the results. As mentioned earlier, competent entities within the bank were asked to evaluate the algorithmic outcomes. This was mostly true for automatic summarization and keyword extraction, whereas the reference extraction did not necessarily require domain specific understanding to be evaluated.

4.1 Reference Extraction Results

The relatively small corpus of documents enabled the creation of a golden standard that contained all the referencing policies for each policy document. This golden standard held the entire collection of true values, against which values the algorithmic results were evaluated. Thus, based on what the outcome was, we were able to compute the accuracy of the algorithm with the use of *precision* (P), *recall* (R) and *F-measure* (F) [30]. P and R have been regularly used to measure the performance of information retrieval and information extraction systems [31], by determining the success rate of the algorithm. As such, P (also known as *confidence*) determines how many of the retrieved values are indeed correctly predicted. Whereas R, also known as *sensitivity*, determines how many values from the gold standard are correctly predicted by the algorithm. Nevertheless, the scientific community decided to combine these performance indicators under a single measure of performance, thus introducing F, which is an equally weighting equation of both P and R. Additionally, researchers such as [32] argue that in IE experiments related to Machine Learning and Computational Linguistics, more importance should be put on determining how confident one can be with the rules or classifier. This means that when measuring the accuracy of the algorithm, more weight should be put on P. In such cases, a variation of F is used, namely the F_β formula. This variation of the F formula usually takes two values as β. It takes a value of 0.5 when more weight should be put on P, and a value of 2 when more weight should be put on R. Hence, this study evaluates the results based on the normal weighted F formula and $F_{0.5}$ formula. The results from these calculations are given in Table 1, together with all the relevant details for each policy. According to these equations, the algorithm managed to achieve a 89% accuracy in extracting relevant information, when both of the performance indicators were weighted equally, and a 94% accuracy when the performance was measured with the F_β formula. Both of these results showed that the algorithm was highly capable to extract relevant information. It should be mentioned that in some cases the algorithm extracted imperfect chunks. These imperfect chunks, mostly, missed a part of the information, since it contained entities expressed in a language different than English. In order to determine how to deal with such imperfect chunks, the literature was consulted. This, showed that the concept of a mistake/negative (i.e. partially correct) is not properly defined and it is a subject of change depending on researcher's perception [33]. Thus in the initial evaluation, the imperfect information were considered as entirely incorrect (negative) values. Furthermore, a second evaluation took place, where these partially correct chunks were considered as correct. This evaluation managed to yield an accuracy of 95 and 97%, when calculating F and F_β respectively. This confirms that the constructed artefact performed impressively.

4.2 Summary Extraction Results

Given the domain specific dictionary that is used to compile these documents, the evaluation of the summaries required matching domain knowledge to assess them. Henceforth, domain experts were charged with the responsibility to evaluate these summaries. The experts were asked to assess the summary and to comment on issues such as whether the summary covered the main aspects of the document, whether the summary was a good representation of the policy, what did the summary miss, and other comments of the similar nature. As mentioned earlier, such a form of evaluation followed a somewhat Black-Box evaluation approach where a comment was provided regarding the accuracy of the outcome. Through an evaluation form the experts could read the generated summary for the respective policy, and provide a comment about its ability to convey the main message of the document. Although the algorithm was capable to construct summaries, the expert evaluation showed that not always the generated outcome was adequate. In some scenarios the generated output had some data quality issues, like having the tendency to generate a more detailed summary than necessary. Nevertheless, next to it, there were also summaries that were a good representation of the policy, as well as a moderate representation of the policy. These comments were equally dispersed across the evaluated corpus, enabling us to sum them up into three main thematic representations:

- *The summary covers the mains aspect of the policy* found in 7 evaluations (30.4%)

Table 1 Reference extraction results

Title	Total	Correct	Part. Corr.	Miss	Extra	P	R	F	Fβ
Policy A	15	11	3	1	0	1	0.73	0.84	0.93
Policy B	15	15	0	0	0	1	1	1	1
Policy C	13	8	4	1	0	1	0.61	0.76	0.88
Policy D	11	4	1	6	0	1	0.36	0.53	0.73
Policy E	13	12	1	0	2	0.85	0.92	0.88	0.86
Policy F	18	13	4	1	0	1	0.72	0.83	0.92
Policy G	14	12	2	0	1	0.92	0.85	0.88	0.90
Policy H	11	11	0	0	0	1	1	1	1
Policy I	27	21	5	1	0	1	0.77	0.87	0.94
Policy J	31	26	4	1	0	1	0.83	0.9	0.96
Policy K	6	6	0	0	0	1	1	1	1
Policy L	20	14	6	0	0	1	0.7	0.82	0.92
Policy M	13	11	2	0	0	1	0.84	0.91	0.96
Policy N	13	13	0	0	0	1	1	1	1
Policy O	24	22	2	0	0	1	0.91	0.95	0.98
Policy P	17	14	0	3	0	1	0.82	0.90	0.95
Policy Q	8	8	0	0	0	1	1	1	1
Policy R	10	10	0	0	0	1	1	1	1
Policy S	11	10	1	0	0	1	0.90	0.94	0.97
Policy T	77	74	1	2	1	0.98	0.96	0.97	0.97
Policy U	29	21	6	2	0	1	0.72	0.83	0.92

(continued)

Table 1 (continued)

Title	Total	Correct	Part. Corr.	Miss	Extra	P	R	F	Fβ
Policy V	22	16	2	4	0	1	0.72	0.84	0.92
Policy W	17	16	0	1	0	1	0.94	0.96	0.98
Total	435	368	44	23	4	22.75	19.3	20.61	21.63
Average	18.91	16	1.91	1	0.17	0.99	0.84	**0.89**	**0.94**

- *The summary is a moderate representation of the policy* found in 8 evaluations (34.7%)
- *The summary is too detailed, thus does not cover the aspect of the policies* found in 8 evaluations (34.7%).

4.3 Keyword Extraction Results

The extracted keywords also required some domain knowledge to be properly evaluated. Thus, together with the summaries, the experts also evaluated the keywords that the algorithm generated for each policy. The experts annotated the algorithmic keywords either as relevant or irrelevant. At the same time, they were advised to add potential keywords or key-phrases that the algorithm had failed to recognize as such. Although in essence, it followed the same evaluation logic as with the summaries, when it came to evaluating keywords, this method enabled to translate the qualitative evaluation into quantitative representations. Such a transformation enabled to measure the performance of the algorithm with the use of P, R and F. The results from such an evaluation are provided in Table 2, together with all the relevant details for each policy. Furthermore, it showed that the algorithm managed to reach an average accuracy of 83% in extracting relevant keywords or key-phrases.

Nevertheless, considering that our list of keywords ranged from 8 up to 51 potential keywords per policy, one may argue that this is beyond a reasonable amount. Thus, looking at similar studies, it was noticed that most of them extracted 9 up to 15 keywords per policy [21, 34–36]. Henceforth, from our lists of keywords, the ones that exceeded this range, were capped at a maximum of 15 most representative keywords. These changes called for a re-evaluation of the algorithm, which showed that if only the 15 highest ranking entities are concerned, the algorithm can reach an extraction accuracy of 82%, indicating that there is not much of a difference between the two calculations. To get a better understanding on how the algorithm performed in these two cases, we benchmarked the generated outcome with other similar studies [21, 34–38]. This benchmarking is shown in Table 3 and it indicates that the outcome of this experiment represents one of the highest accuracy levels that have been achieved when extracting keywords.

5 Case Study Implications

What started as an attempt to determine the use of TM on business policies, more specifically bank policies, evolved into a novel approach of processing such text-heavy, domain-specific documents. Even though the available publications on policy documents gave little to no indication about which techniques should be used for the case study requirements, the rich body of literature in other domains pinpointed

Table 2 Keyword extraction results

Title	Total	Correct	Incorrect	Suggested	P	R	F
Policy A	8	4	4	2	0.50	0.67	0.57
Policy B	13	8	5	0	0.62	1	0.76
Policy C	20	18	2	3	0.90	0.86	0.88
Policy D	29	23	6	2	0.79	0.92	0.85
Policy E	16	12	4	7	0.75	0.63	0.69
Policy F	14	10	4	2	0.71	0.83	0.77
Policy G	10	8	2	1	0.80	0.89	0.84
Policy H	19	11	8	0	0.58	1	0.73
Policy I	29	24	5	3	0.83	0.89	0.86
Policy J	51	43	8	10	0.84	0.81	0.83
Policy K	14	10	4	6	0.71	0.63	0.67
Policy L	13	10	3	2	0.77	0.83	0.80
Policy M	12	11	1	0	0.92	1	0.96
Policy N	13	11	2	0	0.85	1	0.92
Policy O	35	29	6	0	0.83	1	0.91
Policy P	13	12	1	0	0.92	1	0.96
Policy Q	10	7	3	0	0.70	1	0.82
Policy R	11	9	2	6	0.82	0.60	0.69
Policy S	17	16	1	0	0.94	1	0.97
Policy T	48	40	8	0	0.83	1	0.89
Policy U	13	12	1	0	0.92	1	0.96
Policy V	19	16	3	0	0.84	1	0.91
Policy W	11	9	2	0	0.82	1	0.90
Average	19	15	3.6	1.91	0.79	0.87	**0.83**

techniques that would attain the study objectives. This literature showed that when dealing with unstructured data, the processing can rely on a single or ensemble of TM techniques, depending on the intended outcome. Such was the nature of the case study at hand, where the set of requirements called for the use of multiple TM techniques. Looking thoroughly at the adaption of these techniques into programmable artefacts and the execution of the artefact on the study corpus, the algorithmic results did not fail to impress. The ability of the algorithm to reach an over 89% accuracy in extracting relevant information, hints strongly towards the potential of such techniques to process policy documents. A distinctive feature of this artefact was its ability to miss a relatively small portion of relevant information from the corpus. On a similar note, the automatic keyword extraction technique proved to be highly capable of recognizing and extracting the most distinctive words of a document, thus outperforming similar studies.

Table 3 A comparative overview of keyword extraction studies

Study	P	R	F
Automatic Keyword extraction from individual documents (RAKE) [34]	33.7%	51.5%	37.2%
TextRank: Bringing order into text [21]	31.2%	43.1%	36.2%
Improved automatic keyword extraction given more linguistic knowledge [35]	22.5%	51.7%	33.9%
Unsupervised approach for automatic keyword extraction using meeting transcripts [37]	NA	NA	19.6%
A study on automatically extracted keywords in Text Categorization [36]	92.89%	72.94%	81.72%
Automatic keyword extraction from documents using conditional random fields [38]	66.3%	41.9%	51.25%
This study	79.1%	89.3%	**83.2%**
This study (max. 15)	79.5%	87.3%	82.3%

Although the summarization of the documents did not share the same level of success, it still managed to reach a moderate success. Furthermore, its tendency to generate more detailed summaries than necessary shows that the issues in this approach are mostly data quality related rather than methodological. The results of this study are the first of its kind when it comes to internal business policies. While most of the literature studies have yet to execute their conceptual frameworks and ratify their claims, the results of this study and their statistical evaluation validate the use of such an approach on business policies and make it distinguishable from other approaches. Considering the novelty of this approach, when it comes to bank policies, and at the same time its validation by experts and statistical methods, a new framework for processing internal business policies is derived. Figure 2 provides a visual representation of this framework. Here, all the utilized techniques are combined into a single representation which provides a step-by-step description of the followed approach. It guides the user from the data preparation phase towards actions that concern reference extraction, keyword extraction and automatic summarization. At the same time, such phases can be recognized in the framework as the main activities, each having a collection of sub-activities. In addition to these, the *Graph* activity is a shared entity among keyword extraction and automatic summarization.

Furthermore, the design of this framework draws upon the meta-algorithmic representation of the approach followed in the case study. Spruit and Jagesar [11] define meta-algorithmic modelling (MAM) as an "[…] engineering discipline where sequences of algorithm selection and configuration activities are specified deterministically for performing analytical tasks based on problem-specific data input characteristics and process preferences". This approach relies on the activity recipes of the case study, which are modelled with the use of method–engineering notations. Method-engineering is defined as "the engineering discipline to design construct and adapt methods, techniques and tools for the development of information systems"

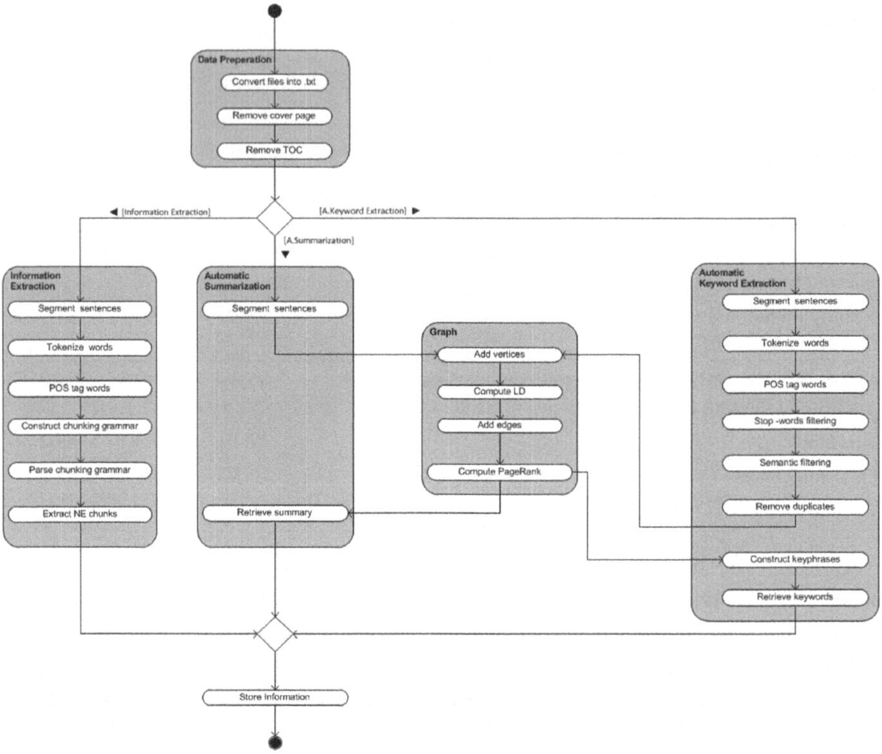

Fig. 2 Meta-algorithmic model for processing business policies

[39]. By following this approach, the framework is depicted as a Process Deliverable Diagram (PDD), which uses a UML activity diagram to represent the processes and a UML class diagram to represent the deliverables [40]. In general, an activity diagram falls in the behavioral class of UML diagrams, thus being a representation that depicts the behavior of a module. Whereas a class diagram belongs to the structural class of UML diagrams, thus representing the structure of the entities in the module. Nevertheless, this framework only represents the processes, thus composed of only a UML activity diagram.

6 Conclusion

Motivated by the fact that Text Mining (TM) is being used extensively across industries, this study has focused on determining the use of such a discipline on business policies, and the benefits that it brings for such documents. This investigation was initiated with a review of the current state of literature that has treated such concepts. Nevertheless, the literature review showed that the use of TM to process policies was far from what it was anticipated to be, which answers RQ1. In a time when the use of TM on disciplines such as biomedicine, national security, finance and law is flourishing, the use of TM on internal business policies, and policies in general, still falls short on both qualitative and quantitative aspects. The amount of publications that treat such concepts was quite limited and failed to match the prominence of other domains. In addition to this, the available publications only considered privacy policies in their study, also known as "Terms and Conditions". This also revealed that the implementation of such techniques on business policies, had only scarcely been addressed yet by other researchers. Furthermore, this small corpus of publications introduced numerous conceptual frameworks, nevertheless, however the majority of them lacked a full scientific validation. This is also reflected in the existence of systems that use TM techniques to process policy documents. Even though most of the literature revolved around the idea of creating such systems, only a small portion embodied the existing frameworks into working modules. Furthermore, as far as the qualitative aspect is concerned, the processing of policy documents often failed to go beyond text categorization/classification, Information Extraction, and Topic Modelling techniques, which answers RQ2. Thus, besides providing an overview of the current state of literature, that has investigated the use of TM on business policies, this study also indicates the gaps in the literature where it contributes. The biggest contribution comes from the introduced Meta-Algorithmic Modelling (MAM) framework that has been implemented and fully validated on bank policies. Besides utilizing a set of unprecedented techniques on policy documents, it also gives a step-by-step recipe of how our Top-3 TM techniques (information extraction, automatic summarization and automatic keyword extraction) can be implemented and what benefits they yield, which answers RQ3.

References

1. Spence, D.: Data, data everywhere: a special report on managing information. Economics, 1–10 (2010)
2. Grimes, S.: Unstructured data and the 80% rule. Clarabridge Bridge (2008)
3. Witten, I.H.: Text mining. Int. J. Comput. Biol. Drug Des. **198** (2004)
4. Friedman, C., Johnson, S.B., Forman, B., Starren, J.: Architectural requirements for a multipurpose natural language processor in the clinical environment. Proc. Symp. Comput. Appl. Med. Care, 347–351 (1995)
5. Haug, P.J., Ranum, D.L., Frederick, P.R.: Computerized extraction of coded findings from free-text radiologic reports. work in progress. Radiology **174**(2), 543–548 (1990)
6. Bholat, D., Hansen, S., Santos, P., Schonhardt-Bailey, C.: Text mining for central banks. Cent. Cent. Bank. Stud. Handb. **33**, 1–19 (2015)
7. Fan, W., Wallace, L., Rich, S., Zhang, Z.: Tapping the power of text mining. Commun. ACM **49**(9), 76–82 (2006)
8. Zhao, Y.: Analysing twitter data with text mining and social network analysis. In: Proceedings of the 11th Australasian Data Mining and Analytics Conference (AusDM 2013) (2013)
9. Anton, A.I., Earp, J.: The Lack of Clarity in Financial Privacy Policies and the Need for Standardization, no. August, pp. 1–12 (2003)

10. Anton, A., Earp, J.: A requirements taxonomy for reducing Web site privacy vulnerabilities. Requir. Eng. **9**, 169–185 (2004)
11. Spruit, M., Jagesar, R.: Power to the people! Meta-algorithmic modelling in applied data science. In: Proceedings of the 8th International Joint Conference on Knowledge Discovery, Knowledge Engineering and Knowledge Management, vol. 1, no. IC3K, pp. 400–406 (2016)
12. Wohlin, C.: Guidelines for snowballing in systematic literature studies and a replication in software engineering. In: 18th International Conference on Evaluation and Assessment in Software Engineering (EASE 2014), pp. 1–10 (2014)
13. Spruit, M., Lytras, M.: Applied data science in patient-centric healthcare: adaptive analytic systems for empowering physicians and patients. Telemat. Inf. (2018)
14. Hevner, A.R., March, S.T., Park, J., Ram, S., Ram, S.: Research essay design science in information. MIS Q. **28**(1), 75–105 (2004)
15. Copeland, L.: A Practitioner's Guide to Software Test Design. Artech House (2003)
16. Sanner, M.F. et al.: Python: a programming language for software integration and development. J. Mol. Graph Model **17**(1), 57–61 (1999)
17. Bird, S., Klein, E., Loper, E.: Natural Language Processing with Python: Analyzing Text with the Natural Language Toolkit. O'Reilly Media, Inc. (2009)
18. Voutilainen, A.: Part-of-speech tagging. The Oxford handbook of computational linguistics (2003)
19. Marcus, M.P., Santorini, B., Marcinkiewicz, M.A.: Building a large annotated corpus of English: The Penn Treebank. Comput. Linguist. **19**(2), 313–330 (1993)
20. Larson, M.: Automatic summarization **5**(3) (2012)
21. Mihalcea, R., Tarau, P.: TextRank: bringing order into texts. Assoc. Comput. Linguist. (2004)
22. Page, L., Brin, S.: PageRank: bringing order to the web. Stanford Digit. Libr. Work. Pap. **72** (1997)
23. Levenshtein, V.I.: Binary codes capable of correcting deletions, insertions, and reversals. Sov. Phys. Dokl. **10**(8), 707–710 (1966)
24. Soukoreff, R.W., MacKenzie, I.S.: Measuring errors in text entry tasks, 319 (2001)
25. Spruit, M.R.: Measuring syntactic variation in Dutch dialects. Lit. Linguist. Comput. **21**(4) (2006)
26. Heeringa, W., Nerbonne, J., Van Bezooijen, R., Spruit, M.R.: Geography and population size as explanatory factors for variation in the Dutch dialectal area. Tijdschr. Voor Ned. Taal-en Lett. **123**(1) (2007)
27. Renz, I., Ficzay, A., Hitzler, H.: Keyword extraction for text characterization. In: 8th International Conference on Application Natural Language to Information Systems, pp. 228–234 (2003)
28. Wilson, R.C., Hancock, E.R.: Levenshtein distance for graph spectral features. In: Proceedings of the International Conference on Pattern Recognition, vol. 2, no. C, pp. 489–492 (2004)
29. Rajaraman, A., Ullman, J.D.: Data mining. Min. Massive Datasets **18**(Suppl), 114–142 (2011)
30. Sasaki, Y.: The truth of the F-measure. Teach Tutor mater, 1–5 (2007)
31. Makhoul, J., Kubala, F.: Performance measures for information extraction, 249–252 (1999)
32. Powers, D.M.W.: Evaluation: from precision, recall and F-measure to roc, informedness, markedness & correlation. J. Mach. Learn. Technol. **2**(1), 37–63 (2011)
33. Hripcsak, G., Rothschild, A.S.: Agreement, the F-measure, and reliability in information retrieval. J. Am. Med. Informatics Assoc. **12**(3), 296–298 (2005)
34. Rose, S., Engel, D., Cramer, N., Cowley, W.: Automatic keyword extraction from individual documents. Text Min. Appl. Theory, 1–277 (2010)
35. Yang, K., Chen, Z., Cai, Y., Huang, D.P., Leung, H.: Improved automatic keyword extraction given more semantic knowledge. **9645**, 112–125 (2016)
36. Hulth, A., Megyeesi, B.B.: A study on automatically extracted keywords in text categorization. In: Proceedings of the 21st International Conference on Computational Linguistics and the 44th Annual Meeting of the Association for Computational Linguistics, vol. 1, pp. 353–360 (July 2006)

37. Liu, F., Pennell, D., Liu, Y.: Unsupervised approaches for automatic keyword extraction using meeting transcripts. In: Proceedings of Human Language Technologies: The 2009 Annual Conference of the North American Chapter of the Association for Computational Linguistics, pp. 620–628 (2009)
38. Zhang, C., Wang, H., Liu, Y., Wu, D., Liao, Y., Wang, B.: Automatic keyword extraction from documents using conditional random fields. J. Comput. Inf. **43**, 1169–1180 (2008)
39. Brinkkemper, S.: Method engineering: Engineering of information systems development methods and tools. Inf. Softw. Technol. **38**, no. 4 SPEC. ISS., pp. 275–280 (1996)
40. van de Weerd, I., Brinkkemper, S.: Meta-modeling for situational analysis and design methods. Handb. Res. Mod. Syst. Anal. Des. Technol. Appl. **35** (2008)

Handing a Hybrid Multicriteria Model for Choosing Specialists to Analyze Application Management Service Tickets

Karla Krupskaia Ferreira Lima, Placido Pinheiro and Marum Simão Filho

Abstract The operational routine in an Application Management Service (AMS) provider begins with the acknowledgment of a customer request for maintenance or support in an application. A ticket then is created to register the customer request. That ticket must be analyzed by one of the AMS specialists, for that it is necessary to identify which specialist is the most appropriated for the task. To achieve better results in choices about alternatives prioritization, it is common to apply decision support methods. Verbal Decision Analysis (VDA) is a framework that consists of a series of multicriteria methods for both classification and ordering alternatives. This paper presents a model constructed to optimize the choice of the AMS specialist to perform the analysis of an AMS ticket. It was collected information of the operational process of an AMS provided by a multinational company. An investigation was conducted to identify the criteria that influences the designation of specialists to a ticket, then a model was constructed based on the hybridization of VDA methods to prioritize the specialists, the ORCLASS method for classification and ZAPROS III-i for ordering. Two reals scenarios were considered to validate the model. The application of the ORCLASS method resulted in two groups of specialists, the preferable and the non-preferable, for the preferable group it was applied the ZAPROS III-i method, resulting in a ranked list of specialists. In the end, the results were satisfactory according to the evaluation of the professionals involved in the AMS studied, which agreed with the prioritization obtained.

K. K. F. Lima (✉) · P. Pinheiro (✉) · M. S. Filho (✉)
University of Fortaleza, Fortaleza, Brazil
e-mail: karla_krupskaia@hotmail.com

P. Pinheiro
e-mail: placido@unifor.br

M. S. Filho
e-mail: marumsimao@gmail.com

M. S. Filho
7 de Setembro University Center, Fortaleza, Brazil

A. Visvizi and M. D. Lytras (eds.), *Research & Innovation Forum 2019*,
Springer Proceedings in Complexity,
https://doi.org/10.1007/978-3-030-30809-4_33

1 Introduction

An Application Management Service (AMS) provider in Brazil has a vast number of contracts with customers of various segments, located in different cities and countries. As soon as an AMS employee acknowledges a customer requirement for support or maintenance in an application, he registers a new *ticket*[1] in the AMS requirements tool. The company has many analysts to deal with the tickets, which can be settled in different cities, have different specialties, varying degrees of professional maturity, distinct cultures, work in different time zones, among other aspects that differentiate them from one another at the AMS environment.

The AMS tickets management involves several processes; one of them is the choice of a specialist (or analyst) to analyze the requisitions described in a ticket. That choice involves criteria evaluated in subjective way, usually by an AMS team leader, the effects reflect throughout the whole service, if the decision is inappropriate it can cause negative impacts for both, customer and service provider.

The diversity of criteria with multiple alternatives of choice and the dependence of the judgment of a person makes up a scenario of decision-making problem with degree of subjectivity adherent to the use of multicriteria methods through qualitative analysis, such like Verbal Decision Analysis (VDA) methods.

This paper reports the construction of a model structured in hybrid multicriteria methods to, through classifying and ordering choice options, optimize the process of AMS specialist's choice to perform AMS tickets analyses.

Before applying the VDA methods, we investigated the operation of the AMS, its particularities, the criteria influencing the determination of an analyst to be the responsible for a ticket; we identified the alternatives of choice and preferences. With the information raised, we built a model with the application of the ORCLASS method to classify the analysts and the ZAPROS III-i method for ordering them.

The organization of this paper is as follows. Section 2 presents an overview of process to designate specialists to perform the analysis of AMS tickets. Section 3 introduces the VDA framework focusing on ORCLASS and ZAPROS III-i methods. Section 4 resumes the model construction. Section 5 brings the results of the model application on two real scenarios from the AMS studied. Section 6 outlines the conclusions and suggestions for future work.

2 Designation of Specialists to Analyze Application Support and Maintenance Tickets

Application Management Services (AMS) refers to services provided to companies that need to outsource their applications management.

[1] A ticket is the unique identification for the customer support solicitation.

This research was carried out from a case study in an AMS operation provided by a multinational IT company in Brazil, with customers in South America.

The AMS operation begins with a customer request for support or maintenance in an application, whenever that happens it reports the key information to identify the situation, such like the application, the area, the subarea, the busyness process involved the type of request, and the criticality. An AMS employee registers the customer's solicitation in the AMS governance tool, generating for that solicitation an AMS ticket. Based on the business process and criticality informed by the client the tool determines the priority and level of service. Thus, a ticket is in general characterized by: customer identification, application, area-subarea, process, priority, ticket type, criticality, and the situation description.

The AMS studied does not have a formal decision model to support the choice of specialists to perform the ticket analyses, and the existing subjectivity on the decision can cause damage to the service. Whenever a choice is performed without the correct criteria evaluation that can implicates unsatisfactory results in the AMS operation, undertaking leading up to negative impacts for both, customer's and AMS provider, as:

- Overload of some AMS analysts and idleness of others;
- Delay for ticket solution;
- Low quality in the final solution of the ticket;
- Additional costs for the customer, not raised on the analysis moment;
- Additional costs not foreseen for the AMS provider;
- Customer dissatisfaction.

The good decision allows tickets resolution within deadlines and with definitive solution, leaving the customer satisfied, with no additional costs.

3 Verbal Decision Analysis Frameworks

Verbal Decision Analysis (VDA) is an approach to support verbal solutions for multicriteria problems, there is no numerical conversion involved.

There are various multicriteria methods elaborated with VDA principles, they are classified into two main groups according to their objectives: to classify or to order the alternatives [1].

The choice of a multicriteria method must meet the characteristics of the problem in question, it is important to evaluate the decision objects and the information available to decide for the most appropriated method [2], the decision-maker's way of thinking and their knowledge of the problem are important in the choice of the method [3] as well. In this study, to address the problem of choice of the specialist who will analyze an AMS ticket, were elected by decision makers involved in the process—a group of specialists from the AMS study—the methods ORCLASS to classify the alternatives and ZAPROS III-i to order them.

3.1 Hybridization of Verbal Decision Analysis Methods

The hybridization of VDA methods consists of applying the criteria and alternatives to a classification method and applying the same criteria and alternatives already classified to an ordering method.

Several articles published involving VDA structured in the classification of alternatives before submitting them to an ordination methodology, showed effectiveness in reducing the complexity of application of the ordering method. The results of some surveys are presented in [2, 4, 5].

The ORCLASS method [6] classifies the alternatives in two groups, preferable and not preferable, reducing the number of alternatives to be ordered, by an ordering method, such as ZAPROS III-i. The result is the ranking of a small number of alternatives in relation to the whole. In this way, the hybridization of VDA methods is interesting for problems with a high number of alternatives.

4 Model Construction

The AMS tickets resolution involves some processes that depend on subjective decisions. In agreement with the AMS specialists of this study, we chose to work on the following process: designation of an analyst to analyze a ticket. We identified that the alternatives of choice vary according to the characteristics of the ticket, and the criteria for the decision are dynamic, they also vary according to the characteristics of the ticket. This way, each ticket features a different decision problem.

To make that decision less subjective, more systematic and assertive we propose a model, illustrate in Fig. 1, organized in three sequential dependent main stages detailed as follow:

1. Problem structuring: the first stage is dedicated to the problem structuration, we begin with the problem definition, to get to know what problem to work on. Once we have the problem defined it is possible to delimitate it, characterizing it with bordered features, what permits to follow to next step, to identify the criteria, criteria values, and the alternatives. The number of alternatives identified will determine how the second stage will be performed;
2. VDA methods application: the second stage involves the application of the methods ORCLASS and ZAPROS III-I, the information raised on the previous stage are inserted to an appropriated tool to apply the methods. As shown on Fig. 1, the ORCLASS method is applied only if the number of alternatives is greater them six, in that case, firstly the criteria is inserted to the ORCLASS tool, secondly the alternatives are inserted, thirdly the rules of classification are constructed, and them a result is obtained. The result will be two classes of alternatives, the preferred one will be considered to the ZAPROS III-i application. However, if the ORCLASS method is not considered, them we go straight to the ZAPOS III-i application, at the appropriated tool we insert the criteria, elicitate the preferences,

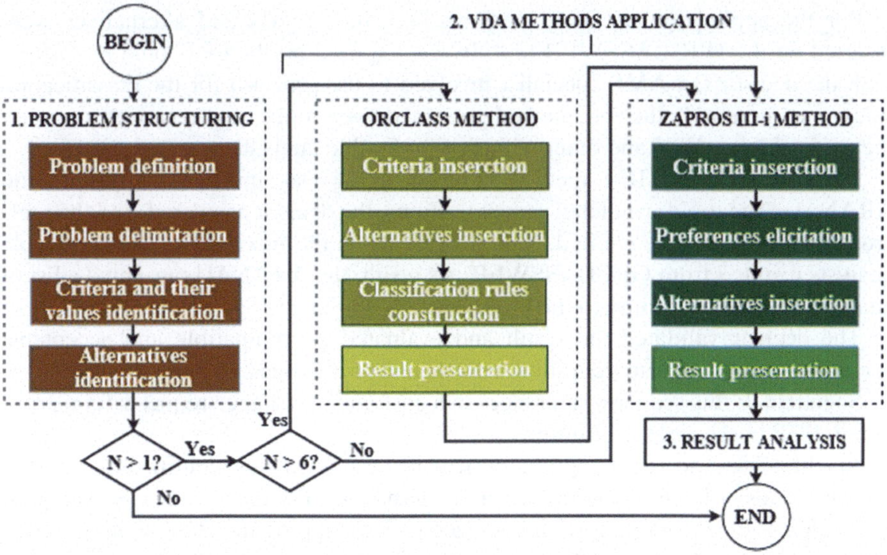

Fig. 1 The model flow

insert the alternatives (the preferred ones resulted from ORCLASS method if it was applied, otherwise, all the alternatives raised on stage of problem structuring), the result is given as a list of the alternatives ordered by the decision maker preferences in accordance with the criteria;

3. Result analysis: the third and last stage is about an analyze performed, on the result obtained from ZAPROS III-i, by the responsible involved on the process, which will choose one alternative from the first to the last one from the list.

5 Model Application

Were approached two real and distinct scenarios to demonstrate the application of the constructed model, one with ten alternatives and the other with six alternatives of choice. For each scenario, we considered the stages of the model indicated on Fig. 1.

Scenario 1: this scenario references to a customer requisition for solution to an error in the ERP application at the time of saving a product receiving.

With problem structuring stage it was identified the criteria, criteria values in order of preferences, and alternatives. We faced a situation with ten analysts as alternatives of choice, which figures an appropriate scenario for applying the ORCLASS and ZAPROS III-i methods as indicates the model flow (Fig. 1).

For the application of ORCLASS method, the criteria and alternatives were inserted on the ORCLASSWEB tool. It was required by the tool some interaction with the decisor (an AMS specialist involved to the process) for the classification rules construction. In the end, the tool presented two groups of alternatives, one with the preferable analysts and other with non-preferable analysts.

For the ZAPROS III-i method submission, it was inserted the criteria on ARANAU. The tool formulated some questions, the decisor answered, and the preferences were this way, elicited. The next step was the insertion of the preferable analysts resulted from ORCLASSWEB. As result, the ARANAU presented a list of analysts ordered according to the preference.

The decisor validated the result and evaluated as compatible and acceptable according to the problem definition, criteria and preferences.

Scenario 2: this scenario references to a customer assistance requests for a problem in ERP invoice report printing.

Just like Scenario 1, at the problem structuring stage it was identified the criteria, criteria values ordered by preference, and alternatives. For the alternatives, we faced a situation with six analysts, what managed us for skipping the classification step in preference groups as directs the model flow (Fig. 1).

Were submitted the criteria to the ZAPROS III-i method trough ARANAU tool. The tool formulated some questions, once again, the decisor answered, and the preferences were elicited this way. Then, the analysts were inserted. As result, the ARANAU presented a list of analysts ordered by preference.

The decisor validated the result and evaluated as compatible and acceptable according to the problem definition, criteria and preferences.

6 Conclusion

One of the recurrent problems in companies offering Application Maintenance Services (AMS) is related to the difficulty in choosing one of the team professionals to perform the technical analysis of a customer request. Typically, this choice is performed in a subjective way by the team leader, and eventually by the first level attendance or even by the analyst himself. That practice can lead to inefficiencies and losses in the service operation.

Considering the systematic work of the AMS provided by a multinational IT company, a structured model was developed based on Verbal Decision Analysis (AVD) methods, the ORCLASS for classification, and ZAPROS III-i for sorting, intending, through multicriteria judgments, to optimize the choice of the AMS specialist who will analyze an AMS ticket). The result was satisfactory according to the validation of the involved in the process, who evaluated that the model makes the decision process in question less intuitive, more systematic and aligned with the company objectives.

As future works, we intend to:

- To develop a tool, based on the model proposed in this paper, integrated to the existing AMS tickets governance application, the ORCLASSWEB tool, and ARANAU tool, with self-learning strategy for dynamic criteria;
- To implement this at the AMS studied.

References

1. Simão, F., Pinheiro, P., Albuquerque, A.: Task allocation in distributed software development aided by verbal decision analysis. In: Proceedings of the 5th Computer Science On-line Conference 2016 (CSOC2016), Software Engineering Perspectives and Application in Intelligent Systems Series, Zlín, vol. 2, pp. 127–137 (2016)
2. Simão Filho, M., Pinheiro, P.R., Albuquerque, A.B.: Verbal decision analysis applied to the prioritization of influencing factors in distributed software development. In: Developments and Advances in Intelligent Systems and Applications. Studies in Computational Intelligence, vol. 718 (2017)
3. Bouyssou, D., Marchant, T., Pirlot, M., Perny, P., Tsoukiàs, A., Vincke, P.: Evaluation and Decision Models: A Critical Perspective. Springer US, vol. 32 (2000)
4. Pinheiro, P.R., Tamanini, I., Pinheiro, M.C.D., Albuquerque, V.H.: Evaluation of the Alzheimer's disease clinical stages under the optics of hybrid approaches in verbal decision analysis. Telemat. Inform. **35**, 776–789 (2018)
5. Leite, M.L.S., Pinheiro, P., Filho, M.S., Araújo, M.L.S.: Ordering assertive strategies for corporate travel agencies: verbal decision analysis model. In: Computational and Statistical Methods in Intelligent Systems. CoMeSySo 2018. Advances in Intelligent Systems and Computing, vol. 859 (2019)
6. Larichev, O.I., Moshkovich, H.M.: Verbal Decision Analysis for Unstructured Problems. Kluwer Academic Publishers, Boston (1997)
7. Tamanini, I.: Hybrid Approaches of Verbal Decision Analysis Methods. Fortaleza (2015)
8. Pinheiro, P.R., Machado, T.C.S., Tamanini, I.: Dealing the selection of project management through hybrid model of verbal decision analysis. Proc. Comput. Sci. **17**(1), 332–339 (2013)
9. Machado, T.: Towards aided by multicriteria support methods and software development: a hybrid model of verbal decision analysis for selecting approaches of project management. Fortaleza (2012)
10. Hanif-ur-Rehman, Bamma, H.K., Nazir, S., Shahzad, S., Hodosi, T.: A sourcing decision model for application maintenance services. Bandung, Indonesia (2017)
11. Ferreira, C., Nery, A., Pinheiro, P.: A multi-criteria model in information technology infrastructure problems. Proc. Comput. Sci. **91**, 642–651 (2016)
12. Vasconcelos, M., Pinheiro, P., Simao Filho, M.: A multicriteria model applied to the choice of a competitive strategy for the printed newspaper. Cybern. Approaches Intell. Syst. **1**, 206–215 (2017)
13. Barbosa, P., Pinheiro, P., de Vasconcelos Silveira, F., Simão Filho, M.: Applying verbal analysis of decision to prioritize software requirement considering the stability of the requirement. Adv. Intell. Syst. Comput. **575**, 416–426 (2017)
14. Pinheiro, P., Machado, T., Tamanini, I.: Verbal decision analysis applied on the choice of educational tools prototypes: a study case aiming at making computer engineering education broadly accessible. Int. J. Eng. Educ. **30**, 585–595 (2014)
15. Simão Filho, M., Pinheiro, P., Albuquerque, A.: Applying verbal decision analysis in distributed software development-rank ordering the influencing factors in task allocation. In: Proceedings of the 11th Iberian Conference on Information Systems and Technologies (CISTI'2016), vol. I, pp. 205–210 (2016)

Data Strategy Framework in Servitization: Case Study of Service Development for a Vehicle Fleet

Jukka Pulkkinen, Jari Jussila, Atte Partanen and Igor Trotskii

Abstract Many companies have been adopting a service business strategy especially in mature industries in order to differentiate their offering and enhance a customer engagement already decades. Therefore, the servitization has been global trend especially among many traditional manufacturing companies and there is a lot of research done in this are but on the other side small or medium size enterprises (SME) has got a less attention in literature. In this paper, the data strategy framework will be presented to develop Proof of Concept (POC) in software development for SMEs companies. Special attention will be paid to take into account the business requirements before software implementation. The purpose of POC is to demonstrate the feasibility of new concept to improve service operation. The service operation improvement is new differentiation possibility to service providers. Lean service development method is utilized in the development of POC to ensure the fulfilment of business requirements. The presented data strategy framework is applied to enhance the service operation efficiency for vehicle fleet manager. The market for the vehicle fleet manager is increasing due to the on-going change in a transport systems. Mobility as a Service (MaaS) is becoming more popular model for different transport methods and this opens new market for the operator who manages whole vehicle fleet. The winner in the new growing market is the player who can manage their operation most effectively and therefore utilizing data is critical success factor.

1 Introduction

Companies have been adopting a service business model especially in mature industries in order to differentiate their offering and enhance a customer engagement already decades. Therefore, the servitization has been global trend among many traditional manufacturing companies and in last years this trend has occurred even in the countries with low production costs; e.g. in China from 1% of servitized enterprises

J. Pulkkinen (✉) · J. Jussila · A. Partanen · I. Trotskii
Häme University of Applied Sciences, 13100 Hämeenlinna, Finland
e-mail: jukka.pulkkinen@hamk.fi

A. Visvizi and M. D. Lytras (eds.), *Research & Innovation Forum 2019*,
Springer Proceedings in Complexity,
https://doi.org/10.1007/978-3-030-30809-4_34

in 2007 to 20% in 2011 [1]. This development means that differentiation through servitization will be more difficult in the future. On the other side, the big data has been increasing sharply due to the development of ICT technology and this provides new differentiation strategy by using data to optimize service operation [2], service delivery and quality decisions [3].

The market for the vehicle fleet manager is increasing due to the on-going change in a transport systems. Mobility as a Service (MaaS) is becoming more popular model for different transport methods and this opens new market for the operator who manages whole vehicle fleet. The winner in the new growing market is the player who can manage their service operation most effectively and therefore utilizing data for efficiency improvement purposes is a critical success factor. The service providers that manage the vehicle fleet are typically small or medium size enterprises (SMEs) in Finland and they are also young companies because the market is relatively new. On the other side, the modern vehicles have a lot of data which could be used to improve the service operation, but naturally this has not been optimally used yet due to the new technology and small enterprises on the market who have limited capabilities to utilize the latest technology.

We argue that the data driven service operation can be used to significantly improve the service performance by having right data strategy in their practical implementation. Recent studies [1, 2, 4] emphasize servitization as one key differentiation factor specially for manufacturing companies by bundling the product and services together into one solution. Nevertheless, less attention has been paid to SMEs, which are not manufacturing companies, and how they can differentiate using the data on their market place.

Our research question is: *How to create the data strategy and implement data driven service operation in SMEs in order to reach significant competitive advantage on their market place.*

In this paper, the data strategy framework will be presented to develop Proof of Concept (POC) in software development where service operation is enhanced with the help of data. Special attention will be paid to take into account the business requirements before software implementation. The purpose of POC is to demonstrate the feasibility of new concept to improve service operation. The service operation improvement with the help of data is new differentiation possibility to service providers. Lean service development method is utilized in the development of POC to ensure the fulfilment of business requirements.

2 Theoretical Background

2.1 Data as Driver for Service Operation

The literature review presented in this chapter is aimed to provide a high-level understanding of service development and how useful they are in the context of SMEs. A special attention will be paid to utilizing latest technology development to integrate software embedded intelligence in the service operation.

The most famous strategic program to improve manufacturing industries competitiveness by using data is the program called Industry 4.0 which has its origin in Germany and aims to upgrade Germany's industrial capabilities with the help of smart factory concept [5]. Actually Germany creates the biggest share (31%) of industrial value in the EU and far behind come Italy (13%), then France (10%) and Spain (7%) [5]. Therefore, Industry 4.0 has got a lot of attention in many countries and similar strategic programs have been initiated in many countries. Industry 4.0 means 4th Generation Industrial Revolution, where "software embedded intelligence is integrated in industrial products and systems" [2]. Thus Industry 4.0 has been discussed a lot in the literature [2, 5–8] but all these are focused on the industrial manufacturing companies and how to create competitive advantage on their market place and they have very limited experience in the area of SMEs outside manufacturing environment.

Many product firms especially in developed countries have already adopted a service strategy and on average service revenue of product firms is about 30% of their total revenue [4]. On the top of the servitization, the data driven strategy and different processes to use data to improve their competitive position are presented in the literature [1, 2, 4, 9]. Actually there are many good elements in the presented processes which can be utilized also to SMEs. But we want to state that the improvements to service performance has often been relatively modest in the literature and this depends at least partly on generalized approach in servitization and they have not taken into account the business requirements well enough. Naturally the big companies face a huge challenge to make transformation from product driven organization to service driven operation model so a change management is very critical in their success. Therefore we argue that there is a specific need to more research in the area how the SMEs can gain benefits from the data driven service operation. The transformation to new operation model is not so big challenge to these companies due to smaller size but on the other side, their investment capability is smaller and they expect very fast return on investment.

The Internet of Things (IoT) technology has been developing very fast in the recent years and its one main reason to an increased amount of available data nowadays. This has been in the focus of the several studies [6, 10, 11]. Typically major part of these are focusing on technical implementation to collect data but less attention has been paid to business requirements and how IoT can be used to improve the competitive advantage. In addition to this, less focus has been paid to service development with the help of IoT technology. But some papers already exist in this area to discuss how IoT based solutions are a cost effective method for creating new competitive advantage in different areas of service operation [2].

We argue that the data driven service operation can be used to significantly improve the service performance in SMEs by having right data strategy in their practical implementation. In the current literature little attention has been paid to SMEs, which are not manufacturing companies by nature, and how they can differentiate using the data on their market place. Therefore, we present the data strategy framework developed for SMEs but we assume that it can be applied also to bigger companies.

2.2 Lean Service Development

The alignment of work processes and supporting software systems has been an ongoing research topic for several decades [12–14]. The business and information technology (IT) research stream builds on the premise that business and IT performance are tightly coupled [13, 15, 16] and enterprises cannot be competitive if their business and IT strategies are not aligned [13].

In order to ensure that the development of supporting software systems will yield business benefits several development methodologies have been introduced, such as lean startup [17], lean service creation [18], lean software development [19], and agile software development [20]. Lean thinking [21] underlies these development methods as a way to organize human activities to deliver more benefits to society and organizations and value to individuals while eliminating waste. A summary of ways to eliminate waste in the focal development methods is introduced in Table 1, compiled from [18, 22].

Poppendieck [22] translated the seven wastes of manufacturing to software development in contrast to operating with mass production paradigm: "if your company writes reams of requirement documents (equivalent to inventory), spends hours upon hours tracking change control (equivalent to order tracking), and has office which defines and monitors the software development process (equivalent to industrial

Table 1 A summary of eliminating waste in manufacturing, software development and service creation

Seven wastes of manufacturing	Seven wastes of software development	Seven wastes of service creation
1. Overproduction	1. Extra features	1. Features not validated by customers
2. Inventory	2. Requirements (e.g. story cards detailed only for current iteration)	2. Features that do not solve a customer problem or fix, eliminate or reduce customer's pain
3. Extra processing steps	3. Extra steps	3. Work not aligned to business objectives and context
4. Motion	4. Finding information	4. Missing feedback
5. Defects	5. Defects not caught by tests	5. Unlovable product
6. Waiting	6. Waiting, including customers	6. Building minimum
7. Transportation	7. Handoffs	7. Self-deception

engineering), you are operating with mass production paradigms. Think 'lean' and you will find a better way." However, criticism has been also leveled towards agile software development from the lean startup community: "agile is very effective on transferring wrong idea to product" [23]. Lean service creation is one approach that combats developing the wrong idea by expanding outside functional silos, involving multidisciplinary teams, maximizing realism to overcome self-deception (of developers), and always validating (learn, measure, build) [18]. In contrast to lean manufacturing and lean/agile software development, lean service creation aims to (1) develop only features that are validated by customers; (2) features that solve customer's customer problem, fix/eliminate/reduce customer's pain and/or that amplify positive emotions that the customer has in getting needs and problems solved; (3) align software development to business objectives and context; (4) get customer feedback and co-create with customers; (5) build lovable products (not just error free), (6) build the minimum lovable product to reduce waiting time for performing experiments and getting customer feedback; (7) avoid self-deception and working on incorrect assumptions by validating and co-designing with customers.

We define "lean service development" as a development approach that combines lean principles, lean software development and lean service creation, where we start from business requirements following lean principles and using lean service creation methods and move towards digital services using lean software development methods in order to create tangible competitive advantages for the enterprise.

3 Methodology

3.1 Case Study

The market for a transport is changing very fast and new business models to provide new possibilities to make transports have been developed in recent years. One example of new transport method is Mobility as a Services (MaaS) where consumers doesn't own the vehicle but transport is provided to them by transport service provider. Naturally this opens new market in the many areas. One new market is a vehicle fleet management because the companies providing transport services has not always a capability to manage their vehicle fleet effectively. The core competence for the transport service providers is different than required for managing the fleet technically. So there are companies focusing on this new niche market providing the vehicle fleet management and this type of company is use case for data strategy framework development. This company is called a service provider in this paper.

Service provider has different tasks like a vehicle maintenance, fixing daily problems, moving vehicles to right places etc. so that vehicle fleet is available in good condition to users. The winner in the new growing market is the player who can manage their operation most effectively and therefore utilizing data is critical success factor. In this case study, the operator is called service provider.

Lean service development is the selected development method in this empirical case study because it combines lean thinking and principles, lean service creation thinking and attitude, and agile philosophy. Lean principle fits nicely to approach starting from the business requirements and ending to digital services which creates tangible competitive advantages.

The research was implemented in five workshops. Before workshop an interview was done to understand business requirements and prepare the workshops. In the first workshop the lean service development was used to collect data from service personnel. In the last workshop the final POC was discussed and fine-tuned. The workshops between these were used to dig deeper to specific area in the data strategy framework to understand better how the data could be used in service provider's work process.

The workshop result were analysed to create data strategy framework for service provider. In practise, the results is Proof of Concept (POC) which will be used to develop further the solution called Service Execution System (SES). SES will instruct service personnel in their daily operation to make all required tasks effectively and ensure a good quality of vehicle fleet. The efficient service operation with high quality will create service provider's competitive advantage

4 Results

The results of our research is a data strategy framework in service development which is used in the software development to make Proof of Concept (POC). First the data strategy framework will be presented in general level in Sect. 4.1 and then the result of our research for service provider is presented in Sect. 4.2.

4.1 Data Strategy Framework

In order to use the data to enhance the service operation, we need to translate the data into useful information. This task is more complex due to larger, faster, and more varied data which is available in different data sources nowadays but to which are not automatically accessible by users. The data strategy consists of two phases.

Phase 1: from business requirements to work process
Phase 2: from data to actions, POC

The Phase 1 pays a lot of attention to business requirements and user experience which are key success factors to ensure efficiency improvement in service operation. This phase consists of three steps which are presented in the Fig. 1:

Step 1—Business requirements: Understanding the nature of the service business is crucial in the beginning. There are a lot of different type of service businesses which requires a different approach in data strategy. One category of the service

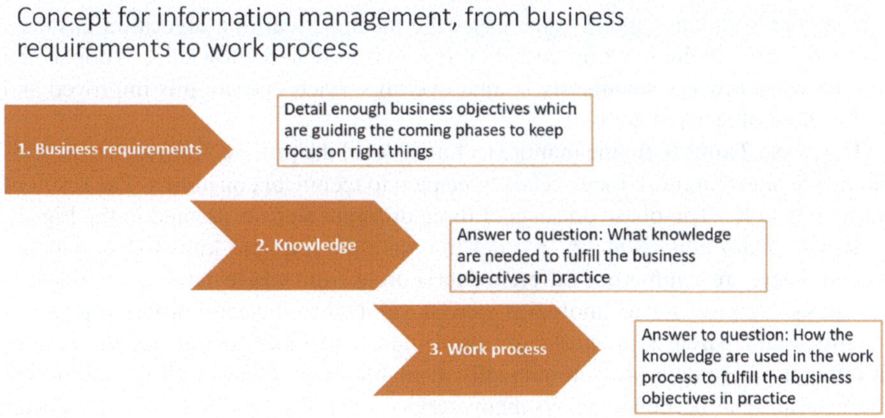

Fig. 1 Three steps model in data strategy framework to create the knowledge starting from business objective and ending to work process

business is combining product and services and the role of service element is to increase the overall value of complete solution to customers. Other type of service business is pure services provided by service vendor to customer and the target is to improve the efficiency with the help of data. Then it might be that the object of service operation is called user but it's not the service vendor's customer who finally pays the bill. This can easily end to situation to enhance the service efficiency at the expense of service quality. Therefore it's very important to understand the nature of the service business in detail level and define the concrete business objectives how to improve the service operation in practice. These business objectives will guide all coming steps in the process.

Step 2—Knowledge: This step aims to create the knowledge which are needed to improve the service operation according to the business objectives. Actually this step needs to be done from the operational point of view so understanding the service operation in detail level is crucial and what data and knowledge would help to improve the concrete operation. The required knowledge can vary a lot like e.g. process knowledge to improve process performance in process industry, how customer is using their product to create knowledge about customer behavior to be used in specific marketing, how service stuff can make their job more effectively, diagnostic information about machines to make predictive maintenance etc. So there are numerous different types of knowledge which are possible to create but the most important is to create knowledge in the line with business requirements defined in the previous step.

Step 3—Work process: In this step the created knowledge is connected into a work process. There are different types of services available in the different businesses environments but all services have one common factor which is people. People has traditionally been a center of service business and this is still the case today in many times. But machine learning algorithms are more and more taking a bigger role in many environments and in some cases the machine learning algorithm has even replaced people in the work process. In this step the created knowledge is connected into the work process seamlessly so that overall service operation is improved and the business objectives are met.

The phase 2 aims to define in more technical level the process from data to actions and this means that more focus needs to be paid to technical constrains. The result of phase to is POC. The phase consists of three different steps presented in the Fig. 2.

Step 1—Data management: In order to create the required knowledge, data are needed. There are a numerous different data sources from where data can be collected nowadays. New sensor technologies were developed to measure different physical quantities and wireless networks are used more and more to transfer the data to an internet where data are automatically available everywhere. Exiting data in the internet can also be beneficial to many service operation and it's easy to transfer into different applications by using API connections. Nowadays so called open data are more and more available and many governments are working to require public organizations to open their data sources to be available to other organizations. Data needs to be cleaned and stored to be effectively available into analytics, machine learning and applications.

Step 2—Analytics and machine learning: This step aims to create the knowledge in practice which are needed to reach the business objectives. The data which are collected from different sources and presented in correct form to right people can already create additional value in many cases. On the other side, the machine learning is typically needed when we aim to reach drastically new level of competitive advantage by changing completely the traditional model of doing services. One example is the case where machine learning is used to replace some current work done by people. Analytics can be classified between the data visualization and the machine learning where collected data is analyzed to figure out new knowledge which are

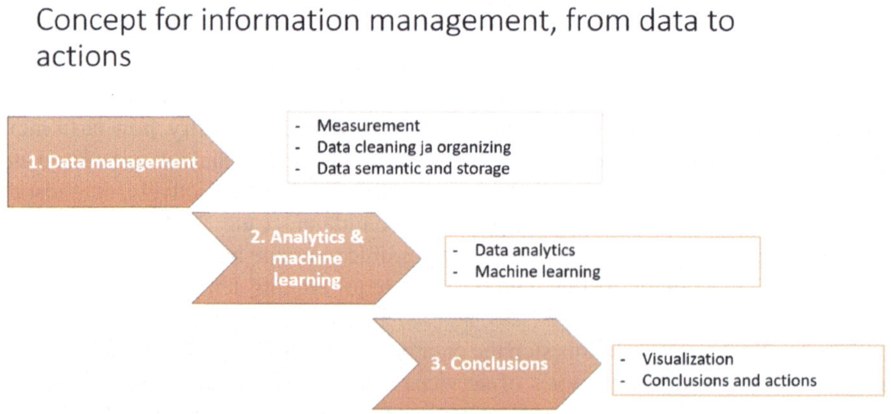

Fig. 2 Three steps model in the data strategy framework to make the actions based on the collected data and data analytics and machine learning

used in service operation. When machine learning is used, the special attention is paid to firstly use the correct machine learning algorithm to fulfill the problem which is defined in the line with the business objectives, and to secondly use comprehensive enough data which are needed to selected machine learning algorithm.

Step 3—Conclusion: Finally in the last step the created knowledge need to be connected into the work process. Actually this is very crucial and a poor implementation can destroy the whole value of data strategy. This happens if accessing to the data and using it in the work process require more effort than an efficiency is improved in some other places and this happens very easily with poor design and implementation. Therefore a user experience is very critical in this phase and how the data is used in the work process need to be defined in a good co-operation with final users of the system. The information can be exchanged between the system and a user through mobile phone, tablet or computer display and a right media needs to be selected and the user interfaces need to be designed to fit for an intended purposes. Even if the machine learning is used to replace a current manual work done by people, the new work process based on the machine learning algorithm needs to be connected to an overall system and all interfaces needs to be designed carefully to ensure the fulfillment of the overall business objectives.

4.2 Data Strategy Framework Implementation in Case Study

In this section the data strategy framework will be presented for specific service provider. In the first phase, we create a so called deviations from the business requirements and then deviations generate work order which are connected to work process according to Fig. 3. So the deviation means that the status of object related to deviation is not within a normal operation range and some actions are needed to

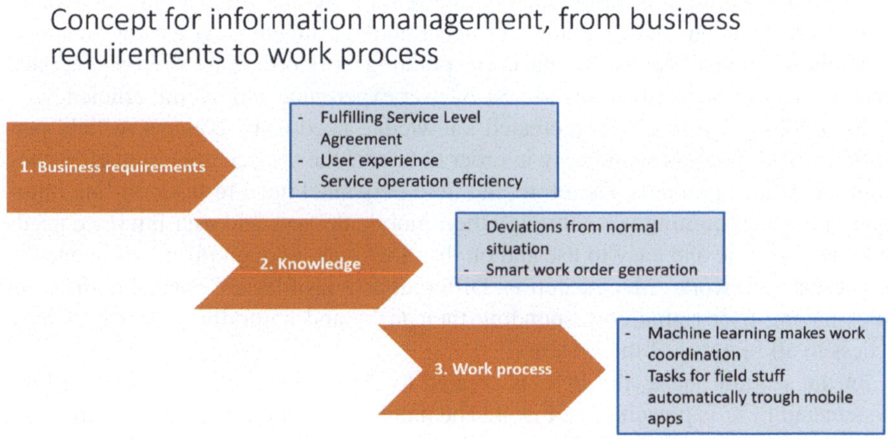

Fig. 3 The data strategy framework for vehicle operator to create the knowledge starting from business objective and ending to work process

keep overall system in required situation. Therefore, we can state that the deviations create solid basis to data strategy for this specific case.

Business requirements: Service Level Agreement (SLA) defines a service scope and a quality requirements for service provider. Typical examples of service scope is periodical maintenance tasks for each vehicle and quality requirements are an availability of bike and how long a bike station can be empty or full to mention few of them. Therefore we can state that SLA defines the most important business requirements for the service operation. Naturally a user experience is also important to service provider to have positive image and the user experience means that all vehicles are in good conditions and available for users when they need them. On the other side, we can state that one purpose of SLA is to define the minimum quality level for the user experience. An efficiency of service operation is also critical business requirement from the service provider's profitability point of view. We can summarize the business requirements by maximizing the efficiency of service operation under the constraint of SLA and user experience.

Step 2—Knowledge: The actual status of all vehicles and stations against SLA requirements are critical knowledge for the service provider. In practice, this means that if the bike station is becoming close to either empty or full it needs to be known by service provider in order to make corrective actions before the station is either empty or full. This situation is called a deviation which needs to be recognized and all deviations need to initiate a corrective work order. Naturally the corrective actions and corresponding deviations must be prioritized from the time point of view which means that the high priority deviations need to be done immediately and low priority deviations have longer time for the corrective actions. So we can state that the deviations are the basis for the data strategy and they can be created from the data collected from vehicles and stations automatically or they can be created from the scheduled maintenance tasks.

On the top of the deviations, work order generation for the field personnel is also very critical knowledge especially from the service operation efficiency point of view. By making this automatically taking into account the overall status of all vehicles and stations have significant impact on the efficiency. At the end, we want to emphasize that all deviations and corresponding work orders are strictly generated from the business requirements like SLA, user experience and overall efficiency.

Step 3—Work process: The created knowledge need to be connect to field personnel's work process seamlessly in order to ensure the service operation efficiency improvement in practice. Therefore, the field personnel need to have on-line information for their coming tasks through their mobile devices and user interface needs to be very simple and easy to use and on the other side, enough information need to be presented to work order execution. Different user profiles are created to different roles having user rights corresponding their tasks and naturally master user have access to all features of the system.

In the second phase, the POC is created to demonstrate main features of final system and POC is presented in Fig. 4. The major part of POC is done with live data but some part are only static displays. The purpose of POC is to provide to service personnel a realistic understanding of functionalities of final system and it verifies

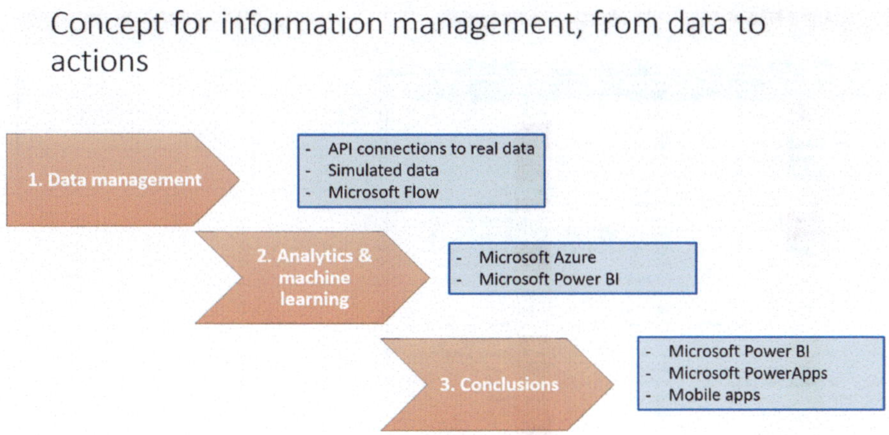

Fig. 4 Proof of concept for data management from data to actions

the correctness of data strategy before system implementation. The system have main displays to manage service operation like scheduled maintenance tasks, deviations and task manager and then data storage for historical data to present different reports and dashboards. Example of displays for deviation is presented in Fig. 5 and for task manager in Fig. 6. With the help of these displays the correctness of POC was verified in the workshop with service personnel.

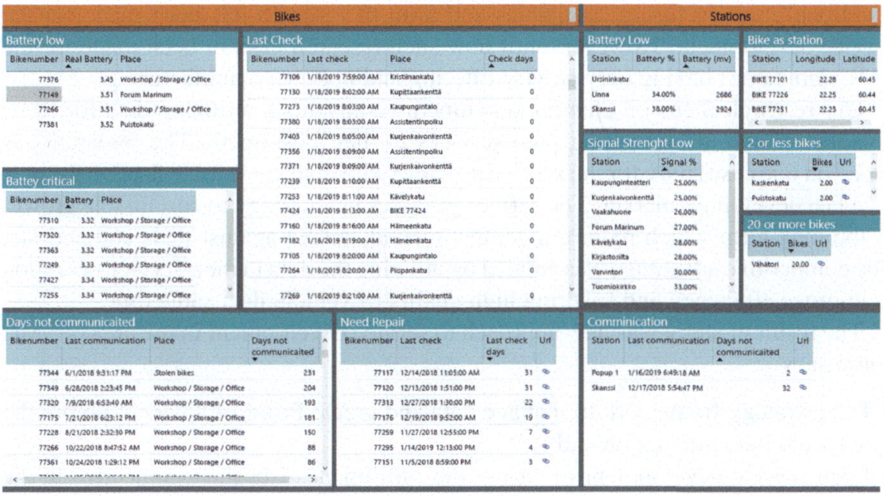

Fig. 5 Proof of concept display for deviations

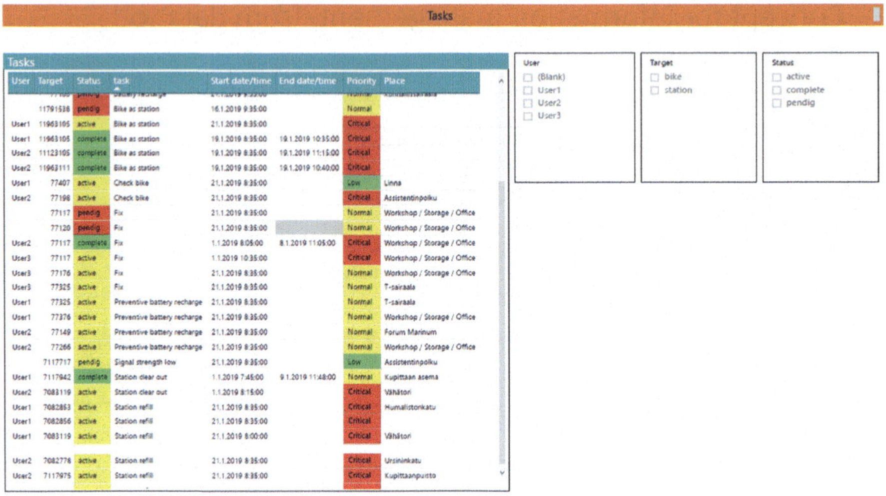

Fig. 6 Proof of concept display for task manager

5 Conclusion

Many companies have been adopting a service business strategy especially in mature industries in order to differentiate their offering and enhance a customer engagement already decades. Therefore, the servitization has been global trend among many traditional manufacturing companies and a lot of research has been made for the service development. However, until now such research has not been made for the SMC companies having only service offering and not own manufacturing.

Our research focus on creating a competitive advantage utilizing data for SMC companies who are providing pure services to their customers. The research was based on one case study for the service provider who manage vehicle fleet on the new nice fast developing market. The service provider has strategy to sustainable growth in the long term which require a competitive advantage against their competitors. The competitive advantage was created by utilizing the data in their service operation to improve efficiency and ensuring high quality of vehicle fleet same time.

The result of study was new approach to service development in the area of SMC companies.

- Data strategy framework to manage data and create Proof of Concept before the software implementation and
- Lean service development to ensure the fulfillment of business objective in the data management.

The presented approach ensures that Proof of Concept fulfill the business requirements and same time all unnecessary features are excluded according the lean service development principle. By excluding the unnecessary features is very important to reach streamlined service operation and in this way maximizing the efficiency.

The presented model can be applied also to different environment and not necessarily to SMC companies but more research are needed in this area.

References

1. Opresnik, D., Taisch, M.: The value of big data in servitization. Int. J. Prod. Econ. **165**, 174–184 (2015)
2. Rymaszewska, A., Helo, P., Gunasekaran, A.: IoT powered servitization of manufacturing–an exploratory case study. Int. J. Prod. Econ. **192**, 92–105 (2017)
3. Miller, H., Mork, P.: From data to decisions: a value chain for big data. It Prof. **15**, 57–59 (2013)
4. Vedrell-Herrero, F., Bustinza, O.F., Parry, G., Georgantzis, N.: Servitization, digitization and supply chain interdependency. Ind. Mark. Manag. **60**, 69–81 (2017)
5. Heng, S.: Industry 4.0-Upgrading of Germany's industrial capabilities on the horizon (2014)
6. Lee, J., Kao, H., Yang, S.: Service innovation and smart analytics for industry 4.0 and big data environment. Proc. Cirp. **16**, 3–8 (2014)
7. Schmidt, R., Möhring, M., Härting, R.-C., Reichstein, C., Neumaier, P., Jozinović, P.: Industry 4.0—potentials for creating smart products: empirical research results. In: International Conference on Business Information Systems, pp. 16–27. Springer, Berlin (2015)
8. Wang, S., Wan, J., Li, D., Zhang, C.: Implementing smart factory of Industrie 4.0: an outlook. Int. J. Distrib. Sens. Networks. **12**, 3159805 (2016)
9. Gilbert, R.J.: E-books: a tale of digital disruption. J. Econ. Perspect. **29**, 165–184 (2015)
10. Gubbi, J., Buyya, R., Marusic, S., Palaniswami, M.: Internet of Things (IoT): a vision, architectural elements, and future directions. Futur. Gener. Comput. Syst. **29**, 1645–1660 (2013)
11. Abdul-Qawy, A.S., Pramod, P.J., Magesh, E., Srinivasulu, T.: The Internet of Things (IoT): an overview. Int. J. Eng. Res. Appl. **1**, 71–82 (2015)
12. Maes, R., Rijsenbrij, D., Truijens, O.: Redefining business-IT alignment through a unified framework (2000)
13. Aversano, L., Grasso, C., Tortorella, M.: A literature review of business/IT alignment strategies. Proc. Technol. **5**, 462–474 (2012)
14. Luftman, J., Brier, T.: Achieving and sustaining business-IT alignment. Calif. Manage. Rev. **42**, 109–122 (1999)
15. Kearns, G.S., Lederer, A.L.: A resource-based view of strategic IT alignment: how knowledge sharing creates competitive advantage. Decis. Sci. **34**, 1–29 (2003)
16. de Leede, J., Looise, J.C., Alders, B.C.M.: Innovation, improvement and operations: an exploration of the management of alignment. Int. J. Technol. Manag. **23**, 353 (2002)
17. Ries, E.: The Lean Startup: How Today's Entrepreneurs Use Continuous Innovation to Create Radically Successful Businesses. Crown Books (2011)
18. Sarvas, R., Nevanlinna, H., Pesonen, J.: Lean service creation. The Handbook V1.8. Futurice (2017)
19. Poppendieck, M., Poppendieck, T.: Lean Software Development: An Agile Toolkit: An Agile Toolkit. Addison-Wesley (2003)
20. Abrahamsson, P., Salo, O., Ronkainen, J., Warsta, J.: Agile Software Development of Mobile Information Systems. VTT publication 478, Espoo (2002)
21. Womack, J.P., Jones, D.T.: Lean Thinking—banish waste and create wealth in your corporation. J. Oper. Res. Soc. **48**, 1148 (1997)
22. Poppendieck, M.: Principles of lean thinking. IT Manag. Sel. **18**, 1–7 (2011)
23. Rossi, A.: Lean Startup. For real!? In: TUT SW Startup day, Tampere (2013)

Big Data Dimensionality Reduction for Wireless Sensor Networks Using Stacked Autoencoders

Muneeba Sirshar, Sajid Saleem, Muhammad U. Ilyas, Muhammad Murtaza Khan, Mohammed Saeed Alkatheiri and Jalal S. Alowibdi

Abstract A typical wireless sensor network (WSN) consists of spatially dispersed sensor nodes for monitoring and gathering data at a central location. Over time, this data accumulates and it becomes difficult to transmit this large volume of data. Owing to the recent advancements in internet of things (IoTs), more complex data is generated. This high-dimensional data must be reduced to a lower dimensional representation before transmission. Principal component analysis (PCA) is a common method for data aggregation. However, PCA performs best for the cases that require linear mapping. Therefore, a method for linear and/or non-linear data mapping is considered in this work. Autoencoders are neural networks that can learn about the internal structure of the data and reduce its dimensionality. Stacked autoencoders are a class of deep feed forward neural networks that perform real learning of the data. Principal component analysis assumes that the underlying data can be fully described by its mean and variance, whereas, autoencoders are adaptive to different data distributions. In this paper, we compare the performance of PCA and autoencoder

M. Sirshar · S. Saleem (✉) · M. U. Ilyas · M. M. Khan
School of Electrical Engineering and Computer Science (SEECS), National University of Sciences and Technology (NUST), Islamabad 44000, Pakistan
e-mail: sajid.saleem@seecs.edu.pk; sssaleem@uj.edu.sa
URL: http://seecs.nust.edu.pk/

M. Sirshar
e-mail: msirshar.msee15seecs@seecs.edu.pk

M. U. Ilyas
e-mail: usman.ilyas@seecs.edu.pk; milyas@uj.edu.sa

M. M. Khan
e-mail: muhammad.murtaza@seecs.edu.pk; mkhan@uj.edu.sa

S. Saleem · M. U. Ilyas · M. M. Khan · M. S. Alkatheiri · J. S. Alowibdi
College of Computer Science and Engineering,
University of Jeddah, Jeddah 23890, Saudi Arabia
e-mail: msalkatheri@uj.edu.sa
URL: http://www.uj.edu.sa/

J. S. Alowibdi
e-mail: jalowibdi@uj.edu.sa

A. Visvizi and M. D. Lytras (eds.), *Research & Innovation Forum 2019*,
Springer Proceedings in Complexity,
https://doi.org/10.1007/978-3-030-30809-4_35

in terms of their respective ability to reduce the reconstruction error in the observed high-dimensional data from a WSN. The results of numerical analysis demonstrate that the autoencoder reduces the reconstruction error by up to 10 times as compared to an ordinary PCA-based scheme.

1 Introduction

The term Big Data was coined by John Mashey in 1987, as he used it to quantify huge volume of information [1]. Later in 2001, Doug Laney, updated the notion of big data by adding the concepts of velocity and variety. Since then, these three factors, i.e, volume, velocity and variety are known as the 3 V's of big data [2].

Big data applications may include diverse data such as demographic data, weather record, time series, etc. [3]. Big data analytics is an important constituent of internet of things (IoT) deployment. IoT mainly consists of deployed sensors, data storage and data analytics. Such systems shall generate enormous amount of data and hence fall in the jurisdiction of big data. Processing and transmission of this data is not feasible without application of dimensionality reduction techniques.

2 Dimensionality Reduction Techniques

Higher-dimensional data-sets are often reduced to lower dimensions because of the 'curse of dimensionality' associated with these data-sets [4]. Dimensionality reduction involves projecting the high-dimensional data to a lower-dimensional space such that vital information stays intact. Principle component analysis (PCA) is the most widely used technique for dimensionality reduction [5]. To cater for the drawbacks associated with PCA, extensions of PCA such as kernel PCA, probabilistic PCA [6] and oriental PCA were proposed and tested [7]. Dimensionality reduction techniques can be broadly divided into convex and non-convex optimization techniques. Convex techniques include algorithms such as Isomap, diffusion maps, multi-dimensional scaling, maximum variance unfolding, Hessian eigenmaps, local linear embedding (LLE) [8–10], where as non-convex techniques include Sammon mapping [11], manifold charting [12] and wavelet compression [13].

2.1 Principal Component Analysis (PCA)

Principal component analysis is a widely used form of analysis that relies on transforming the existing coordinate system to new axes that capture the maximum data variance [14]. These new axes are called principal components. PCA provides an insight into the latent factors of the data as the generated principal components are in the order of decreasing variance. This means that the first principal component captures the most significant portion of the variance followed by the second, and so on. The first few components, in most cases, sufficiently capture the data variance, and hence, these few components may be used to reconstruct the data. These few principal components provide a reduced dimensional representation of the data. However, PCA assumes that the data set follows the principle of linearity. It is due to this assumption that the PCA's representation of data, in terms of an approximate linear basis, can frequently be very restrictive. The higher-dimensional complex systems are often non-linear and PCA may prove to be insufficient to represent such data in lower-dimensional space. Local linear approximation is used in these cases because higher-order terms often vanish for larger variations. The noise in the data itself can be reduced by finding the latent factors in the data. The estimated noise covariance for this work follows the probabilistic PCA model from Tipping and Bishop [15]. Redundancy must also be minimized for effective implementation of the dimensionality reduction scheme. To reduce redundancy, the goal is to process the data in such a manner that the covariance matrix is diagonalized.

Let us consider a wireless sensor network (WSN) that consists of randomly deployed sensor nodes, $X_1, X_2, \dots, X_n$, where each node records some observation and transmits it to the cluster head [16]. In each cluster, the number of cluster members are about the same. All the sensor readings are subjected to two data similarity checks:

1. *Data Magnitude Similarity*:

$$c = D * kk, \tag{1}$$

and

2. *Data Correlation*:

$$Corr(X_i, X_j) \leq \epsilon_c, \forall i, j \in \{1, 2, \dots, n\}, \tag{2}$$

where D is the difference in readings of any two sensors, say X_1 and X_2, kk is a scaling factor, c is the defined similarity threshold, and similarly, ϵ_c is the the similarity threshold for correlation. Note that $Corr(\cdot, \cdot)$ denotes the correlation of two sensor signals.

Figure 1 shows the average data distribution across the ensemble of sensors from which the readings are available. Once clustering has been done, cluster head performs data aggregation using principal component analysis [17]. However, linearity supports the changing of basis for PCA transformation. A lot of research (such as kernel PCA) has been done on applying non-linearity prior to PCA. It also assumes that the data is fully described by the mean and variance, whereas, the only zero mean probability distribution that is fully described by its variance is the Gaussian distribution. This assumption also employs that SNR and covariance matrix fully characterize the noise and redundancies in the data-set. These drawbacks can be addressed by other dimensionality reduction schemes such as the autoencoders.

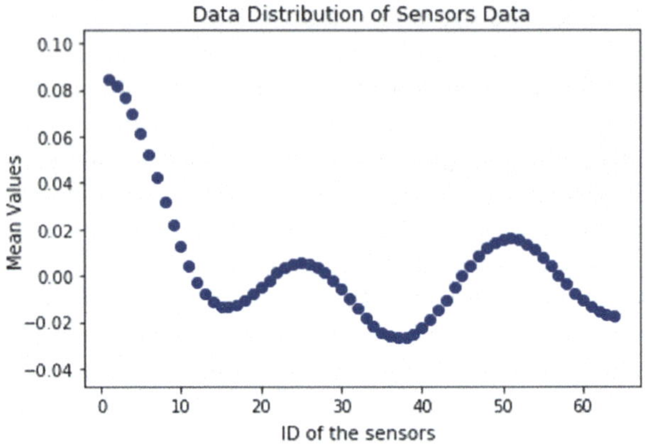

Fig. 1 Data distribution in sensor nodes

2.2 *Autoencoders*

Autoencoders are a class of neural networks that are designed with the purpose of copying the input to the output with the goal of performing dimensionality reduction in the hidden layers. This is generally achieved by having a hidden layer that is smaller than the input layer. Typically, an autoencoder is designed with an odd number of hidden layers between the fully-connected input and output layers. The number of neurons in the input layer is equal to the number of neurons in the output layer. The neurons in the hidden layers define the level of compression or reduction that is allowed. To develope an understanding for autoencoders, lets begin by considering a single layer neural network with n inputs, B outputs and a linear activation function. The i-th output of this layer is given as follows:

$$y_i = \sum_{j=1}^{n} w_{ij}x_j, \forall i \in \{1, 2, \ldots, B\} \text{ and } j \in \{1, 2, \ldots, n\}, \tag{3}$$

where w_{ij} is the weight that links the j-th node of a layer to the i-th node of the next layer. Now, determining the best weights is a supervised learning task in which we learn the mapping from input to output using multiple training examples T. We need a set of weights that satisfy the system of linear equations defined as follows:

$$x_1^{\alpha} w_{i1} + x_2^{\alpha} w_{i2} + \cdots + x_B^{\alpha} w_{iB} = d_i^{\alpha}, \tag{4}$$

where d_i is the desired output, $\alpha \in \{1, 2, \ldots, T\}$, and T represents the number of training examples. This solution fails if there are no set of weights that satisfy all the equations simultaneously. So, typically, a least-mean-square based error minimization approach is adopted:

$$Error = \sum_{\alpha=1}^{T} \sum_{i=1}^{B} (d_i^{\alpha} - y_i^{\alpha})^2. \tag{5}$$

To achieve the desired dimensionality reduction, let us now consider an autoencoder network with a hidden layer and a non-linear activation function σ_k [18]. Now, the least-mean-squared error at the reduced dimensional inner layer can be defined as

$$E = \sum_{k=1}^{m} (d_k - \sigma_k)^2, \tag{6}$$

where m is the number of neurons in the hidden layer, assumed to be significantly less than n. For an autoencoder, the number of nodes in the input and output layer are always equal. Using the back-propagation learning algorithm, the optimized change in weights at each step is given as

$$\Delta w_{ji} = \epsilon \delta_j \sigma_i, \tag{7}$$

where ϵ represents the learning rate and δ_j, depending upon on the encoding layer, can be computed as follows:

For the output unit,

$$\delta_j = (d_j - \sigma_j)\sigma_j(1 - \sigma_j), \tag{8}$$

and for the nodes in the hidden layer,

$$\delta_j = \{\sum_k \delta_k w_{kj}\}\sigma_j(1 - \sigma_j), \tag{9}$$

where Δw_{ji}, for the output unit, is the same as least-mean-square (LMS) with a non-linear output.

The autoencoder presented above is trained to minimize the mean-squared error between the input and the output. Data compression is performed in the encoding layer. A sigmoid activation function is generally used at both the input and the output whereas for the hidden layers, linear transformation is usually employed. However, the autoencoder has a high number of connections and hence it converges slowly. It is also likely that the back propagation approach gets stuck in the local minima due to higher number of interconnections. To address this issue, a stacked autoencoder may be used.

3 Stacked Autoencoders

Stacked autoencoder consist of a multi-layered structure as shown in Fig. 2. At the first hidden layer, dimensionality reduction is performed as the number of nodes is less than the number at input layer. However, the number of nodes at this layer is

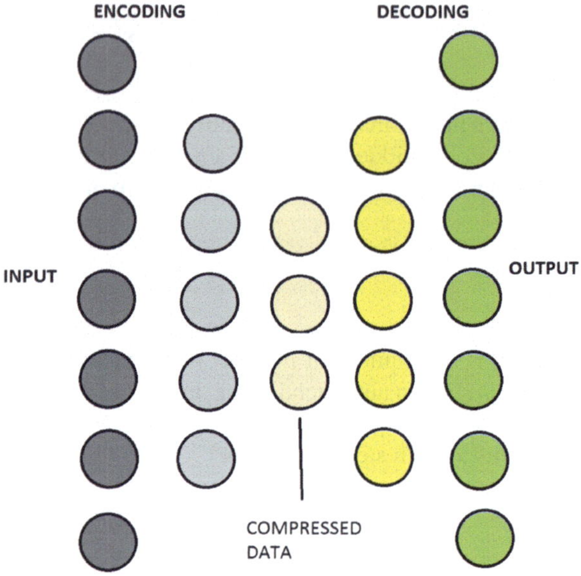

Fig. 2 General stacked autoencoder layout. Stacked autoencoder may consist of more than one hidden layers along with an input and an output layer. Note that the number of neurons in the input and output layers are equal

greater than the number of nodes at the innermost layer. For this work, the stacked autoencoder's design was generated using TensorFlow package. Dense neural layers have been constructed with rectified-linear unit (Relu) function as the activation function of neurons in the network. He-initializer has been used for the initialization of weights and l_2-regularization has been used as kernel for regularization. At the first hidden layer, the dimensionality is reduced to the number of dimensions equal to k_{opt}. At the innermost layer, the desired dimensionality reduction is achieved.

The error between the input data and the reconstructed data, after compression, is called reconstruction error. The reconstruction error can be minimized by regularizing the cost function. As a design choice, mean-squared error can be used as a cost function followed by identity function as a linear transformation function.

4 Comparison and Analysis

These schemes are tested on seven data-sets. Five of these data-sets are synthetic data-sets with parameters shown in Table 1. The other two data-sets are real data-sets taken from UCI's Machine Learning Repository [19]. The five synthetic data-sets, indexed by γ, are used for the evaluation of the proposed method. These data sets are based on the superposition of γ Gaussian distributions multiplied by a damping exponential and a cosine factor.

Table 1 Parameters for synthesized data-sets. The mean vector, covariance matrix, frequency and damping factor are included for all the distributions in the Gaussian mixture model

k	μ_k	Σ_k	f_k (Hz)	δ_k
1	[3, 2]	[10, 2]; [2, 9]	0.1	10
2	[8, 14]	[8, 2]; [2, 10]	0.2	15
3	[13, 14]	[9, 1]; [1, 8]	0.05	8
4	[4, 2]	[7, 3]; [3, 10]	0.3	12
5	[10, 15]	[11, 3]; [3, 10]	0.4	7

$$x_\gamma(i, j, t) = 10^3 \sum_{k=1}^{\gamma} (g_{\mu_k, \Sigma_k}(i, j) e^{-\frac{t}{\delta_k}} \cos(2\pi f_k t)), \tag{10}$$

where i and j represent the physical location of the sensor on the grid, $\gamma \in \{1, 2, \ldots, 5\}$, t is the time instant of an observation, Σ_k is the covariance matrix and μ_k is the mean vector of the k-th two dimensional Gaussian random vector.

The real data-sets used for this paper are taken from UCI's Machine Learning Repository. These data-sets include the Ozone level detection data-sets [19] and gas drift concentrations data-set [20]. Figure 3 shows the performance of the PCA with increasing number of principal components for a mixture of Gaussians. The horizontal axis shows the number of Gaussian distributions in the mixture, i.e., γ, and the vertical axis represents the noise variance. The results in Fig. 3 verify that as we increase the dimensionality of the compressed signal by allowing more PCA components for representation, a decrease in noise variance and an increase in data variance is observed. Another important trend that can be seen is that the optimum value of components that carry a sufficient data variance depends on the number of Gaussian spatial modes in the signal.

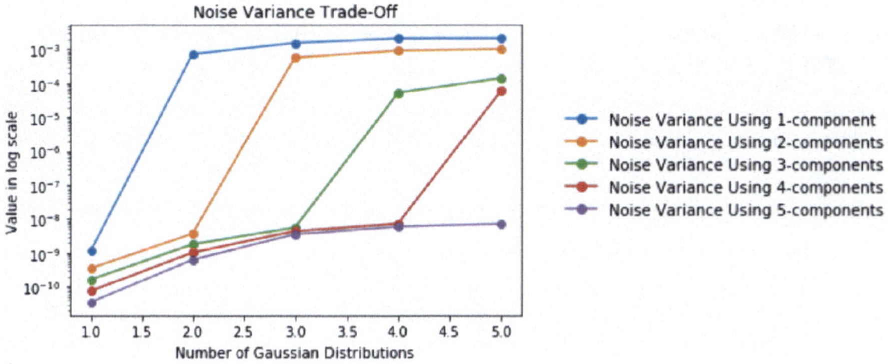

Fig. 3 Captured noise variance using different number of principal components

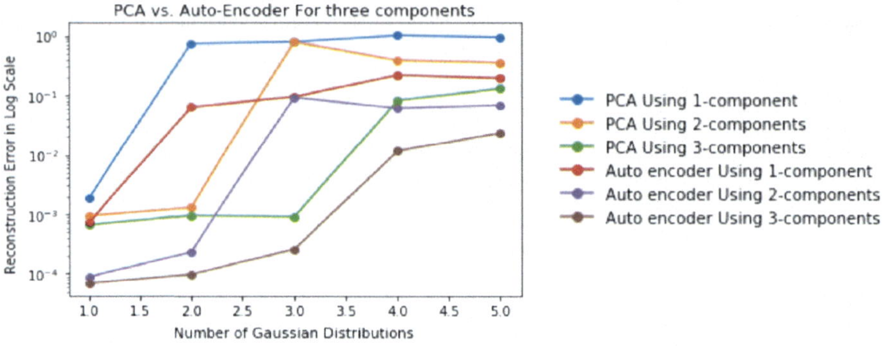

Fig. 4 Reconstruction error of PCA versus auto encoder with different number of Gaussian spatial modes. The reconstruction error for autoencoder is about ten times less than that of PCA

It is interesting to observe from Fig. 4 that the autoencoder reduces the reconstruction error 10 times as compared to a PCA-based scheme. The success of autoencoders lies in its learning and optimization characteristics. It is also evident that the stacked autoencoders can further reduce the reconstruction error by 10 times as compared to an autoencoder. Therefore, autoencoders reduce the reconstruction error significantly due to their self-learning nature.

The success behind the proposed solution lies in the fact that autoencoders with linear neurons that are trained to minimize mean-squared error (MSE) will simply perform PCA. An autoencoder can learn non-linear transformations, unlike PCA, with a non-linear activation function and multiple layers. Therefore, an autoencoder provides a compact representation at the output of each layer, and having multiple representations of different dimensions is very useful.

Figures 5 and 6 show the compression results for real data-sets of ozone concentration and gas drift, respectively. The horizontal axis represents the number of units

Fig. 5 Results of ozone level detection data-set. Increasing number of dimensions allow better description of data variance

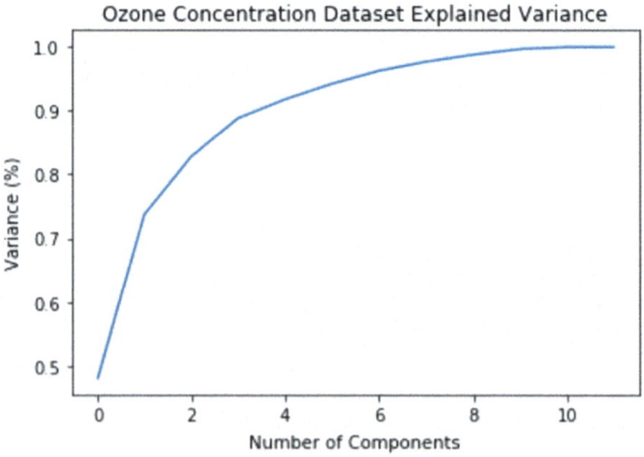

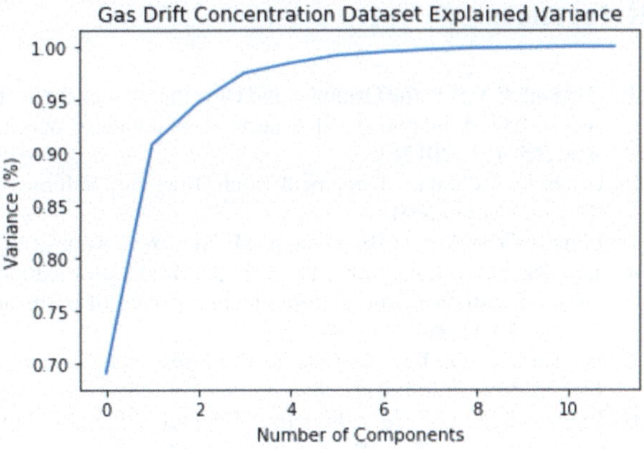

Fig. 6 Results of gas drift concentration data-set. Increasing number of dimensions yield better description of data variance

in the hidden layer, i.e., the reduced dimensional representation using the stacked autoencoder. Similarly, the vertical axis represents the percentage of variance of the data described using a specific number of components. It is clear that increasing number of components capture more variance of data, whereas, the gains are diminishing with increasing number of components. These results justify the idea that the higher-dimensional data can be represented very effectively in a lower dimension using fewer components.

The results in this section establish that autoencoders perform better than other dimensionality reduction techniques because of their self-learning nature, which is independent of the linearity/non-linearity constraint. Therefore, autoencoders are generally robust to any kind of data variations.

5 Conclusion

In this paper, we designed autoencoders for dimensionality reduction of data. This may be beneficial for processing and transmitting huge amounts of data, which shall be generated in the context of IoT framework. It was demonstrated that a simple autoencoder consisting of one hidden layer performs better than PCA, with upto ten times less reconstruction error. For a fair comparison, the number of nodes in the hidden layer of this autoencoder were considered equal to the number of principal components used for data reconstruction from PCA. It was also demonstrated that data can be compressed more efficiently if a stacked autoencoder is used, with reconstruction error reducing further to one-tenth as compared to that of a simple autoencoder.

References

1. Diebold, F.X.: On the Origin(s) and Development of the Term 'Big Data'. PIER Working Paper No. 12-037. Available at SSRN https://ssrn.com/abstract=2152421 or https://doi.org/10.2139/ssrn.2152421 (2012)
2. Laney D: 3-d data management: controlling data volume, velocity and variety. META Group Research Note (2001)
3. Lohr, S.: The Age of Big Data, p. 11. New York Times (2012)
4. Jimenez, L.O., Landgrebe, D.A.: Supervised classification in high-dimensional space: geometrical, statistical, and asymptotical properties of multivariate data. IEEE Trans. Syst., Man Cybern. **28**(1), 39–54 (1997)
5. Pearson, K.: On lines and planes of closest fit to systems of points in space. Philiosophical Mag. **2**, 559–572 (1901)
6. Roweis, S.T.: EM algorithms for PCA and SPCA. In: Advances in Neural Information Processing Systems, vol. 10, pp. 626–632 (1997)
7. Saul, L.K., Weinberger, K.Q., Ham, J.H., Sha, F., Lee, D.D.: Spectral methods for dimensionality reduction. In: Semisupervised Learning. The MIT Press, Cambridge, MA, USA (2006)
8. Burges, C.J.C.: Data Mining and Knowledge Discovery Handbook: A Complete Guide for Practitioners and Researchers, chapter Geometric Methods for Feature Selection and Dimensional Reduction: A Guided Tour. Kluwer Academic Publishers (2005)
9. Venna, J.: Dimensionality reduction for visual exploration of similarity structures. Ph.D. thesis, Helsinki University of Technology (2007)
10. Lee, J.A., Verleysen, M.: Nonlinear Dimensionality Reduction. Springer, New York (2007)
11. Torgerson, W.S.: Multidimensional scaling I: theory and method. Psychometrika **17**, 401–419 (1952)
12. Brand, M.: Charting a manifold. In: Advances in Neural Information Processing Systems, vol. 15, pp. 985–992. The MIT Press, Cambridge, MA, USA (2002)
13. Guo, C., Nelson, J.D.B.: Compressive imaging with complex wavelet transform and turbo amp reconstruction. In: 23rd European Signal Processing Conference (EUSIPCO), pp. 1751–1755 (2015)
14. Hotelling, H.: Analysis of a complex of statistical variables into principal components. J. Educ. Psychol. **24**, 417441 (1933)
15. Tipping and Bishop: Pattern Recognition and Machine Learning by C. Bishop, 12.2.1 p. 574 (1999)
16. Chen, Z., Yang, S., Li, L., Xie, Z.: A clustering approximation mechanism based on data spatial correlation in wireless sensor networks. In: Wireless Telecommunications Symposium (WTS), p. 17 (2010)
17. Li, J., Guo, S., Yang, Y., He, J.: Data aggregation with principal component analysis in big data wireless sensor networks. In: 12th International Conference on Mobile Ad-Hoc and Sensor Networks (2016)
18. LeCuns, Y.: Online Available at http://yann.lecun.com/exdb/lenet/index.html
19. Fan, W., Zhang, K., Yuan, X.J.: UCI Machine Learning Repository. Online Available http://archive.ics.uci.edu/ml
20. Vergar, A.: UCI Machine Learning Repository. Online Available http://archive.ics.uci.edu/ml

Medical Informatics

BYOD, Personal Area Networks (PANs) and IOT: Threats to Patients Privacy

Samara Ahmed

Abstract The passage of FISMA and HIPPA Acts have mandated various security controls that ensure the privacy of patients' data. Hospitals and health-care organizations are required by law to ensure that patients' data is stored and disseminated in a secure fashion. The advent of Bring Your Own Devices (BYOD), mobile devices, instant messaging (such as WhatsApp) and cloud technology however, have brought forth new challenges. The advent of Internet of Things (IOT) have complicated the matters further as organizations are not fully cognizant to the all facets of threats to data privacy. Physicians and health care practitioners need to be made aware of various new avenues of data storage and transmission that need to be secured and controlled. In this paper we look at various threats and challenges that IOT, Bring Your Own Device (BYOD) and Personal Area Networks (PANs) technologies pose to the patients' privacy data. We conclude the paper by providing the results of a survey that gauge the depth of understanding of healthcare professionals regarding the emerging threats to patients' privacy.

1 Introduction

The networked nature of healthcare environment has allowed medical and healthcare practitioners to exchange patients' information across various platforms. The promise of web3.0 [1] brings a plethora of opportunities to glean useful information from historical data that is available to hospitals and healthcare organizations. Work done in [2] shows the potential to detecting mental health symptoms by scavenging through twitter data. Authors in [3] provide an overview of various efforts that have been carried out in this vein.

A smart healthcare environment has evolved where healthcare services are delivered seamlessly. Not only, the environment provides seamless access to information to healthcare professionals and the patients, but also provides a platform to help

S. Ahmed (✉)
College of Medicine, King Abdulaziz University, Jeddah, Saudi Arabia
e-mail: samraa2018@gmail.com

A. Visvizi and M. D. Lytras (eds.), *Research & Innovation Forum 2019*,
Springer Proceedings in Complexity,
https://doi.org/10.1007/978-3-030-30809-4_36

monitor patients remotely and at times also deliver medications automatically [4, 5]. The use of storage and network technologies is a prerequisite for such environment. However, such technologies put the data that is being stored at risks and in turn increases the risk of putting patients' privacy at risk. The Health Insurance and Portability Accountability Act (HIPAA) was passed in 1996 [6] that articulated various regulations that would help safeguard patients' privacy. The Federal Information Security Management Act of 2002 (FISMA) [7] was passed to augment the guidelines set out in HIPAA. Both the laws complimented each other in terms of putting various security controls in place.

The FISMA legislation was passed in 2002 and the dramatic change witnessed by the FISMA act witnessed many areas that were not covered. Beginning with the mobile devices and the control they provided, sensors were introduced in medical devices and even human body for exchange of sensitive patient data. These technologies pushed the bolstering of the FISMA initiative with Cyber Enhancement Act of 2014 [8]. The act further refined the security controls that the federal agencies must put in place. The Food and Drug Administration (FDA) published non-binding requirements for medical devices [9] and also supported the Mitre corporation effort that specifically focusses on making medical devices more secure [10].

Securing the medical devices is a small part of the security equation however. The study conducted in [11] provided results on lack of understanding on part of people on the working of internet and the various components that are involved and how majority of the people are unaware of the underlying risks in the new healthcare environment. The growth of Internet of Things (IoT) paradigm complicated the matters further as it allowed various mobile devices and sensors to monitor heterogeneous systems and humans in real-time. While practitioners can now practice preventive medicine via IoT platform and at times even deliver medication [12, 13], little thought has been given to ensuring a secure exchange of such information. The underlying architecture for such enabling technologies further complicates the comprehension of the vast landscape that underscores a smart health care environment. This paper aims at addressing the following:

1. Describe a typical smart healthcare architecture and its components
2. Describe legislations and other efforts that put policies and procedures in place
3. Challenges provided by PAN and BYOD devices
4. Initial surveys.

2 Typical Smart Healthcare Architecture

As mentioned in [11], one of the biggest challenges when dealing with patients' data and the underlying privacy is the lack of understanding as to how the information is transmitted and stored via a network. Figure 1 taken from [5] shows a typical smart healthcare environment.

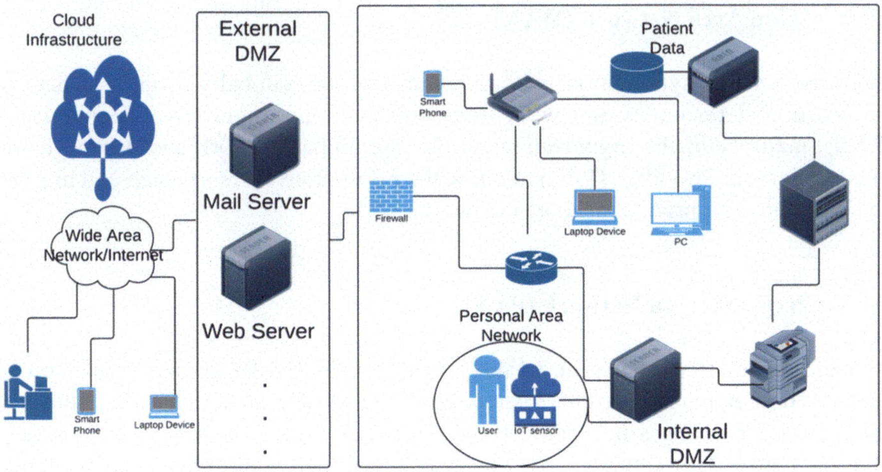

Fig. 1 A typical smart health-care architecture

The study grouped the threats to patients' data under the following two categories namely (1) Network and (2) Technology. For our work, we will focus on the Network layer only.

2.1 *Network Layer*

The Network layer can be further divided into the following three areas.

1. Local Area Network (LAN)
2. Wide Area Network (WAN)
3. Personal Area Network (PAN)

We briefly describe and explain the LAN and WAN networks as this topic has been discussed in detail in literature and traditional security solutions have focused on such technologies. We will expand upon the PAN aspect as it introduces more vulnerability to the patients' data.

2.1.1 Local Area Network (LAN)

The LAN environment is the internal network of an organization and is typically hidden from outside the enterprise boundaries. Employees can connect to the corporate LAN within the organization or employ techniques such as Virtual Private Network when connecting over the Internet (encrypting the link that allows the user to access the digital assets remotely). A VPN is considered relatively safe as the organization sets it up per the best practices usually and is difficult to hack.

2.1.2 Wide Area Network (WAN)

WAN has become synonymous with the word Internet. Globally people can access various digital resources and while some corporations might have setup private WANs, majority of the organizations utilize the global network and depend upon Internet Service Providers (ISPs) to allow their corporate users to reach and use the resources on their private local area networks.

2.1.3 Personal Area Network (PAN)

A much smaller network in size, A PAN is limited to within 10–20 m of the person or device. These networks are characterized by a sensor and an actuator. The sensor's job is to notice changes in certain stimuli (e.g., heart rate of a person, temperature of a room etc.). The actuator is the physical device that would move once a trigger is activated. For example, the thermostat would cause the AC to run off once the temperature falls below a certain range and activate a trigger.

The presence of a PAN is usually dependent upon either a mobile device with the user or a device that is within the proximity of the patient. As pointed out by [14], there are primarily four standards that help in forming a PAN.

1. Bluetooth
2. IEEE 802.1.5.4 Low rate WPAN
3. IEEE 802.15.3 High rate WPAN
4. IEEE 802.15. 6 Body Area Networks

National Institute of Science and Technology (NIST) [15] summarizes the risks that the mobile devices can cause to the corporate networks in SP800-46. These include:

1. Lack of Physical Security Controls: This refers to the fact that the device used by the end user that can connect to the LAN and PAN over the Internet can easily be stolen and accessed by a malicious user
2. Unsecured Networks and Man in The Middle (MITM) attacks: The device used by the healthcare practitioners can easily be sniffed by someone on both the WAN and the PANs. The security level on the Internet cannot be controlled by the enterprise while the PANs are relatively new and can be compromised with a much more ease by the malicious user
3. Infected Devices: The mobile device used by the healthcare practitioner can be infected with malicious software. The device having access to the PANs allow the malicious user to access the PAN directly and in turn get access to the patient data.

Furthermore, most of the mobile devices are connected to the cloud which can possible have lax security controls in place this endangering the patient private data.

3 Related Legislation

Various laws have been formulated by the United States Congress that mandate the privacy requirements for patients' data. While the jurisdiction of such laws is limited to the US, similar laws are being adopted globally. We briefly present the applicable laws below.

3.1 Applicable Laws

3.1.1 Hipaa

The HIPAA law [6] for the first time discussed how the privacy of patient needs to be preserved and the patient need to be informed of how his information is handled and disseminated. One of the corollaries of the HIPAA law was the effect it had on information stored in IT systems and the safe storage and transmission of the patients' information.

3.1.2 Federal Information Security Management Act of 2002

The FISMA act (also known as the E-Government Act of 2002) zeroed in on the electronic services provided by the government and established the governance structure to propagate the importance of information security and privacy [7]. Given the networked nature of the various IT systems in place and the associated security controls needed in place, FISMA delegated the assigned the task of developing various standards and policies to National Institute of Science and Technology (NIST). Such standards and policies are paramount to put information security controls in place. Work done in [16] sheds light on the privacy aspect of electronic health recors and the challenges posed.

3.1.3 Cyber Enhancement Act 2014

The Cyber Enhancement Act (CEA) of 2014 [8] augments the FISMA 2002 Act and emphasizes further areas of focus. The act addresses the physical aspect of information security and also briefly addresses cyber-physical systems (backbone of the Personal Area Networks) [14]. The CEA Act also recognizes the nascent nature of the recent standards and thus emphasizes further research in this area. This work is a step in this direction.

3.2 NIST Recommendations

NIST is a non-regulatory body under the US Department of Commerce and is tasked with developing policies and recommendations for information security controls. NIST publishes two set of documents namely Federal Information Processing Standards (FIPS) series and the Special Publications 800 (SP-800) series. The FIPS series is a list of mandatory standards for federal organizations while the SP-800 series recommends various controls that can be implemented. Initial Survey Results for Healthcare Professionals.

Given the above and similar to [11], we carried a brief survey of healthcare professionals. Please note the following:

1. Before proceeding to get the ethics approval, we did an informal survey on survey monkey that was completely anonymous
2. The sample size was relatively small but our goal was to see whether the idea warrants further exploration
3. The sample size was not specific to a particular institution but rather spread across many institutions
4. The sample size included a range of healthcare professionals and included 70% physicians and 30% nursing staff
5. The data was based in one city
6. The data came from three hospitals.

We got the following initial results.

Number of participants	50
Awareness of cloud risks	6%
Using WhatsApp application to discuss patient info	69%
Perpetual bluetooth device usage	82%

Looking at the above data, we gathered the following:

1. Majority of the healthcare professionals are using instant messaging technology to discuss patients' cases
2. Healthcare professionals are unaware of the risk posed by the Bluetooth technology
3. Healthcare professionals are unaware of the amount/type of data that is being stored in the cloud.

4 Initial Survey Results for Healthcare Professionals

The challenges to preserving the patients' data security and privacy are quite a few. However, given the explosion in the network and mobile technology has given rise to unprecedented challenges. In this paper, we have provided an overview of such technologies and more importantly we have set the platform for conducting a more through survey to gauge the healthcare professionals' insight into the threats of patients' privacy and the effects of HIPAA and other related laws on the usage of emerging technologies. We hope to expand on this work and explore this topic further.

References

1. Lassila, O., Hendler, J.:Embracing "Web 3.0". IEEE Int. Comput. **11**(3) (2007)
2. Rajput, A., Ahmed, S.: Making a Case for Social Media Corpus for Detecting Depression. arXiv preprint arXiv:1902.00702 (2019)
3. Rajput, A.E., Ahmed, S.M.: Big Data and Social/Medical Sciences: State of the Art and Future Trends. arXiv preprint arXiv:1902.00705 (2019)
4. Rajput, A.: Natural language processing, sentiment analysis and clinical analytics. In: Lytras, M. et al. (eds.) Innovation in Health Informatics: A Smart Healthcare Primer. Elsevier, Amsterdam, Netherlands (available as arXiv preprint arXiv:1902.00679) (2019)
5. Ahmed, S., Rajput, A.: Threats to patients privacy in smart healthcare environment. In: Lytras, M., et al. (eds.) Innovation in Health Informatics: A Smart Healthcare Primer. Elsevier, Amsterdam, The Netherlands (2019)
6. https://www.gpo.gov/fdsys/pkg/BILLS-104s1028is/pdf/BILLS-104s1028is.pdf
7. https://csrc.nist.gov/topics/laws-and-regulations/laws/fisma
8. https://www.gpo.gov/fdsys/pkg/PLAW-113publ283/pdf/PLAW-113publ283.pdf
9. https://www.fda.gov/downloads/MedicalDevices/DeviceRegulationandGuidance/GuidanceDocuments/UCM623529.pdf
10. https://www.mitre.org/publications/technical-papers/medical-device-cybersecurity-regional-incident-preparedness-and
11. Kang, R., Dabbish, L., Fruchter, N., Kiesler, S.: "My data just goes everywhere:" user mental models of the internet and implications for privacy and security. In: Symposium on Usable Privacy and Security (SOUPS), pp. 39–52. USENIX Association, Berkeley, CA (2015)
12. Hussain, A., Wenbi, R., da Silva, A.L., Nadher, M., Mudhish, M.: Health and emergency-care platform for the elderly and disabled people in the smart city. J. Syst. Softw. **110**, 253–263 (2015)
13. Doukas, C., Metsis, V., Becker, E., Le, Z., Makedon, F., Maglogiannis, I.: Digital cities of the future: extending@ home assistive technologies for the elderly and the disabled. Telematics Inform. **28**(3), 176–190 (2011)
14. Rajput, A., Brahimi, T.: Characterizing IOMT/personal area networks landscape. In: Lytras, M. et al. (eds.) Innovation in Health Informatics: A Smart Healthcare Primer. Elsevier, Amsterdam, The Netherlands (earlier version available as arXiv preprint arXiv:1902.00675) (2019)
15. www.nist.gov
16. Haas, S., Wohlgemuth, S., Echizen, I., Sonehara, N., Müller, G.: Aspects of privacy for electronic health records. Int. J. Med. Inf. **80**(2), e26–e31 (2011)
17. Huang, L.C., Chu, H.C., Lien, C.Y., Hsiao, C.H., Kao, T.: Privacy preservation and information security protection for patients' portable electronic health records. Comput. Biol. Med. **39**(9), 743–750 (2009)

18. Fernández-Alemán, J.L., Señor, I.C., Lozoya, P.Á.O., Toval, A.: Security and privacy in electronic health records: a systematic literature review. J. Biomed. Inform. **46**(3), 541–562 (2013)
19. Anwar, M., Joshi, J., Tan, J.: Anytime, anywhere access to secure, privacy-aware healthcare services: issues, approaches and challenges. Health Policy Technol. **4**(4), 299–311 (2015)
20. Xu, L., Jiang, C., Wang, J., Yuan, J., Ren, Y.: Information security in big data: privacy and data mining. IEEE Access **2**, 1149–1176 (2014)
21. Yüksel, B., Küpçü, A., Özkasap, Ö.: Research issues for privacy and security of electronic health services. Futur. Gener. Comput. Syst. **68**, 1–13 (2017)
22. Arias, O., Wurm, J., Hoang, K., Jin, Y.: Privacy and security in internet of things and wearable devices. IEEE Trans. Multi-Scale Comput. Syst. **1**(2), 99–109 (2015)
23. Farooq, M.U., Waseem, M., Khairi, A., Mazhar, S.: A critical analysis on the security concerns of internet of things (IoT). Int. J. Comput. Appl. **111**(7) (2015)
24. Sadan, B.: Patient data confidentiality and patient rights. Int. J. Med. Inf. **62**(1), 41–49 (2001)
25. Sahi, A., Lai, D., Li, Y.: Security and privacy preserving approaches in the eHealth clouds with disaster recovery plan. Comput. Biol. Med. **78**, 1–8 (2016)
26. Wu, R.: Secure Sharing of Electronic Medical Records in Cloud Computing. Arizona State University (2012)
27. Sweeney, L.: k-anonymity: a model for protecting privacy. Int. J. Uncertain. Fuzziness Knowl.-Based Syst. **10**(05), 557–570 (2002)
28. https://csrc.nist.gov/publications/fips
29. https://csrc.nist.gov/publications/sp800

Towards the Early Diagnosis of Dementia in People with HIV/AIDS Using Bayesian Networks

Placido Pinheiro, Luana Ibiapina Cordeiro, Maria Lúcia Duarte Pereira, Wilson Correa de Abreu, Mirian Caliope Dantas Pinheiro, Pedro Gabriel Caliope Dantas Pinheiro and Egidio Gomes Filho

Abstract The high prevalence of people with HIV in some countries can lead to the development of dementia syndromes (HIV-associated dementia). Besides the natural aging process, there is an action of the virus on the Central Nervous System. The neuroinflammation may led to cognitive and functional decline, related to motor disorders. This neurocognitive disorders associated with HIV, are usually called HIV-associated dementia, which implies severe alterations on two or more cognitive domains and mark functional interference with daily living, for example, performance activities, attention deficit, psychomotor slowing, dysdiadochokinesia and spastic gait. Its progression can be rapid and interfere with the motor sphere, due to the myelopathy and peripheral neuropathy, even with the use of the highly active antiretroviral therapy. This study proposes the construction of a model using Bayesian networks to assist specialists in the search for the early diagnosis of dementia in people with HIV. To generate the Bayesian Networks the Netica J-API was

P. Pinheiro (✉) · M. C. D. Pinheiro (✉) · P. G. C. D. Pinheiro (✉) · E. G. Filho (✉)
University of Fortaleza, Fortaleza, CE, Brazil
e-mail: placido@unifor.br

M. C. D. Pinheiro
e-mail: mcaliope@uol.com.br

P. G. C. D. Pinheiro
e-mail: pedrogcdpinheiro@hotmail.com

E. G. Filho
e-mail: egidio.filho@gmail.com

L. I. Cordeiro (✉) · M. L. D. Pereira (✉)
State University of Ceará, Fortaleza, CE, Brazil
e-mail: luanaibiapina1@hotmail.com

M. L. D. Pereira
e-mail: luciad029@gmail.com

W. C. de Abreu (✉)
School of Nursing of Porto, Porto, Portugal
e-mail: wjabreu@esenf.pt

A. Visvizi and M. D. Lytras (eds.), *Research & Innovation Forum 2019*, Springer Proceedings in Complexity,
https://doi.org/10.1007/978-3-030-30809-4_37

used. Results show that the Bayesian Network is an accurate and reliable method for study and support the assistance of HIV infected patients.

1 Introduction

Around 36.9 million people living the Human Immunodeficiency Virus (HIV) worldwide. The HIV infection, due to its particular characteristics of transmission, represents a serious public health problem. It is considered by the World Health Organization [1, 2] as a pandemic, whose form of occurrence in the different regions of the world depends, among several determinants, on individual and collective human behavior. Since its emergence more than 78 million people have been affected and 39 million have died worldwide [3].

With the introduction of current antiretroviral treatments into clinical practice, there was a significant drop in the high morbidity and mortality rates associated with the syndrome. From 1996, with the emergence of highly active antiretroviral therapy (HAART) significant successes were achieved in the treatment of human immunodeficiency virus (HIV) infection and the conditions were provided for it to be considered a chronic disease [4].

One main goal stated by the United Nations Member States was end the AIDS epidemic by 2030; and the investment of billion of dollars, millions of health-care workers, social workers, community-based organizations and researchers are working towards this goal [5]. However, for reasons that are due to poor socioeconomic conditions, public health problems and health inequalities, there are fluctuations in the incidence rates between countries and even in the same country over time. This reality, a major concern to the societies, ask for new strategies to identify both incidence increases and other associated diseases. The success in saving lives has not been matched with equal success in reducing new HIV infections. New HIV infections are not falling fast enough. HIV prevention services are not being provided on an adequate scale and with sufficient intensity and are not reaching some groups of people who need them [5].

Despite these advances, neurocognitive disturbances from the HIV virus still persist as complex and clinically important problems [6] due to the action of the virus, the therapeutic regimen or even social isolation. The prevalence of the HIV and the neurocognitive disturbances reinforces the likely-hood for the development of Dementia Syndromes. In addition to the natural aging process, there is an action of the virus on the Central Nervous System (CNS) [7, 8]. Thus, neuroinflammation, on the occasion of the primary infection, may trigger different cognitive and motor declines called HIV-Associated Neurocognitive Disorders (HAND), which are classified according to the results of the neuropsychological assessments and the impact of the disease to perform daily activities; as well as by clinical diagnosis and analysis of cerebrospinal fluid and plasma, whose methods promote a more detailed knowledge of CNS changes during HIV infection and may differentiate HAND from other CNS disorders [9, 10].

According to the Centers for Disease and Control and Prevention (CDC), the most likely determinant of HAND is the degree of immunosuppression of the infected individual, whose values are nadir of LTCD4 + <350 cells/mm^3 or LTCD4 + >350 cells/mm^3. Among the risk factors, the age group is over 50 years old [10].

The most common clinical conditions of HAND are asymptomatic neurocognitive disorder (ANI); mild to moderate neurocognitive disorder (MND), and HIV-associated dementia (HDD), which is considered its most severe form, as it implies severe alterations of 02 or more cognitive domains and the presence of marked functional interference in instrumental activities of daily life, such as attention deficit, psychomotor slowing, dysdiadochokinesia and spastic gait. In addition its progression, can be considered to be rapid and lead to intense motor impairment due to the possible development of myelopathy and/or peripheral neuropathy, even with the use of Highly Active Antiretroviral Therapy (HAART) [11, 12].

Many health professionals reveal the need for educational and/or technological support to assist them in decision-making, since some of them affirm that they do not feel sure about practices to help people living with HIV (PLHIV) [13, 14].

The use of technological strategies, such as Information and Communication Technologies (ICTs), can be considered important tools for the detection and early diagnosis of public health problems. In recent years, a field of computer science has been gaining prominence in health sciences, Artificial Intelligence (IA) [15].

Some studies have sought to improve the performance of Specialist Systems by applying Multicriteria Decision Support Methodologies, such as: the diagnosis of Alzheimer's disease [16]. It should be emphasized that decision support models do not replace the professional's knowledge, but they can be helpful in guiding safe, quality and effective care to patients, and they will work successfully in the context of clinical problems [16–18]. The start question of this study was: How can a multicentric model of decision supported by Bayesian networks predict an early diagnosis for dementia in people who are infected with HIV?

It is expected, through the early diagnosis of dementia, the optimization in the diagnosys and implementation od adequate therapeutic interventions, which can have repercussions in the favorable clinical and cognitive prognosis.

The objective of this study is to propose the construction of a model using Bayesian networks to assist specialists in the search for the early diagnosis of dementia in people with HIV.

2 Methodology

It is a descriptive, applied and evaluative research carried out from October 2017 to December 2018. The database is composed of 111 people with HIV/AIDS, aged 50 years or more and with a level of mild to moderate dementia that are followed in hospital wards and community services. Data were collected by the following instruments: a sociodemographic questionnaire, that includes health and risk behaviors; the Simplified Medication Adherence Questionnaire (SMAQ); the International Dementia Scale to detect the risk for dementia in HIV patients; the Barthel Index and the Lawton and Brody Scale to assess the degree of functional capacity or decline. Finally, the Social Support Scale for People Living with HIV/AIDS. Sociodemographic and health data, as well as data on the application of the SMAQ, the Barthel index and the international dementia scale, were compiled and analyzed using the statistical program Statistical Package for the Social Sciences (SPSS 24). With all probabilities calculated, a priori and a posteriori probabilities, the Bayesian Network (RB) generated by the tool, API Netica-J, is saved in a ".dne" file format, since it allows the visualization and manipulation of RB in Netica software. Knowledge in Bayesian Networks are represented and manipulated based on fundamental probabilistic principles, represented both qualitatively and quantitatively [19]. These probabilities have been used to analyze the attractiveness of each parameter, in relation to the level of influence (little or too much influence) for the diagnosis of HAND.

3 Results and Discussion

Because it is a probabilistic method, the Bayesian Network (RB) will provide positive results capable of assisting decision making and attention to clinical care. Thus, RB corresponds to an important tool in the representation of knowledge and inference under conditions of uncertainty [19]. In this model proposed by the network the nodes were composed of two groups: sociodemographic profile; signs and symptoms through the physical and psychological manifestations of dementia in people with HIV.

According to the data obtained it is possible for the expert to add or not, information into the network. After performing the inference process, the professional obtains the conditional probability of the other nodes of the network.

The nodes of the network represent evaluation criteria for the early diagnosis of dementia. The nodes that offered the greatest contribution, in this study, have been marital status and nervousness. However, the others also contributed to the final results, even with different weights.

The structure of the Bayesian network is the specification of marginal and conditional probabilities. Such probabilities were obtained through an automatic learning process from a database.

The marginal probabilities are shown without parents. These probabilities correspond to the prevalence of dementia in medical care with PLHIV. It can be identified in Table 1.

Another provision is one of the conditionals, most of the network nodes related to their parents. Table 2 represents the distribution of conditional probabilities of patients with high risk to cognitive disorders given the symptoms that occur:

Table 1 Marginal probabilities to nodes without parents

Table of marginal probabilities (n = 111)			
	Never	Sometimes	Always
Nervousness	8.1	45.0	46.8
Memory loss	18.9	52.3	28.8
Annoyance	25.2	41.4	33.3
Appetite loss	39.6	48.6	11.7
Feel alone	34.2	45.0	20.7
Sadness	18.0	55.9	26.1
Persecutory delusion	40.5	36.0	23.4

Source Prepared by the researchers

Table 2 Conditional probabilities of dementia

Probability to develop dementia presenting particular symptom (n = 111)			
	Never	Sometimes	Always
Nervousness	7.8	44.7	47.5
Memory loss	18.5	51.6	29.2
Annoyance	24.5	42.7	32.8
Appetite loss	38.3	50.1	11.6
Feel alone	32.4	47.4	20.2
Sadness	17.7	55.9	26.4
Persecutory delusion	40.2	36.1	23.7

Source Prepared by the researchers

In this context, two Bayesian Networks were built: the first with seven behavioural factors (as seen in Fig. 1), and the second with fifteen clinical and social factors, Fig. 2.

4 Conclusion

With the prolonged life expectancy of patients aged over 50 years, age is a factor in HIV-related cognitive disturbances but other variables can also interfere. An estimated grow of people with HIV-associated neurocognitive disorders (HANDs) cause significant morbidity and mortality, accelerate the memory decline and can facilitate the spread of the virus and the progression of HIV seropositivity.

Bayesian networks were used in this study to recognize and validate the neurocognitive disorders on the basis of a set of symptoms, with the support of a probabilistic model. In this study, we showed only a part of the wide research, which includes more variables. This model enabled to determine probabilities of causes on the basis of

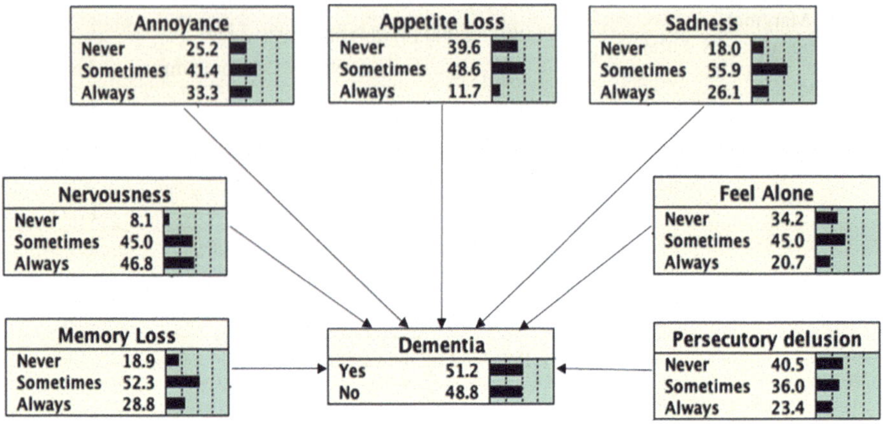

Fig. 1 Behavioural factors Bayesian network

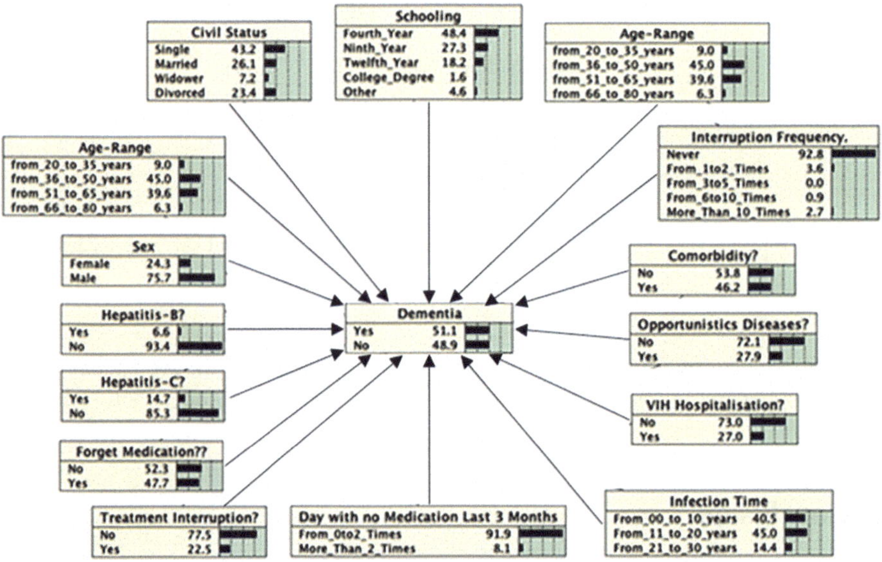

Fig. 2 Clinical and social factors Bayesian network

probabilities of results. The results with RB were very similar to the results of the literature, thus showing a compatibility between human and computational reasoning. The operation of the network shows that in certain cases this type of modeling can used profitably, especially when it has a large number of parents, and when the parents have characteristics in common. Other studies can consider different Bayesian classifiers and evaluate the network using real cases and the corresponding expert assessment. This assessment will provide different views on parts of the network and will contribute to further deployments. It would be clinically relevant to perform practical experiments with the network, sensitivity analysis, and the development of a user interface and the corresponding application. RB can be used as a decision support tool (29) for health professionals, and its use in clinical routines can improve patient care, as it can help the professionals at the moment of decision making about which modality is more appropriate in that particular population.

References

1. WHO, World Health Organization: HIV/AIDS Data and statistics (2018). Homepage, http://www.who.int/hiv/data/en/
2. WHO, World Health Organization: Global Health Sector Strategy on HIV 2016–2021 (2016). Homepage, http://apps.who.int/iris/bitstream/10665/246178/1/WHO-HIV-2016.05-eng.pdf?ua=1. ISBN 9789241501651
3. UNAIDS Gap Report (2014). Homepage, http://files.unaids.org/en/media/unaids/contentassets/documents/unaidspublication/2014/UNAIDS_Gap_report_en.pdf. ISBN 978-92-9253-062-4
4. Seidl, E.M.F., et al.: Pessoas vivendo com HIV/AIDS: variáveis associadas à adesão ao tratamento anti-retroviral. Cadernos de Saúde Pública **23**(10), 2305–2316 (2007)
5. UNAIDS Gap Report (2018). [Consult. 24 June 2018]. Homepage, http://www.unaids.org/sites/default/files/media_asset/unaids-data-2018_en.pdf
6. Watkins, C., Treisman, G.J.: Cognitive impairment in patients with Aids—prevalence and severity. HIV/AIDS Res. Palliat. Care **7**, 35–47 (2015)
7. Deus, D.M.V., Possas, C.C.: Updates on cortical dementia in HIV-infected individuals. Humanae Controversial Issues Contemp. World **12**(1), 1–11 (2018)
8. UNAIDS. Joint United Nations Programme on HIV/Aids. Relatório global do UNAids sobre prevenção destaca avanços e desafios da resposta à Aids no Brasil. UNAids Brasil - Prevention Gaz Report (2016)
9. Becker, B.W.: Longitudinal change in cognitive function and medication adherence in HIV-infected adults. AIDS Behav. **15**(8), 1888–1894 (2011)
10. Barbosa, A., Marques, A., Sousa, L., Nolan, M., Figueiredo, D.: Effects of a psycho-educational intervention on direct care workers communicative behaviors with residents with dementia. Health Commun. **31**(4), 453–459 (2015)
11. Sacktor, N.: Changing clinical phenotypes of HIV-associated neurocognitive disorders. J. Neurovirology **24**(2), 141–145 (2018)
12. Saylor, D., et al.: HIV-associated neurocognitive disorder—pathogenesis and prospects for treatment. Nat. Rev. Neurol. **12**(4), 234–248 (2016)
13. Lee, S., Abbott, P.A.: Bayesian networks for knowledge discovery in large datasets: basics for nurse researchers. J. Biomed. Inform. **36**(1), 389–399 (2003)
14. Serrano-Villar, A.R.S., Gutierrez, F., Mirall, E.S.C: HIV as a chronic disease: evaluation and management of non-AIDS defining, Management of HIV-Associated Disease in HIV. **3**(2), 1–15 (2016)
15. Tamanini, I., Pinheiro, P.R., Santos, C.N.: A hybrid approach of verbal decision analysis and machine learning. Lect. Notes Comput. Sci. **7413**(1), 126–131 (2012)
16. Castro, A.K.A., Pinheiro, P.R., Pinheiro, M.C.D., Tamanini, I.: Towards the applied hybrid model in decision making: a neuropsychological diagnosis of Alzheimer's disease study case. Int. J. Comput. Intell. Syst. **4**(1), 89–99 (2011)
17. Gomes, L.F.A.M., Gomes, C.F.S.: Managerial decision making: multicriteria approach. In: Atlas, 4th edn. Atlas, São Paulo (2012)

18. Pinheiro, P.R., Tamanini, I., Pinheiro, M.C.D., Albuquerque, V.H.C.: Evaluation of the Alzheimer's disease clinical stages under the optics of hybrid approaches in Verbal Decision Analysis. Telematics Inform. **35**, 776–789 (2018)
19. Sampieri, R.H., Collado, C.F., Lucio, M.P.B.: Research methodology. In: Penso, 5th edn. Porto Alegre (2013)

A Hybrid Model to Guide the Consultation of Children with Autism Spectrum Disorder

Luciano Comin Nunes, Placido Pinheiro, Mirian Caliope Dantas Pinheiro, Monica Pompeu, Marum Simão Filho, Rafael Comin-Nunes and Pedro Gabriel Caliope Dantas Pinheiro

Abstract Considering that psychological disorders have specific causes and symptoms, the conventional way to establish diagnosis consists of analysing the behavioural reactions of a human being to events in their daily lives. The manifestation of psychological disorders varies from person to person and the degree of severity. Like 'personality,' 'consciousness' and 'intelligence,' the expression 'abnormal behaviour' is difficult to define due to sociocultural subjectivities intrinsic to the human being. Social interactions are essential factors in the search for solutions to the problems of the information society. However, neurodevelopmental disorders may affect some people from birth, hampering their healthy growth and consequently their participation in social interactions. In this context, autistic spectrum disorder is one of the primary disorders of neurodevelopment, which manifests itself before the age of three. Despite the existence of diagnostic criteria for autism spectrum disorder, there is a lack of system with models containing algorithms that help the families and health professionals in the early diagnosis of this disorder. The objective of this study is to make proactive the decision-making process for the establishment of the early

L. C. Nunes (✉) · P. Pinheiro · M. C. D. Pinheiro · M. Pompeu · M. S. Filho · P. G. C. D. Pinheiro
Graduate Program in Applied Informatics, University of Fortaleza, Fortaleza, CE, Brazil
e-mail: lcominn@uol.com.br

P. Pinheiro
e-mail: placido@unifor.br

M. C. D. Pinheiro
e-mail: caliope@unifor.br

M. Pompeu
e-mail: monicarios@unifor.br

M. S. Filho
e-mail: marumsimao@gmail.com

P. G. C. D. Pinheiro
e-mail: pedrogcdpinheiro@hotmail.com

R. Comin-Nunes
Federal University of Ceara, Fortaleza, CE, Brazil
e-mail: rafael23.labomar@gmail.com

A. Visvizi and M. D. Lytras (eds.), *Research & Innovation Forum 2019*,
Springer Proceedings in Complexity,
https://doi.org/10.1007/978-3-030-30809-4_38

diagnosis of autism spectrum disorder. For this, the research presents a hybrid model proposal composed of a specialist system structured in decision-making methodologies (Multi-Criteria Decision Analysis) and structured representations of knowledge in production rules and probabilities using Artificial Intelligence. In this context, the research aims to show one of the potentials of information technologies as a way to generate knowledge that adds value and serves as a support in medical decision making.

1 Introduction

Organisation's survival in a globalised world requires competitiveness and sustainability. Since organisations rely heavily on people, they continually seek healthy, qualified, ethic, innovative and proactive professionals to build work teams. Therefore, each one adopts a selection and recruitment process based on their models or existing methodologies.

Regarding applicants' healthiness, it is common to use selection models that exclude the ones who suffer specific issues such as Neurodevelopmental Disorders listed on Fig. 1, with emphasis on the Autism Spectrum Disorder (ASD) ones, contributing factors to disqualification on selection and production process.

It is common in workplaces for team members to react sceptically with another member suffering from psychopathology. A reason for this collective behaviour is misinformation among the other team members. Consequently, one of the more drastic consequences to professional suffering from psychological disorders is lack

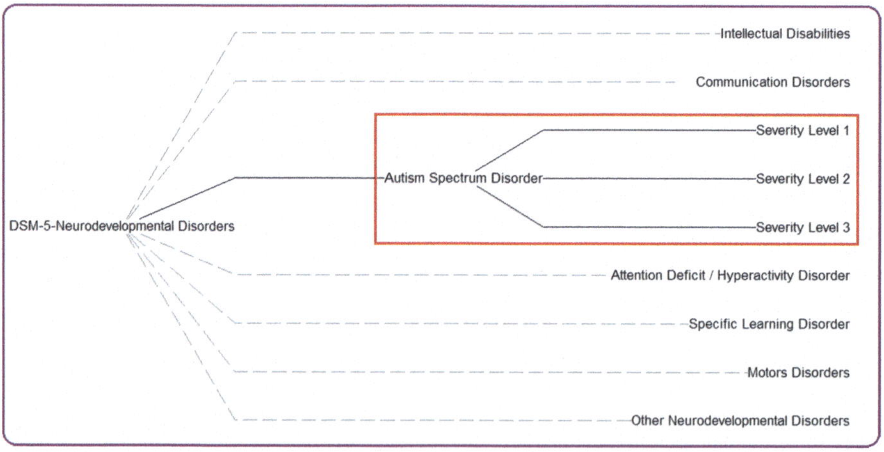

Fig. 1 Neurodevelopmental disorders categorisation according to the American Psychiatric Association (APA)'s Diagnostic and Statistical Manual of Mental Disorders (DSM–5)

of trust in their capacity. Even if this worker strives to maintain his/her reputation, prejudice and damage are destructive.

Under those circumstances, professionals on the autism spectrum will have lower chances of being selected and recruited. In a whole socioeconomic perspective, this paper presents a model proposal for early ASDs detection and diagnosis of children younger than two years old. It is important to note how this allows for steps towards lowering the exclusion effects on individual's socioeconomically current ages.

1.1 Summary of the Article's Proposed Model

The present study proposes a model for early ASDs detection and diagnosis manifested on children up to two years old. The model comes from the hybridisation of a multicriteria methodology for decision aiding and a specialised system based on production rules, probabilities and artificial intelligence usage. Hybrid models are applied to decision making assistance in the health field, aiming to diagnose illness. For instance, Castro developed a hybrid model to aid in decision making towards a neuropsychological diagnosis of Alzheimer's Disease [1]. In another study, Menezes presents a hybrid model proposal for early type 2 diabetes diagnostic, using decision aiding methods [2]. Filho et al. developed a Heterogeneous Methodology to Support the Early Diagnosis of Gestational Diabetes, with data organized by the multicriteria methodology, applying Multiattribute Utility Theory (MAUT) methods and information structured in the knowledge base of a specialist system [3].

The difference between the three studies' models and the one presented in this paper is the integrated usage of artificial intelligence techniques within a Specialized System with multicriteria analysis method to support diagnostic training. The technologies integration aims to increase accuracy towards best alternatives indication of ASD. This study was based on the American Psychiatric Association (APA)'s Fifth edition of the Diagnostic and Statistical Manual of Mental Disorders (DSM-5), due to its synthesis and overall configuration.

1.2 Article's Structure

This paper has been divided into five sections, including the introductory section, which presents the study's rationale and objectives, as well as a categorisation of Neurodevelopmental Disorders. The second section conceptualises ASD and presents the control events that will be used in the proposed model. Section three analyses qualitatively and comparatively ASD's criteria and events, while detailing the proposed model. The fourth section presents a specialist system applied in the ASD diagnostic, using the control events from the third section, while indicating the due confidence degrees in diagnostic construction based on the disorder severity. The final section presents conclusions and possible future works regarding the subject.

2 Autism Spectrum Disorder Characteristics

ASD is part of the set of Neurodevelopmental Disorders. The ASD refers to a range of conditions characterised by impaired social behaviour, as well as by almost non-existent communication and language. Individuals with ASD often have comorbidities, such as epilepsy, depression, anxiety, and attention deficit hyperactivity disorder (ADHD).

Aiming to assist in the diagnosis of such comorbidities, in an integrated way to that of ASD, Nunes et al. present a proposed model of diagnosis of psychological disorders [4].

Throughout the world, people with ASD are often subject to stigma, discrimination and human rights violations. World estimates show one autistic child for every 160 [5]. In Brasil, there is the ratio of 1/360. Although considered underestimated, these data from Brazil show that there is a significant demand for specialised care [6].

2.1 Control Events that Characterise the ASD

According to the DSM-5, an individual with ASD has persistent deficits in communication and social interaction in multiple contexts, including non-verbal communication behaviours used in skills to develop, maintain and understand relationships. In addition to deficits in social communication, the individual demonstrates restricted and repetitive patterns of behaviours, interests or activities [7]. There is a compromise of the individual's communicative abilities, that is, those forms of visual contact of the individual with the objects and events around him, through coordinated gestures and verbalisations [8, 9]. Cited compromises are characterised as possible markers for the diagnosis of ADS [10].

On the other hand, studies indicate that the genetic factor contributes to the incidence of ASD [11]. Table 1 presents the main control events for ASD diagnosis. These events were extracted from the ASD foundations described in DMS-5, as well as from studies and observations of specialists in the psychopathology in children and adolescents. The numerical values in Table 1 represent the influence of each event in establishing the diagnosis of ASD. These values are assigned by the scale described in Table 2.

Figure 2 shows the main control events related to ASD in its most severe form (Severity Level 3). The numerical values in Fig. 2 represent the influence of the event in establishing the diagnosis for this disorder. These values constitute the application of a scale of confidence factor of each event for the occurrence of the disorder. Mentioned scale, broken down in Table 2, uses diagnostic criteria of ASD, described in DSM-5, as well as studies of specialists in the psychopathology of children and adolescents.

Table 1 Principals events of ASD control

Control event	SL-1[a]	SL-2[a]	SL-3[a]	Description
Def.Social-Emotional	50	75	100	Deficits in social emotional reciprocity
Def.Nonverbal Commun	50	75	100	Deficits in nonverbal communicative behaviors
Def.Developing Relat	50	75	100	Deficits in developing, maintaining, and understanding relationships
Repetitive Motor Mov	50	75	100	Stereotyped or repetitive motor movements, use of objects or speech
Ritualised Patterns	50	75	100	Insistence on sameness, ritualised patterns of verbal or nonverbal behaviour
Restricted Interests	50	75	100	Highly restricted interests that are abnormal in intensity or focus
Reactivity to Sensor	25	50	75	Hyper- or hypo-reactivity to sensory input or unusual interest in sensory aspects of the environment
Sleep Disturbance	25	50	75	Sleep phase delay; Rapid Eyes Movement sleep behavioural disorder
Apraxia	25	50	75	Speech apraxia or difficulty in producing some phonemes
Comorbidity	25	50	75	Simultaneous occurrence of disorders
Genetics	25	25	50	Hereditary predisposition
Environment	25	25	25	Environmental factors
Lonely	50	75	100	Individual isolating himself from others
Following Attention	75	50	25	Following the attention
Drive Attention	75	50	25	Drive attention
Name Orientation	75	50	25	Orientation to the name
Face Orientation	75	50	25	Orientation to the face of the other
Object Orientation	75	50	25	Object orientation with others
Social Touch Aversion	75	50	25	Aversion to social touch
Unbalanced Serotonin	25	25	50	Low-level neurotransmitter Serotonin
Incapacity	25	50	75	Inability to perform activities for the age group, requiring support

[a]Refers to the ASD Severity Level: 1, 2, and 3

Table 2 The scale of the influence of the ASD control events

0	25	50	75	100
Indifferent	Little influence	Moderate influence	Much influence	Decisive influence

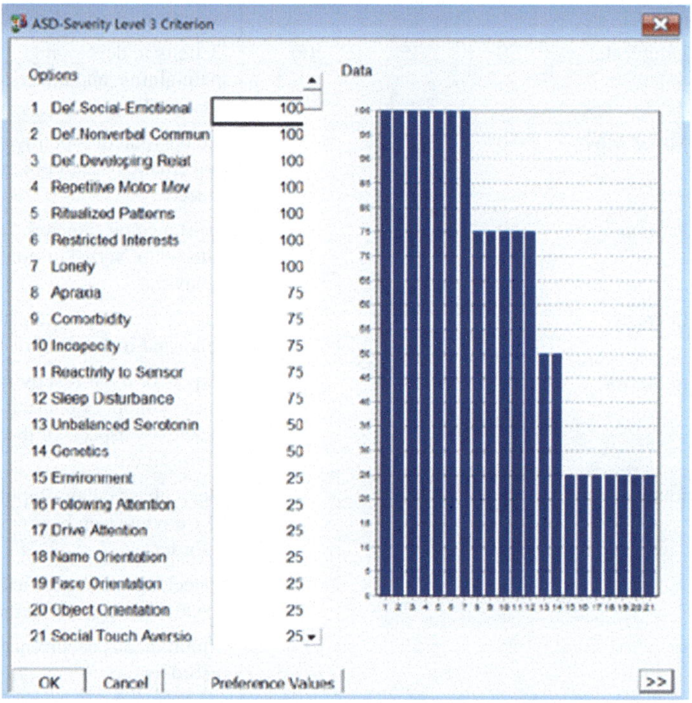

Fig. 2 The ASD-Severity Level 3 control events

3 Proposed Model of Support in the Diagnosis of ASD

Methodologies to support multicriteria decision making provide the decision maker with techniques and tools that allow structuring the control events, as well as hierarchize these events for the proper classification in order of the degree of importance of each one in the decision-making process.

Figure 3 presents the algorithm of the proposed model in the form of a graph containing the integrations between the technologies used in the present study.

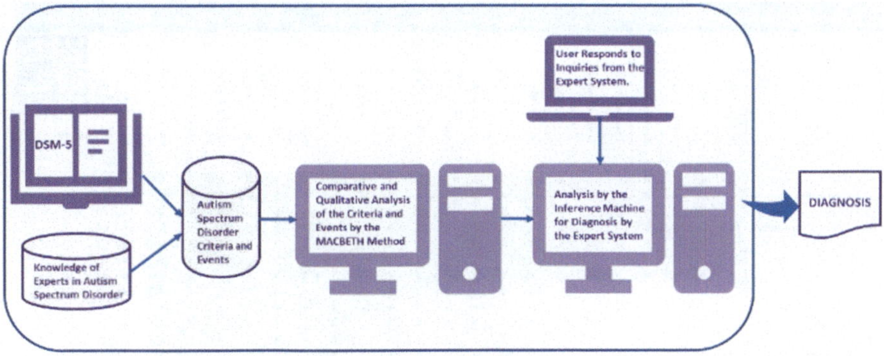

Fig. 3 Proposal of the model to support the early diagnosis of autism spectrum disorder

3.1 MACBETH Method Application

Technique is a multicriteria decision support approach that allows assigning a value to each alternative through a peer comparison. Given two alternatives, the decision maker should say which is the most attractive and the degree of confidence of this attractiveness in a semantic scale that corresponds to an ordinal scale [12].

In the MACBETH method, the decision maker makes value judgments about the alternative in each situation due to the attractiveness of this alternative. This task is defined by the construction of a criterion function vj, such that [13]:

- For $a, b \in A$, $v(a) > v(b)$, if and only if, for the evaluator, a is more attractive than b (aPb);
- Any positive difference, $v(a) > v(b)$, numerically represents the value difference between a and b, with aPb always regarding a fundamental point of view $j(PVFj)$, or criterion j. Then, for each $a, b, c, d \in A$, with a being more attractive than b e c being more attractive than d, we see that $v(a) - v(b) > v(c) - v(d)$, if and only if, "the difference in attractiveness between a and b is greater than the difference in attractiveness between c and d".

The critical question of the MACBETH method is: Given the impacts ij (a) and ij (b) of two potential actions a and b, from a fundamental point of view $PVFj$, being judged more attractive than b, the difference in attractiveness between a and b is judged to be "null," "very weak," "weak," "moderate," "strong," "very strong," or "extreme."

Figure 4 shows the degrees of attractiveness between events in Table 1. The degrees of attractiveness between events form the Current Scale of confidence factors, indicating "Consistent Judgments" or qualitative consistency in peer-to-peer comparison.

The model proposes that the control events and respective confidence factors of each level of ASD severity of Fig. 1, come to compose the knowledge base of the Expert System described in Sect. 4.

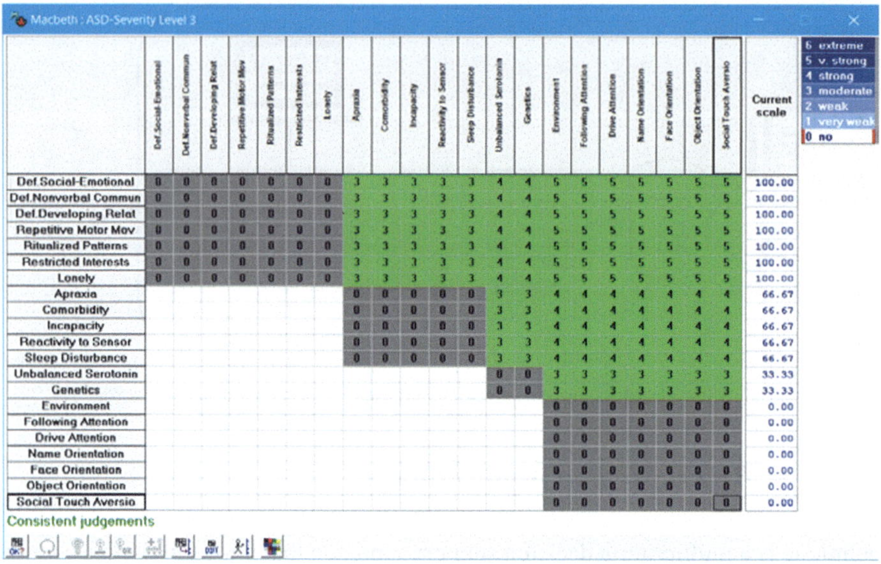

Fig. 4 The matrix of value judgment and attractiveness difference between ASD events

Regarding the confidence factors, Fig. 5 shows how the decision maker can adjust these degrees of attractiveness among the control events, pointing in a ruler the limits between the different degrees of the attractiveness of these events. If the decision maker modifies these values, the new degrees of attractiveness indicated in the rule are reflected in the current scale of the matrix, as well as adjustments in the confidence factors of the control events.

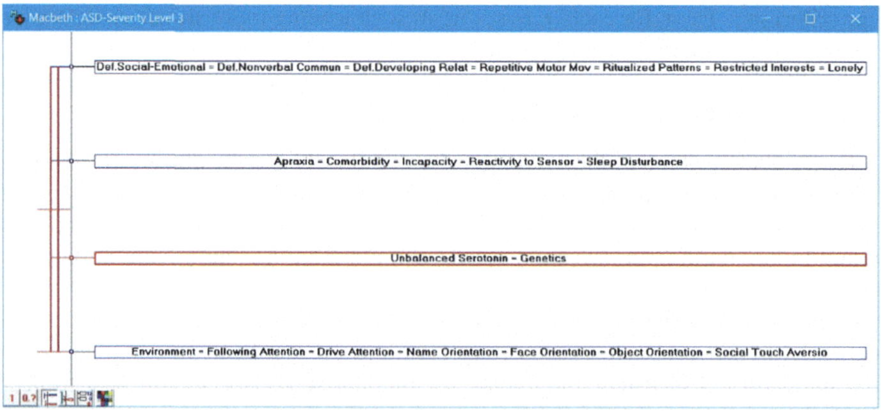

Fig. 5 Control events with new confidence factors within permissible intervals

4 An Expert System for ASD Diagnosis

The mobility of Information Technology (IT) resources enables the portability of the expert systems, making them more attractive and popularly more accessible. Providing tools of this type goes beyond the provision of graphics and tables to the user. This new IT helps decision-makers in identifying their needs, simulating scenarios and enabling speed and quality in the solutions to their problems. Expert Systems are associated with the expression of Artificial Intelligence (AI) and are useful in the generation of knowledge, which is why they can assist in decision making processes in the health area. The most common architecture of expert systems involves production rules structured in a set of conditions in style IF… THEN…, with the possibility of including logical connectives relating the attributes in the scope of knowledge and the use of probabilities.

For the construction of the Expert System of this study, the ExpertSinta tool was used, which applies AI techniques for automatic generation of expert systems, using a knowledge representation model based on production rules and probabilities.

According to Fig. 3, the control events described in Table 1 and submitted to the influence scale values described in Table 2 are analysed and compared in pairs by the MACBETH method, generating the values of the attractiveness differences necessary for the judgment of the matrix of values, as exemplified in Fig. 4. After this, the information of these control events indirectly feeds the Specialist System.

It is important to emphasise that, because the diagnosis of ASD is mostly clinical, it requires anamnesis with the participation of parents and caregivers, by observing the child's behaviour. To do so, the Specialist System uses information from the control events, as well as questions on screening scales such as Modified-Checklist for Autism in Toddlers (M-CHAT), Autism Behavior Checklist (ABC), among others. Thus, said Specialist System assists in the early diagnosis, constituting a useful tool for Primary Health Care Professionals. The following is the sequence of steps for the configuration and use of the Specialist System.

(a) Define the precedence of the logical operators that will be used by the Inference Machine, indicating one of the following structures: '(*A and B*) *or C*'; '*A and* (*B or C*)'.
(b) Define the variables from the control events submitted to the MACBETH method (see Fig. 6).
(c) Define the objective-variables that will indicate the final diagnosis (see Fig. 7).
(d) Define the minimum value for the confidence factor.
(e) Create a password for the database.
(f) Create an interface for user/system interaction. Figure 7 illustrates how the questions that are presented to the user are created. At this point, it is indicated if, together with the response, the user must inform the confidence factor (field 'CNF' in Fig. 8) of the occurrence of the event that is the subject of the question.
(g) In addition to the variables, objectives and interfaces, the Expert System requires logical rules, from which it will indicate the diagnosis. In the course of executing the expert system, the user interacts through graphic interfaces as shown in

Fig. 6 Variables definition

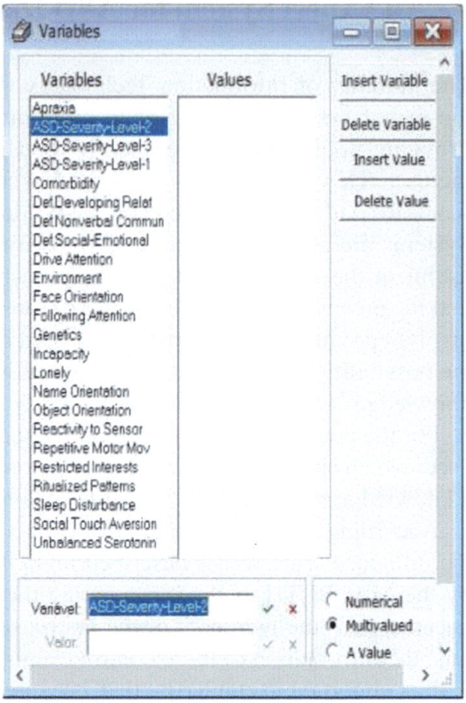

Fig. 9. The interaction through these interfaces enables the collection of values that feed the variables and trust factors used by the expert system.

(h) The expert system also offers a research tree that is the path of logical reasoning conducted by the specialist. This track assists in the analysis of the results after obtaining the diagnosis.

5 Conclusion

Despite advances in the Information Technology field, there is a great need for improvements regarding decision process automation, especially when there is a multicriteria analyse associated to incomplete knowledge, which is usually in the healthcare field.

In the field of learning, some studies with "Serious games in K-12 education" show positive results, such as increased attention, motivation, independence, autonomy and self-esteem in students with special educational needs [14]. The results of the present study may be, in future works, integrated into the surveys with "Serious games in K-12 education".

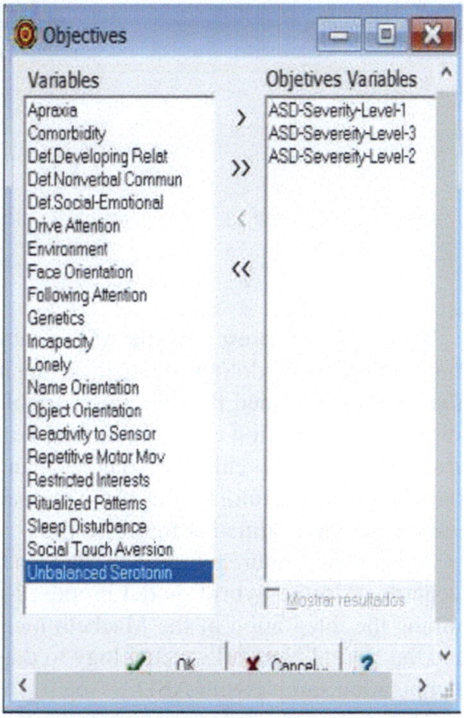

Fig. 7 Objectives definition

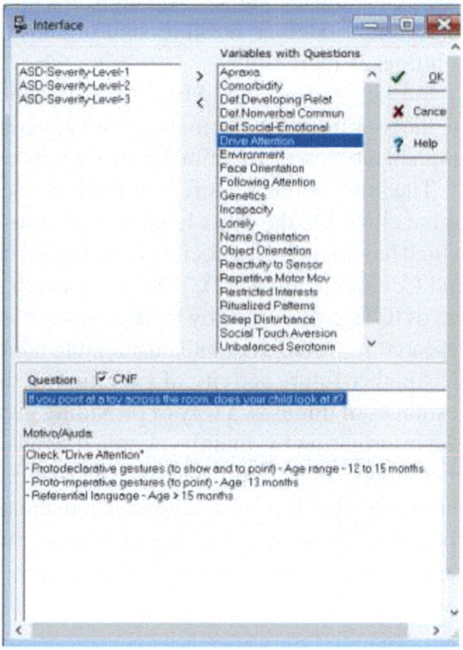

Fig. 8 Interfaces definition

Fig. 9 Data entry via a user interface

This essay has presented the ASD control events, categorised and analysed in a multicriteria methodology, being those the input to a specialised system. The system after that transformed the data into variables that made possible interface creation towards the collection of the additional necessary information for ASD early diagnostic indication on children younger than two years old, synthesising this into its knowledge base. Automation of the transaction process between the methodology and the system is aimed at future works.

A significant future advance for this research is the insertion of Neural Networks technology in the hybrid model proposed in this study. That is a new model containing the integration of the Macbeth method with the Expert System technology and the Neural Networks technology to diagnose and establish behavior patterns of a population sample with ASD.

This study's research process implies that related entities should encourage more researches on the subject, resulting in more concrete findings. Also, the research on hybrid models is feasible, and organisation interest in using specialised system solutions is increasing.

Further studies should be done to enhance the model. We propose enhancements on the user interface, formatting and generation of generic questionnaires, as well as implementation of functionalities such as exporting and importing of data files.

The innovations required by modern society accelerate the construction of intelligent cities. On the other hand, society must adopt an intelligent educational model, which has an inclusive social participation in this new era of Information Technology and Communication (ITC). Thus, research should explore the resources of this new era of ICT, as discussed by Lystras et al. in an article that addresses social networking research for intelligent and sustainable education [15].

Another future activity of evolution of the present study is the use of the Internet resources of things as a way of providing greater interaction, exchange of experiences and information by, mainly, people diagnosed with ASD through hyper connectivity and networking. In this context, the Makori research [16] presents some aspects of innovation that propitiate dynamism in the academic environment where people with special needs must be included that have much to contribute with the solution of the problems of the current and future society.

References

1. de Castro, A.K.A., Pinheiro, P.R., Pinheiro, M.C.D.: A hybrid model for aiding in decision making for the neuropsychological diagnosis of Alzheimer's disease. In: Lecture Notes in Computer Science, Rough Sets and Current Trends in Computing, vol. 5306, pp. 495–504 (2008)
2. Menezes, A.C., Pinheiro, P.R., Pinheiro, M.C.D., Cavalcante, T.P.: Towards the applied hybrid model in decision making: support the early diagnosis of type 2 diabetes. In: 3rd International Conference on Information Computing and Applications (ICICA), LNCS, vol. 7473, pp. 648–655 (2012)
3. Filho, E.G., Pinheiro, P.R., Pinheiro, M.C.D., Nunes, L.C., Gomes, L.B.G.: Heterogeneous methodology to support the early diagnosis of gestational diabetes. IEEE Access **1**, 1 (2019)
4. Nunes, L.C., Pinheiro, P.R., Pinheiro, M.C.D., Filho, M.S., Nunes, R.E.C., Pinheiro, P.G.C.D.: Automatic detection and diagnosis of neurologic diseases. IEEE Access **1**, 1 (2019)
5. WHO Homepage, https://www.who.int/news-room/fact-sheets/detail/autism-spectrum-disorders. Last accessed 27 Jan 2019
6. Paula, C.S., Ribeiro, S.H., Fombonne, E., Mercadante, M.T.: Brief report: prevalence of pervasive developmental disorder in Brazil: A pilot study. J. Autism Dev. Disord. **41**(12), 1738–1742 (2011)
7. DSM-5, Diagnostic and Statistical Manual of Mental Disorders. Fifth Edition, New School Library, American Psychiatric Association, Washington-DC, London-England (2013)
8. Dawson, G., Toth, K., Abbott, R., Osterling, J., Munson, J., Estes, A.: Defining the early social attention impairments in autism: social orienting, joint attention, and responses to emotions. Dev. Psychol. **40**, 271–283 (2004)
9. Delinicolas, E., Young, R.: Joint attention, language, social relating, and stereotypical behaviours in children with autistic disorder. Autism **11**(5), 425–436 (2007)
10. Mundy, P., Sigman, M.: Specifying the nature of the social impairment in autism. In: Dawson, G. (ed.) Autism: New Perspectives on Nature, Diagnosis, and Treatment, pp. 3–21. Guilford, New York (1989)
11. Astling, D.P., Sikela, J.M., Heft, I.E., Jones, K.L.: High-resolution measurement of DUF1220 domain copy number from whole genome sequence data. BioMed Central Genomics **18**, 614 (2017). https://doi.org/10.1186/s12864-017-3976-z
12. Bana e Costa, C.A., Corte, J.M.D., Vansnick, J.C.: MACBETH, LSE-OR Working Paper, 56 (2003)
13. de Souza, G.G.C.: A Multicriteria Model to Produce a Journal, f. 71, the University of Fortaleza (2003). www.unifor.br
14. Papanastasiou, G., Drigas, A., Skianis, C., Lytras, M.D.: Serious games in K-12 education: benefits and impacts on students with attention, memory and developmental disabilities. Program **51**(4), 424–440 (2017). https://doi.org/10.1108/PROG-02-2016-0020
15. Lystras, M.D., Visvizi, A., Daniela, L., Sarirete, A., De Pablos, P.O.: Social networks research for sustainable smart education. Licensee MDPI, Basel, Switzerland. This article is an open access article distributed under the terms and conditions of the Creative Commons Attribution (CC BY) license (2018). http://creativecommons.org/licenses/by/4.0/
16. Makori, E.O.: Promoting innovation and application of internet of things in academic and research information organizations. Libr. Rev. **66**(8/9), 655–678 (2017). https://doi.org/10.1108/lr-01-2017-0002

Use of EEG Technology with Based Brain-Computer Interface to Address Amyotrophic Lateral Sclerosis—ALS

Nighat Mir, Akila Sarirete, Jehad Hejres and Manar Al Omairi

Abstract Amyotrophic lateral sclerosis (ALS), also stated as "Lou Gehrig's Disease," is an advanced neurodegenerative disease that disturbs motor nerve cells in the brain and in the spinal cord. Motor neurons act a message carrier starts from the brain to the spinal cord and from the spinal cord to the all the muscles throughout the body. There are about 40 million ALS patients in the world, and it mostly affects patients of age (40–70). An early symptom is a painless, weakness in a limb, difficulty with speech and difficulty in walking. However, this weakness quickly extends to other parts of the body, producing progressive paralysis of the trunk and diaphragm, at that point patients may elect to have a tracheotomy so their lungs can be ventilated mechanically. This research is about suggesting a solution for ALS patients using EEG brain-computer interface (BCI), bionics and Emotiv system through smart phone application for their basic communication. It is aimed to help the ALS patients to control their environment and to communicate with their family members or care givers via computer interface.

1 Introduction and Background

Technology has been playing a vital role in transforming the health into smart healthcare by providing efficient and effective solutions. Human health being a global concern is revolutionizing with the integration of technological devices and communication technologies for a reliable and faster mechanism presented in [1]. Several

N. Mir (✉) · A. Sarirete · J. Hejres · M. Al Omairi
Effat University, Jeddah, Saudi Arabia
e-mail: nmir@effatuniversity.edu.sa

A. Sarirete
e-mail: asarirete@effatuniversity.edu.sa

J. Hejres
e-mail: jhajres@effatuniversity.edu.sa

M. Al Omairi
e-mail: malomairi@effatuniversity.edu.sa

A. Visvizi and M. D. Lytras (eds.), *Research & Innovation Forum 2019*,
Springer Proceedings in Complexity,
https://doi.org/10.1007/978-3-030-30809-4_39

solutions have been presented for various problems and still there are various areas to be addressed from different angles.

Success factor of these smart health care solutions can be seen by the devices already in practice all over the world, some of these are used as remote monitoring systems as shown in [2], similarly for patients at home and emergency rooms in [3] and some are using the social media platform to monitor and track the patients in [4].

Amyotrophic lateral sclerosis (ALS) patient have to be cleaned regularly and in the last stage when it becomes very difficult to chew food, and they are fed using tubes which are inserted into their stomach to confirm a balanced supply of nutrients for their body. One of the most difficult moments comes when the lungs begin to fail and the patient is unable to communicate. Then it is decided whether they should be connected to the ventilator or not. But in the contemporary world of technology, where the use of technology is proliferating in every field from medical science to support forces the technology is overwhelming human capabilities. Brain-computer interface (BCI), bionic and emotiv devices which work on the principle of electroencephalography (EEG)are to record the cognitive activities of human have mostly practiced with the entertainment activities such as movies, cartoon and games. However, these can be integrated for the smart health care application to help Amyotrophic lateral sclerosis (ALS) patient and their family to be taken care.

An Emotiv-EPOC and the EGG devices have fourteen sensors in addition of two reference sensors for tuning electrical signals which are generated in the brain. These are for the detection of thought patterns, patterns of expressions and feelings in the real time [5]. However, it is based on an assumption that this device is able to receive this kind of information which can be used for any occasion like playing a particular game using this interface or in the control of any common action like using an application in a computer without resorting to a mouse or keyboard, or in performing a physical action by using Robotics in synchronization with EPOC. Making use of this type of data precisely and the combination of a robotic interface like the Arduino can create prototypes for testing concepts of the functionality, thereby expanding scope for utilization in larger applications which have greater influences on several scientific and commercial environments.

Making use of two devices, EPOC is used as a data input device and Arduino as an output device. This study is an attempt of combining the two devices to develop an application as an interface taking brain or neural readings of an ALS patients whose nervous system is effected, in a controlled environment, hence a brain interface device with an application to interpret the thoughts. ALS patients due to nervous damage suffer from progressive disconnection with their body and brain cells which makes it difficult for any physical activity. This research is an attempt to develop a communication interface between these patients and their family members or care takers.

The physical defects related to amyotrophic lateral sclerosis (ALS) are easily deducible from its name. The word "Amyotrophic" which means that the muscles have lost their nutrition; and they become smaller and weaker. "Lateral" designates that the disease affects the sides of the spinal cord, where the motor nerve are located.

Lastly, "sclerosis" means that the diseased part of the spinal cord develops hardened or scarred tissue in place of healthy nerves. This progressive degeneration of the motor neurons in ALS finally leads to the inability of the brain to control muscle movement reported in [6].

Amyotrophic lateral sclerosis (ALS) is a neurodegenerative disease [7] at advanced level that damage nerve cells in the brain and the spinal cord. Usually motor neurons are the carriers that act as massager from the brain to the spinal cord and from the spinal cord to the all the muscles throughout the whole body. And any damage to these nerve originate amyotrophic lateral sclerosis (ALS) which finally leads to death. If the motor nerve die, the brain function are effected hardly and the brain loses control over the human body. Those muscle which act voluntary, their actions are very much effected and these cause the patient to become wholly paralyzed body have different kind of nerves. In brain there are different types of nerves; some to help in thinking, some which help in memory and for feeling the sensation such as hot, cold, sharp, dull etc. There are also nerves which help in hearing, talking and seeing. But the damage to motor nerve cause ALS, which basically controls the body actions and movements. Brain-computer interfaces (BCIs) may help to re-establish communication to people severely disabled by amyotrophic lateral sclerosis (ALS) presented in [8].

Functionality and reliability of EEG and BCI for ALS patients at a controlled home environment and showed how BCI can support communication for patients who are severely disabled and find it very useful to the given extent of use [9].

2 Proposed Solution

This research for Amyotrophic lateral sclerosis (ALS) patients is tested using the Emotive device that is connected to the patient and communicates data to a computer application. This application transforms and manipulates the given instructions such as controlling appliances or any other messages. Figure 1 shows a prototype of the given scenario of how information is communicated.

3 Emotive System

Emotiv Systems is an Australian electronics company for developing brain computer interfaces (BCI) based on electroencephalography (EEG) technology. It is used to detect the energy produced with a brain activity. The data it produced is compared with the pre-defined data on a computer and thus a system is trained to detect the feelings. It tracks the signal in the brain work using the principle of electroencephalography (EEG). The brain wear have 16 electrodes which help in detection and recording of the signals produced by the brain for different activities. These signal are sent to

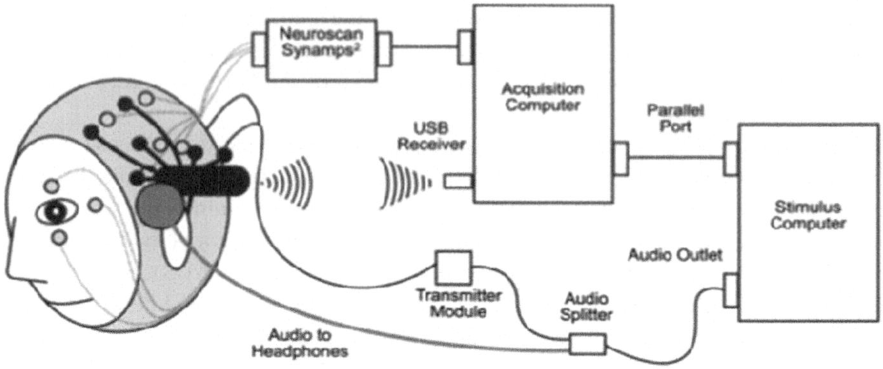

Fig. 1 Prototype of proposed solution

a computer which has a pre-developed software, for analysis of data. And that data is recorded, which is used in future for controlling the activities with mind.

Emotive system has three parts Emotiv EPOC, Emotiv EEG, and Emotiv Insight.

The EPOC is a neuro signal gaining and handling wireless headset which uses 16 sensors to capture the electric signals produced by your brain. When put on, detects the human thoughts and acts an interface for computer and human. It uses electrical signals produced by the brain to detect thoughts, feelings and expressions. Where the high resolution multiport EEG is having all the features of the Emotiv EPOC. Emotive insight measures the brain performance, which has 5 EEG sensors with other two reference sensors to help cover the complete coverage of the main sites around the cerebral cortex.

For this research, cocos2dx smart phone application has been developed using the Android platform, where data signals are received from ALS patient through emotiv device and are communicated to the application in a computer, which uses HTTP protocol to communicate this to a mobile application and hence allows a family member or care taken to reach the patient. At this stage system is tested for controlling the home appliances such as light, TV, fan, calling family and sending this information as a message, which can be seen in Fig. 2, it also shows the main parts of the system such as emotive device, smart phone, and personal computer.

Figure 3 shows the activities of the system where a patient's thoughts are captured by Emotive device and a respective Emo-key is generated for the API where the commands are instructions for trained thoughts are stored, a responsive notification is sent to the care giver or a family member and accordingly the patient receives the assistance.

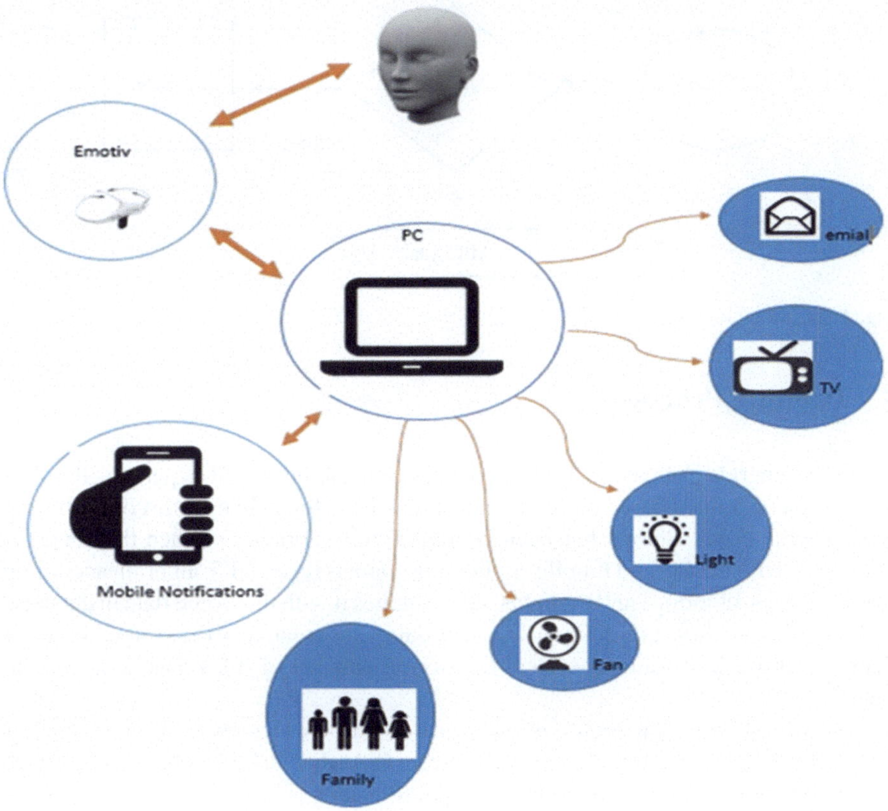

Fig. 2 System design for proposed solution

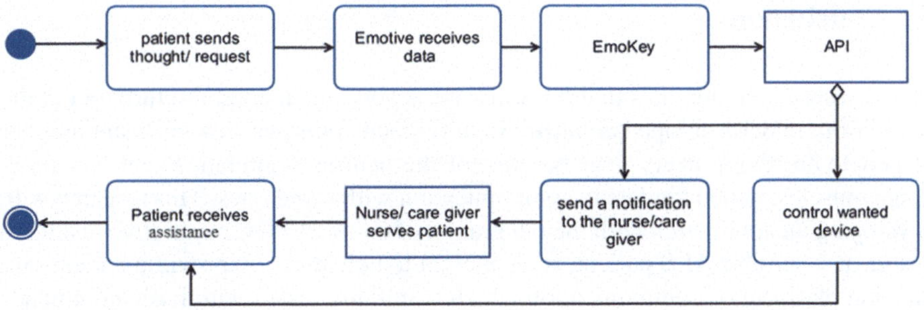

Fig. 3 Activity of system

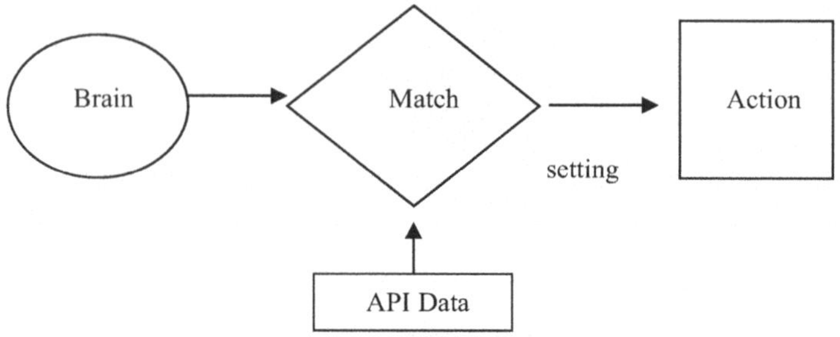

Fig. 4 Training process

4 Training Process

Electrical energy changes in brain depending on the action of a person, the BCI system can be trained based on brain scan or how EEG looks like while concentrating on a specific point. Emotiv has been connected to the brain, and then the device is connected with the PC and mobile application. Data is received from brain according to the feelings, emotions and thoughts, API matches it with the stored data in database and according to these pre-defined data, new data set is created by the system. Figure 4 depicts a process of training system for storing new set of rules, based on present data.

Accordingly once the match is found, data will be executed to form a signal or instruction to take an action. Language is preset to Arabic on both ends, user interface and mobile application, for getting the required results.

5 Conclusion

This research demonstrates that ALS disease is very much common throughout the world from underdeveloped countries to developed countries. A significant number of people are dying every year because of the neuron degradation and less functional muscles, which hinders in communication with loved ones. This research with developing an application with the integration of emotiv EPOC or emotive insight to design a system for ALS patient, is an attempt to help them by providing a communication platform of communication which is initially tested with basic commands of controlling the home appliances and calling their family members or caregivers.

References

1. Shamim Hossain, M., Xu, C., Li, Y., Bilbao, J., El Saddik, A.: Advances in next-generation technologies for smart healthcare. IEEE Commun. Mag. (2018) (13 April)
2. Park, K., Park, J., Lee, J.: An IoT system for remote monitoring of patients at home. Appl. Sci. **7**, 260 (2017)
3. Rojas, E., Sepúlveda, M., Munoz-Gama, J., Capurro, D., Traver, V., Fernandez-Llatas, C.: Question-driven methodology for analyzing emergency room processes using process mining. Appl. Sci. **7**, 302 (2017)
4. Kuang, S., Davison, B.D.: Learning word embeddings with chi-square weights for healthcare tweet classification. Appl. Sci. **7**, 846 (2017)
5. EMOTIV Homepage, https://www.emotiv.com/. Last accessed 2018
6. Cipresso, P., Meriggi, P., Carelli, L., Solca, F., Meazzi, D., Poletti, B., Lulé, D., Ludolph, A.C., Riva, G., Silani, V.: The combined use of Brain Computer Interface and Eye-Tracking technology for cognitive assessment in Amyotrophic Lateral Sclerosis. In: 5th International Conference on Pervasive Computing Technologies for Healthcare (PervasiveHealth) and Workshops). IEEE (2011)
7. Deagle, K.: Critical review: Can brain-computer interface (BCI) systems facilitate communication in-persons with Amyotrophic Lateral Sclerosis (ALS)? In: Proceedings, PubMED (2010)
8. Mccane, L., Sellers, E.W., Mcfarland, D., Mak, J.N.: Brain-computer interface (BCI) evaluation in people with amyotrophic lateral sclerosis. Informa Healthcare (2014)
9. Wolpaw, J., Bedlack, R.S., Reda, D.J., Ringer, R.J.: Independent home use of a brain-computer interface by people with amyotrophic lateral sclerosis. Neurology (2018)

Big Data-Assisted Word Sense Disambiguation for Sign Language

Luis Naranjo-Zeledón, Antonio Ferrández, Jesús Peral and Mario Chacón-Rivas

Abstract Automatic word sense disambiguation (WSD) from text is a task of great importance in various applications of natural language processing, for example, in machine translation, question answering, automatic summarization or sentiment analysis. There are different approaches to finding the meaning of a word within a context, whether using supervised, unsupervised, semi-supervised or knowledge-based methods. Several studies have been conducted to automatically translate from text to sign language, reproducing the result of the translation with a signing avatar, in a way that deaf users have access to informative contents that otherwise are highly inaccessible, because sign language is their mother tongue. The many proposals that have been made look forward to minimize these informative and communicative barriers. Sign languages, however, do not have as many words as the spoken languages, so an automatic translation must be as accurate and free of ambiguities as possible. In this paper, we propose to evaluate the use of public access big data resources, as well as appropriate techniques to access this type of resources for WSD tasks, illustrating their effects in a translation system from text in Spanish to Costa Rican Sign Language (LESCO). The architecture of the actual system incorporates the use of a folksonomy, from which the disambiguation process will benefit. When an exact word is not found for a given detected sense in the source text, the ontology will be fed back with a new relationship of hyperonymy, to alert the curator on the need to propose a new sign in that category, thus promoting an enrichment in a key component of the architecture. As a result of the evaluation, the most appropriate big data public resources and techniques for WSD for sign language will be elucidated.

L. Naranjo-Zeledón · A. Ferrández · J. Peral (✉)
University of Alicante, Crtra. San Vicente, Alicante, Spain
e-mail: jperal@dlsi.ua.es

L. Naranjo-Zeledón
e-mail: lnaranjo@alu.ua.es; lnaranjo@itcr.ac.cr

A. Ferrández
e-mail: antonio@dlsi.ua.es

L. Naranjo-Zeledón · M. Chacón-Rivas
Inclutec, Instituto Tecnológico de Costa Rica, Cartago, Costa Rica
e-mail: machacon@itcr.ac.cr

A. Visvizi and M. D. Lytras (eds.), *Research & Innovation Forum 2019*,
Springer Proceedings in Complexity,
https://doi.org/10.1007/978-3-030-30809-4_40

1 Introduction

Machine translation (MT) is a part of natural language processing (NLP), and consists of translating from a language into another, by means of algorithms and data. Bel-Enguix and Jiménez [1] claim that there is a lack of coherent explanations or models for NLP, turning it into a complex system. The most widely used techniques of MT can be grouped into five large groups, as depicted in Fig. 1.

Rule based MT maps source and target words and their structures, in order to relate them and produce the translation. This requires a basis of linguistic resources (dictionaries and grammars), curated by experts. Statistical approaches use the principle that a document can be translated by using a probability distribution, relying on conditional probability (Bayes Theorem). Example-based uses a large knowledge base of parallel texts, exploiting analogies. Hybrid approaches usually refer to combinations of statistics and rules. Machine learning has used neural networks, support vector machines, bayesian networks and genetic algorithms.

Kumar et al. [2] summarize the application areas of WSD: in Information Retrieval, WSD improves the accuracy of indexing and searching. Information extraction focuses on the recognition, tagging and extraction of key elements of information from a large document database, requiring WSD to identity those elements. An automated online assistant or answering machine must deliver an answer within a valid context, requiring accurate information if the words in the user question are ambiguous. In speech recognition an acoustic signal must be translated into some language, requiring accurate results in case of ambiguity. Finally, MT must disambiguate in order to produce a correct output from text.

The lack of proposals for sign languages motivates for the use of big data-assisted WSD, because of its inductive nature, used to obtain information from massive data sets [3]. Knowledge discovery takes place within any level of structure, discovering patterns and discrepancies. The work by Simonini and Guerra [4] is one of the few

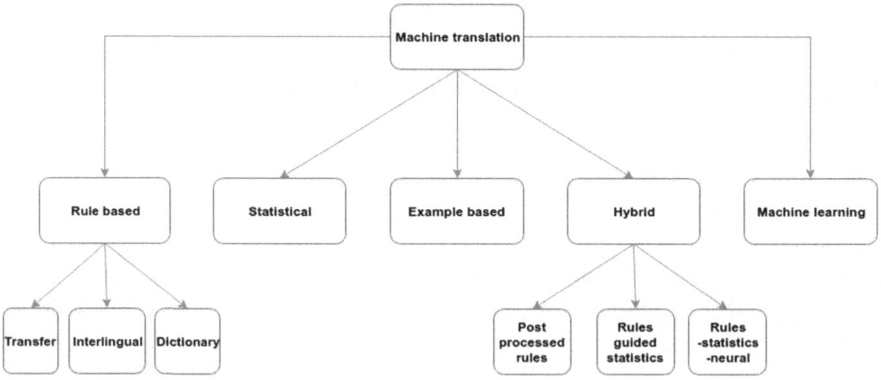

Fig. 1 Machine translation techniques

dealing with big-data WSD, by extracting corpora and metadata from the Web as input to an index of sense clusters (iSC) without targets.

In the next section, we review the previous works. Then, in Sect. 3, we present our architecture that is suitable for sign language WSD. Finally, Sect. 4, shows the conclusions, explaining our contribution to WSD.

2 Background

A general classification of supervised, unsupervised and knowledge-based WSD methods is provided by Kumar et al. [2] as shown in Fig. 2. The supervision level plays a key role in the study and implementation of WSD.

Supervised methods rely on sense tagged corpora, achieving very high accuracy in particular domains. This accuracy is costly, though, in terms of time and effort. Creating corpora manually for all languages in all domains is not reasonable [5]. Unsupervised methods infer potential categories from unlabelled data [6] and do not need sense tagged corpora.

Siddiqui [7] comments on commonly used features for supervised machine learning: part of speech (POS), collocation vector, neighboring words, and co-occurrence vector. These features are the basis to construct rules and list them in a list of value-sense-score decision.

According to Belsare and Akarte [5] unsupervised systems have gained widespread interest since the accuracy of supervised methods has reached a plateau. Yarowsky [8] proposed an unsupervised algorithm that accurately disambiguates in a large untagged corpus, using properties of human language in an iterative bootstrapping setup, and extracting classes of words by from an explicit concept hierarchy. K-nearest neighbors is one of the simplest machine learning algorithms to learn context in a training set. The test set is matched with the learned context and the k most similar contexts identified in a training set are selected.

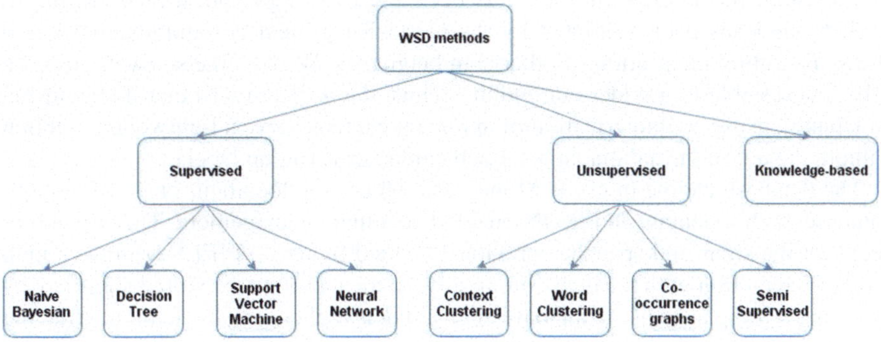

Fig. 2 Supervised, unsupervised and knowledge-based WSD methods classification

An article by Rawat and Chandak [9] identifies algorithmic basis for unsupervised learning (Support Vector Machines, Neural Networks and Decision Trees) as well as clustering algorithms (K-Means, Self-Organizing Maps, and hierarchical clustering). A co-occurrence graph is an alternative to vectors, with nodes representing words and edges representing syntactic relations.

Knowledge based systems are characterized by making use of knowledge resources, and may be confused with supervised methods, but the former learn from information on structured data, whereas the latter learn from examples [5].

Belsare and Akarte [5] divide knowledge-based into three categories: Lesk algorithm, conceptual density and Walker algorithm. Lesk algorithm looks for overlaps between ambiguous senses of a word and those of its surrounding words. Some variants, like that of Guthrie et al. [10] use subject-dependent co-occurrence neighborhoods (subject classifications from LDOCE). Cowie, Guthrie and Guthrie [11] algorithm simulates annealing with good results. Vasilescu et al. [12] determined simplified version that works better.

Conceptual density measures how a concept that a word represents is related to the concepts of other words in its context. This measure can be obtained directly from WordNet.

Walker algorithm [10] works with subject categories from Roget thesaurus codes, based on the assumption that the sense of a word derives from the categories assigned to it. Words with more than one category would have more than one sense.

3 The Proposed Architecture

As a contribution of this article we propose a design complementary to an existing platform for editing and reproducing sign language by means of a signing avatar. The architecture of our platform for sign languages (PIELS, Spanish for International Platform of Sign Language Editing), contemplates a big data component, in order to disambiguate words within a domain, as shown in Fig. 3.

Currently, the PIELS platform has been used for sign and speech editing in LESCO, and has been validated by the Costa Rican deaf community. Its design allows for future integration of other sign languages. We use free software to make PIELS accessible to a wider community. Thus, the decision was that it should run on Ubuntu, using containerization of applications, to achieve a lightweight solution without the aid of virtual machines. Each container is run on Docker.

The database engine in use is MongoDB, which is a document-oriented NoSQL database, with schemata and JSON-enabled document management. This repository keeps all the signs and speeches previously edited by users. PIELS requires a high level of interaction with the deaf community, so the nature of the stored signs is very dynamic, forcing us to use a database engine that is highly scalable without affecting the consistency of already saved data.

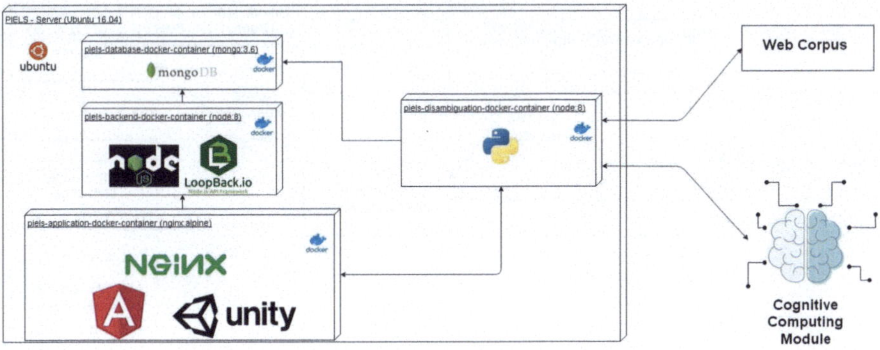

Fig. 3 PIELS architecture and its big data assisted disambiguation engine

For the back-end we use Node.js, adhering to the "JavaScript everywhere" paradigm. LoopBack.io is the used framework, because of its high simplicity to develop new APIs. These two components run together in a Docker.

Another Docker contains three components for the front-end, where the edition and visual display takes place: Angular, Nginx and Unity. Angular is used for communication and primary validations. Unity is used specifically for the deployment of the signing avatar. Nginx is key in balance loading of the activities of the whole platform, because the level of expressiveness and response time in the display of signs is a key human-computer interaction point.

Finally, we propose to use a Docker for the disambiguation module, coded in Python. It will be tested using three different approaches: rule-based with access to WordNet, unsupervised machine learning and hybridization. In the latter two cases, different corpora will be crawled from the Web (big data), much in the way suggested by Simonini and Guerra [4] and IBM Watson will be the cognitive computing module (an algorithmic intermediary), but we'll also explore a solution without intermediary, as explained below.

By zooming into the disambiguation component of the architecture, we can further decompose the tasks needed to produce gains from using big data crawled from the Web and constructing the iSC, as shown in Fig. 4.

We use esDBpedia, the Spanish version of DBpedia that has generated semantic information from Wikipedia since 2011, with a website and a SPARQL Endpoint. The extraction process produces 100 million RDF Triples from the Spanish Wikipedia. All the RDF Triples are available in the SPARQL endpoint [13].

A pre-processing process is generally a good idea when dealing with text, particularly when it comes in a large size and there is no certainty about its structure. Therefore, we will use WebCorpus, and open-source project to extract statistics from the text. It acts as a pipeline of various Hadoop MapReduce jobs, and can be used to perform the following actions:

- Stop-words removal
- Stemming

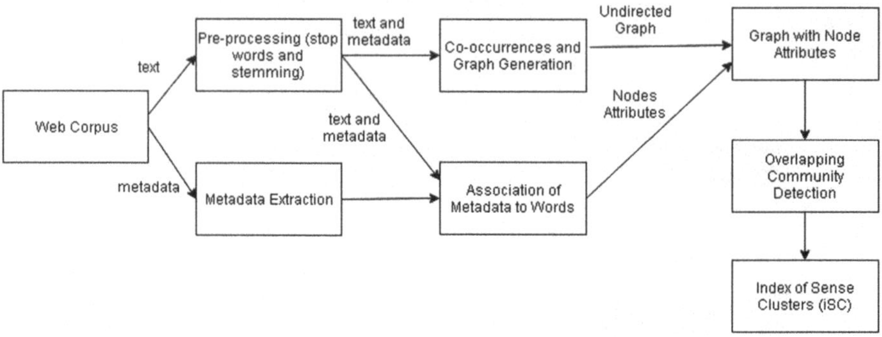

Fig. 4 Construction of the iSC of the disambiguation module. Adapted from Simonini and Guerra [4]

- Attributes extraction from metadata (categories)
- Words co-occurrence computation
- Similarity scores (strength of the co-occurrences).

By extracting metadata, we can introduce an intuitive idea of the semantics of words in text. There is an important gain in this process, because it can help to better detect communities. Simonini and Guerra [4] contribution to traditional WSD techniques consists mainly in taking in account this information, and not only the relations among nodes of co-occurrence graphs.

In our case, the use of esDBpedia greatly facilitates the extraction of metadata, as it is designed for this type of task, as well as the search for relationships. The project uses the SPARQL query language, which offers the advantage of being based on a widely documented query method.

Graph-based algorithms perform a partitioning of the graph from a list of target words, to produce a set of non-overlapping sense clusters of words, each one representing a context for a specific target word. If the dataset is of a highly dynamic nature, then there is need to execute the graph-construction and sense cluster generation again. This motivates for a more efficient means of graph construction, since we are dealing with dynamic content.

An interesting possibility is to deal with a set of different domains of application, selecting the attributes accordingly with those domains to construct iSCs. This provides for flexibility, as well as ulterior comparative studies across domains.

The detection of overlapping communities is of particular importance, providing for the data-centric determination of senses. The similarity measure of election will be log-likelihood, as suggested by Moore [14] since it allows for a greater recall. The task at hand consists in finding the parameter θ that maximizes the log-likelihood of an observed sample ζ, thus maximizing the likelihood function $L(\theta; \zeta)$, because the natural logarithm is a monotone increasing function.

For the construction of the inverted index, it is necessary to find all the possible communities of affiliation where for each word. The chosen overlapping community detection algorithm is CESNA, because its method considers both nodes relationships inside communities, as well as dependency between communities and attributes. It can also detect overlapping, hierarchical and nested communities.

According to Yang et al. [15] CESNA is based upon the following intuitive ideas: nodes belong to the same communities are likely to be connected to each other; nodes in the same community will likely share common attributes; communities may overlap, since nodes may belong to more than one single community; if two nodes belong to common communities, they are more likely connected, even more than if they are only part of a single community.

4 Conclusions

Available literature on WSD account for the state of the art on disambiguation, but it is difficult to find specific applications of WSD in sign language, which can become a serious problem when there are no equivalent words in the correct sense between the source language and the target language. Sign languages are most sensitive to this situation, because their lexicon is generally smaller than those of spoken languages.

We have reviewed well-known WSD algorithms and their relationship to data, in order to discover how to adopt them for sign language synthesis. To the best of our knowledge, there is a lack of WSD techniques specifically designed and empirically tested for sign languages. Also, and although the techniques for WSD are well documented, very little literature on disambiguation assisted by big data has been detected, from which great benefits could be achieved, hence deserving an important effort in terms of proposals, experimentation and documentation of results.

Feeding back into lexical resources is certainly a challenging field of research, particularly by recurring to big data. In this article we have proposed a robust architecture to assist in the subprocess of disambiguation, in order to translate from Spanish to LESCO, by means of a hybrid approach of a corpus curated by experts, as well as big data-enabled machine learning.

Acknowledgements The authors thank the School of Computing and the Computer Research Center of the Technological Institute of Costa Rica for the financial support, as well as ONICIT (Consejo Nacional para Investigaciones Científicas y Tecnológicas), Costa Rica, under grant 290-2006. This work was supported in part by the Spanish Ministry of Science, Innovation and Universities through the Project ECLIPSE-UA under Grant RTI2018-094283-B-C32, the Project RESCATA under Grant TIN2015-65100-R, the Project PROMETEO/2018/089, and the Lucentia AGI Grant.

References

1. Bel-Enguix, G., Jiménez, M.: Language as a Complex System: Interdisciplinary Approaches. Cambridge Scholars Publishing (2010)
2. Kumar, R., Khanna, R., Goyal, V.: A review of literature on word sense disambiguation. Res. Cell Int. J. Eng. Sci. **6** (2012)

3. Lomotey, R., Deters, R.: Towards knowledge discovery in big data. In: IEEE 8th International Symposium on Service Oriented System Engineering (2014)
4. Simonini G., Guerra, F.: Using big data to support automatic word sense disambiguation. In: International Conference on High Performance Computing & Simulation (HPCS) (2014)
5. Belsare, R., Akarte, S.: A review on strategies of word sense disambiguation. Int. J. Sci. Eng. Appl. **5**(4) (2016)
6. Kumar, A., Kumar, S.: Word sense disambiguation using association rules: a review. Int. J. Innov. Res. Stud. **2**(2) (2013)
7. Siddiqui, T.: Review on word sense disambiguation techniques. Aryabhatta J. Math. Inform. **7**(2) (2015)
8. Yarowsky, D.: Unsupervised word sense disambiguation rivaling supervised methods. In: Proceedings of the 33rd Annual Meeting of the Association for Computational Linguistics, Cambridge, MA, pp. 189–96 (1995)
9. Rawat, S., Chandak, M.: Word sense disambiguation and classification algorithms: a review. Int. J. Comput. Sci. Appl. **8**(1) (2015)
10. Guthrie, J., Guthrie, L., Wilks, Y., Aidinejad, H.: Subject dependent co-occurrence and word sense disambiguation. In: Proceedings of the Annual Meeting of the Association for Computational Linguistics. ACL (1991)
11. Cowie, J., Guthrie, J., Guthrie, L.: Lexical disambiguation using simulated annealing. In: Proceedings of the International Conference on Computational Linguistics, COLING (1992)
12. Vasilescu, F., Langlais, P., Lapalme, G.: Evaluating variants of the Lesk approach for disambiguation words. In: Proceedings of Language Resources and Evaluation Conference (LREC), pp. 633–636 (2004)
13. DBPedia Homepage, http://es.dbpedia.org/. Last accessed 12 Jan 2018
14. Moore, R.: On log-likelihood-ratios and the significance of rare events. In: EMNLP (2004). http://aclweb.org/anthology/W04-3243
15. Yang, J., McAuley, J., Leskovec, J.: Community detection in networks with node attributes (2012). https://arxiv.org/pdf/1401.7267.pdf

Emerging Issues: Technology and Society

Surfing the Waves of Digital Automation in Spanish Labor Market

Josep Lladós-Masllorens

Abstract The emerging technologies based on artificial intelligence and robotics are challenging the landscape for human labor. Revolution 4.0 is progressively questioning some business models, changing the balance between creation and destruction of jobs, accelerating the transformation of occupational and skills requirements and affecting the distribution of income, because these technologies can replace both manual and cognitive skills. Our research analyzes how digital revolution is affecting labor market in Spain. We mainly focus on the new employment created since the outbreak of the recent financial crisis. We confirm a skill-biased technological change, although the new occupations are mostly demanding a limited set of complex skills. This is a consequence of the routine task-intensive economic structure in Spain. Although digital technologies are inciting a task-biased technological change, decreasing the demand for workers performing routine tasks, this is not the case in Spain. The predominance of routine (but not repetitive) tasks in many services linked to serving, attending, health and care is limiting the scope of automation. The immediate consequence is a growing polarization in employment opportunities and incomes. In addition, a progressive de-skilling effect is emerging, with high-skilled workers moving down the occupational ladder, whereas the wage premium is declining.

1 Introduction

1.1 Theoretical Background

Technology is playing a role in the new configuration of global production and consequently in the composition of employment. Emerging digital technologies are rapidly changing people's jobs and lives [1]. Significantly, the academic debate about the extent of job destruction linked to digital automation has been considerable.

J. Lladós-Masllorens (✉)
Open University of Catalonia (Universitat Oberta de Catalunya), Av. Tibidabo 39-43, 08035 Barcelona, Spain
e-mail: jlladosm@uoc.edu

A. Visvizi and M. D. Lytras (eds.), *Research & Innovation Forum 2019*, Springer Proceedings in Complexity,
https://doi.org/10.1007/978-3-030-30809-4_41

Although empirical evidence has shown that the deep transformation of production processes has not resulted in a significant increase in the aggregate levels of technological unemployment, the debate about the consequences of technological change on the labor market still has not come to an end [2–5].

The most relevant effects are found in the composition, rather than in the level, of employment. The digital technologies are skill-biased because they complement skilled workers, increasing their productivity and the relative demand for their labor services [6–9]. In many developed and developing countries, this skill-biased technical change is also generating a skill premium [10–15].

In addition, robotics and artificial intelligence are also leading to employment polarization, with a reduction of middling occupations. At all skill levels, workers perform a variety of tasks. A seminal work [16] classified work tasks along two main dimensions: their degree of routinization and their cognitive or manual nature. Technology is predicted to improve relative demand for workers that perform non-routine tasks, because they are not easily automated and replaced by labor-saving technologies.

While non-routine cognitive tasks are typical of skilled jobs, non-routine manual tasks are essentially unskilled jobs. With regard to routine cognitive or manual tasks, in which the technology has the potential to replace human labour, both of them are characteristic jobs performed by middle-skilled workers. This routine-biased technological change is generating employment polarization and, in some economies, a reduction in the skill-premium [17–24].

However, it is necessary to keep in mind that the recent evolution of the labour market and the past influence of determining factors do not necessarily will decide the future course of employment, because the potential disruptive effects of these emerging technologies [25, 26].

1.2 Research Design

We focus our analysis on the labor market in Spain, one of the countries more dramatically affected by the recent financial turmoil. We use the information of the Labor Survey to study how the digital revolution is affecting the composition of employment. The sample provides information about more than 160,000 individuals.

Using quantitative research methods, we concentrate our research essentially on the new employment created since the outbreak of financial crisis. In particular, we analyze the changes in the composition of employment between the first quarter of 2014 and the end of 2018. We quantify the consequences on the distribution of employment and the evolution of wage gains, according to the skills level and the type of occupation, and examine how new jobs are organized.

2 Analysis of Results

The evolution of Spanish labor marketshows a skill-biased technological change, confirming previous research [27]. Digital technologies tend to complement better the most skilled-workers, so the demand for skills has increased significantly. As the risk of automation declines as educational attainment level rise, probably 4.0 technologies primarily threaten low-skilled works (Fig. 1).

However, the new occupations are mostly demanding a limited set of complex skills. This is a consequence of the routine task-intensive economic structure in Spain, relative to the rest of European Union. Although, in many countries, digital technologies are instigating a task-biased technological change and the demand for workers performing routine tasks has perceptibly decreased, this is not exactly the case in Spain. In fact, the predominance of routine (but not repetitive) tasks in many services linked to serving, attending, health and care is limiting the scope of automation, although most of them are low-skilled jobs.

The immediate consequence is a growing pattern of job polarization by skill level. This segmentation affects employment opportunities and labor incomes, whereas the wage premium, both for educational attainment and skills complexity, is clearly

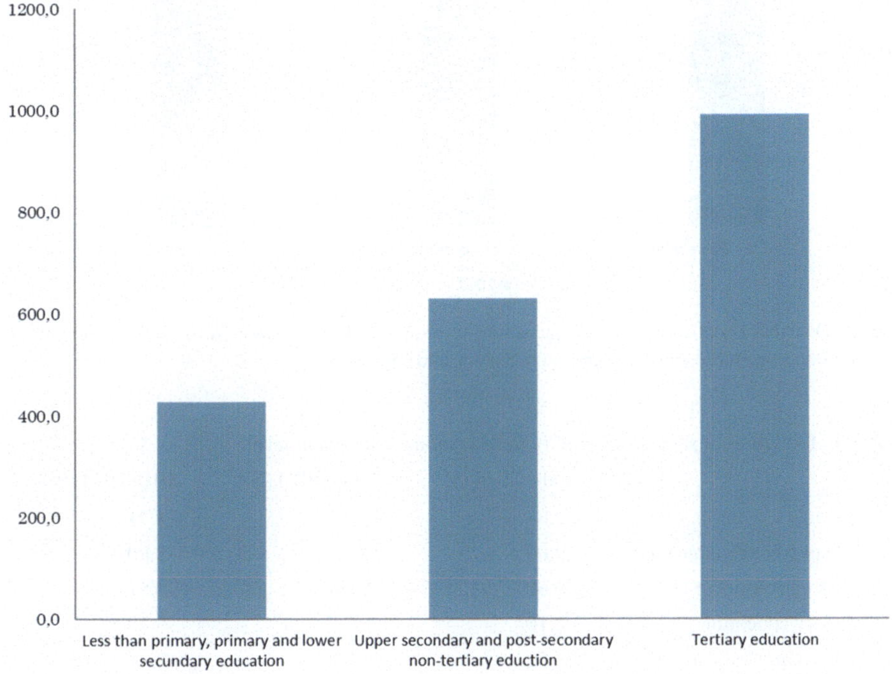

Fig. 1 Evolution of employment according to educational attainment level (2014–2018). *Source* Compiled by the author based on data from the Active Population Survey

shrinking. This reduction occurs at the same time that the average level of wages is decreasing in Spain (Fig. 2; Table 1).

In addition, a progressive de-skilling effect is emerging in the whole labor market, with high-skilled workers moving down the occupational ladder. This is also the case of mid-level education workers because most of new jobs appear in economic activities that are demanding a limited set of skills, so they begin to perform jobs traditionally performed by lower-skilled workers.

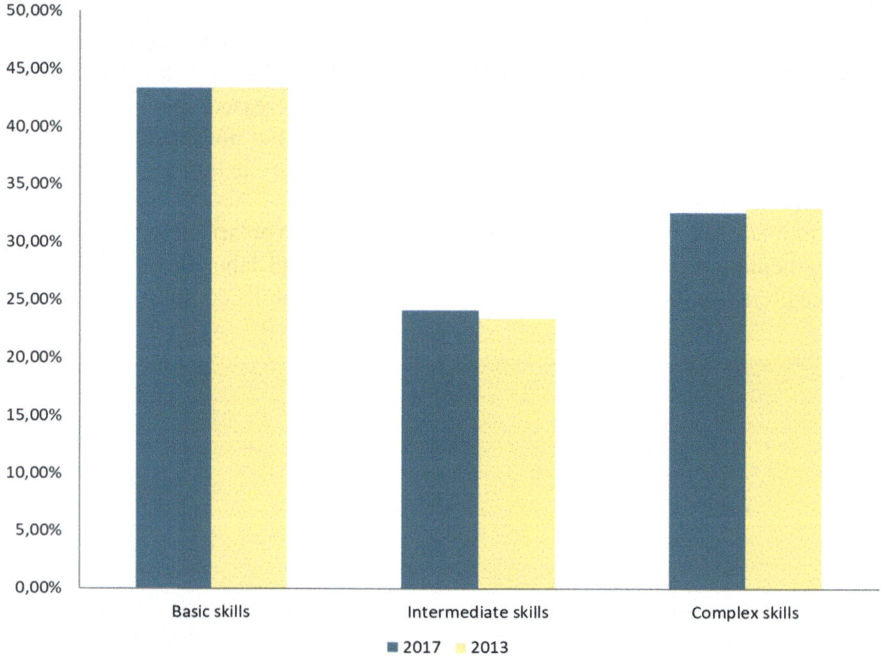

Fig. 2 Distribution of employment according to required skills. *Source* Compiled by the author based on data from the Active Population Survey and ISCO-08

Table 1 Evolution of salaries according to educational attainment level

	2006–2014 (%)	2006–2010 (%)	2010–2014 (%)
Average	9.84	15.27	−4.71
Less than primary education	9.00	12.13	−2.80
Primary education	0.60	9.96	−8.51
Secondary education	2.10	9.31	−6.60
Post-secondary education	12.52	15.22	−2.34
Tertiary education	10.33	13.53	−2.81

Source Compiled by the author based on data from the Active Population Survey and Wage Structure Survey

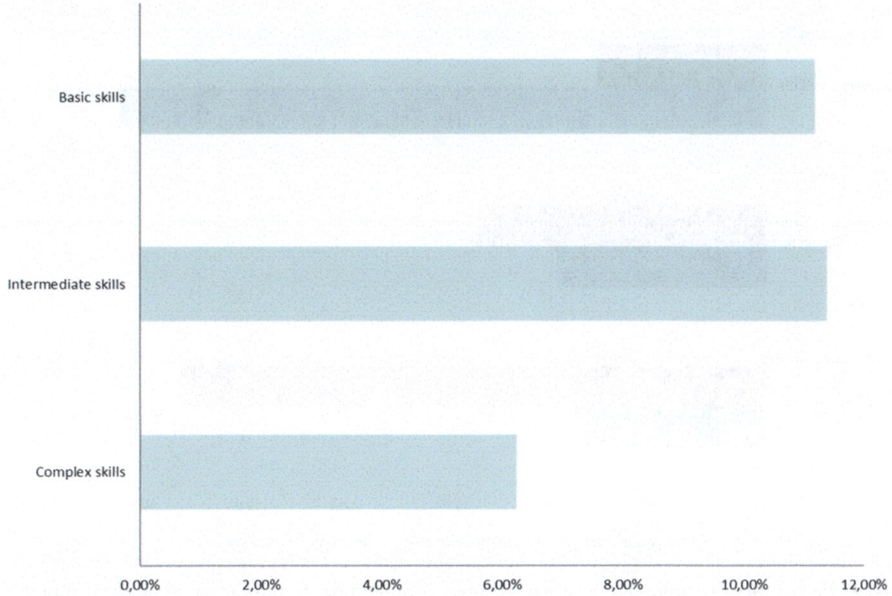

Fig. 3 Evolution of salaries according to the required skills. *Source* Compiled by the author based on data from the Active Population Survey and ISCO-08

The cognitive skills connected with creative and social intelligence, which are also needed to solve non-standard problems, probably will continue to benefit from a high premium. Nevertheless, many highly skilled workers cannot take advantage of their skills due the poor demand for them in the labor market (Fig. 3).

Part of this de-skilling process could be the consequence of technological maturity. As ICT are general-purpose technologies, once they are widely adopted, the demand for complex skills drops [28]. A similar result could emerge from the digital automation that breaks down complex operations into simple tasks. This de-skilling process in Spain is accelerating over time, causing a displacement and substitution of low-skilled workers despite the persistent demand for routine tasks and basic skills (Figs. 4 and 5).

3 Concluding Discussion

Digitalization will have a key influence on the future of work in Spain. However, the economic structure of production in this country, with the dominance of jobs performing routine tasks and demanding non-complex skills, is strongly affecting the current composition of employment. Consequently, an obvious skill-biased effect on employment opportunities goes with the use of emerging technologies, but the digital transformation is advancing slowly in this economy, probably because it is

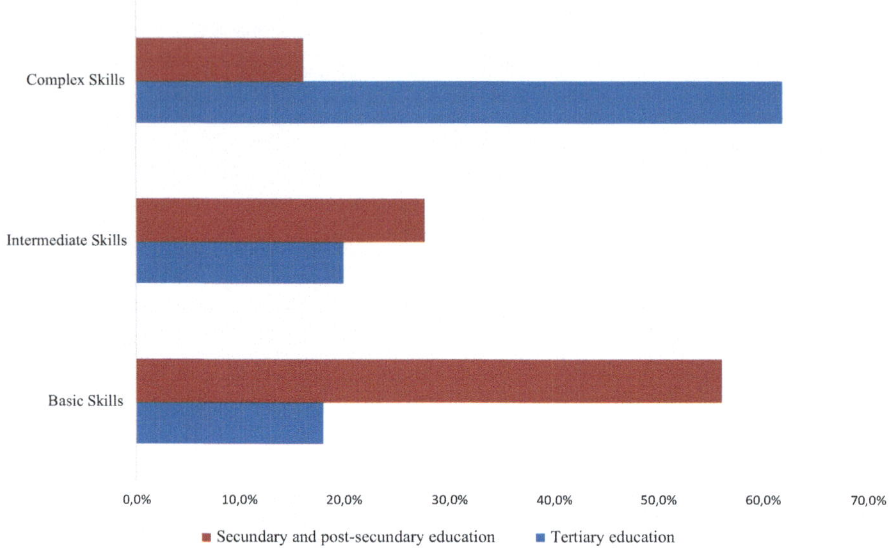

Fig. 4 Probability of employment (2018). *Source* Compiled by the author based on data from the Active Population Survey and ISCO-08

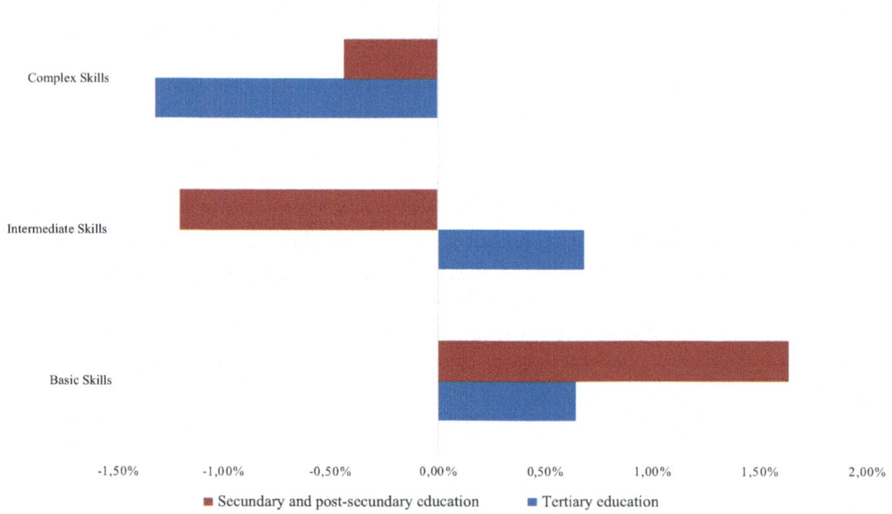

Fig. 5 Change in the probability of employment (2014–2018). *Source* Compiled by the author based on data from the Active Population Survey and ISCO-08

facing a strong competition from business models based on low wages and poor working conditions.

A de-skilling effect is detected, resulting from the mismatch between workforce qualifications and the requirements of labor demand. A polarization in labor incomes is emerging while high-skilled workers move down the occupational ladder, whereas their wage premium is progressively declining. The process of de-skilling is especially harmful for low-skilled workers that are already in a vulnerable position in the Spanish labor market. Probably, the impact of the financial crisis on Southern Europe economies has aggravated this growing asymmetry in the labor market. The interaction between both effects has reduced job opportunities for less qualified work [29, 30].

Therefore, this unequal distribution of the risk of digital automation raises the stakes involved in policies to prepare workers for changing job requirements. Probably, education, adult learning, research, work organization and competition policies, affecting both supply and demand of labor, seem to be the most appropriate strategies to transform this challenge into an opportunity for social progress.

Future work could be carried out in this research about the evolution of employment in Spain according to the skill composition of labor and educational attainment in different regions. It would be significant to analyze the influence of technological change in regions with different economic specialization, as tourism or high-tech manufacturing.

References

1. OECD: Putting faces to the jobs art risk of automation. Policy Brief on the Future of Work, March (2018)
2. Frey, C.B., Osborne, M.: The future of employment: how susceptible are jobs to computerization? Working Paper, Oxford Martin School, University of Oxford (2013)
3. Autor, D.H.: Why are there still so many jobs? The history and future of workplace automation. J. Econ. Perspect. **29**(3), 3–30 (2015)
4. Vivarelli, M.: Innovation, employment and skills in advanced and developing countries: a survey of economic literature. J. Econ. Issues **48**(1), 123–154 (2014)
5. Gregory, T., Salomons, A., Zierahn, U.: Racing with or against the machine? Evidence from Europe. In: Conference paper. Technological Change, Versatility and Macro-Labor Models, B18-V2 (2016)
6. Berman, E., Bound, J., Griliches, Z.: Changes in the demand for skilled labor within US manufacturing: evidence from the annual survey of manufacturers. Q. J. Econ. **109**(2), 367–397 (1994)
7. Autor, D.H., Katz, L.F., Krueger, A.B.: Computing inequality: have computers changed the labor market? Q. J. Econ. **113**(4), 1169–1213 (1998)
8. Bresnahan, T.F., Brynjolfsson, E., Hitt, L.M.: Information technology, workplace organization, and the demand for skilled labor: firm-level evidence. Q. J. Econ. **117**(1), 339–376 (2002)
9. Acemoglu, D., Autor, D.: Skills, tasks and technologies: implications for employment and earnings. In: Card, D., Ashenfelter, O. (eds.) Handbook of Labor Economics, vol. 4b, pp. 1043–1171 (2011)
10. Gera, S., Gu, W., Lin, Z.: Technology and the demand for skills in canada: an industry-level analysis. Can. J. Econ./Revue canadienne d'economique **34**(1), 132–148 (2001)

11. Gregory, M., Zissimos, B., Greenhalgh, C.: Jobs for the skilled: how technology, trade, and domestic demand changed the structure of UK employment, 1979–90. Oxford Econ. Pap. **53**(1), 20–46 (2001)
12. Piva, M., Santarelli, E., Vivarelli, M.: The skill bias effect of technological and organisational change: evidence and policy implications. Res. Policy **34**(2), 141–157 (2005)
13. Autor, D.H., Katz, L.F., Kearney, M.S.: Trends in US wage inequality: revising the revisionists. Rev. Econ. Stat. **90**(2), 300–323 (2008)
14. Fajgelbaum, P.D., Khandelwal, A.K.: Measuring the unequal gains from trade. Q. J. Econ. **131**(3), 1113–1180 (2016)
15. Meschi, E., Taymaz, E., Vivarelli, M.: Globalization, technological change and labor demand: a firm-level analysis for Turkey. Rev. World Econ. **4**(152), 655–680 (2016)
16. Autor, D.H., Levy, F., Murname, R.J.: The skill content of recent technological change. Q. J. Econ. **118**(4), 1279–1333 (2003)
17. Autor, D.H., Katz, L.F., Kearney, M.S.: The polarization of the U.S. labor market. Am. Econ. Rev. Pap. Proc. **96**(2), 189–194 (2006)
18. Goos, M., Manning, A.: Lousy and lovely jobs: the rising polarization of work in Britain. Rev. Econ. Stat. **89**, 118–133 (2007)
19. Goos, M., Manning, A., Salomons, A.: Explaining job polarization: routine-biased technological change and offshoring. Am. Econ. Rev. **104**(8), 2509–2526 (2014)
20. Arntz, M., Gregory, T., Zierahn, U.: The risk of automatization for jobs in PECD countries. A comparative analysis. OECD Social, Employment and Migration Working Papers, 189 (2016)
21. Arntz, M., Gregory, T., Zierahn, U.: Revisiting the risk of automatization. Econ. Lett. **159**, 157–160 (2017)
22. Marcolin, L., Miroudot, S., Squicciarini, M.: Routine jobs, employment and technological innovation in global value chains. OECD Science, Technology and Industry Working Papers, 2016/01. OECD Publishing (2016)
23. Graetz, G., Michaels, G.: Robots at Work CEP Discussion Paper No. 1335 (2015)
24. Parro, F.: Capital-skill complementarity and the skill premium in a quantitative model of trade. Am. Econ. J. Macroecon. **5**(2), 72–117 (2013)
25. Brynjolfsson, E., McAfee, A.: The Second Machine Age: Work, Progress and Prosperity in a Time of Brilliant Technologies. WW Norton & Company, New York (2014)
26. Mokyr, J., Vickers, C., Ziebarth, N.L.: The history of technological anxiety and the future of economic growth: is this time different? J. Econ. Perspect. **29**(3), 31–50 (2015)
27. Aguirregabiria, V., Alonso-Borrego, C.: Occupational structure, technological innovation, and reorganization of production. Labour Econ. **8**(1), 43–73 (2001)
28. Charles, K.K., Hurst, E., Notowidigdo, M.J.: The masking of the decline in manufacturing employment by the housing bubble. J. Econ. Perspect. **30**(2), 179–200 (2016)
29. Miguel, A.M.: Labour Market Measures in Spain 2008–13: The Crisis and Beyond. International Labour Office (2015)
30. Visvizi, A.: The crisis in Greece and the EU-IMF rescue package: determinants and pitfalls. Acta Oerconomica **64**(4) (2014)

Impact of Linguistic Feature Related to Fraud on Pledge Results of the Crowdfunding Campaigns

Wei Wang, Yenchun Jim Wu and Ling He

Abstract In order to achieve the pre-set funding goal, some entrepreneurs may engage in malicious fraud, that is, using fraudulent textual descriptions to attract monetary contribution from crowd. Thus, fraud is inevitable in the online financial market. Fraudulent texts are not strictly equivalent to low-quality campaigns, but fraudulent content can jeopardize users' perceptions of project quality. Thus, the fraudulent text has great drawbacks for the development of crowdfunding model, leading investors lose confidence in this newborn financing model. Through text mining, 4 indicators are adopted to measure the linguistic feature related to fraud. And 126,593 campaigns from Kickstarter is employed to estimate the impact of linguistic feature related to fraud on the fundraising outcomes. Multi text levels are selected as the study objects include abstract, detailed description and the reward narratives. The results show that in general, lower linguistic feature related to fraud attracts the investors to contribute more money, the predictive model also validates this conclusion. However, some fraud indicators have no significant negative impacts on financing, or even show positive influences. Moreover, the detailed delivery terms in the reward text, the higher ratio of successful funding. This study provides a guideline for the founders to generate attractive description for the crowdfunding campaigns.

W. Wang · L. He
College of Business Administration, Huaqiao University, 362021 Quanzhou, China
e-mail: wayswang@gmail.com

L. He
e-mail: helingqz@163.com

Y. J. Wu (✉)
Graduate Institute of Global Business and Strategy,
National Taiwan Normal University, Taipei City, Taiwan
e-mail: wuyenchun@gmail.com

A. Visvizi and M. D. Lytras (eds.), *Research & Innovation Forum 2019*,
Springer Proceedings in Complexity,
https://doi.org/10.1007/978-3-030-30809-4_42

1 Introduction

As a new Internet financial model, crowdfunding market volume has grown 100% every year in recent years. Generally, entrepreneurs need to raise enough money to support the creative idea, some founders may resort to all means including fraud to get support. To pledge enough money, there must be some fraudulent information on the crowdfunding market. Due to information asymmetry and the cluster effect of the online behavior [1], some founders may fraud investors by immature ideas or fraudulent description. In addition, in order to attract more investors, the creators may exaggerate the stories deliberately to make the text description seems unrealistic. Therefore, the studying on the impact of fraudulent texts on crowdfunding financing has both theoretical and practical merit.

In the field of text mining, some methods have been proposed to detect fraud, many of which are based on linguistic features, such as machine learning [2]. From the perspective of text analysis, linguistic features indicate the fraudulent cues [3], while from the perspective of psycholinguistics, the fabricated stories are different from the true stories significantly [4], which provides a guideline for the fraudulent cues detection from the description for the crowdfunding campaigns. The campaigns with more fraudulent cues in the description can be considered as poor quality, because these projects are likely to cannot honor their commitments, therefore these campaigns are more likely to be funded failed.

Extant studies generally agreed that the more fraudulent cues in the text, the lower quality of the text description, as fraudulent descriptions in financial reporting [5] and fraudulent content in insurance field [6]. The general idea is that the text contains more fraudulent cues indicates that the project is less reliable, therefore, the funding successful ratio of the campaigns will be lower. However, the opposite view is widespread. The investment in crowdfunding projects comes from the expectation on the reward of the project. This subjective belief is achieved by the attractiveness of the textual description generated by the entrepreneurs. If entrepreneurs describe the project more attractive, the greater the expectation of investors on the project, thus, the greater the willingness to invest in [7]. Existing study does not attempt to explore the impact of fraudulent cues on crowdfunding, which prompts us to investigate the new issue.

We can't simply assume that fraudulent cues hinder financing, because the simple assertion may face the following questions: (1) It is a vague indicator itself for the fraudulent cues, thus, how to measure it for the online user generated content (UGC)? (2) Extant studies do not involve what type of fraudulent cues would influence the investment intention and whether there are differences between indicators? (3) How the fraudulent cues affect investment decisions and fundraising outcomes? In this study, we attempt to use text analysis to identify fraudulent cues from online text. And we employ real data from Kickstarter to estimate the influence of fraudulent cues on fundraising outcomes. We expect our results to provide implications for entrepreneurs and crowdfunding platforms both theoretically and practically.

2 Literature Review

Crowdfunding is often referred to as Crowdfinancing or Crowdinvesting. The "Crowdfunding" is defined as the process of obtaining needed services, ideas, or content by soliciting monetary contributions from a large group of people, especially from an online community, rather than from employees or suppliers [8]. Nowadays, crowdfunding is often performed via Internet-mediated registries, but the concept can also be executed through mail-order subscriptions, benefit events, and other channels. Crowdfunding is a form of alternative finance, which has emerged outside of the traditional financial system. At present, there have been many discussions on online crowdfunding, mainly focusing on the factors affecting successful financing. Among them, the fraud of the project is a common concern of investors.

Benford's Law is a law on digital frequency, that is, the probability of the number that appears to follow the Benford distribution. In detail, in a large amounts figures, the frequency of 1 at the beginning of the number is not 1/9, but 30.1%, and the frequency of 2-beginning figures is 17.6%, while the lowest frequency is 9 with 4.6% [8]. Benford's law is not generated from a strict mathematical proof, but obtained from many statistics. It can be employed to detect fraudulent activity, which suggests that the statistical analysis can identify fraudulent cues [9], namely, the description can reveal fraudulent cues which provides a channel for our study to detect fraudulent content.

Analysis from psycholinguist, there are significant differences between the fabricated story and the real story on the syntax [4], which provides an implication for the detection of fraudulent cues for crowdfunding projects. In a study of telephone fraudulent detection on insurance industry, scholars drawn the conclusion that the real stories tend to use the exact words while the fraudulent content always use many specious and ambiguous vocabularies [10]. The study shows that it is able to identify fraudulent cues from the text itself, as linguistic features can be detected by text mining [11]. In the description of crowdfunding projects, founders always expect a fascinating story to convince investors to invest in the project, therefore, these projects contain many fraudulent cues in the description can be considered as poorly because they include more fabricated content.

3 Experimental Data and Methodology

3.1 Experimental Data

Kickstarter is launched in 2009, and the experimental data we used also started from that time as previous study did [12]. A total of 136,309 raw projects were collected during the observed window, but these data include projects that be funding during data collection. This part of the project must be removed from the sample because it is impossible to judge whether this part of the project can be successful funded, which are not suitable for the study.

Another type of project that deserves special discussion is the projects that have been cancelled. Figure 1 shows an example of a cancelled project on Kickstarter. The project has no clear information on the status of success or failure. The example shows that during the fundraising, the entrepreneur canceled the project (the scheduled funding deadline is February 28, 2014, but during the process of financing in February 21, 2014, the founder canceled the project when the project has raised $244,923 to reach 24% of the preset target). Even if the project has been funded successfully, it may be cancelled due to a change in the founder's plan, copyright reasons or a vulnerability in the business plan.

Among the raw data, there are 9791 cancelled projects, accounting for about 7%. Since the object of this study is to estimate the impact of fraudulent cues on the fundraising results, the cancelled projects are treated as follows: (1) If the project is canceled when the fundraising progress has reached or exceeded 100%, then the project is considered as a successful because the actual funded money has reached the preset target; (2) If the project is cancelled when the fundraising progress does not reach 100%, then we cannot judge the final fundraising outcomes for this part projects, so we remove these projects from the corpus.

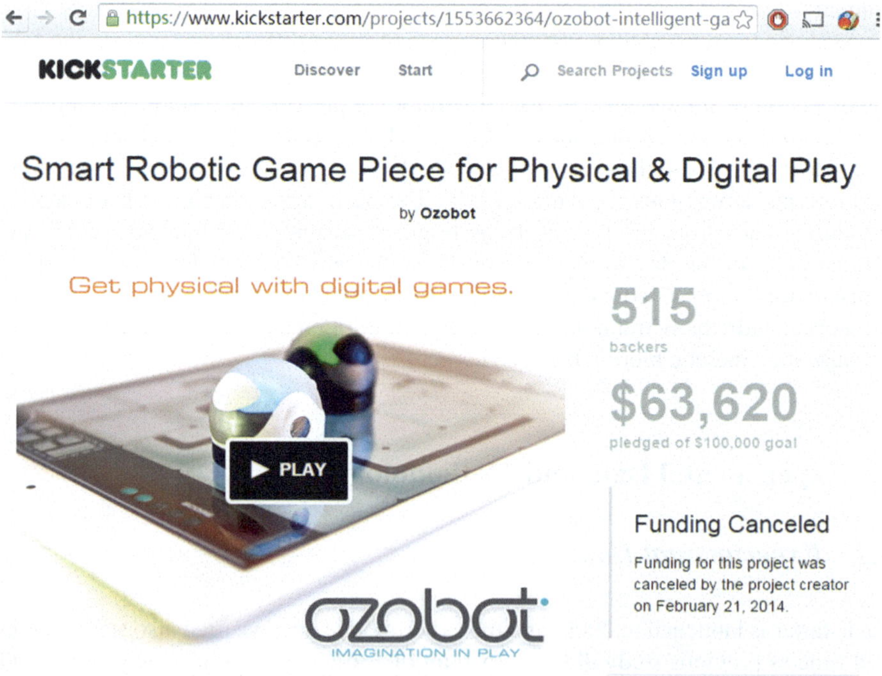

Fig. 1 An example of a project that has been canceled on Kickstarter

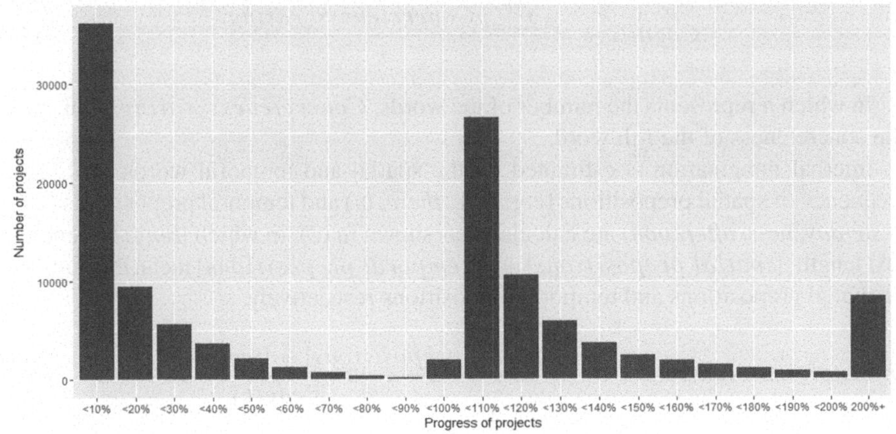

Fig. 2 The trend of the sample

After the above processing steps, the final experimental data includes 126,593 campaigns. According to the official data, the average success rate is about 40%, and the success rate in this study is about 52%. This difference comes from the fact that some projects cannot attract a supporter, so we cannot collect these projects. However, as a balanced data, a study on about half of the failed campaigns in the sample can reveal the impact of linguistic feature related to fraud on the project's fundraising outcomes; another reason for eliminating the cancelled projects without affecting the conclusions is that there are major problems in some aspects inevitable if a project was cancelled passively or actively, which may not be related to linguistic features. Retaining this part of the projects may have a negative impact on the results of text analytics.

Figure 2 shows the overall trend of the sample. Generally speaking, most projects have attracted less than 20% of the funding goal, and most successful projects have pledged ranging from 100 to 120% of the funding goal. The result shows the trend for investors to invest in projects: once a project is funded successful, they are less willing to invest in.

3.2 *Text Mining Model*

Machine learning and text mining provide methods for mining fraudulent cues from the text description [13]. Based on existing studies, cognitive load, internal imagination, dissociation, and negative emotion are employed to estimate the linguistic feature related to fraud [14].

Cognitive load is estimated by the mean of concreteness of all words [15]. Equation (1) shows the calculation for cognitive load.

$$Cognitive = \frac{\sum_{i=1}^{n} Concreteness\_rating_i}{n} \quad (1)$$

In which n represents the number of the words, $Concreteness\_rating_i$ indicates the concreteness of the i-th word.

Internal imagination is estimated by the spatial and temporal words [16]. The frequency of spatial prepositions (e.g. *here*, *there*, *in*) and temporal prepositions (e.g. *yesterday*, *now*, *afternoon*) are calculated as shown in (2), in which $|length|$ denotes text length; $|spatial\ prepositions|$ and $|temporal\ prepositions|$ mean the number of spatial prepositions and temporal prepositions respectively.

$$internal\_imagination = 1 - \frac{|spatial\ prepositions| + |temporal\ prepositions|}{|length|} \quad (2)$$

For dissociation, the fraudulent text often uses non-first person pronouns to distinguish the fictional figures from reality (e.g. *he*, *him*, and *her*). And due to the psychological and moral guilt from deception (lying), authors of fraudulent text need to release negative sentiment in writing. Thus, frequency of non-first person pronouns and sentiment are adopted to estimate the dissociation and negative emotion respectively.

Figure 3 shows the framework of the study, we use text mining to identify fraudulent cues in the textual description of crowdfunding projects, including the blurbs, detailed texts, and reward statements. Then, we obtain the impact of fraudulent cues on the fundraising outcomes using the econometric model.

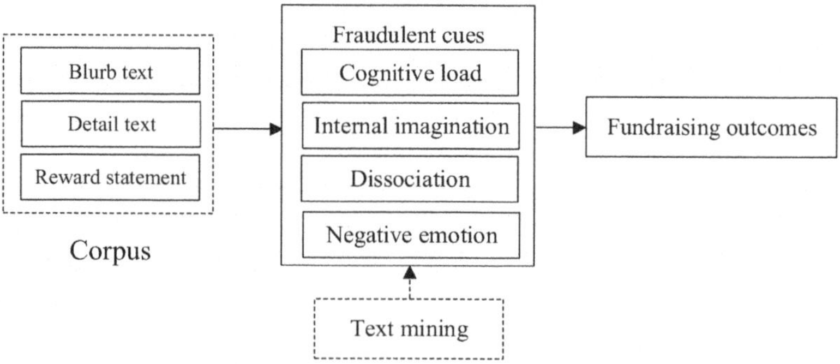

Fig. 3 Framework of the study

4 Results and Discussion

Table 1 shows the results of fraudulent cues detection. Firstly, in terms of cognitive load, the least concrete description is reward statement which suggests that founders tend to keep reward terms simple. And the text with the highest cohesion is the detailed description, since founders employ a great many conjunctions to improve the readability of such long narratives. In terms of internal imagination, there are many temporal words and spatial words in reward description, which makes it is the least imaginary narrative. But it should be noted that reward description is the most isolated and fragmented description. Finally, in terms of negative emotions, there is no significant difference among blurb, detailed description and reward statement.

Table 2 shows the impacts of linguistic cues to fraud on crowdfunding campaigns' fundraising outcomes. Firstly, in terms of cognitive load, concise blurb and detailed description promotes the success of a campaign, whereas reward terms should be clear about certain conditions and detail reward terms instead of concise description.

In terms of cohesion, the cohesion of blurb, detailed description and reward text promotes the success of a campaign significantly. Therefore, the cohesion of project description should be improved. In terms of internal imagination, imaginary blurb

Table 1 The text mining results for fraudulent cues detection

Text level	Aspect	Indicator	Min.	Max.	Median	Avg.	SD.
Blurb text	Cognitive load	Concreteness	0	5	2.639	2.656	0.305
		Cohesion	0	2.600	0.120	0.129	0.085
	Internal imagination	Internal imagination	0	3.000	0.087	0.094	0.072
	Dissociation	Dissociation	0	2.000	0.150	0.162	0.106
	Negative emotion	Negative emotion	0	2.000	0.000	0.012	0.032
Detailed text	Cognitive load	Concreteness	0	4.870	2.512	2.515	0.159
		Cohesion	0	4.000	0.144	0.144	0.034
	Internal imagination	Internal imagination	0	2.000	0.097	0.098	0.025
	Dissociation	Dissociation	0	2.000	0.181	0.182	0.041
	Negative emotion	Negative emotion	0	2.000	0.014	0.016	0.013
Reward text	Cognitive load	Concreteness	0	4.930	2.700	2.730	0.222
		Cohesion	0	0.667	0.122	0.122	0.044
	Internal imagination	Internal imagination	0	0.546	0.079	0.081	0.037
	Dissociation	Dissociation	0	0.690	0.190	0.189	0.061
	Negative emotion	Negative emotion	0	0.417	0.007	0.013	0.018

Table 2 The impacts of linguistic cues to fraud

Aspects	Indicator	Blurb level	Detail text	Reward text
Cognitive load	Concreteness	0.049***	0.075***	−0.030***
	Cohesion	0.098***	0.461***	0.218***
Internal imagination	Internal imagination	0.024	−0.206***	0.406***
Dissociation	Dissociation	0.071***	−0.345***	0.085***
Negative emotion	Negative emotion	−0.123**	−0.851***	−0.454

$*p < 0.1$, $**p < 0.01$, $***p < 0.001$

has no significant impact on the success of a campaign, imaginary detailed text has negative impact, while imaginary reward text has positive impact. The results may be contrary to common knowledge, but it makes sense in crowdfunding campaigns because it needs to specify reward terms with specific time and manner, hence, it contains a great amount of temporal words. In terms of dissociation, the result is like internal imagination, which shows that the dissociation of blurb and reward has positive impact on fundraising outcomes while that of the detailed text has negative influence. The result indicates that founders should employ more first-person pronouns rather than non-first-person pronouns. In terms of negative emotions, blurb and detailed text should express positive sentiments to attract investors. Besides that, negative emotions in reward text are not significantly correlated with the success of a campaign, which suggests that negative emotions in reward terms have little impact on investment intention.

5 Conclusions

The impact of linguistic cues to fraud on the successes of crowdfunding campaigns are analysed by lexical-based approaches in this study. The description of a crowdfunding campaign is divided into blurb, detailed description and reward statement. Through empirical study, the following main conclusion is obtained: (1) most linguistic cues to fraud are negatively correlated with success of fundraising outcomes, so it is necessary to lessen linguistic cues to fraud, (2) however, there are some linguistic cues to fraud have no significant impact on the successes of campaigns. Therefore, entrepreneurs should consider many aspects when creating text descriptions.

Although a preliminary study is conducted, there are some insufficiencies in this study. Further study will be conducted to tackle the following issues: (1) the dependent variable in the empirical model is the dummy variable, it is reasonable due to that Kickstarter adopts “Nothing-or-More” funding model. However, there are other funding models adopted by other platforms, for example, Rockethub employs a “All-and-More” funding model, which means founders can dominate the raised money regardless whether it reaches the pledge goal or not, thus dummy variable is not applicable in this scenario, (2) natural language processing develops rapidly

[17]. Recent text mining researches provide some new perspectives, such as keyword extraction, text format recognition and depth grammar recognition, such comprehensive approaches can be employed to analyze the text of crowdfunding campaigns in the future.

Acknowledgements This work is partially supported by the NSFC Grant (71601082), Natural Science Foundation of Fujian Province (2017J01132), Huaqiao University's High Level Talent Research Start Project Funding (16SKBS102) and Teaching development reform project for Huaqiao University teachers (17JF-JXGZ17), Ministry of Science & Technology, Taiwan (MOST 106-2511-S-003-029-MY3).

References

1. Herzenstein, M., Dholakia, U.M., Andrews, R.L.: Strategic herding behavior in peer-to-peer loan auctions. J. Interact. Mark. **25**(1), 27–36 (2011)
2. Nizamani, S., et al.: Detection of fraudulent emails by employing advanced feature abundance. Egypt. Inform. J. **15**(3), 169–174 (2014)
3. Bertrand, R., et al.: A system based on intrinsic features for fraudulent document detection. In: 12th International Conference on Document Analysis and Recognition (ICDAR). IEEE, Washington, DC, USA (2013)
4. Pennebaker, J.W., Mehl, M.R., Niederhoffer, K.G.: Psychological aspects of natural language use: our words, our selves. Annu. Rev. Psychol. **54**(1), 547–577 (2003)
5. Wells, J.T.: ... And nothing but the truth: uncovering fraudulent disclosures. J. Accountancy **192**(1), 47 (2001)
6. Viaene, S., et al.: Strategies for detecting fraudulent claims in the automobile insurance industry. Eur. J. Oper. Res. **176**(1), 565–583 (2007)
7. Mollick, E.: The dynamics of crowdfunding: an exploratory study. J. Bus. Ventur. **29**(1), 1–16 (2014)
8. Raimi, R.A.: The peculiar distribution of first digits. Sci. Am. **221**(6), 109–121 (1969)
9. Diekmann, A.: Not the first digit! Using Benford's Law to detect fraudulent scientific data. J. Appl. Stat. **34**(3), 321–329 (2007)
10. Leal, S., et al.: You cannot hide your telephone lies: Providing a model statement as an aid to detect deception in insurance telephone calls. Legal Criminological Psychol. **20**(1), 129–146 (2015)
11. Wang, W., et al.: The impact of sentiment orientations on successful crowdfunding campaigns through text analytics. IET Softw. **11**(5), 229–238 (2017)
12. Guo, L., et al.: Distance diffusion of home bias for crowdfunding campaigns between categories: insights from data analytics. Sustainability **10**(4), 1251 (2018)
13. Wang, W., Tan, G., Wang, H.: Cross-domain comparison of algorithm performance in extracting aspect-based opinions from Chinese online reviews. Int. J. Mach. Learn. Cybernet. **8**(3), 1053–1070 (2017)
14. Kirkos, E., Spathis, C., Manolopoulos, Y.: Data mining techniques for the detection of fraudulent financial statements. Expert Syst. Appl. **32**(4), 995–1003 (2007)
15. Seufert, T.: The interplay between self-regulation in learning and cognitive load. Educ. Res. Rev. **24**, 116–129 (2018)
16. Devitt, A.L., Addis, D.R., Schacter, D.L.: Episodic and semantic content of memory and imagination: a multilevel analysis. Mem. Cogn. **45**(7), 1078–1094 (2017)
17. Young, T., et al.: Recent trends in deep learning based natural language processing. IEEE Comput. Intell. Mag. **13**(3), 55–75 (2018)

Sustainable Economic Development: Some Reflections on Access to Technology as a Matter of Social Engagement

Isabel Novo-Corti, Xose Picatoste and Diana Mihaela Țîrcă

Abstract Technological advances have been crucial for economic development as well as for increasing social and individual welfare; however, it is not uncommon for imbalances and inequalities to appear simultaneously and may give rise to new social groups at risk of exclusion. In the age of technology, there is still the risk of technological exclusion and the digital divide is still present. The concern for sustainable development in its three aspects: environmental, economic and social has led to the search for socio-political environments that minimize the risk of exclusion. The objective of this paper is to analyze the access to technology as a social and human rights problem, which transcends individual needs to become a key issue in public policies. Primary data have been collected, through a survey with the objective of evaluating the opinion of citizens on the relevance of implementing public policies targeted to achieve technology universal access, as a central issue in public agendas. Different social groups' profiles have shown divergent opinions.

I. Novo-Corti (✉)
Universidade Da Coruña, 15071 A Coruña, Spain
e-mail: isabel.novo.corti@udc.es

Faculty of Economy & Business, Campus Elviña S/N, 15071 A Coruna, Spain

X. Picatoste
Department of Economics, University of A Coruna, Coruna, Spain
e-mail: xose92@gmail.com

Faculty of Economics and Business, Calle Francisco Tomás y Valiente 5, 28949 Canto Blanco, Spain

D. M. Țîrcă
Constantin Brancusi University, 210007 Târgu Jiu, Romania
e-mail: diana.mihaela.tirca@gmail.com

Faculty of Economics Sciences, Tineretului Street, no. 4, 210185 Târgu Jiu, Romania

A. Visvizi and M. D. Lytras (eds.), *Research & Innovation Forum 2019*,
Springer Proceedings in Complexity,
https://doi.org/10.1007/978-3-030-30809-4_43

1 Introduction

The concept of sustainable development is linked to the future from it is in its own nature [1]. Since the future is on the new generations, their own concerns have become a key issue for social and global sustainability.

The industrialization, globalization and urbanization economic process have originated a continuous migration from the countryside to the cities and the number of inhabitants of urban areas has been increasing whilst the rural population has being decreasing. From this perspective, the urban areas have to face the challenge of providing education, health services, housing, employment, etc. for making possible people's lives in a sustainable basis compatible with technological advancements and environmental preservation. The smart cities try to provide all this requirements.

Knowing the opinion of population about the importance of technology and environment could help to policymakers to drive adequate policies [2] to give response to their claims.

2 The Knowledge Society, the Digital Divide and Technology Access

The digital divide represents a difficulty for social sustainability [3], particularly for the vulnerable groups [4, 5]. The extent of the digital society and all the social improvements arisen from its implementation have put the knowledge society and the technology access at the core of the concerns about sustainable societies.

The Smart Cities play an important role for as promotor of social sustainability are closely linked to the necessity of break down the digital barriers as well as facilitate the access to digital society, from technical, educational and economic access perspectives. Two are the key variables in this field: the ICTs access and the perception of the importance of the Development of Knowledge Society for social development in a sustainable way.

3 Objectives

The main objective of this work is an exploratory analysis to identify the social awareness about the relation between social sustainability and ICTs in the context of Smart cities and Knowledge society. Taking account the key arguments pointed above, two are the main perspectives for facing this analysis: one from the perspective of access to technology (TA) and the other form the importance given to the framework of the Knowledge societies (KS) for balanced social development (BSD). Two are the research questions to answer:

– RQ1: What are the main factors explaining the importance of TA for BSD?
– RQ2: What are the main factors explaining the importance of KS for BSD?

4 Method and Data Analysis

The linear regression method is commonly accepted as adequate to explain the causal relationships in economics and social sciences [6, 7], so it was chosen to analyze the relationship between the variables in this model. The data were collected from a specific survey conducted at European level on November and December, 2018, by means of the snowball procedure, with the google drive support. The social networks were used to get the data. The objective public were the so called "millennials", and this survey was mainly asked to young people, under 30 and more than three hundred valid responses were gotten for the sample, where 186 come from women and 128 from man from different European countries.

The asked questions were assessed by a 5 points Likert scale, where respondents should inform about their agreement or disagreement with several statements about the main social issues stablished by the European Agenda 2030 in the context of the United Nations Sustainable Development Goals.

The applied methodology was a regression analysis was carried out by a "stepwise" procedure by means of the use of the support of IBM SPSS statistic package software, 21 version. To answer the RQ1 the dependent variable was "Importance given to the TA for SBD" and for answering the RQ2 the dependent variable was "Importance given to KS for SBD".

In this linear regression analysis can be considered as a "multiple linear regression", since more than one are the independent variables, in fact, ten variables were introduced as the explanatory ones for explaining the importance given to TA for SBD and nine for the importance given to KS for SBD. For the construction of the adequate equation is based on the selection of the variables one by one, that is, "step by step". The objective is to find among all the possible explanatory (independent) variables those that more and better explain the dependent variable without any of them being a linear combination of the rest. According to this procedure, three assumptions are implied: (a) in each step only that variable that meets entry criteria is introduced; (b) once introduced, in each step it is assessed if any of the variables meet exit criteria; and (c), in each step the goodness of fit of the data to the linear regression model is evaluated and the parameters of the model verified in said step are calculated. The action starts without any independent variable in the regression equation and the process concludes when there is no variable outside the equation that satisfies the selection criterion (it guarantees that the selected variables are significant) and/or the elimination criterion (which means a guarantee that a selected variable is not redundant) [8].

5 Results

Although the same database was the raw font, the explanatory variables were very different for each of the Research Questions.

5.1 *The Importance of Technology Access for Sustainable Balanced Development*

The stepwise regression analysis gave the results shown in Tables 1 and 2. The classification of those variables according the three pillars of sustainable development (for both TA and KS) are shown in Table 5.

Results in Table 1, as well as Table 3, are shown how the R^2 coefficient is increasing at the same time that the steps go forward because one more variable is introduced in the analysis (which is statistically significant). Since the added variables are introduced from the higher to the lower explanatory capacity it is possible to assess the convenience of considering the introduction one more independent variable or not.

The results in Table 1 reveal the great influence of the variables 1 and 2: "Education" and "Knowledge society and Technology" for the explanation of the "Importance given to the TA for SBD since the R^2 value achieves 0.286, that is to say, both of them are able to explain almost the thirty percent of the variability of the dependent variable, whilst the other eight additional variables rise up the value of R^2 in 0.12 points.

Nevertheless, for taking account of all relevant variables, all of them should be considered. As it is shown in Table 2, all of them are statistically significant, so all of them are relevant and should be well thought-out, particularly if the policymakers try to be very refined, smart and well targeted in their policies design, because sometimes

Table 1 Variables for explaining the TA and successive R^2 scores

Step	Introduced variables (dependent variable: TA)	R^2
1	Education	0.187
2	Knowledge society and technology	0.286
3	Demography	0.316
4	Economic development	0.336
5	End poverty in all its forms everywhere	0.350
6	Ensure sustainable consumption and production patterns	0.371
7	Social protection	0.386
8	Youth specific policies	0.393
9	Social values	0.401
10	Take urgent action to combat climate change and its impacts	0.407

Table 2 Regression coefficients

Independent variables	Non-standardized coefficients		Typified coefficients	t	Sig.
	β	Standard error	Beta		
(Constant)	0.065	0.240		0.270	0.787
Education	0.332	0.053	0.251	6.255	0.000
Knowledge society and Technology	0.233	0.039	0.247	5.952	0.000
Demography	0.103	0.031	0.123	3.351	0.001
Economic development	0.179	0.035	0.196	5.177	.000
End poverty in all its forms everywhere	−0.136	0.035	−0.158	−3.851	0.000
Ensure sustainable consumption and production patterns	0.171	0.039	0.178	4.419	0.000
Social Protection	0.119	0.037	0.129	3.238	0.001
Youth specific policies	−0.088	0.038	−0.100	−2.325	0.020
Social Values	0.118	0.043	0.114	2.767	0.006
Take urgent action to combat climate change and its impacts	−0.092	0.037	−0.102	−2.491	0.013

Variables for explaining the TA

Table 3 Variables for explaining the KS and successive R^2 scores

Step	Introduced variables (dependent variable: KS)	R^2
1	Youth specific policies	0.283
2	Technology access and Knowledge Society	0.379
3	Build resilient infrastructure, promote inclusive and sustainable industrialization and foster innovation	0.401
4	Inequality	0.419
5	Good governance, democracy, the rule of law and human rights	0.440
6	Health	0.445
7	Democracy	0.450
8	Gender issues	0.454
9	Equal opportunities and access to the labour market	0.460

one particular variable can improve the results more than it was initially expected due its influence on the other.

The specific characteristics of each situation should give a clue about how many variables could be taken as more relevant. Other important issue is the relation between the cost (not only from an economic perspective but also in terms of management, time, etc.) of taking account each of those last added variables and the expected influence of the action on that variable.

5.2 *The Importance of Knowledge Society for Sustainable Balanced Development*

The s stepwise regression analysis gave the results shown in Table 3. The "Youth specific policies" is the most explanatory variable for KS, since the asked group is constructed by "millennials", and its explanatory capacity goes near the thirty percent of the KS.

The results in Table 3 indicate that the most important variables are the first four, because from the fifth and onwards, the R^2 is increasing at a decreasing rate. The "youth specific policies" variable itself explains almost thirty percent of the variability of the dependent variable (Importance given to the KS for SBD) and the addition of the second variable "Technology access and Knowledge Society" increases this value to almost forty percent (37.9%). Although the other additional variables gives some explanatory value, they are less relevant since it is necessary the addition of 6 more variables to increase the R^2 in 0.06 points (see Table 4).

5.3 *TA, KS and the Three Pillars for Sustainable Development*

See Table 5.

6 Discussion and Conclusions

The interaction between the three pillars has showed up for both aspects: TA and KS. This was an expected result, since academic literature and empirical studies point to it. The most interesting result is the one revealing that there is only one independent variable shearing for explaining the two analyzed aspects: it is the "generalist" one related to "*Youth specific policies*".

This underlines established difference among TA and KS and the importance of analysing them in a separate way. Nevertheless the results indicates that youth are

Table 4 Regression coefficients

Independent variables	Non-standardized coefficients		Typified coefficients	t	Sig.
	β	Standard error	Beta		
(Constant)	0.464	0.217		2.135	0.033
Youth specific policies	0.366	0.034	0.392	10.696	0.000
Technology access and Knowledge Society	0.291	0.036	0.275	8.014	0.000
Build resilient infrastructure, promote inclusive and sustainable industrialization and foster innovation	0.189	0.038	0.184	4.952	0.000
Inequality	−0.163	0.039	−0.165	−4.188	0.000
Good governance, democracy, the rule of law and human rights	0.175	0.039	0.168	4.545	0.000
Health	0.112	0.042	0.108	2.695	0.007
Democracy	−0.096	0.044	−0.089	−2.199	0.028
Gender issues	−0.080	0.032	−0.097	−2.493	0.013
Equal opportunities and access to the labour market	0.095	0.039	0.094	2.412	0.016

Variables for explaining the KS

clamming for their own space into the public policies, revealing that they are aware of the specific differences of this group related to other social groups.

In addition, whilst this variable itself is able to explain 0.283 (see Table 3) percent of the variability of the dependent variable KS, its contribution to the explanation of the TA could be understood as the difference between R2 after and before its introduction in the analysis (see Table 1), that is to say $0.386 - 0.393 = 0.007\%$. From our point of view, this is because the TA is a wider variable, which should be taking account and solved previously to the KS. In fact, if there is not a whole access, from technical, economical and skills perspectives, the successful implementation and exploitation of KS for improving people's lives will not be possible. This is a widener result affecting all social groups, which goes beyond the specific youth policies.

Summarizing, the answer for our research questions "RQ1: Which are the main factors explaining the importance of TA for BSD?" and "RQ2: Which are the main factors explaining the importance of KS for BSD?" points out to the examination of

Table 5 Relation between TA, KS and the three pillars for sustainable development

Explaining factors (independent variables)	TECH perspective	Sustainability pillars			TECH
		EC	S	EN	
Education	TA1		X		
Knowledge society and technology	TA2				X
Demography	TA3		X		
Economic development	TA4	X			
End poverty in all its forms everywhere	TA5	X	X		
Ensure sustainable consumption and production patterns	TA6	X			
Social Protection	TA7		X		
Youth specific policies	*TA8*	*X*	*X*		
Social values	TA9		X		
Take urgent action to combat climate change and its impacts	TA10			X	
Youth specific policies	*KS1*	*X*	*X*		
Technology access and Knowledge Society	KS2	X	X		X
Build resilient infrastructure, promote inclusive and sustainable industrialization and foster innovation	KS3	X		X	X
Soc & EC: inequality	KS4	X	X		
Good governance, democracy, the rule of law and human rights	KS5		X		
Health	KS6		X		
Democracy	KS7		X		
Gender issues	KS8		X		
Equal opportunities and access to the labour market	KS9	X	X		

Note: EC = Economy, EN = Environment, S = Society, TECH = Technology

the three pillars of Sustainable Development and to the specific differences among these two aspects of Technology Access and the successful of Knowledge Society.

For this simple approximation to the proposed issues, KA and KS, the main conclusions are three. The first one is that all the three pillars of sustainable development are closely related to the Technology and all of them need to be taking account, particularly for analysing the possible effects of a new technology implementation. This is becoming more important as time goes by since the innovative process and its influence on the environment as well as social and economic life [9, 10] go faster and faster and it influence unbalanced to different countries and social and economic groups [11].

The second one is that the KS as a key factor for improving people's lives is strictly linked to the whole access to technology, and this is a previous stage to get the right framework for implementing the KS, and then the Smart Cities' goals and success [12]. The urban environment in the sustainable context of a modern smart city helps people to improve their lives by means of new opportunities.

And, finally, it is clear that each social group features and necessities should be analysed under their own specific characteristics and they need specific policies to address their main problems with lasting solutions.

References

1. WCED Brundtland Commission: Our common future. World Commission on Environment and Development, vol. 1 (1987)
2. Bongers, A., Díaz-Roldán, C.: Stabilization policies and technological shocks: towards a sustainable economic growth path. Sustainability **11**(1), 205 (2019). https://doi.org/10.3390/su11010205
3. Novo-Corti, I., Barreiro-Gen, M.: Public policies based on social networks for the introduction of technology at home: demographic and socioeconomic profiles of households. Comput. Hum. Behav. **51**, 1216–1228 (2015). https://doi.org/10.1016/j.chb.2014.12.040
4. Novo-Corti, I., Varela-Candamio, L., Teresa Garcia-Alvarez, M.: Breaking the walls of social exclusion of women rural by means of ICTs: the case of 'digital divides' in Galician. Comput Hum Behav. **30**, 497–507. https://doi.org/10.1016/j.chb.2013.06.017 (2014)
5. Novo-Corti, I., Barreiro-Gen, M.: Additional barriers to access to labour market for prisoners due to digital isolation. In: Kelly Chandler-Olcott, K.A.H. (eds.) Journal of Adolescent & Adult Literacy, pp. 391–396. Blackwell Publishing Ltd, Hoboken (2017). https://doi.org/10.1007/978-3-319-46068-0_51
6. Zhao, F., Collier, A., Deng, H.: A multidimensional and integrative approach to study global digital divide and e-government development. Inf. Technol. People **27**(1), 38–62 (2014). https://doi.org/10.1108/ITP-01-2013-0022
7. Perez-Estebanez, R., Urquía-Grande, E., Rautiainen, A.: Technological and economic factors determining ICT level: evidence from rural micro-businesses in democratic Republic of Congo. J. Int. Dev. **30**(1), 118–133 (2018). https://doi.org/10.1002/jid.3281
8. Rodríguez, M., Mora, R.: Estadística informática: casos y ejemplos con el SPSS. Publicaciones Universidad de Alicante, España (2001)
9. Saiprasad, N., Kalam, A., Zayegh, A.: Triple bottom line analysis and optimum sizing of renewable energy using improved hybrid optimization employing the genetic algorithm: a case study from India. Energies **12**(3), 349 (2019). https://doi.org/10.3390/en12030349
10. Dao, V., Langella, I., Carbo, J.: From green to sustainability: information technology and an integrated sustainability framework. J. Strateg. Inf. Syst. **20**(1), 63–79 (2011). https://doi.org/10.1016/j.jsis.2011.01.002
11. Picatoste, J., Pérez-Ortiz, L., Ruesga-Benito, S.M.: A new educational pattern in response to new technologies and sustainable development. Enlightening ICT skills for youth employability in the European Union. Telemat. Inform. **35**(4), 1031–1038 (2018). https://doi.org/10.1016/j.tele.2017.09.014
12. Bibri, S.E., Krogstie, J.: On the social shaping dimensions of smart sustainable cities: A study in science, technology, and society. Sustain. Cities Soc. **29**, 219–246 (2017). https://doi.org/10.1016/j.scs.2016.11.004

Using Crowdsourcing to Identify a Proxy of Socio-economic Status

Adil E. Rajput, Akila Sarirete and Tamer F. Desouky

Abstract Social Media provides researchers with an unprecedented opportunity to gain insight into various facets of human life. Researchers put a great emphasis on pinpointing socioeconomic status (SES) of individuals as they can use to it to predict numerous outcomes of interest. Crowdsourcing is a term coined that entails gathering intelligence from a user community online. In order to group online users into a common conversation, researchers have made use of hashtags that will label users and user content into tags that can be easily searched for. In this paper, we propose a mechanism to group a group of users based on their geographic background and build a corpus for such users. Specifically, we have looked at online discussion forums for commercial vehicles where the website has established forums for different geographic areas to share information, have discussions, and provide additional information about the vehicle of interest. From such a discussion, it was possible to glean the vocabulary that these group of users adhere to. We compared the corpus of different communities and noted the difference in the choice of language. This provided us with the groundwork for predicting a proxy of SES of such communities. More work is underway to take words and emojis out of vocabulary (OOV) and assessing the average score as special cases.

1 Introduction

Social media platforms provide an unprecedented opportunity for research in the social sciences and online behavior. It provides a forum for users to express their opinions in a disinhibited fashion as they need not disclose their real identity. While many social media platforms today do require the end users to confirm their real identity to a certain degree, the process is not guaranteed 100%. Furthermore, social media companies are bound by federal regulations to safeguard the real identity of

A. E. Rajput (✉) · A. Sarirete · T. F. Desouky
College of Engineering, Effat University, Jeddah, Kingdom of Saudi Arabia
e-mail: aiilahi@effatuniversity.edu.sa

A. Visvizi and M. D. Lytras (eds.), *Research & Innovation Forum 2019*,
Springer Proceedings in Complexity,
https://doi.org/10.1007/978-3-030-30809-4_44

the end user. The US presidential campaign of 2004 made the concept of online campaigning popular and propelled certain authors to study the effectiveness of such campaigns [29].

Researchers in social and medical sciences have started paying attention to looking at the plethora of data that is available to them. While the data available cannot be used to detect and diagnose the problems being faced by a certain individual per se, the data available can provide a basis for detecting various symptoms that can prove an indication for the onset of certain mental health issues [18]. The techniques developed in the Natural Language Processing (NLP) domain can prove invaluable in processing, segmenting and clustering the available text data per the various segmentation techniques needed by the social and medical sciences practitioners. The choice of corpus is one of the essential requirements to perform these series of steps. A corpus is defined as "a collection of naturally occurring text, chosen to characterize a state or variety of a language" [20]. Generally speaking, development of corpus entails looking at a text specific to the problem at hand and deriving keywords, bigrams and at times trigrams (two and three-word phrases) that are heavily used in a particular domain. As an example, authors in [25] argue for the need of establishing a corpus that would help mental health practitioners detect depression among users given a certain group. The authors looked at hashtag #depression on twitter and gleaned the keywords and established that such words were part of the vocabulary of depression patients [19]. Once such a corpus is established, researchers would look at a random text and predict with a certain assurance whether the words used by a given person show the same frequency as those in the corpus.

One of the factors that mental health practitioners and sociologists look at is the socio-economic status (SES) of a given individual. SES status is used to predict the potential issues the individual might face. As an example, Collins [6] discusses the comorbidity of SES and alcoholism. One of the key factors that determine the SES of an individual is the level of education. This in turn affects the writing style of an individual, which eventually could be used as an indicator of their level of education. The Flesch-Kincaid test [14] developed a formula that combines the following ratios of a text to determine the grade level of the text at hand, as follows:

1. The ratio of total words to total sentences in a given text
2. The ratio of total syllables to total words in a given text

The grade level obtained from the above formula described above correlates directly to the school grade level of the text at hand. For example, a grade level of 12 indicates that a student in grade level 12 of school could comprehend the text at hand while a grade level of 14 would indicate that the text is written at the level of a student in the second year of University.

We looked at a public online discussion forum where users who bought a particular car posted their thoughts and impressions. The discussion forum was segmented already by various regions in the USA and Canada (e.g., Northeast, South, Western Ontario etc.). We scraped the forum and ran various analyses to see whether individuals from different regions differ in Flesh-Kincaid grade level.

The paper will present the following:

1. Establish a corpus that represents the thoughts and expressions by users from different regions of the USA and Canada,
2. Compute the Flesh-Kincaid grade level score for the users and compute the average,
3. Compare the scores across the regions to see whether there is any difference.

2 Literature Review

SES is an important composite index used to measure the social standing of an individual or groups of individuals along socially constructed group identities or along geographic locations. It is typically composed of an individual's (1) education, (2) income, and (3) occupation [2, 3]. Education can measured by number of formal schooling years completed or along a Likert scale. Income, on the other hand can, be measured using objective data such as annual income or more subjective measures such as economic stress and ability to pay immediate bills [5]. Finally, occupation can be measured by job titles as well as standardized lists. A rank order of prestige is derived from opinion polls and other organizations. However, some researchers suggest alternative ways of measuring occupation as subjective in nature by allowing respondents to self-compare to others in a social hierarchy [7].

Numerous fields in the social sciences use SES as a significant predictor of important variables of interest. For example, SES has been shown to have a medium to strong relationship with academic achievement [22, 30] as well as causal effects on health [1]. For instance, the association between SES and obesity is well-established. An update [17] to the original review by Sobal and Stunkard [23] has shown consistent results. Interestingly, SES has even been shown to be a predictor of brain function (specifically language and executive functions) [13]. From studies looking at child development [4] to how minority groups experience and cope with stress [10], SES is an important variable to consider.

Some studies use measures of SES as a proxy for individual characteristics when data is incomplete, and the validity of doing so has been questioned [12]. An aggregate proxy (e.g., median household income) may inflate effects shown in statistical analysis [11, 24]. However, our study is unique in that we do not use SES as a proxy, but rather individual scores on a readability test as a reflection of educational attainment, which is a core factor in SES. By looking at a proxy measure of education via online posts, information can be collected to identify and perhaps target these individuals in a more appropriate and customized manner.

One method to look at education is by measuring indirect constructs associated with it. As previously stated, SES has an effect on the human brain by affecting language attainment and expression as well as core executive functions [13]. Therefore, a measure of language expression, via writing, can potentially reflect an individual's SES. One common method used by the United States Military [14] as well in word processing software is the Flesch-Kincaid Readability Tests [26]. The Flesch Reading Ease Test rates a text on a 100 point scale with higher scores indicating easier readability [9]. The Flesch-Kincaid Grade Level Test standardizes the score to the U.S. grade school levels [15]. For example, a score of 12.0 indicates that twelve grader can comprehend a given text. Some researchers have begun to use such measures in order to start adopting to the realm of big data [8]. Much research is being conducted on readability with some measures superior to the Flesch-Kincaid test [21]. However, even though progress has been made, some core problems still persist in terms of consistency [16, 27].

Crowdsourcing techniques, on the other hand, are used in various scenarios to help quickly collect information about large groups of people in social networks. Wazn [28] reviewed the definition of crowdsourcing, crowdsourcing taxonomies, crowdsourcing research, regulatory and ethical aspects, including some prominent examples of crowdsourcing. The author concluded that crowdsourcing has the potential to be extremely promising, in particular in healthcare as it has the ability to collect quickly and the information in a cost-effective way. In order to group users online into communities, researchers have made use of hashtags that will collect the interest of a community of users.

In this paper, we propose a mechanism to group a group of users based on their geographic background and build a corpus for such users. Specifically, we will look at discussion forums for commercial vehicles. From such a discussion, we can collect the vocabulary that these group of users adhere to. We compare the corpus of different communities and note the difference in the overall readability of their posts. Finally, we compare these groups with median household income to validate our approach.

3 Experimental Setup

3.1 Assumptions

To evaluate the effectiveness of our corpus for social media, we focus on a public discussion forum of a given vehicle. Specifically, we capitalize on the fact that the discussion forum has been set up with various regions of the USA and Canada. The following assumptions hold true:

1. There is no way to confirm whether the users are posting in the right region. For example, a user from Northeast USA can post to another region. Even though users report their respective cities, there is no way for the forum to confirm whether the data is accurate.
2. We focus on corpora in the English language only and hence our results will only apply to the English speaking community.
3. We make sure that we do not count a duplicate post.
4. The posts by various users differ in size and we do not normalize the size. Rather, we compute the Flesch-Kincaid measure.

3.2 *Data Sources and Data Gathering*

One of the biggest challenges when gathering data is ensuring the legality of using the data [31]. The data we gathered originates from a public source. Various vehicle manufacturers have forums on the net for their customers to discuss issues. Representatives of the company monitor these forums so that potential problems can be addressed effectively. We did not store any user credentials. Instead we anonymized them by giving fictitious identifiers. Moreover, we use the following open source packages:

1. *textstat package*[1]
2. *beautifulsoup package*[2]

3.3 *Pre-processing and Processing Data*

The following process was implemented for pre-processing and processing of data:

1. We used the built-in 'urllib' functionality of python to handle all url functionality;
2. We used the beautifulsoup package to scrape the data that is present in the forum;
3. We anonymized the user data to ensure that the privacy of individuals is not violated (even though the data is public);
4. We computed the Flesch-Kincaid grade level for each text within a forum;
5. We computed the average of each region and store it in the database.

We implemented the above on a standard Dell running Ubuntu Linux and Python3 program with a 16G RAM. Given that we did not have any performance requirements, the program can be ported to any platform that supports Python3.

4 Evaluation

After setting up the experiment, we ran the following experiments:

1. We computed the individual Flesch-Kincaid score for each post;
2. We computed the Flesch-Kincaid grade level for each post;
3. We tabulated the results and computed the average for each region along with the standard deviation using the textstat package (see footnote 1);
4. A difference of 1 on the results meant a whole school grade difference so it was deemed of practical significance.

---

[1] https://pypi.org/project/textstat/.

[2] https://pypi.org/project/beautifulsoup4.

5 Results and Discussion

We present the tabulated results below.

Note the following:

1. Given that the number of posts in various regions are different, we use both the average and standard deviation to compare the results. As we can see from Table 1 the Northeast, Mid-Atlantic, South, Midwest, and Canada-East regions have similar standard deviations. The total number of posts range from 22 to 101.
2. The actual data for each of the regions mentioned in 1 above along with Southwest regions contain 10–15% posts with a score of zero. Upon manual inspection, the authors noted that such posts either contained emojis or acronyms used on internet blogs such as "lol"—"laugh out loud".
3. While the Canada-West region shows the highest Flesch-Kincaid grade level, we ignore the results as the number of posts in that region are only 5.
4. The South and Northwest regions showed the highest average Flesch-Kincaid grade levels of 8.5 and 7.5 respectively. The Southwest region was not that behind with an average score of 7.3.
5. The Northeast, Mid-Atlantic. Midwest and Canada-East regions have similar average Flesch-Kincaid grade levels around 6.5.

The South and Southwest regions discussed in point 4 are characterized by wages that are less compared to the regions discussed in 5. Thus, people posting in regions discussed in points 4 show a writing level much less than this discussed in 4 and

Table 1 Flesch-Kincaid grade level for each of the North American regions

Region	Real median family income (Personal Income) 2017[a]	Total number of posts	Average Flesch Kincaid score	Standard deviation
Northeast	$83,641	66	6.62	3.07
Mid Atlantic	($33,116)	101	6.51	3.60
South	$69,278 ($30,107)	22	8.41	3.46
Midwest	$78,118 ($31,428)	92	6.70	3.44
Southwest	–	41	7.32	4.47
Northwest	–	12	7.50	2.71
West	$78,346 ($32,365)	36	5.58	2.47
Canada-East	–	56	6.50	3.09
Canada-West	–	5	12.40	2.88

[a]Census Regions (Northeast, South, Midwest, and West) https://fred.stlouisfed.org/categories/32043

combining it with the average income could give us valuable insights into the SES of the people buying this vehicle.

6 Conclusion and Future Work

In this paper, our goal was to establish a framework through which we can start gathering information that can give researchers clues about the SES status of the people posting on various sites including social media. Such status can be helpful to medical and social scientists, practitioners and researchers. We established a corpus using the posts on a forum that was designed for a particular vehicle. The vehicle manufacturer divided the forum into six regions into the US and two regions in Canada. After gathering the posts, we computed the average Flesch-Kincaid grade level for each of the regions and compared the regions. We found that the South and Southwest regions showed the highest average grade level (we ignored the Canada-West region as it only had 5 posts).

We intend to do the following in the future:

1. Refine the work by considering Out of Vocabulary (OOV) words and emojis and evaluating the average score by handling them as special cases;
2. Looking at other forums that have a huge number of posts;
3. Computing a similar score for Twitter posts.

References

1. Adler, N.E., Ostrove, J.M.: Socioeconomic status and health: what we know and what we don't. Ann. N. Y. Acad. Sci. **896**(1), 3–15 (1999)
2. American Psychological Association: Measuring socioeconomic status and subjective social status. Public Interest Directorate, Socioeconomic Status Office, Resources and Publication (2016)
3. Baker, E.H.: Socioeconomic Status, Definition. The Wiley Blackwell Encyclopedia of Health, Illness, Behavior, and Society, 2210–2214. https://doi.org/10.1002/9781118410868.wbehibs395 (2014)
4. Bradley, R.H., Corwyn, R.F.: Socioeconomic status and child development. Annu. Rev. Psychol. **53**(1), 371–399 (2002)
5. Chen, E., Paterson, L.Q.: Neighborhood, family, and subjective socioeconomic status: how do they relate to adolescent health? Health Psychol. **25**(6), 704 (2006)
6. Collins, S.E.: Associations between socioeconomic factors and alcohol outcomes. Alcohol Res. Curr. Rev. **38**(1), 83–94 (2016)
7. Diemer, M.A., Mistry, R.S., Wadsworth, M.E., López, I., Reimers, F.: Best practices in conceptualizing and measuring social class in psychological research. Anal. Soc. Issues Public Policy **13**(1), 77–113 (2013)
8. Flaounas, I., Ali, O., Lansdall-Welfare, T., De Bie, T., Mosdell, N., Lewis, J., Cristianini, N.: Research methods in the age of digital journalism: massive-scale automated analysis of news-content—topics, style and gender. Digital J. **1**(1), 102–116 (2013)

9. Flesch, R.: A new readability yardstick. J. Appl. Psychol. **32**(3), 221–233 (1948). https://doi.org/10.1037/h0057532
10. Gayman, M.D., Cislo, A.M., Goidel, A.R., Ueno, K.: SES and race-ethnic differences in the stress-buffering effects of coping resources among young adults. Ethn. Health **19**(2), 198–216 (2014)
11. Geronimus, A.T., Bound, J.: Use of census-based aggregate variables to proxy for socioeconomic group: evidence from national samples. Am. J. Epidemiol. **148**(5), 475–486 (1998)
12. Geronimus, A.T., Bound, J., Neidert, L.J.: On the validity of using census geocode characteristics to proxy individual socioeconomic characteristics. J. Am. Stat. Assoc. **91**(434), 529–537 (1996)
13. Hackman, D.A., Farah, M.J.: Socioeconomic status and the developing brain. Trends Cogn. Sci. **13**(2), 65–73 (2009)
14. Kincaid, J.P., Fishburne Jr, R.P., Rogers, R.L., Chissom, B.S.: Derivation of new readability formulas (automated readability index, fog count and Flesch reading ease formula) for navy enlisted personnel (1975)
15. Kincaid, J.P., Fishburne, R.P., Rogers, R.L., Chissom, B.S.: Derivation of new readability formulas (automated readability index, fog count, and flesch reading ease formula) for Navy enlisted personnel. Research Branch Report 8–75. Chief of Naval Technical Training: Naval Air Station Memphis (1975)
16. Mailloux, S.L., Johnson, M.E., Fisher, D.G., Pettibone, T.J.: How reliable is computerized assessment of readability? Comput. Nurs. **13**, 221 (1995)
17. McLaren, L.: Socioeconomic status and obesity. Epidemiol. Rev. **29**(1), 29–48 (2007)
18. Rajput, A., Ahmed, S.: Big Data and Social/Medical Sciences: State of the Art and Future Trends. IACHS 2018 Available at arXiv preprint arXiv:1902.00705 (2018)
19. Rajput, A., Ahmed, S.: Making a case for Social Media Corpus to detect Depression. IACHSS 2018 Available at arXiv preprint arXiv:1902.00702 (2018)
20. Schvaneveldt, R.W., Meyer, D.E., Becker, C.A.: Lexical ambiguity, semantic context, and visual word recognition. J. Exp. Psychol.: Hum. Percept. Perform. **2**(2), 243 (1976)
21. Si, L., Callan, J.: A statistical model for scientific readability. In: Proceedings of the Tenth International Conference on Information and Knowledge Management, pp. 574–576. ACM, New York (2001, October)
22. Sirin, S.R.: Socioeconomic status and academic achievement: a meta-analytic review of research. Rev. Educ. Res. **75**(3), 417–453 (2005)
23. Sobal, J., Stunkard, A.J.: Socioeconomic status and obesity: a review of the literature. Psychol. Bull. **105**(2), 260 (1989)
24. Soobader, M.J., LeClere, F.B., Hadden, W., Maury, B.: Using aggregate geographic data to proxy individual socioeconomic status: does size matter? Am. J. Public Health **91**(4), 632 (2001)
25. Sparck Jones, K.: A statistical interpretation of term specificity and its application in retrieval. J. Doc. **28**(1), 11–21 (1972)
26. Stockmeyer, N.O.: Using Microsoft Word's readability program. Mich. Bar J. **88**, 46 (2009)
27. Wang, L.W., Miller, M.J., Schmitt, M.R., Wen, F.K.: Assessing readability formula differences with written health information materials: application, results, and recommendations. Res. Soc. Adm. Pharm. **9**(5), 503–516 (2013)
28. Wazny, K.: "Crowdsourcing" ten years: a review. J. Glob. Health **7**(2), 020602 (2017)
29. Weinberg, B.D., Williams, C.B.: The 2004 US Presidential campaign: impact of hybrid offline and online 'meetup' communities. J. Direct Data Digit. Mark. Pract. **8**(1), 46–57 (2006)
30. White, K.R.: The relation between socioeconomic status and academic achievement. Psychol. Bull. **91**(3), 461 (1982)
31. Youyou, W., Kosinski, M., Stillwell, D.: Computer-based personality judgments are more accurate than those made by humans. Proc. Natl. Acad. Sci. **112**(4), 1036–1040 (2015)

IoT-IaaS: A New Public Cloud Service to Unleash Innovation in the IoT Space

Muhammad U. Ilyas, Muhammad Murtaza Khan, Sajid Saleem and Jalal S. Alowibdi

Abstract IoT solution developers today are hampered by the regulatory permit burdens and high up-front deployment and maintenance costs, particularly in the public domain. Essentially, it keeps the development of IoT solutions requiring large deployments in the purview of deep-pocketed players, to the exclusion of millions of small developers who may have figured out novel solutions to problems, but lack the resources to pilot them. This is akin to the situation of cellphones before the release of SDKs for Apple iOS and Android smartphones; phones with relatively capable hardware were often restricted to running the dozen-or-so apps they shipped with. Together, smartphone SDKs and public cloud services, which were becoming available at around the same time, less well-resourced developers were able to bring their ideas for apps to the public, thus launching the smartphone revolution. In this paper we conducted a survey of studies that describe various pieces of the technology puzzle that could contribute towards democratizing access to public IoT resources.

M. U. Ilyas (✉) · S. Saleem
Department of Computer and Network Engineering, College of Computer Science and Engineering, University of Jeddah, Jeddah 23890, Saudi Arabia
e-mail: milyas@uj.edu.sa; usman.ilyas@seecs.edu.pk
URL: http://seecs.nust.edu.pk/

S. Saleem
e-mail: sssaleem@uj.edu.sa; sajid.saleem@seecs.edu.pk
URL: http://seecs.nust.edu.pk/

M. M. Khan · J. S. Alowibdi
Department of Computer Science and Artificial Intelligence, College of Computer Science and Engineering, University of Jeddah, Jeddah 23890, Saudi Arabia
e-mail: mkhan@uj.edu.sa; muhammad.murtaza@seecs.edu.pk
URL: http://www.uj.edu.sa/; http://seecs.nust.edu.pk/

J. S. Alowibdi
e-mail: jalowibdi@uj.edu.sa
URL: http://www.uj.edu.sa/

M. U. Ilyas · M. M. Khan · S. Saleem
School of Electrical Engineering and Computer Science, National University of Sciences and Technology, Islamabad 44000, Pakistan

A. Visvizi and M. D. Lytras (eds.), *Research & Innovation Forum 2019*, Springer Proceedings in Complexity,
https://doi.org/10.1007/978-3-030-30809-4_45

1 Introduction

The National Institute of Standards and Technology formally defines cloud computing as "*a model for enabling ubiquitous, convenient, on-demand network access to a shared pool of configurable computing resources* (e.g., *networks, servers, storage, applications, and services) that can be rapidly provisioned and released with minimal management effort or service provider interaction,*" [12, 14, 18].

Cloud computing grew out of the early concept of grid computing, the idea that computing services should be available like any other utility in our daily lives, i.e. ubiquitously, on-demand, and in virtually unlimited quantity.

Presently, there are three major service models in cloud computing:

The **Infrastructure-as-a-Service** (IaaS) model provides users scalable, on-demands access to virtual computing infrastructure including networks, servers and storage.

The **Platform-as-a-Service** (PaaS) model provides users with on-demand access to software development tools, libraries and services while freeing consumers from the management of infrastructure resources (operating systems, hardware, configuration, etc.) on which they are hosted.

The **Software-as-a-Service** (SaaS) model provides users access to applications hosted on cloud infrastructure, thereby freeing them from managing applications or the cloud infrastructure they are hosted on.

Public cloud service models gave developers with limited resources the ability to build and run highly available and scalable services and applications without incurring prohibitive capital expenditure (CapEx) necessary to own required hardware and software resources.

In this paper we describe a new cloud service model that we call *Internet-of-Things Infrastructure-as-a-Service* (IoT-IaaS), whose time may have come with growing IoT application deployments and the imminent deployment of 5G mobile networks that will give these further impetus. Software Development Kits (SDKs) for smartphone operating systems took app development out of the exclusive domain of phone manufacturers and put them in the hands of millions of developers across the world.

Wide area and metropolitan area scale IoT applications have high deployment costs and require scaling significant regulatory hurdles that keep IoT solution development in the realm of public institutions and resourceful private sector organizations. Small developers, the kind that ushered in the era of smart-phone apps, are presently still unable to translate their ideas for large scale systems into prototypes.

In this paper we examine the various ways by which access to IoT infrastructure in the public domain could be made more democratic.

Opening up otherwise inaccessible IoT-infrastructure to a large number of creative developers, in the same way that smartphone SDKs and cloud computing made app development and service scalability accessible to everyone, has the promise of unleashing an order of magnitude larger wave of innovation in the IoT space. We also summarize various recent proposals and architectures that provide building blocks by which this might be achieved.

Paper Organization: Section 2 covers the major proprietary solutions on offer by industry at this point with regards to IoT-cloud integration. Section 3 describes OpenFog, an industry-academia consortium that has developed a broad, open architecture for fog computing systems. Section 4 discusses recently proposed clean-slate architecture approaches. Section 5 discusses recently proposed approaches that are mindful of the existing eco-system of public clouds and IoT. Section 6 concludes the paper.

2 Proprietary Solutions

Fog computing [3, 13] or edge computing [17] pushes out computational resources from the data center at the core out to the edge of the network, where devices interacting with the physical world may be controlled or data may be collected from using sensors.

It has the advantage of leveraging underutilized compute resources at the edge, reduce latency between data acquisition and processing, and/or reduce the need for bandwidth necessary to transmit high volume data back all the way to the network core.

Cisco Systems is among network equipment vendors that offers fog computing solutions [19]. Cisco considers fog computing an extension of the cloud computing paradigm, which extends the core of the network to the edge of the network.

However, Cisco's fog computing solution is proprietary.

Amazon Web Services (AWS) Greengrass [4] is another proprietary fog/edge computing solution offered by a cloud service provider. Greengrass allows customers to connect their own devices to VM/container instances hosted on AWS, while using edge nodes for sensing and/or compute functions. However, customers need to deploy their own IoT infrastructure, and there is no sharing of that infrastructure by multiple tenants. Edge node devices must run either Amazon FreeRTOS or IoT Device SDK in order to connect to AWS Greengrass services.

3 OpenFog: The Open Alternative

On the other hand, OpenFog is an industry-academia consortium that describes its mission as "*To drive industry and academic leadership in fog computing and networking architecture, testbed development, and interoperability and composability deliverables that seamlessly bridge the cloud-to-things continuum.*"

The OpenFog Architecture Workgroup developed a reference architecture [5] in which it defines fog computing as "*A horizontal, system-level architecture that distributes computing, storage, control and networking functions closer to the users along a cloud-to-thing continuum.*"

A key difference between OpenFog's and AWS's fog architectures is that the former is open, whereas the latter is proprietary.

OpenFog's reference architecture allows for a great deal of flexibility.

It allows for varying degrees to which the fog may be extended outward from the data center, as well as a single or multiple tenants using edge devices, various virtualization technologies, mobile or wired sensors, etc.

4 Clean-Slate Architecture Approaches

In this section we identify proposals that integrate IoT into cloud data centers but do not consider any pre-existing legacy of relevant open or closed projects, i.e., clean-slate architecture approaches.

Renner et al. [16] proposed the *Device-cloud.* The Device-cloud is middleware that provides the ability to share IoT devices in what the authors describe as a hybrid IaaS-PaaS mashup. They also developed a simple proof-of-concept implementation, but the level of detail provided is sparse. However, the proposed Device-cloud architecture is still limited in its features because it lists multitenancy as an untackled challenge, as well as others like QoS and service execution environment. Their work also appears to be largely unaware of the existence of OpenStack.

Kliem and Kao [11] proposed a similarly named *Device Cloud.* Their Device Cloud of sensors and devices integrates IoT devices with the cloud. Physical IoT resources become an integral part of the cloud resource pool and are shared and provisioned like other IaaS cloud resources. However, their approach relies on the use of gateways which in turn connect to IoT devices. The IoT devices in their Device Cloud do not support multitenancy or edge computing capabilities. However, they did not report any reference implementation or results to demonstrate the Device Cloud architecture.

Gupta and Mukherjee's proposal [6, 7] is also sparse on details. Like Khan et al. [10] they too share sensors among multiple users by representing physical sensors as virtual sensors through an API that application developers can use in their solutions. The approach is alternatively described as IaaS and SaaS. The IoT devices they used were running tinyOS and simply respond to requests for sensor data. Therefore, they do not support application deployment on the IoT devices or multitenancy.

Khan et al. [9, 10] proposed the virtualized Wireless Sensor Networks which offers users WSN IaaS. Node sharing either sequentially (via round robin sharing) or simultaneously (via context switching). They propose the use of a middleware layer that virtualizes physical sensors attached to IoT devices as virtual sensors. Data from those virtual sensors is made accessible to all users. Physical sensors attached to IoT devices are not directly addressable and are accessible indirectly through an IP-enabled gateway.

Hammoudi et al. [8], proposed a data storage service that leverages the spare storage capacity available in IoT devices connected to a cloud. The organization is quite straightforward: IoT devices are grouped into clusters, each managed by an assigned cluster head node, according to the file types they will store. Files are replicated across devices in the same cluster for redundancy.

5 Evolutionary Approaches

OpenStack and OpenFog are two open source projects most relevant to an IoT-IaaS cloud service model. However, among the studies that have proposed the integration of IoTs into cloud computing there are relatively few that acknowledge these.

Merlino et al. [15] built their solution using OpenStack components. They made the case for a more comprehensive solution that goes beyond connecting to things/devices that feature a network interface. They made the case that "sensing and actuation devices should be handled along the same lines as computing and storage abstractions in traditional Clouds, i.e., virtualized and multiplexed over (scarce) hardware resources." Their prototype implementation used Arduino boards with Linux as stand-ins for IoT devices. However, their prototype does not offer multitenancy using VMs or containers on IoT devices. In a sense, the inability to run applications or wholly or partially on IoT devices means the resulting solution is very different from using IoT data from a data service provider providing access to a live data stream. Merlino et al. proposed the use of of AMQP to transfer data from Arduino boards to compute nodes. This means the flow of information is one way only (sensing) and there may not be support for having actuators. Moreover, the applications appear to be limited to low bandwidth sensing applications.

Alam et al. [2] proposed an evolutionary solution and used an OpenStack cloud deployment together with a Cloud Foundry to demonstrate two applications sharing IoT infrastructure. It is similar to many of the above proposals in the sense that it too describes a middleware layer that virtualizes physical IoT devices. However, it departs from them because it goes beyond applications that use IoT devices for sensing-only tasks, but also considers IoT devices with actuators like drones that require a reverse path. The proposal is quite detailed and also describes the mechanism by which IoT devices' capabilities will be published. The IoT devices used in the prototype implementation included Advanticsys Sky Tmote sensors running Contiki OS and a Virtenio Preon32 running JVM based Virtenio OS.

In our earlier work Ahmad et al. [1] we proposed an evolutionary approach that is the greatest departure from all preceding proposals. We developed the virtualized-IoT (vIoT) testbed that demonstrates IoT infrastructure integration with an OpenStack deployment. Unlike all the above proposals, however, it does not build its own middleware or require the use of gateway nodes to serve as intermediaries between the data center and the IoT devices. Ahmad et al. [1] used Raspberry Pi 2 and 3 computers as stand-ins for IoT devices. All Pi computers were running Ubuntu OS, OpenStack Nova and LXD containers. The deployment of OpenStack Nova directly on the Pi computers makes them a logical part of the data center resources. Furthermore, the deployment of LXD containers enables multitenancy on each Pi, i.e., multiple users

Table 1 Matrix of features supported by IoT-cloud integration proposals

Study	Cleanslate Architecture	Edge Computing	Edge OS	Multitenancy	Sensor virtualization
Renner et al. [16]	✓	X		X	X
Kliem and Kao [11]	✓	X		X	X
Gupta and Mukherjee [6, 7]	✓	X	tinyOS	X	✓
Khan et al. [10]	✓	X		X	✓
Hammoudi et al. [8]	✓	X		X	N/A
Merlino et al. [15]	OpenStack	X	Arduino	X	✓
Alam et al. [2]	OpenStack	X	Contiki OS, JVM/ Virtenio OS	X	✓
Ahmad et al. [1]	OpenStack, LXD	✓	Ubuntu	✓	✓

can deploy parts of their applications at the very edge (the fog region) of the cloud. Table 1 gives a summary of all prior approaches.

6 Conclusion

In this paper we motivated the integration of IoT systems into cloud computing and the introduction of a new IoT-IaaS service model for cloud computing. In particular, we reviewed the state-of-the-art in various efforts that offer promising pieces to solving this puzzle. We broadly categorized works into two categories; those taking a clean-slate approach independent of pre-existing cloud computing deployments and evolutionary approaches that are integrated with relevant pre-existing open or closed sourced projects, e.g., OpenStack, Cloud Foundry, LXD containers, etc. With the exception of Ahmad et al., all proposed solutions use intermediary gateways to connect IoT devices to the cloud and use some form of sensor virtualization to share sensors for multiple users. Ahmad et al. [1] stands out as the only solution that truly integrates the IoT end-devices into the cloud infrastructure by configuring them as Nova compute nodes of the OpenStack cloud, capable of hosting a number of LXD containers. This way, their approach is the only one that allows users to deploy and execute a part of their application on the IoT device itself in true fog/ edge computing fashion.

References

1. Ahmad, M., Alowibdi, J.S., Ilyas, M.U.: viot: A first step towards a shared, multi-tenant iot infrastructure architecture. In: 2017 IEEE International Conference on Communications Workshops (ICC Workshops), pp. 308–313. IEEE (2017)
2. Alam, M.N., Glitho, R.H.: An infrastructure as a service for the internet of things. In: 2018 IEEE 7th International Conference on Cloud Networking (CloudNet), pp. 1–7. IEEE (2018)
3. Bonomi, F., Milito, R., Zhu, J., Addepalli, S.: Fog computing and its role in the internet of things. In: Proceedings of the First Edition of the MCC Workshop on Mobile Cloud Computing, pp. 13–16. ACM (2012)
4. Greengrass, A.: https://aws.amazon.com/greengrass/, https://aws.amazon.com/greengrass/
5. Group, O.C.A.W., et al.: Openfog reference architecture for fog computing. OPFRA001 **20817**, 162 (2017)
6. Gupta, A., Mukherjee, N.: Implementation of virtual sensors for building a sensor-cloud environment. In: COMSNETS, pp. 1–8 (2016)
7. Gupta, A., Mukherjee, N.: Can the challenges of IOT be overcome by virtual sensors. In: 2017 IEEE International Conference on Internet of Things (iThings) and IEEE Green Computing and Communications (GreenCom) and IEEE Cyber, Physical and Social Computing (CPSCom) and IEEE Smart Data (SmartData), pp. 584–590. IEEE (2017)
8. Hammoudi, S., Aliouat, Z., Harous, S.: A new infrastructure as a service for IOT-cloud. In: 2018 14th International Wireless Communications and Mobile Computing Conference (IWCMC), pp. 786–792. IEEE (2018)
9. Khan, I., Belqasmi, F., Glitho, R., Crespi, N., Morrow, M., Polakos, P.: Wireless sensor network virtualization: early architecture and research perspectives. IEEE Netw. **29**(3), 104–112 (2015)
10. Khan, I., Errounda, F.Z., Yangui, S., Glitho, R., Crespi, N.: Getting virtualized wireless sensor networks' iaas ready for paas. In: 2015 International Conference on Distributed Computing in Sensor Systems (DCOSS), pp. 224–229. IEEE (2015)
11. Kliem, A., Kao, O.: The internet of things resource management challenge. In: 2015 IEEE International Conference on Data Science and Data Intensive Systems (DSDIS), pp. 483–490. IEEE (2015)
12. Liu, F., Tong, J., Mao, J., Bohn, R., Messina, J., Badger, L., Leaf, D.: Nist cloud computing reference architecture. NIST Spec. Publ. **500**, 292 (2011)
13. Mahmud, R., Kotagiri, R., Buyya, R.: Fog computing: a taxonomy, survey and future directions. In: Internet of Everything, pp. 103–130. Springer (2018)
14. Mell, P., Grance, T., et al.: The nist definition of cloud computing (2011)
15. Merlino, G., Bruneo, D., Distefano, S., Longo, F., Puliafito, A.: Stack4things: Integrating IOT with openstack in a smart city context. In: 2014 International Conference on Smart Computing Workshops (SMARTCOMP Workshops), pp. 21–28. IEEE (2014)
16. Renner, T., Kliem, A., Kao, O.: The device cloud-applying cloud computing concepts to the internet of things. In: 2014 IEEE 11th International Conference on Ubiquitous Intelligence and Computing, and 2014 IEEE 11th International Conference on Autonomic and Trusted Computing, and IEEE 14th International Conference on Scalable Computing and Communications and Its Associated Workshops (UTC-ATC-ScalCom), pp. 396–401. IEEE (2014)
17. Shi, W., Cao, J., Zhang, Q., Li, Y., Xu, L.: Edge computing: vision and challenges. IEEE Internet Things J. **3**(5), 637–646 (2016)
18. Subashini, S., Kavitha, V.: A survey on security issues in service delivery models of cloud computing. J. Netw. Comput. Appl. **1**(34), 1–11 (2011)
19. Systems Inc., C.: Fog Computing and the Internet of Things: Extend the Cloud to Where the Things Are, https://www.cisco.com/c/dam/en_us/solutions/trends/iot/docs/computing-overview.pdf

The Climb to Success: A Big Data Analysis to Find Out Why Huawei Has Conquered the Market

Orlando Troisi, Mara Grimaldi, Francesca Loia and Gennaro Maione

Abstract Today's electronic devices are in common use for most people, who use them to perform a multitude of activities. Technical change and new product proliferation have made this industry extremely dynamic. The work aims to highlight the reasons why millions of consumers prefer the purchase and use of products offered by Huawei, one of the most important companies in the sector. In this regard, a Big Data-oriented approach is followed. In particular, social media analytics is applied on Twitter by taking into consideration people's opinions expressed about Huawei over a five months period. Overall, over one million tweets were selected, collected and analyzed. In particular, social media analytics are applied on Twitter by taking into consideration people's opinions expressed about Huawei over a five months period. Overall, over one million tweets were selected, collected and analyzed. Afterward the collected data were subjected to a sentiment analysis, a word cloud analysis and a cluster analysis. Results show the existence of numerous aspects able to affect the smartphone consumers' behavior. In fact, in addition to traditional factors (such as price, value for money and so on) it is surprising that many consumers choose Huawei based on additional elements, such as design, quality, and brand loyalty.

O. Troisi · M. Grimaldi (✉) · G. Maione
University of Salerno, Fisciano 84084, ITA, Italy
e-mail: margrimaldi@unisa.it

O. Troisi
e-mail: otroisi@unisa.it

G. Maione
e-mail: gmaione@unisa.it

F. Loia
University of Rome "La Sapienza", Rome 00185, ITA, Italy
e-mail: francesca.loia@uniroma1.it

A. Visvizi and M. D. Lytras (eds.), *Research & Innovation Forum 2019*,
Springer Proceedings in Complexity,
https://doi.org/10.1007/978-3-030-30809-4_46

1 Introduction

For just over a decade, the vast ecosystem of technology has undergone a radical transformation, characterized by a rapidity of innovation unprecedented in the recent history, marking the contemporary and future evolution of society in all its aspects [28, 32, 59]. This change was determined by various factors but above all by the spread of electronic devices (especially smartphones, tablets, laptops, smartwatches, etc.) able to satisfy any kind of need. Nowadays, in fact, there is no person who does not use or, at least, know what these electronic devices are and how they are made.

Today's electronic devices are in common use for most people, who use them to perform a multitude of activities. Specifically, smartphones, tablets and laptops can be employed for recreational reasons (for instance, to take pictures or record videos, connect to social networks, send instant messages to friends and relatives, listen to music, etc.), for motives related to study or work (such as to write or read documents, consult e-mails, organize or join a videoconference, etc.), to move easily (using the device as a satellite navigator), to search for a restaurant or hotel for a certain period of time, to book a trip by plane or train) and so forth [31, 29, 60].

Thus, thanks to the possibility offered by new technologies (enhanced connectivity, faster and immediate interactions, and easier purchase modalities) the pervasive use of mobile devices redefines users' lives and, consequently, their purchasing behavior. On the other hand, companies have the opportunity to monitor and collect users' feedback 24 h per day to understand their opinions and/or predict their behaviors.

Despite the significant impact of Big Data on contemporary marketing evolution, extant research does not employ big data analysis and analytics to shed light on customer decision-making processes and behaviors [16]. Literature on big data analysis and consumer behavior can be integrated to explore how the new technologies can help companies to observe consumer's attitude, intentions and behaviors.

In this scenario, the work aims to provide an empirical evidence of the main reasons of the success of a well-known company operating in the development, production and marketing of electronic products, digital systems, network, and telecommunications solutions [30, 51, 52].

In particular, the research seeks to highlight the reasons why millions of consumers prefer the purchase and use of products offered by Huawei, one of the most important companies in the sector, founded in China but present in more than 140 countries with over 180,000 employees. The scientific articles aimed at investigating the reasons for the success of Huawei are rather few and, among other things, none of them performs a quali-quantitative investigation by considering a sample with dimensions capable of allowing for an adequate generalization of the results.

In order to meet the research goals and to address the gaps identified in extant research, the work follows a qualitative-quantitative approach, based on a big data analysis, carried out with regard to the people's opinions expressed on a specific digital platform (social network) in a chosen period.

The work is structured in 6 sections. It opens with the analysis of the theoretical background, reporting the scientific evidence emerging from the literature on big data analysis and on its role in understanding electronic devices' consumer behavior, focusing on the contributions dedicated to the company "Huawei". Subsequently, the research design used to collect and analyze data is described; later, through the use of a series of tables and graphs, the results arisen from the analysis are summarized; after that, the findings obtained are discussed to offer a plausible interpretation of their meaning; following, the potential implications of the work are highlighted, both from a theoretical-scientific and a practical-managerial point of view. Finally, within the conclusions, the limits of the work are underlined and some suggestions for possible future research are offered.

2 Theoretical Background

2.1 Smartphone Industry and Consumer Behavior

Smartphones, i.e. mobile phones with advanced computing capabilities and connectivity than the regular ones, came into the consumer market in the late 90s, gaining a mainstream popularity with the introduction of Apple's iPhone in 2007. Starting from the launching of the first iPhone, Apple definitely defined a new category of product, accelerating the convergence of traditional mobile telephony, Internet services and personal computing into a new industry. In this regard, Park and Chen [43] show that the behavioral intention to use the smartphone was largely influenced by perceived usefulness and ease of use toward using it. Today, becoming the standard configuration among different types of mobile devices, the mobile phone industry represents a very innovative segment within the ICTs sector. Technical change and new product proliferation made (and keep on making) this industry extremely dynamic, even if market shares are highly concentrated in the hands of few companies.

Recent trends in the smartphone industry expanded previous conceptions of the industry and its boundaries. For instance, the increasing importance of Internet and cloud-based services that in many ways lie outside the control of the physical device, operating system, and even the cellular network, seems to be changing the roles and strategies of key firms in the ecosystem [45]. However, smartphone sellers shipped 355.6 million units worldwide during the third quarter of 2018, how depicted in Fig. 1, resulting in a 5.9% decline when compared to the 377.8 million units shipped in the third quarter of 2017. The drop marks the fourth consecutive quarter of year-over-year declines for the global smartphone market.

This result, as The Economist writes (https://www.economist.com), shows the market saturation and the need of specific marketing strategies to attract consumers. In this mean, successful innovation lies not in bowing down to consumer resistance, but in understanding the causes and developing a marketing strategy to attack them [46]. Indeed, since the smartphone market has reached a saturation state, device

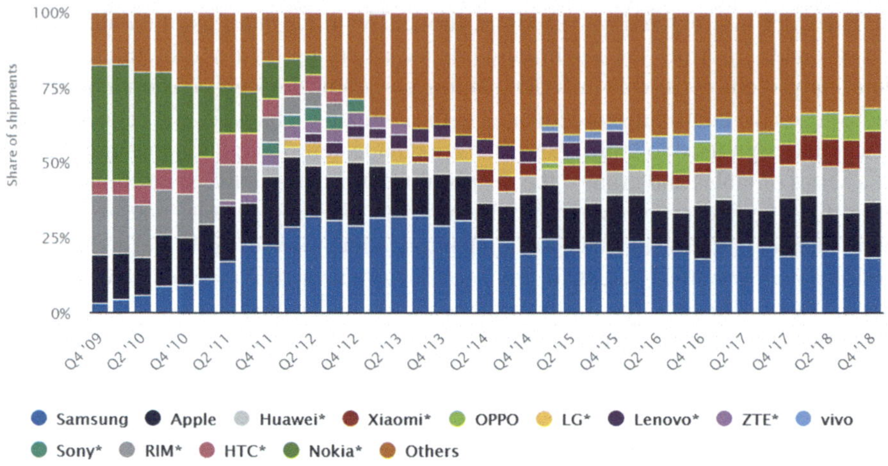

Fig. 1 Global smartphone vendors from 4th quarter 2009 to 4th quarter 2018. *Source* https://www.statista.com

manufacturing companies have to re-focus their resources and capabilities to enhance customer loyalty in order to retain existing customers and attract new ones.

That said, there are only few studies on customer loyalty of smartphones. Kim et al. [22] analyze the Korean smartphone market, showing that customer satisfaction and switching barriers (alternative attractiveness and switching cost) have significant impacts on customer loyalty. The device features (functions, usability and design) and corporate factors (customer support and corporate image) significantly influence customer satisfaction. Usage characteristics (relationship length and usage experience) moderate some of the links in the research model. According to a report by IDC (https://www.idc.com), the world's five biggest smartphone companies are Samsung, Huawei, Apple, Xiaomi, and Oppo. Samsung remains the leader, while the Chinese smartphone maker, Huawei, has surpassed Apple. Huawei, consequently, by maintaining its second position based on global market share, surpassed Apple for the second consecutive quarter as well as continues to lead the China smartphone market with 13.4% YoY growth in Q3 2018. Also, Huawei's P20/P20 Pro series found strong demand in the US$600 < $800 price segment, helping Huawei build a high profile in the global market while its extended distribution network remained the driver to push its presence in the domestic market.

This complex and dynamic scenario makes central the role of the consumers and their purchasing behavior, as today they are increasingly selective in product choice. Simultaneously, product life cycles are shortening, competition is intensifying and the new product failure rate is growing. Understanding the consumer buying process, indeed, can make the difference between success and failure in smartphone industries.

The new technologies and the subsequent innovative features of "digitalized" products introduce a revolution in contemporary markets that leads to new ways of understanding consumer behavior and formulating marketing strategy.

To understand new attitude, trends or gaps in services [16], a strategic approach to the use of new technologies and big data is needed to enhance customers' satisfaction and loyalty thanks to the continuous collection of their evaluation, complaints and feedback on products, services and brands.

Big Data consumer analytics allow the extraction of hidden insight about consumer behavior and the exploitation of that insight through advantageous interpretation. For this reason, Big Data permit to exploit the possibilities offered from new technologies to gain insights about consumer's behaviors and translate those insights into market advantage.

2.2 Big Data: A Scientometric Approach

In order to identify the scientific profile (authors, contributions, topics most frequently treated and relationships between them) of the studies carried out in the period 1990-today by Business, Management and Accounting (BMA) on the theme of Big Data, we followed a scientometric approach and, in this context, the methodologies of bibliographic mapping and clustering. It was used VosViewer (Visualization of Similarities) by Van Eck and Waltman [57, 58] software developed with the specific goal to construct, display and make publicly available bibliometric maps. It provides distance-based maps, or graphical representations, in which the importance of a term is represented by its size on the map and the distance between two terms reflects the strength of the relationships between them: the smaller the distance, the more intense the relationship binds them.

The question was addressed to Scopus, the largest database of citations and abstracts of peer-reviewed literature. Concretely, Scopus returned 2242 references.

Subsequently, through VosViewer, a bibliometric map relative to the bibliographic coupling of the census references was created. VosViewer highlighted 1106 linked contributions, grouped into four different clusters, as shown in Fig. 2.

The first cluster (the green one) is focused on "Data Mining techniques for Big Data". Big Data has becoming increasingly central, generating a new era in data exploration and utilization [8]. The "mass digitization" [14] led to a rapid expansion of large amounts of data, characterized by large-volume, complexity, multiple and autonomous sources. This phenomenon is continuously evolving, as reported by the "3Vs" model of Laney, which highlights the increasing of the volume, velocity and variety of generated data [3, 67, 68]. Later, other concepts as veracity and value have been attributed to this model, with the aim to highlight respectively the quality across datasets and the capacity to generate useful output for industry challenges and issues [56]. In this scope, the Data mining techniques permit extracting useful information from large datasets or streams of data [17] and can reveal insights, supporting decision making. Indeed, from the data mining perspective, the data-driven model involves demand-driven aggregation of information sources, mining and analysis, user interest modeling, and security and privacy considerations [44, 63].

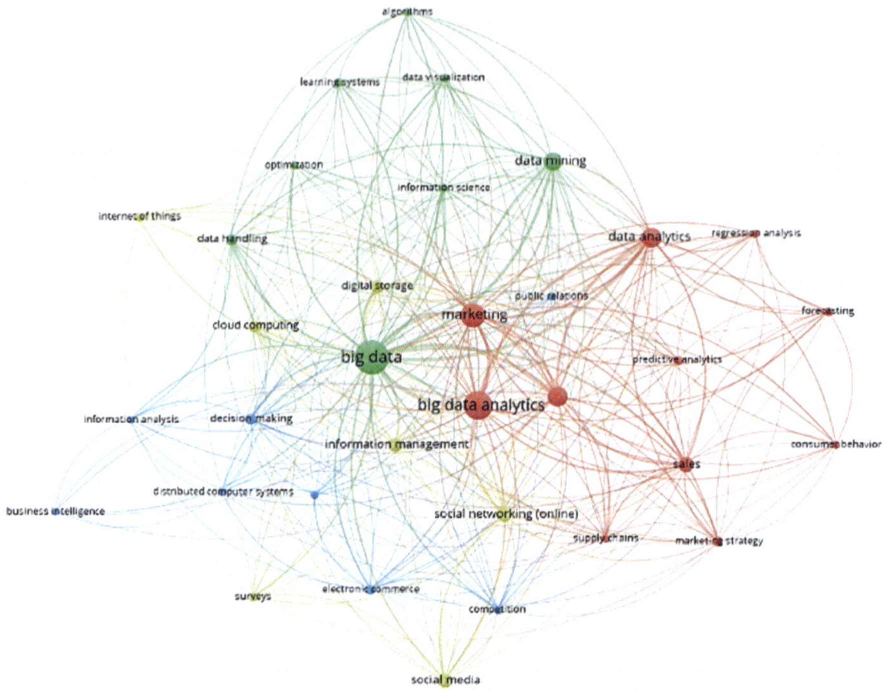

Fig. 2 Big data network visualization in business, management and accounting field. *Source* Authors' elaboration

The second cluster (the red one) regards "Big Data Analytics for Marketing". Undoubtedly, consumer analytics is at the epicenter of a Big Data revolution. New analysis techniques helps capture rich and plentiful data on consumer phenomena in real time, enabling the process of collecting, storing, extracting and utilizing consumer insight to enhance companies' dynamic and adaptive capabilities [16]. Therefore, marketing practice is always more influenced by data culture, by relying on the processes of data mining and A/B testing rather than human intuition [20]. The aim is to carry out consumer behavior analytics offering a look inside the "black box" of markets by providing meaningful information able to support the decision-making process and to improve of various business services to significantly improve customer experience as well as value creation for organizations [2].

The third cluster (the yellow) is about the "Big Data generation". The social networking phenomenon permit to create, modify, share, and discuss Internet content in a very dynamic way [1, 13]. Thanks to user-generated content, such as text posts or comments, digital photos or videos, and data generated through all online interactions, the online social networks grow by connecting a user's profile with those of other individuals or groups [39]. Therefore, social networking Web sites are amassing vast quantities of data and computational social science is providing tools to process this data. Especially, with announcements of growing data aggregation by both Google and Facebook, the need for consideration of these issues is becoming urgent [40]. Indeed, leveraging the social network paradigm could enable a level of collaboration to help solve big data processing challenges [49]. On the other hand, also the phenomenon of Internet of Things (IoT), relying on physical objects interconnected between each other, creates a mesh of devices producing large quantities of information. The sensors are surrounding our environment (e.g., cars, buildings and smartphones) and continuously collect data about our living environment [6]. Over the last few years, all such solutions capture large amounts of data pertaining to the environment as well as their users. The IoT's goal is to learn more and better serve system users.

The fourth cluster (the blue one) is about the "Decision-making and Business Intelligence in Big Data era". Therefore, decisions will increasingly be based on data and analysis rather than on experience and intuition [33]. Organizations are looking for ways to harness the power of big data to improve their decision making, taking advantage through an evolutionary process in which the gradually understanding of the potential of big data and the routinization of processes plays a crucial role [21]. Indeed, with exponential growth in data, enterprises must act to make the most of the vast data landscape, by applying multiple technologies, carefully selecting key data for specific investigations, and innovatively tailor large integrated datasets to support specific queries and analyses. All these actions will flow from a data value chain a framework to manage data holistically from capture to decision making and to support a variety of stakeholders and their technologies [34]. Connected to the impact of data-approach in contemporary business organizations is the Business Intelligence field [8]. Numerous companies already foresee the enormous business effects that analytical scenarios based on big data can have, and the impacts that it will hence have on advertising, commerce, and business intelligence [61].

3 Research Design

To achieve the research goal the work follows a both qualitative and quantitative approach, based on big data analysis, realized taking into consideration people's opinions expressed on a specific digital virtual platform (social network) in a chosen time span. In detail, the "social search" was used. It is based on the mining and analysis of online data [5] to provide summary information, preliminary to reliable considerations about the investigated phenomenon [11]. The social search is increasingly used for research purposes since it offers numerous advantages, especially in terms of the number of metrics that can be generated, identification of the most frequently used key words (hashtags), anticipation of future trends, reports drafting, minimization of business risks, analysis of attitudes, deduction of the Key Performance Indicators (KPIs) in corporate decision-making, better targeting of marketing strategies, analysis of time series, optimization of resources, identification of solutions to complex problems and so forth [36].

Data collection started at the beginning of September of last year and was completed at the end of January of the current year. The virtual environment used for data collection is Twitter, one of the most popular social networks (third after Facebook and YouTube) with over 330 million active users [52]. Among the various possible alternatives, the choice fell on Twitter to exploit the "rule" imposed by the developers of the platform, which allows each user to write posts containing no more than 280 characters. This aspect has made possible to streamline data collection.

Figure 3 represents a screenshot of the Twitter page, made during the search for posts containing the hashtag #Huawei.

At the end of the collection phase, the data were analyzed to be appropriately interpreted. Specifically, the users' posts were taken into account to obtain sufficiently reliable information about the reasons why more and more people prefer Huawei products, contributing to its success all over the world. To avoid conceptual distortions, the extracted text was "cleaned" through a procedure that eliminated all stopwords. Moreover, to facilitate the understanding of the results, a wordcloud was represented in Fig. 4, with terms of different size and color depending on the frequency with which the users used them.

Furthermore, a sentiment analysis was carried out to understand the polarity (positive, neutral or negative) of the comments, i.e. to know how much the users' opinion expressed through their tweets was favorable with respect to Huawei products.

Fig. 3 Screenshot of data searching. *Source* Authors' elaboration

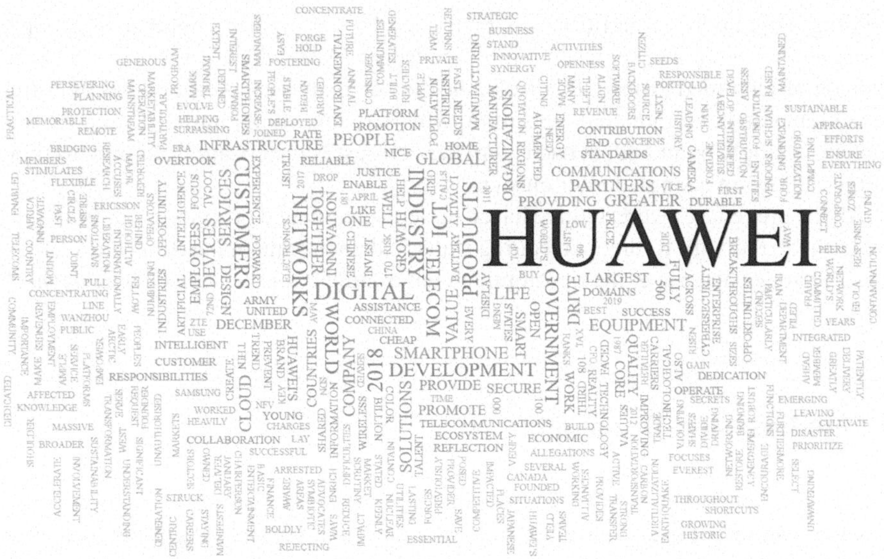

Fig. 4 Wordcloud

This analysis is performed usually to extract information from online sources and texts (reviews, comments, posts, etc.) through the determination of positive, negative or neutral polarity. The most common approaches can be sub-divided into three macro-categories: keyword detection, lexical affinity and statistical methods. Keywords detection allows the classification of the text into some recognisable emotional categories, identified based on the presence of unambiguous emotional words (e.g. happy, sad, and bored). The method of lexical affinity, after the detection of the emotional keywords, assigns arbitrary words to particular emotions based on their semantic affinity.

After mining, the gathered data were subjected to a sentiment analysis to enable the identification of people's perceptions about Huawei's products, allowing understanding the overall polarity of the extracted words: the most frequently found words were submit- ted to the sentiment check against a lexicon annotated with sentiment values in order to establish their potential positive, negative or neutral value. The check returned values in the [0–1] range, which represents the words' positivity, negativity or neutrality, whose total sum is 1.

Next, a cluster analysis was performed to identify the main "group of elements" considered by Huawei's consumers. Introduced by Tryon [54], it is based on the use of multivariate analysis techniques aimed at the selection and grouping of homogeneous elements in a dataset. Clustering techniques are based on measures related to the similarity between the elements, expressed in terms of distance in a multidimensional space. The clustering algorithms group the elements by considering their mutual distance, which indicates the belonging to a certain set (cluster). To this end, the hierarchical method was chosen for the analysis, by virtue of the fact that it allows building a hierarchy of partitions characterized by a decreasing number of groups, viewable through a tree representation (dendrogram), in which it is highlighted the division of identified groups. Precisely, the cluster analysis was carried out by using an agglomerative hierarchical algorithm, which assumes that, initially, each cluster contains a single point and, at each step, the "closest" clusters are from time to time merged to obtain a single larger cluster.

All the procedures were defined based on semi-automated procedures, defined and operationalized with R, an open source statistical environment based on a programming language and a specific development environment for the statistical analysis of data.

4 Findings and Discussion

The analysis provided useful information about the main features of Huawei products that can orient users' attitude. In particular, the results highlight the existence of multiple factors, grouped in five clusters (Fig. 5), capable of inducing users of electronic devices to prefer Huawei.

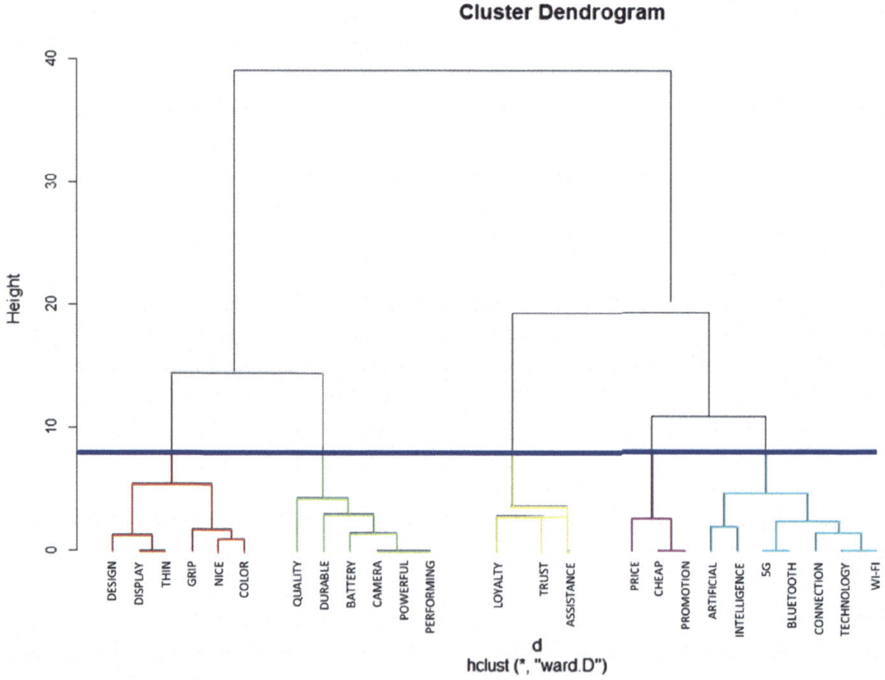

Fig. 5 Cluster analysis

First, from the analysis carried out emerges that the cluster "Design" figures at the first place. Factors as "display", "thin", "grip", "nice" and "color" belong to this group, indicating the consumer attitude to focalize on aesthetic aspects. According to Nanda et al. [38], the aesthetic design of a smartphone has an impact on emotional reaction of people, showing that smartphone is always more perceived as a fashion accessory, as well as a communicating tool. Therefore, consumer's switching behavior from incumbent smartphone to new one can be explained by not merely rational assessment but also affective aspects like aesthetic or pleasure and emotional points like attachment. Consumer's attachment to aesthetic would influence rational evaluation to switch or not to other brands, becoming an important research topic [26]. Another study [50] highlights how aesthetics' primary effect on purchase intention is not direct but rather indirect through perceived sociality and to a lesser extent, perceived emotional value while the importance of aesthetics on perceived functional value is far less. Therefore, these aspects highlights how aesthetics and product design can be used to strengthen purchase intention and influence consumer behavior in terms of both product development and promotional strategies. In this direction, aesthetics' appeal to social and emotional perceived values provides ideas on how to exploit aesthetics in promotional campaigns. Huawei, in particular, starting from the C7100 mobile phone, aimed at the Chinese market, sold more than a third of a million in the three months after its launch in 2008. After, the company carried out the strategy of elaborating technology connected to the European demand for simple aesthetic design. The result was a distinctive design language that is central to Huawei's international strategy. The company now has one of the largest in-house design teams in the world with design centers in the UK, Japan and the United States (https://tangerine.net). Then, Huawei opened its first global research center focused on innovation in aesthetics and design, part of the $600 million investment plan in France (https://huawei.eu). During the 2018, Huawei has announced that the Huawei P20 Pro has become its best-selling high-end device in Western Europe, with a growth of over 300% on an annual basis. Huawei P20 Pro, characterized by particular design, was announced last March 27 in Paris and has been put on the market. In the four weeks that followed, twitter related to aesthetic factors grew and sales rose by 316% compared to last year with P10 Plus (https://huawei.hdblog.it). In addition, Huawei's launched an accompanying Porsche Design Huawei Mate RS, which is a fancied-up P20 Pro. The aim is to offer in the luxury mobile segment a device with great technical performance, as the Chinese house promises, but with the distinctive touch of design of a brand rooted in a European imaginary.

In the second place, "Quality" is ranked. Aspects as "durable", "battery" and "camera" represent the perceived quality of smartphone by consumers. Indeed, according to Parasuraman and Grewal [42], Tsiotsou [55], perceived product quality is expected to influence consumer repurchase intention. Chen and Ann [7] highlights that the attribute of 'battery duration' is very important for customers but offers relatively lower satisfaction, which should be improved as a top priority for the smartphone companies. The longer battery duration has been a common expectation for the smartphone addicts, and smartphone manufacturers need to focus on this aspect. In the future, it is probably that the battery duration will be longer than one day so that people will no longer need to carry around an extra battery charger for normal use. Filieri and Lin [19] shows that the consumers that have already purchased, and then experienced, a smartphone brand can judge its level of quality in terms of its durability, functionality, reliability, and performance. Based on previous usage experience, if smartphone brand is perceived to be of high quality, consumers will repurchase it, otherwise they may switch to another brand. Therefore, a smartphone brand that is considered as durable, functional, reliable and robust will also influence consumers' repurchase intentions. This result can be explained by the fact that in the current society smartphones are used daily to perform multiple and important tasks, requiring an excellent quality to enable individuals to do efficiently and effectively what they want. Presented to the western consumer as a Chinese economic sub-brand, today Huawei earned respect and admiration by consumers. Indeed, the opinions of users on Huawei are mostly positive. In particular, Huawei Consumer Business Group (CBG) South Africa strives on becoming the first brand for consumers in the country after finishing 2nd in the South African Customer Satisfaction Index (SAcsi) report, conducted by Consulta, on mobile phones. Now in its fifth year, the SAcsi for Mobile Handsets offers impartial insights into the South African mobile handset industry by measuring the customer's overall satisfaction. This satisfaction score is based on brands exceeding or falling short of customer expectations and the respondents' idea of the ideal product to achieve an overall result out of 100. The measurement also includes, among other measures, Customer Expectations Index, Perceived Quality Index and a Perceived Value Index (https://mybroadband.co). In addition, thanks to the collaboration with Leica, a famed photography brand known for some of the most iconic portraits and street photography images, Huawei transfers that picture quality to a smartphone, conquering consumers regarding the camera potentialities (https://www.stuff.tv).

The third position is occupied by "Price". "Promotion", "rate" and "cheap" belong to this group, highlighting the importance of the convenient purchase for consumers. Price concern is one of the determinants tested to find out the effects on demand of Smartphone [10]. In addition, a research found that price significantly impact on the purchase intention of Smartphone among young adults in UTAR, Perak, Malaysia [9]. Therefore, the smartphone customer is often influenced by the product quality and its price as well [15]. In terms of pricing strategies, each company follows a particular path. Apple did not follow a wave of low-price handsets strategies, introducing new high level smartphones to create more profit and providing lower selling prices for the old smartphone models. Apple iPhones are more expensive than the others and the decline in their price points are very limited. Still, when Samsung and HTC released high-level smartphones into the market, their price levels fall quite quickly after a few months. Nevertheless, the sales volume of Samsung Galaxy is not as expected possibly due to market saturation of high-level handsets and the fierce competition of low-level handsets coming from China [7]. Huawei faces tough competition from Ericson, Cisco System, ZTE corporation, Apple etc. As the numbers of competitors are more in the telecommunication industry, Huawei follows a competitive pricing strategy in its marketing mix for its products. The reason for this pricing strategy is because buyers (consumers) have more bargaining power and can switch brands easily. Huawei always focuses on providing best quality products to its customers as pricing are nearly same by all competitors. Huawei also charges high prices for its new and innovative products that are not in the competitor's product line. Huawei sometimes follows elastic pricing policy and gives discount on its products mostly sold through ecommerce. For its business division of Enterprise and carrier products, the company charges premium pricing strategy for its innovative products and solutions (https://www.mbaskool.com).

In the fourth place "Brand Loyalty" is ranked. In this group, some factors are "assistance", "trust" and "trend", representing an attitude of loyalty from the consumer perspective towards Huawei brand. Indeed, with the growth and competition of the smartphone industry, developing a better understanding of what drives consumers' loyalty to smartphone brands has become an important issue for academics and practitioners. Yoo et al. [66] investigate the importance of positive reputation from external experience sources for diffusion of smartphones. By categorizing the different reputation sources into four group (personal, expert, consumers and mass media), the authors found that the prior experience of consumer group has the largest importance for the purchasing decision of the potential adopters. Moreover, early adopters and female consumers give more importance on the prior consumers' opinions. The reputation from expert and mass media was relatively lower than it from consumers and personal group. A study in Malaysian shows that 35.6% of the total 1814 respondents reported that the trend in community is the most influential factor for consumers to acquire smartphone instead of the actual needs [41]. In addition, another study on the factors influencing smartphone buying process [27] reveal that, together with the aesthetic value or beauty of design, the brand reputation are positively correlated with the repurchase of smartphones. In order to analyse consumer behaviour in the smartphone market in Vietnam, Wollenberg and Thuong [62] highlights that there is substantial influence of brand perception on customer choice in smartphone market. Yeh et al. [65] show that age enhances the "emotional value-brand loyalty" and "social value-brand loyalty" linkages but weakens the "brand identification-brand loyalty" relationship. Furthermore, gender does not play a moderating role in the determination of smartphone brand loyalty. However, for many years, the brand loyalty was a force for Apple, by leveraging on the ecosystem Apple (iPad, Mac, MacBook, etc.), iOS, after-sales assistance, and trend. Instead, regarding Huawei, diversifying portfolio made harder building brand loyalty, as it is difficult to build up hype around so many different devices. According a report (https://www.brandindex.com) Samsung has an Index score (which measures a range of metrics including quality, value and reputation) of 37.5, Sony follows on 24.3, ahead of Apple's iPhone on 22.7, Nokia on 16.5 and LG on 14.7. Huawei's score is just 6.4. Anyway, the brand Huawei is headed in the right direction. Indeed, its Index score is up by 5.5 points over the past year, a statistically significant increase. In particular, consideration is rising up 1.8 points to 5.2 over the past year, as the purchase intent. Therefore, Huawei should concentrate more on areas such as image building, manufacturing quality products, and ensuring customer satisfactions [23]. As regards the Chinese market, Huawei is perceived better than Apple by consumers despite their preference for global companies (https://www.scmp.com) but this aspect does not influence the analysis considering that Chinese people generally do not use Twitter.

The last cluster is "Innovation". This group contains factors as "artificial intelligence", "5G" and "augmented reality", underlining the importance of innovative technology for smartphone consumers'. Probably, this group figures at the end of the ranking because consumers tend to be resistant to innovation. Indeed, a study [35] identified two factors that influence consumer behavioral intention to resist to innovation like smartphone: perceived risk and perceived complexity. The result indicates that both factors have significant influence on smartphone consumers. In this mean, consumer resistance as consumer intention to adopt or reject the product makes very important difference in success of innovative products. Therefore, behavioural research shows that reasons for and reasons against adopting innovations differ qualitatively, and they influence consumers' decisions in dissimilar ways. This has important implications for theorists and managers, as overcoming barriers that cause resistance to innovation calls for marketing approaches other than promoting reasons for adoption of new products and services [12]. In this regard, the choice of an appropriate branding strategy is a critical determinant of new product success. Truong et al. [53] point out that both earlier and later adopters will favor existing brands to cope with the elevated risk associated with an innovative high technology product. However, innovation image plays an important role for smartphone brands, influencing customer satisfaction [7]. Huawei is distinguished by research and development efforts, investing between $15 billion and $20 billion in Research & Development (R&D). The research involves cloud, Internet of Things (IoT) and 5G technologies, making the company one of the most innovative in the world (https://www.reuters.com). Also, Huawei is working on Augmented Reality (AR) smart glasses that could debut in the next one or two years, potentially putting it in a race against Apple, which is reportedly working on a similar product of its own (https://www.cnbc.com).

5 Theoretical and Managerial Insights

Consistently with the most recent research conducted in the field of management and of consumer behavior [4, 48, 64], the work can be considered as a useful tool for scholars and practitioners, since it offers original ideas both under a theoretical profile and from a practical point of view.

In fact, as concerns its contribution to scientific research, the article spreads awareness about the usefulness of implementing big data analysis to interpret the reasons behind people's purchasing and using choices. Indeed, from the study carried out emerges that big data analysis can bring new opportunities to modern society and challenges to data scientists, by understanding consumers' preferences regarding products and services in order to offer them relevant offers proactively [18, 25]. Consumer analytics through big data technology helps to comprehend consumer behavior, influencing various marketing activities and enabling firms to better exploit its potentialities [16]. Therefore, big data analytics associated with database searching, mining and analysis can be seen as an innovative IT capability that can definitely improve firm performance [24]. Furthermore, the research provides empirical results able to corroborate what emerged in some recent studies dedicated to the theme of consumer behavior in smartphone field, offering the advantage of results generalizability, due to the consideration of a particularly large sample. In particular, enriching the existing literature [37, 65], this work underlines and classifies the factors capable of inducing users of electronic devices to prefer Huawei, highlighting the increasing importance of aesthetic aspect of a smartphone, by having the potential to impact the consumer emotion [47]. Instead, the analysis shows that regarding innovation the consumer tends to be resistance [35]. This has important impact for both theorists and managers, which should overcome the obstacles that cause resistance to innovation by applying different marketing approaches to the promotion and adoption of innovative products and services [12]. With reference to managerial implications, the work could be considered useful by practitioners who want to know how the smartphone market is moving, what are the variables that can be used to attract consumers, what are the assets toward which invest mostly in order to achieve the desired objectives. In particular, the document could encourage entrepreneurs and managers to reflect on factors to be taken into account to take advantage of all the needs connected to the smartphone. In this regard, the work highlights five variables, ranking them in order of occurrences, allowing those who have invested or intend to invest in smartphone sector to know which aspects to pay most attention. Indeed, in a dynamic and saturated market as smartphone sector, always more are required studies in consumer behavior. Tackling this challenge from companies is less difficult if they have a clear understanding of consumers' needs and preferences. In short, the reading of the article could represent a theoretical-empirical basis from which to understand how a company can quickly acquire and retain an increasing number of customers and conquer the market.

6 Conclusions, Limitations and Future Research

The work presents several original elements, traceable both in the research objective and in the approach followed to achieve it. In fact, previous scientific contributions do not analyze the success of Huawei through a big data analysis to provide generalizable empirical evidence.

Moreover, the exploration of users' behavior thanks to Big Data extraction can enhance the understanding of the possibilities offered from new technologies to monitor all the stages of consumer decision-making cycle, including what the consumer does, how it is done, how he can choose between similar products and how he enjoys services. In this way, academic and managerial dialogue between consumer behavior and data mining researchers can be encouraged.

However, the paper presents some limitations, related to the type of technique used to collect and analyze data. In fact, even if big data analysis offers a series of advantages (primarily the possibility of analyzing a huge amount of data in real time), it is characterized by a certain degree of superficiality, since it does not allow going deep into the understanding of the people's opinions [11]. In fact, although the sample was particularly large, the automated collection of people's comments has prevented from going deeper in the analysis of users' complete thought. In this regard, for instance, think of the impossibility of adequately interpreting the sarcastic or ironic statements.

Therefore, the methodological limits of the work are related to general criticalities of big data (veracity, volatility) and on the impossibility to generalize findings obtained through data mining [20]. Big data collection and analysis allow the exploration of what people say without providing an analysis of the motivations underlying their opinions. Complex constructs, such as motivation and attitude can be only inferred from Big Data sources [16]. Big Data sets do not permit to store all the factors influencing consumer decisions. For instance, users can express opinions only on given features of a service and the opinions expressed by users that already enjoyed a service cannot be distinguished from those of the users that have never experienced the product.

Such weaknesses could induce to perform a further analysis about the same topic to compare the results emerged from this study with the findings that could arise by using a qualitative approach (such as in-depth interviews).

References

1. Aghaei, S., Nematbakhsh, M.A., Farsani, H.K.: Evolution of the world wide web: From WEB 1.0 TO WEB 4.0. Int. J. Web Semant. Technol. **3**(1), 1–10 (2012)
2. Balar, A., Malviya, N., Prasad, S., Gangurde, A.: Forecasting consumer behavior with innovative value proposition for organizations using big data analytics. In: 2013 IEEE International Conference on Computational Intelligence and Computing Research (ICCIC), pp. 1–4. IEEE (2013)
3. Beyer, M.A., Laney, D.: The Importance of "Big Data": A Definition. Gartner, Stamford, CT (2012)
4. Cantallops, A.S., Salvi, F.: New consumer behavior: a review of research on eWOM and hotels. Int. J. Hospitality Manage. **36**, 41–51 (2014)
5. Carmel, D., Zwerdling, N., Guy, I., Ofek-Koifman, S., Har'El, N., Ronen, I., Chernov, S.: Personalized social search based on the user's social network. In: Proceedings of the 18th ACM Conference on Information and Knowledge Management, pp. 1227–1236. ACM (2009)
6. Cecchinel, C., Jimenez, M., Mosser, S., Riveill, M.: An architecture to support the collection of big data in the internet of things. In: 2014 IEEE World Congress on Services (SERVICES), pp. 442–449. IEEE (2014)
7. Chen, C.M., Ann, B.Y.: Efficiencies vs. importance-performance analysis for the leading smartphone brands of Apple, Samsung and HTC. Total Qual. Manage. Bus. Excellence **27**(3–4), 227–249 (2016)
8. Chen, H., Chiang, R.H., Storey, V.C.: Business intelligence and analytics: from big data to big impact. MIS Q. **36**(4), 1165–1188 (2012)
9. Chew, J.Q.: Exploring the Factors Affecting Purchase Intention of Smartphone: A Study of Young Adults in UTAR. Unpublished Degree Paper. Universiti Tunku Abdul Rahman, Perak Campus, Malaysia (2012)

10. Chow, M.M.: Conceptual paper: factors affecting the demand of smartphone among young adult. Int. J. Soc. Sci. Econ. Art **2**(2), 44–49 (2011)
11. Ciasullo, M. V., Troisi, O., Loia, F., Maione, G.: Carpooling: travelers' perceptions from a big data analysis. TQM J. (2018)
12. Claudy, M.C., Garcia, R., O'Driscoll, A.: Consumer resistance to innovation—a behavioral reasoning perspective. J. Acad. Mark. Sci. **43**(4), 528–544 (2015)
13. Cormode, G., Krishnamurthy, B.: Key differences between Web 1.0 and Web 2.0. First Monday **13**(6), (2008)
14. Coyle, K.: Mass digitization of books. J. Acad. Libr. **32**(6), 641–645 (2006)
15. Djatmiko, T., Pradana, R.: Brand image and product price; its impact for Samsung smartphone purchasing decision. Procedia Soc. Behav. Sci. **219**, 221–227 (2016)
16. Erevelles, S., Fukawa, N., Swayne, L.: Big data consumer analytics and the transformation of marketing. J. Bus. Res. **69**(2), 897–904 (2016)
17. Fan, W., Bifet, A.: Mining big data: current status, and forecast to the future. ACM SIGKDD Explor. Newsl **14**(2), 1–5 (2013)
18. Fan, J., Han, F., Liu, H.: Challenges of big data analysis. Natl. Sci. Rev. **1**(2), 293–314 (2014)
19. Filieri, R., Lin, Z.: The role of aesthetic, cultural, utilitarian and branding factors in young Chinese consumers' repurchase intention of smartphone brands. Comput. Hum. Behav. **67**, 139–150 (2017)
20. Hofacker, C.F., Malthouse, E.C., Sultan, F.: Big data and consumer behavior: imminent opportunities. J. Consum. Mark. **33**(2), 89–97 (2016)
21. Janssen, M., van der Voort, H., Wahyudi, A.: Factors influencing big data decision-making quality. J. Bus. Res. **70**, 338–345 (2017)
22. Kim, M.K., Wong, S.F., Chang, Y., Park, J.H.: Determinants of customer loyalty in the Korean smartphone market: moderating effects of usage characteristics. Telematics Inform. **33**(4), 936–949 (2016)
23. Kumar, S.S., Menon, R.P.: Brand loyalty of customers in smartphone brands. Indian J. Mark. **47**(3), 8–15 (2017)
24. Kwon, O., Lee, N., Shin, B.: Data quality management, data usage experience and acquisition intention of big data analytics. Int. J. Inf. Manage. **34**(3), 387–394 (2014)
25. Labrinidis, A., Jagadish, H.V.: Challenges and opportunities with big data. Proc. VLDB Endowment **5**(12), 2032–2033 (2012)
26. Lee, W.K., Park, J.H.: Effects of brand attachment and perceived aesthetic on intention to purchase new smartphone. J. Inf. Syst. **23**(4), 147–168 (2014)
27. Leelakulthanit, O., Hongcharu, B.: Factors influencing smartphone repurchase. J. Bus. Econ. Res. (Online) **10**(11), 623 (2012)
28. Loia, V.: Soft computing technology for dynamic web pages categorization. In: Fuzzy Logic and the Internet, pp. 157–171. Springer, Berlin, Heidelberg (2004)
29. Lytras, M.D., Raghavan, V., Damiani, E.: Big data and data analytics research: from metaphors to value space for collective wisdom in human decision making and smart machines. Int. J. Semant. Web Inf. Syst. (IJSWIS) **13**(1), 1–10 (2017)
30. Lytras, M., Visvizi, A.: Who uses smart city services and what to make of it: toward interdisciplinary smart cities research. Sustainability **10**(6), 1998 (2018)
31. Lytras, M.D., Papadopoulou, P. (eds.): Applying Big Data Analytics in Bioinformatics and Medicine. IGI Global (2017)
32. Lytras, M.D., Pouloudi, A., Poulymenakou, A.: A framework for technology convergence in learning and working. Educ. Technol. Soc. J. Int. Forum Educ. Technol. Soc. IEEE Learn. Technol. Task Force **5**(2), 99–106 (2002)
33. McAfee, A., Brynjolfsson, E., Davenport, T.H., Patil, D.J., Barton, D.: Big data: the management revolution. Harvard Bus. Rev. **90**(10), 60–68 (2012)
34. Miller, H.G., Mork, P.: From data to decisions: a value chain for big data. It Prof. **15**(1), 57–59 (2013)
35. Mohtar, S.B., Abbas, M.: Factors influencing consumer resistance to innovation (Smartphone). Middle-East J. Sci. Res. **23**(7), 1390–1394 (2015)

36. Morris, M.R., Teevan, J., Panovich, K.: A comparison of information seeking using search engines and social networks. ICWSM **10**, 23–26 (2010)
37. Nagarkoti, B.: Factors influencing consumer behavior of Smartphone users, degree thesis. Int. Bus. BBA (2014)
38. Nanda, P., Bos, J., Kramer, K.L., Hay, C., Ignacz, J.: Effect of smartphone aesthetic design on users' emotional reaction: an empirical study. TQM J. **20**(4), 348–355 (2008)
39. Obar, J.A., Wildman, S.: Social media definition and the governance challenge: an introduction to the special issue. Telecommun. policy **39**(9), 745–750 (2015)
40. Oboler, A., Welsh, K., Cruz, L.: The danger of big data: social media as computational social science. First Monday **17**(7) (2012)
41. Osman, M.A., Talib, A.Z., Sanusi, Z.A., Shiang-Yen, T., Alwi, A.S.: A study of the trend of smartphone and its usage behavior in Malaysia. Int. J. New Comput. Architectures Appl. **2**(1), 275–287 (2012)
42. Parasuraman, A., Grewal, D.: The impact of technology on the quality-value-loyalty chain: a research agenda. J. Acad. Mark. Sci. **28**(1), 168–174 (2000)
43. Park, Y., Chen, J.V.: Acceptance and adoption of the innovative use of smartphone. Ind. Manage. Data Syst. **107**(9), 1349–1365 (2007)
44. Pllana, S., Janciak, I., Brezany, P., Wohrer, A.: A survey of the state of the art in data mining and integration query languages. In 2011 14th International Conference on Network-Based Information Systems (NBiS), pp. 341–348. IEEE (2011, September)
45. Pon, B., Seppälä, T., Kenney, M.: One ring to unite them all: convergence, the smartphone, and the cloud. J. Ind. Competition Trade **15**(1), 21–33 (2015)
46. Ram, S., Sheth, J.N.: Consumer resistance to innovations: the marketing problem and its solutions. J. Consum. Mark. **6**(2), 5–14 (1989)
47. Shin, D.H.: Cross-analysis of usability and aesthetic in smart devices: what influences users' preferences? Cross Cult. Manage. Int. J. **19**(4), 563–587 (2012)
48. Solomon, M.R., White, K., Dahl, D.W., Zaichkowsky, J.L., Polegato, R.: Consumer behavior: buying, having, and being. Pearson Boston, MA, USA (2017)
49. Tan, W., Blake, M.B., Saleh, I., Dustdar, S.: Social-network-sourced big data analytics. IEEE Internet Comput. **17**(5), 62–69 (2013)
50. Toufani, S., Stanton, J.P., Chikweche, T.: The importance of aesthetics on customers' intentions to purchase smartphones. Mark. Intell. Planning **35**(3), 316–338 (2017)
51. Troisi, O., D'Arco, M., Loia, F., Maione, G.: Big data management: The case of Mulino Bianco's engagement platform for value co-creation. Int. J. Eng. Bus. Manage. **10**, 1847979018767776 (2018)
52. Troisi, O., Grimaldi, M., Loia, F., Maione, G.: Big data and sentiment analysis to highlight decision behaviours: a case study for student population. Behav. Inf. Tech. **37**(10–11), 1111–1128 (2018)
53. Truong, Y., Klink, R.R., Simmons, G., Grinstein, A., Palmer, M.: Branding strategies for high-technology products: the effects of consumer and product innovativeness. J. Bus. Res. **70**, 85–91 (2017)
54. Tryon, R.C.: Cluster analysis: correlation profile and orthometric (factor) analysis for the isolation of unities in mind and personality. Edwards brother, Incorporated, Ann Arbor, MI (1939)
55. Tsiotsou, R.: Perceived quality levels and their relation to involvement, satisfaction, and purchase intentions. Mark. Bull. **16**(4), 1–10 (2005)
56. Uddin, M.F., Gupta, N.: Seven V's of Big Data understanding Big Data to extract value. In: Proceedings of the 2014 zone 1 conference of the American Society for Engineering Education, pp. 1–5. IEEE (2014)
57. Van Eck, N.J., Waltman, : L.: Text mining and visualization using VOSviewer. arXiv preprint. arXiv:1109.2058 (2011)
58. Van Eck, N.J., Waltman, L.: Visualizing bibliometric networks. In: Measuring scholarly impact, pp. 285–320. Springer, Cham (2014)

59. Visvizi, A., Lytras, M.D.: Rescaling and refocusing smart cities research: from mega cities to smart villages. J. Sci. Technol. Policy Manage. (2018)
60. Visvizi, A., Lytras, M. D., Daniela, L.: (Re) defining smart education: towards dynamic education and information systems for innovation networks. In: Enhancing Knowledge Discovery and Innovation in the Digital Era, pp. 1–12. IGI Global (2018)
61. Vossen, G.: Big data as the new enabler in business and other intelligence. Vietnam J. Comput. Sci. **1**(1), 3–14 (2014)
62. Wollenberg, A., Thuong, T.T.: Consumer behaviour in the smartphone market in vietnam. Int. J. Innov. Manage. Technol. **5**(6), 412 (2014)
63. Wu, X., Zhu, X., Wu, G.Q., Ding, W.: Data mining with big data. IEEE Trans. Knowl. Data Eng. **26**(1), 97–107 (2014)
64. Xiang, Z., Magnini, V.P., Fesenmaier, D.R.: Information technology and consumer behavior in travel and tourism: insights from travel planning using the internet. J. Retail. Consum. Serv. **22**, 244–249 (2015)
65. Yeh, C.H., Wang, Y.S., Yieh, K.: Predicting smartphone brand loyalty: Consumer value and consumer-brand identification perspectives. Int. J. Inf. Manage. **36**(3), 245–257 (2016)
66. Yoo, J., Yoon, Y., Choi, M.: Importance of positive reputation for Smartphone adoption. In: 2010 International Conference on Information and Communication Technology Convergence (ICTC), pp. 314–318. IEEE (2010, November)
67. Zaslavsky, A., Perera, C., Georgakopoulos, D.: Sensing as a Service and Big Data. arXiv preprint arXiv:1301.0159 (2013)
68. Zikopoulos, P., Eaton, C.: Understanding Big Data: Analytics for Enterprise Class Hadoop and Streaming Data. McGraw-Hill Osborne Media, New York (2011)

Identification of Factors of Indigenous Ethnic Identity: A Case Study on the Waorani Amazonian Ethnicity

Aldrin Espín-León, Antonio Jimeno-Morenilla, María L. Pertegal-Felices and Jorge Azorín-López

Abstract Latin American Indians have undergone changes in their cultural identity, especially as a result of contact with Western cultures. In order to determine the degree of change in cultural identity, the authors have recently developed an instrument designed from an indigenous perspective that provides information on 30 subdimensions and 5 cultural dimensions and was successfully tested for validity and reliability. In this paper, the instrument is analyzed from an Artificial Intelligence point of view in order to automatically classify the individuals and to provide a subspace of it able to identify the weights of the subcomponents of this instrument in regard to its contribution to the Waorani identity. The systematic application of the instrument together with the AI-based system can provide decision-makers with valuable information about which aspects of their identity are most sensitive to change and thus help design development policies that minimally interfere with their ethnic identity.

1 Introduction

The cultural study of a community is very complex because of the large number of variables involved, including: knowledge, beliefs, art, morals, law, customs, habits and acquired abilities [1]. It also considers ways of life and ways of living in society [2].

A. Espín-León
Faculty of Sociology and Social Work, Central University of Ecuador, Quito, Ecuador

A. Jimeno-Morenilla · J. Azorín-López
Department of Computer Technology, Universidad de Alicante, Alicante, Spain

M. L. Pertegal-Felices (✉)
Developmental and Educational Psychology Department,
Universidad de Alicante, Alicante, Spain
e-mail: ml.pertegal@ua.es

A. Visvizi and M. D. Lytras (eds.), *Research & Innovation Forum 2019*,
Springer Proceedings in Complexity,
https://doi.org/10.1007/978-3-030-30809-4_47

Identity is part of culture, and cannot be determined tangibly or through physical characteristics, nor is the individual considered in isolation from his or her community and is historically defined by various aspects of immaterial character such as language, social relations, ceremonies, rites, beliefs [3]. The culture and identity of peoples are constantly changing or merged by territorial heterogeneity and diversity of cultural expressions, making a system of indicators for analysis difficult and complex [4].

In Latin America, indigenous communities and nationalities have undergone accelerated cultural changes and mergers of identity as a consequence of colonyism by dominant groups [5]. The constant abolitions of their culture were established from the political power and the ethnic domination was maintained, excluding them of all type citizen participation, the state considered that they should be of an only culture, an only language and an only religion [6].

In the case of the Amazonian Indians, there is widespread concern that the inhabitants will lose their identity or, at worst, disappear. The link that exists between individuals and nature is essential in Amazonian societies [7].

1.1 Aim of the Research

Knowing and measuring the identity of indigenous communities is one of the most important steps in protecting their culture. Precisely, this study aims to present tools that allow us to discern the cultural state of an indigenous nationality in order to check whether it is being degraded as a consequence of a process of western acculturation.

The authors developed at an earlier stage an instrument that made it possible to identify factors of indigenous identity through variables and indicators. The instrument itself is not a solution, but the identification of ethnic factors provides valid information to be taken into account in making decisions that help these communities. The objective of this research is to know which of the variables identified in the instrument are those that most influence their cultural identity. This identification would make it possible to know whether certain policies aimed at these peoples cause undesirable effects on their identity. In order to determine this influence, clustering analysis will be carried out using techniques based on artificial intelligence.

2 Background

Measuring identity and culture is complex. In cultural statistics, when there is a subjective assessment or they have ideological components, alternatives must be proposed so that the assessments are objective and can be analyzed quantitatively [8]. If measuring identity has a large qualitative component, quantitative analysis minimizes intuitive and subjective conceptions.

Ethnic identity, as proposed by several authors, is determined by four dimensions: cognitive, evaluative, affective, and behavioral. Instruments often ask about a definition of self, about feelings of belonging to a group, about pride or intention of belonging to it, about the language, religion and knowledge of history of their own group [9, 10].

On the other hand, authors such as Cross and Helms [11, 12], developed theoretical models that contemplate three states referring to ethnic identity: (1) negative valuation of blacks and their preference toward another race (whites); (2) knowledge of and interest in them; and (3) international and group acceptance.

The ethnic identity seen from the acculturation includes questionnaires that use external and internal variables. External variables deal with language, media (tv radio), friendship relations and ethnic traditions, while internal variables contain cognitive, affective and moral issues [13]. Another model of ethnic identity proposed by Phinney and Rotheram [14], includes six components: Self-definition, attitudes towards one's own ethnic group, attitude towards oneself, interest of the ethnic group and commitment to ethnic identity.

However, these models and components are focused from generalist points of view and lack items that specify aspects of indigenous ethnic identity and even less of the Amazon indigenous.

There are works that measure gender identity, related to sexual roles and the adaptation of gender identity towards stereotypes [15, 16]. Muradas and Rodríguez [17], on the other hand, use instruments with items to measure tangible culture such as churches, museums, parks, cultural centers, among others. These instruments would be limited to gender and would only make sense in populations with a Western structure.

A similar study on identity was conducted in the community of Chiapas, Mexico [18]. They worked on identity narratives using the autobiographical multi-methodology [19]. Three groups participated in the experiment: Chiapas indigenous people, students from a rural area, and from a Spanish city. The results obtained were clearly qualitative; however, this study takes the individual as a methodological reference and was applied to different regions with different identity characteristics.

From the previous analysis it can be seen that the identification and measurement of the construct of cultural identity is a complex issue that is difficult to address due to the large number of subjective variables to be taken into account. If we focus on indigenous communities, the scientific baggage is even less, with qualitative studies that do not provide qualitative measurements for the variables described. In the case of the Amazon, with communities in serious danger of extinction, no instruments of this type have been described.

The paper has the following structure: Sect. 3 will briefly describe the instrument developed by the authors in a previous study. In Sect. 4, the instrument is analyzed from an Artificial Intelligence point of view in order to automatically classify the individuals and to provide a subspace of it able to identify the weights of the subcomponents of this instrument in regard to its contribution to the Waorani. Section 5 presents the main conclusions of the study and the lines of future work.

3 Instrument for the Measurement of the Waoroani Ethnic Identity

In the Ecuadorian state, the indigenous population represents 7.0%, of which 78.5% are found in rural areas, while 21.5% are concentrated in urban areas, bringing together 14 nationalities that give rise to 17 peoples. Its population underwent significant changes, rising from 65% in the 18th century to 7.0% in 2010 [20]. However, these demographic data do not demonstrate the qualitative changes in their identity that they may have undergone.

The Waorani Nationality has approximately 13,000 inhabitants, distributed in 22 communities. The territory located in the Amazon (South America) shared by Ecuador and Peru is located in the provinces of Pastaza, Napo and Orellana (see Fig. 1). It is an extensive area of about 790,000 ha whose boundaries range from the Napo River in the north to the Villano and Curaray Rivers in the south. Much of its territory is surrounded by the Yasuní National Park—protected by its exuberant biodiversity—and by an area called intangible in which the Tagaeri and Taromenane inhabit, uncontacted peoples who live in isolation with their original culture without any contact from the West or other cultures.

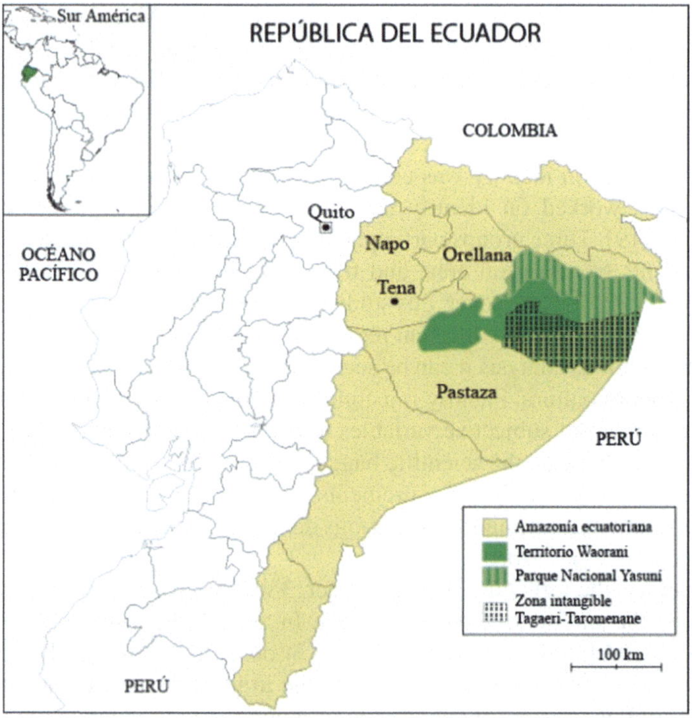

Fig. 1 Geographical location of the Waorani territory

Table 1 Dimensions and subdimensions of the Waorani Identity Instrument

Dimensions		Subdimensions
1. Economic (E)	9	Handicrafts (E1), Exchange of products (E2), Cultivation (E3), Tourism (E4), Work (E5), Mingas (E6)—mingas or nopos, indigenous words that means community work without payment-, Trade (E7), Hunting and fishing (E8), Breeding of animals (E9)
2. Family and reproduction (F)	6	Education (F1), Care and upbringing of children (F2), Medicine (F3), Marriage (F4), Coexistence (F5), Reproduction (F6)
3. Ideological (I)	4	Religion (I1), Beliefs (I2), Spirituality (I3), Rites (I4)
4. Organization (O)	3	Community (O1), Justice (O2), Government (O3)
5. Social (S)	8	Music, dance and songs (S1), Art (S2), Food (S3), Dress (S4), Housing (S5), Culture, ethnicity and identity (S6), Language (S7), Sports and recreation (S8)
Total subdimensions	30	

The instrument used in this study has been elaborated by the authors as a result of a previous research in which the dimensions were obtained from a qualitative study in which the *Atlas.ti* software was used for the identification of such dimensions. The developed instrument consisted of 99 items that provided information about 30 subdimensions grouped in 5 dimensions (see Table 1).

Evidence of reliability was obtained through Cronbach's alpha statistic which gave an excellent overall result of 0.974 for the 99 items. In the case of the 5 constructs the evidences were: Economy 0.896 for 22 items; Family and reproduction 0.935 for 21 items; Ideology 0.906 for 13 items; Organization 0.602 for 6 items and Social 0.931 for 37 items. Evidence of validity was obtained through the judgment of 9 experts using CVR index based on Lawshe [21] for each of the 99 items of the instrument.

4 Experiments

4.1 Experimental Setup

The experiments have been carried out using classic classifiers in order to validate the ability of the instrument presented in the previous section to represent and classify the individuals. Specifically, the Self-Organizing Map (SOM) [22], Supervised Self-Organizing Map (SSOM) [23], Neural GAS (NGAS) [24], Linear Discriminant Analysis (LDA), k-Nearest Neighbour (kNN) [25] and the Support Vector Machines

(SVM) [26] have been used. Moreover, a multiclassifier (MC) boosting designed from the above classifiers has been applied. The MC calculates from an input the most frequent class classified by the mentioned classic techniques.

The data of the instrument have been collected by interviewing 299 people: 88 Waorani individuals, 100 from Quito and 111 from the village of "El Tena". The samples were taken by simple random sampling using a systemic procedure from these three populations. These three populations represent 3 types of identities present in Ecuador: (1) The indigenous Waorani Amazonian Nationality is the target population of the study (it consists of 22 communities, the data collection was carried out in three of the 22 communities: Konipare, Menipare and Gareno). The capital, Quito, was selected because it is a totally westernized city in Ecuador and may represent the opposite extreme to the indigenous community within the country. The city of Tena is one of the cities closest to the Waoranis indigenous communities. It is a westernized village but still retains vestiges of indigenous culture and has frequent commercial exchanges with various indigenous nationalities including communities of Waorani nationality, therefore the inhabitants of Ciudad de El Tena are supposed to be culturally placed between Amazonian Indians and western culture.

They conform the output of the classification process. Since the input data is imbalanced, the Synthetic Minority Over-Sampling Technique (SMOTE) [27] has been applied in order to balance the classes. Additionally, the 30 subdimensions of the Waorani instrument (showed in Table 1) used as the inputs for each classifier have been normalized to the range (0 1). Each dimension of the instrument has been divided by the maximum value for each component.

Finally, a 10-fold cross validation has been performed obtaining Sensitivity and Specificity values, and the ROC curves in order to analyse the performance of classifiers and the instrument from an Artificial Intelligence point of view, to distinguish the class of the individuals (Waorani, Quito or El Tena).

4.2 Results and Discussion

Results of classification accuracy for each classifier of the subdimensions of the instrument are presented in Table 2. The performance criteria is related to the *Sensitivity* (correctly classified positive samples/true positive samples), *Specificity* (correctly classified negative samples/true negative samples) and *Accuracy* (correctly classified

Table 2 Performance results according to different classifiers

Criteria	3-knn	LDA	SOM	SSOM	NGAS	SVM	MC
Sensitivity	0.9298	0.9632	0.9097	0.9331	0.9164	**0.9699**	0.9532
Specificity	0.9649	0.9816	0.9548	0.9665	0.9582	**0.9849**	0.9765
Accuracy	0.9532	0.9755	0.9398	0.9554	0.9444	**0.9799**	0.9688

Bold significance in table means a better result, that is, a better performance in classification positive and negative samples/samples). The best results are achieved using the SVM classifier achieving a 97% of sensitivity, 98% of specificity and 98% of accuracy.

In order to visually illustrate the performance of classifiers for the instrument, a ROC (receiver operating characteristic) curve is used in Fig. 2. It represents the relation between the probability of correctly detect the class of an individual (Sensitivity) and the probability of classify it in another class (1-Specificity). The ROC space shows the excellent performance for all classifiers that are able to correctly identify the ethnic identity above 91% and having less than 5% of false alarms.

Next, confusion matrix for the best classifier SVM is presented in Table 3. Matrix columns represent the actual classes, and rows represent the classifier prediction. As it can be observed, the Waorani individuals are almost perfectly detected with a 98.86% of probability (87/88 individuals). Moreover, only 1 person from Quito and El Tena was identified as Waorani (1/211 individuals), representing less than 0.5% of false Waorani identification.

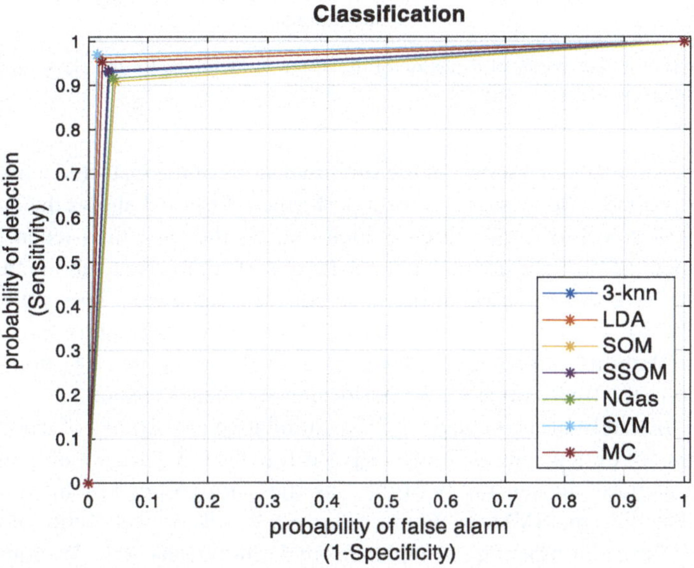

Fig. 2 Classification performance for the different methods to distinguish the classes "Quito", "El Tena" and "Waorani"

Table 3 Confusion matrix for the SVM classifier

		Actual class		
		Waorani	El Tena	Quito
Predicted	Waorani	87	0	1
	El Tena	1	109	5
	Quito	0	2	94

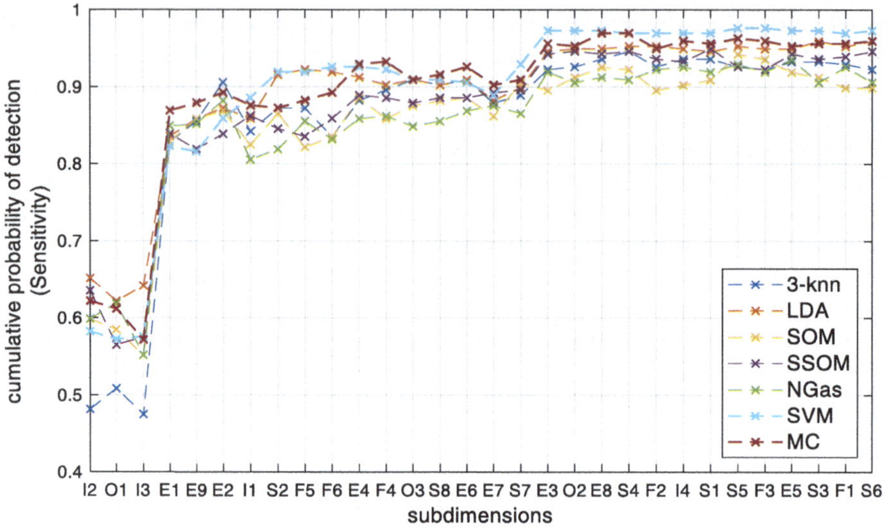

Fig. 3 Cumulative probability of detection of the different identities for the classifiers according to the subdimension of the instrument

Finally, a study about a subset of the subdimensions of the instrument that are able to identify properly the classes has been performed. Figure 3 shows the cumulative probability of detection of the different identities for the classifiers according to the subcomponent of the instrument. As it can be observed, the first subdimension I2 is able to identify about the 60% of the different identities (only the simple 3-knn is below the 50%). The more subdimensions are considered as input of the classifiers, the more probability of detection. However, after using the first 18 subdimensions represented in the figure (axis x), the performance is near constant.

Finally, the study of the behavior of the cumulative probability of detection provides a subset of dimensions as shown in Table 4. In the table, the subdimensions that are able to increase the sensitivity of the classifiers are grouped by the dimensions. The best classifier, the SVM, is able to achieve a 97.7% of probability of detection and 1.2% of false alarms using 9 components of 4 dimensions of the Waorani Identity Instrument. In this case, the Organization dimension is not relevant, being the most important the Economic and Social aspects.

5 Conclusions

This paper analyses an instrument for the measurement of the Waoroani ethnic identity from an Artificial Intelligence (AI) point of view. The instrument composed by 30 subcomponents has been used as the input of several classifiers in order to analyze the performance of an AI system to automatically classify individuals.

Table 4 Subdimensions having an impact in the classification performance

Class.	Economic	Fam.	Ideo.	Org.	Soc.	Sens.	Spec.
3-knn	E1, E2, E3, E8, E9		I2	O1, O3, O2	S4	0.946	0.973
LDA	E1, E2, E3, E9	F5	I2	O2	S2, S4, S3	0.959	0.979
SOM	E1, E2, E4, E8, E9, E8		I2	O2	S7, S5	0.943	0.972
SSOM	E1, E3, E4, E7		I1, I2	O2	S1, S7	0.949	0.975
NGAS	E1, E2, E3, E5, E9	F2	I2, I4	O1	S5	0.936	0.968
SVM	E1, E2, E3	F6	I1, I2		S2, S7, S5	**0.977**	**0.988**
MC	E1, E2, E3, E4, E8, E9	F4	I2			0.969	0.985

Bold significance in table means a better result, that is, a better performance in classification

The results show a high performance both the sensitivity and the specificity of the classifiers being able to provide with high accuracy the ethnic identity of an individual. Moreover, this paper analyses the components of the instrument as a feature selection problem obtaining that a subset of about 10 variables are able to achieve the same performance results.

In particular, the SVM classifier obtained the best results being able to use 4 of the 5 components of the instrument to properly classify the identity of the individuals. Particularly, it used 9 of the 30 original subcomponents representing only the 30% of them. It allows us to reduce the dimensions and practically the number of questions needed to identify them.

The contribution of the AI to the research is twofold. First, it allow us to provide a machine learning method able to automatically identify the identify factors of the Waorani indigenous identity, showing a high performance even compared to close individuals as the as the inhabitants of "La Tena" village. Second, the machine learning method is able to determine the most important dimensions of the instrument to reduce the number of variables needed to identify properly the Waorani indigenous identity and, in consequence, improve the instrument initially provided.

In consequence, as future line a detailed study of the variable selection is proposed. It also will allow us to design distance function (using the subcomponents) able to determine what level of identity is in an individual and whether it can be considered that the original identity of the indigenous nationality has been preserved.

References

1. Goodenaugh, W.H., Kahn, J., Kroeber, A., Malinowski, B., Tylor, E.B., White, L.A.: El concepto de cultura: textos fundamentales. Anagrama (1975)

2. UNESCO: Declaración Universal sobre la Diversidad Cultural una visión una plataforma conceptual un semillero de ideas un paradigma nuevo (2002)
3. González-Varas, I.: Conservación de bienes culturales: teoría, historia, principios y normas. Ed. Catedra (1999)
4. Carrasco-Arroyo, S.: Medir la cultura: una tarea inacabada, pp. 1–2 (2006)
5. Quijano, A.: Colonialidad y modernidad/racionalidad. Perú Indígena **13**(29), 11–20 (1992)
6. Codenpe: Construyendo el Estado Plurinacional. Quito (2011)
7. Vivieros de Castro, E.: Perspectivismo y multinaturalismo en la América indígena. *en Tierra Adentro: Territorio indígena y percepción del entorno*. Lima: International Work Group for Indigenous Affairs (2004)
8. Carrasco, S.: Indicadores culturales: una reflexión. Univ. Valencia Recuperado **21**, 3–4 (1999)
9. Tajfel, H.: Human groups and social categories: Studies in social psychology. CUP Archive (1981)
10. Verkuyten, M.: Self-esteem and the evaluation of ethnic identity among Turkish and Dutch adolescents in the Netherlands. J. Soc. Psychol. **130**(3), 285–297 (1990)
11. Cross Jr., W.E.: The Thomas and cross models of psychological nigrescence: a review. J. Black Psychol. **5**(1), 13–31 (1978)
12. Helms, J.E.: An update of Helm's White and people of color racial identity models.. In: Versions were presented at the Psychology and Societal Transformation Conference, U Western Cape, South Africa, Jan 1994, and at a workshop entitled "Helm's Racial Identity Theory," Annual Multicultural Winter Roundtable, Teachers Coll–Columbia U, New York, Feb 1994 (1995)
13. Espín, J.V., Marín, M.Á., Rodríguez, M., Cabrera, F.: Elaboración de un cuestionario para medir la identidad étnica y la aculturación en la adolescencia. Rev. Educ. Madr. **315**, 227–249 (1998)
14. Phinney, J.S., Rotheram, M.J.: Children's ethnic socialization: pluralism and development. ERIC (1986)
15. Vergara, A.I., Páez, D.: Revisión teórico-metodológica de los instrumentos para la medición de la identidad de género. Rev. Psicol. Soc. **8**(2), 133–152 (1993)
16. Vega, V.C.: Adaptación argentina de un inventario para medir identidad de rol de género. Rev. Latinoam. Psicol. **39**(3), 537–546 (2007)
17. Muradas, I.M., Rodríguez, J.R.O.: Diseño de un instrumento de medida para el análisis comparativo de los recursos culturales tangibles. Número Patrocinado Por, p. 245 (2005)
18. Esteban, M., Vila, I.: Modelos culturales y retratos de identidad. Un estudio empírico con jóvenes de distintos contextos sociodemográficos. Estud. Psicol. **31**(2), 173–185 (2010)
19. Bagnoli, A.: Researching identities with multi-method autobiographies. Sociol. Res. Online **9**(2), 1–15 (2004)
20. Enrique, A.M.: Lucha política y origen de los partidos políticos en el Ecuador, Corporación Editora Nacional, 2da. Ed. Quito Ecuad. (1982)
21. Tristán-López, A.: Modificación al modelo de Lawshe para el dictamen cuantitativo de la validez de contenido de un instrumento objetivo. Avances en medición **6**(1), 37–48 (2008)
22. Kohonen, T.: The self-organizing map. Proc. IEEE **78**(9), 1464–1480 (1990)
23. Papadimitriou, S., Mavroudi, S., Vladutu, L., Pavlides, G., Bezerianos, A.: The supervised network self-organizing map for classification. Appl. Intell. **16**(3), 185–203 (2002)
24. Martinetz, T.M., Berkovich, S.G., Schulten, K.J.: 'Neural-gas' network for vector quantization and its application to time-series prediction. IEEE Trans. Neural Networks/Publication of the IEEE Neural Networks Council **4**(4), 558–69 (1993)
25. Cover, T.M., Hart, P.E.: Nearest neighbor. IEEE Trans. Inf. Theory **13**(1), 21–27 (1967)
26. Hsu, C.W., Lin, C.J.: A comparison of methods for multiclass support vector machines. IEEE Trans. Neural Networks **13**(2), 415–425 (2002)
27. Chawla, N.V., Bowyer, K.W., Hall, L.O., Kegelmeyer, W.P.: SMOTE: synthetic minority oversampling technique. J. Artif. Intell. Res. **16**, 321–357 (2002)

Analysis of Virtual Currencies as Driver of Business Marketing

Higinio Mora and Rafael Mollá Sirvent

Abstract Brand perception is a powerful tool to engage consumers. In a globalized world, the differentiation between products and services is often very difficult. Firms are continuously policing competitors in case they copy their products and services. At the same time, they are looking for characteristic elements to differentiate them over the competition. In this context, new ideas are needed to create a brand image perception among users. Currently, virtual currencies are a novel disruptive technology that is beginning to be known in modern society. In fact, the future course of this new technology and its real effects remain unknown. However, this technology has a modern and innovative feeling that can help firms to gain popularity. This work lies with this emerging issue track at the cross-section of technology and business. Its main contribution is exploring the connection between virtual currency as payment method and marketing strategies of firms.

1 Introduction

The constant advance of Information and Communication Technologies (ICT) has made us face the problems of modern societies in innovative ways. Due to this development of technology and the capabilities it offers, we are witnessing a worldwide revolution. New concepts such as Internet of Things, big data, social network, or cognitive computing are disrupting many areas of industry, society and business. Virtual currencies are one of these new disruptive technologies in modern world [1–3]. A disruptive technology introduces a different set of performance attributes. It can displace an established technology and has the ability to shake up the industry to create an evolution on society [4, 5].

H. Mora (✉) · R. M. Sirvent
Department of Computer Technology and Computation, University of Alicante, Campus de San Vicente del Raspeig, 03690 Alicante, Spain
e-mail: hmora@dtic.ua.es

R. M. Sirvent
e-mail: rmolla@dtic.ua.es

A. Visvizi and M. D. Lytras (eds.), *Research & Innovation Forum 2019*,
Springer Proceedings in Complexity,
https://doi.org/10.1007/978-3-030-30809-4_48

The assimilation of this issues into education and research programs will allow us to make better use of technology and to apply it to the progress of society to keep it moving towards a better world [6].

To understand the scope of virtual currency concept, the next questions-answers are introduced:

What is a virtual currency? it can be defined as a digital representation of value that is not issued by a credit institution nor central bank and it is not backed up by any government. However, it can be used, in some cases, as an alternative to money [7].

How a virtual currency gets its value as money? Virtual currencies acquire their value when used as a means of exchange. If a virtual currency were not used, it would not have intrinsic utility.

Where is using virtual currency? Virtual currencies are mainly used as a payment method although they are also used as a means of speculation.

Currently, virtual currencies are a novel disruptive technology that is beginning to be known in modern society. In fact, the future course of this new technology and its real effects remain unknown. However, this technology has a modern and innovative feeling that can help firms to gain popularity. Beside the inherent advantages of using virtual currencies as payment method, they could be also used as driver of business marketing.

Given this scenario, this paper defends the hypothesis that virtual currencies are used more as marketing tools than as a means of payment.

The remainder of this work is organized as follows: section two presents the role of virtual currencies for business; section three describes the use of virtual currencies as payment method; section four presents the use of virtual currencies as marketing driver; and finally, the conclusions are presented in section five.

2 The Role of Virtual Currencies for Business

To analyze the role played by virtual currencies in business, it would be interesting to first expose the characteristic elements of a virtual currency system. In this way, later, it will be easier to understand the comparison of advantages and disadvantages between traditional and virtual currencies.

The main elements that compose a virtual currency system are the following: the Peer-to-Peer (P2P) network, the blockchain, digital wallets and the currency issuance method [8].

The P2P network is composed of geographic distributed nodes used to propagate transactions or blockchain updates. When a new transaction is created, it is disseminated to all nodes. These nodes are called miners. Miners compete to solve a cryptographic test according to the specification of the transaction. When a transaction is confirmed that update is disseminated to the entire P2P network to update the blockchain.

Blockchain is a database designed to record transaction in a distributed and immutable way. Each node of the P2P network has a copy of the blockchain. The blockchain is public and stores all confirmed transactions by the miners.

Digital wallet is the element whose store the cryptographic key, based on asymmetric cryptography methods. These cryptographic keys allow participants to access their bitcoin address, that is used for making transactions and spending the money. Digital wallets can be stored on a computer, a mobile device or on a physical memory card. There also exist digital wallets in the cloud.

The currency issuance method is based on the validation process of the transactions. The miner nodes compete to calculate valid hashes and are rewarded with new virtual money. Virtual currencies have a limited supply of money. This limitation make them predictable and prevent interferences from central banks.

Once we know the most relevant elements of a virtual currencies system, we can compare them with traditional currencies in terms of use in companies or businesses, to identify the advantages and disadvantages between them. To make this comparative, we will use the Bitcoin as reference virtual currency.

Next, the main advantages of virtual currencies will be exposed [9]:

First, no taxes are paid when receiving virtual currencies. However, for certain types of payments most wallets charge reasonable taxes. These taxes can be higher if the transaction needs to be confirmed sooner.

Second, there is a problem with payments that are later cancelled. For example, using a credit card the buyer can cancel the payment when the seller has already sent the product. This supposes an extra cost for the seller, which affects the final price of the product. With the use of Bitcoin for payments this problem is solved, since Bitcoin payments are irreversible and safe. This protects merchants against fraud in product returns.

Third, the use of Bitcoin for payments offers the possibility of making international payments in a more efficient way. No waiting, no additional charges and no limitations on minimum and maximum amounts that can be sent.

Fourth, in order to make online payments using credit cards, a large number of security measures are necessary, and these must also comply with the PCI (Payment Card Industry) standard. This implies that no costs are applied in the treatment of confidential information of the clients, such as the numbers of the credit cards. Even so, users must keep their wallets safely.

Fifth, the fact that a business starts accepting Bitcoins as a means of payment does give the company some visibility and the possibility of getting new customers.

Sixth, the use of Bitcoin offers the possibility that a payment is only made if a subset of a group of people authorizes the transaction. This feature is known as multi-signature. For example, the board of directors of a company could use this feature to prevent purchases from being made without the sufficient consent of other members.

Finally, the use of Bitcoin allows to offer the greatest possible transparency, since thanks to the blockchain technology, all the transactions made can be consulted. The most indicative case would be the non-profit organizations which could show the collaborators the transactions they receive and even in what the money is used.

Table 1 Advantages and disadvantages of virtual currencies

Advantages	Disadvantages
No taxes are paid when receiving Bitcoins	High volatility of value
For payments, taxes are not proportional to the amount of Bitcoin	Impossibility of use promissory notes
Protection against fraudulent refunds	Not enough laws and regulations
Fast international payments	
No implicit costs for security expenses	
Free visibility	
Protection with multi-signature transactions	
The greatest possible transparency	

About the disadvantages:
First, there is the high volatility of the Bitcoin value with respect to traditional currencies. A year ago, we saw how the capitalization of Bitcoin reached a historical record with a value of 20,089 USD (17/12/2017) and today we see how it is in 4535 USD (21/12/2018) [10]. This supposes a variation of more than 77% value in only one year. This radical change makes businesses distrust Bitcoin. As you will see in the next section of this article, most companies that use Bitcoin to receive payments do so through a currency exchange service to their local currency, to protect themselves against this change in value.

Second, many companies do not have enough cash to keep their activities and finances up to date. Because of this, most businesses use promissory notes at different times to make their purchases. Nowadays, this is not possible to carry out with Bitcoin because the transactions are confirmed almost instantaneously.

Finally, most countries still do not have laws regulating the use of Bitcoin and how to apply the relevant taxes to transactions.

In order to make this comparative more visual, the aforementioned data have been represented in the following Table 1.

The use of Bitcoin offers a series of advantages over traditional currencies, but these advantages are impacted by the lack of legislation by governments and by the impossibility of using promissory notes, which limits to a large extent the use of them, by companies and businesses.

3 Use of Virtual Currencies as Payment Method

Since a few years ago many companies accept Bitcoin as a payment method. These are companies from different places around the world and different levels of billing and size. You can find different websites with extensive listings of companies that accept this method of payment [11, 12], heat maps with the locations of these companies [13] and even search engines that depending on the product you want to buy tell you in which store you can do it with Bitcoin [14]. Some of the most important companies that accept virtual currencies today are Overstock, Microsoft and Expedia [15–17].

However, most companies do not operate directly with Bitcoin, but to make transactions they use currency exchange services, from Bitcoin to their common use currency [18–20]. This is due to the volatility of the Bitcoin and the uncertainty generated in the entrepreneurs by the change of value with respect to the currencies of legal tender. There are several companies offering this service: PayBear.io [21], CoinGate [22], BitPay [23] y Coinbase Commerce [24] which is the most important of the mentioned.

Knowing the modus operandi of the companies regarding the collection of payments in VC it is interesting to observe the data of growth in terms of income and transactions of Coinbase in recent years. The revenues of Coinbase come mainly by the collection of fees in transactions, which depend on the country from which they are made, the amount of the transaction and the currency they use. Taking this into account, it is striking to observe the data of Coinbase's income in 2017 made by the independent study Superfly Insights [25, 26].

It is difficult to determine the exact percentage of companies that accept virtual currencies as a payment method since not all public administrations offer information about how many companies are registered in their territory and every day thousands of new companies are created and others closed. There are websites that offer an estimation of the number of companies registered in the world, although the estimations indicate that there are around 200 million companies in the whole world, we will use the data of the United States of which we have proof [27].

In the United States, there are 63,496,420 registered companies, of which 1075 accept Bitcoin. With this data, it is obtained that the 0.0001693008247% of the companies in the United States accept virtual currencies as payment method or what is the same, 1 of every 59,066 (Data at the end of 2018. The data of companies that accept VC have been obtained from the Coinmap REST API [28], limiting the query to the United States territory). That percentage is too low for what should be analyzed why the companies do not decide to accept Bitcoin.

To date it is very complex to determine if the virtual currency will end up establishing as usual method of payment or if they will disappear completely. The regularization of the use of virtual currency by governments and the proliferation of "stable coins", pave the way for this to happen.

4 Use of Virtual Currencies as Marketing Driver

The hypothesis defined in this work states that virtual currencies are used more as marketing instrument than as a means of payment.

Our reasons to define this assumption is based on the fact that virtual currencies are perceived by citizens as a disruptive technology. In addition, the technology that supports virtual currencies such as blockchain ledger and smart contracts are proposing a disruptive way of doing things [29]. They open a new way to do business

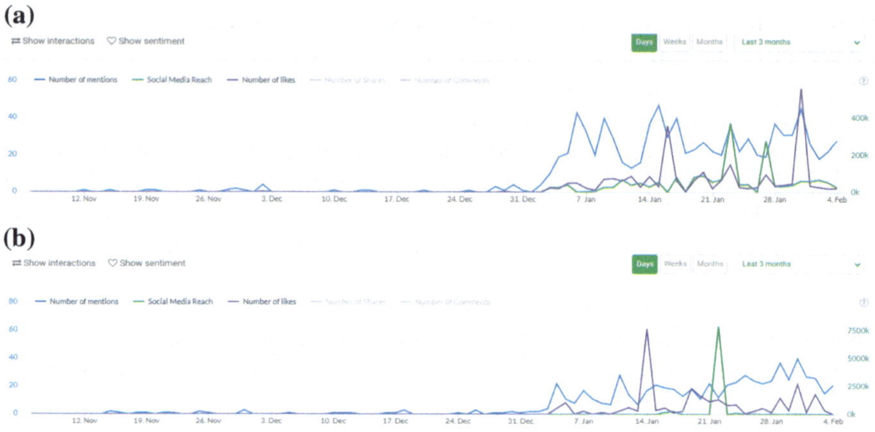

Fig. 1 Impact of social media. **a** McDonalds Canada; **b** KFC Canada. Data retrieved from Brand24 in 04/02/2019

that is in line with modern social business initiatives such as social commerce and collaborative economy [2, 3] and these are also related to sustainable economy [30].

Beyond our assumptions, there are objective facts that point towards this idea: in this work we made a preliminary exploration of social media through a 'Social Media Monitoring Tool' to know the impact of accepting virtual currencies. The tool used is Brand24 (https://brand24.com/). This tool is a reliable, simple and cost-effective social monitoring application used by companies of all sizes and sectors to identify, connect and analyse in real-time online conversations about their brands, products and competitors across the network.

In this work we briefly analyze the companies KFC and McDonalds in Canada. In this country, KFC accepts Bitcoin from last year as payment method [31]. Figure 1 compares the social media interactions between the two companies in Canada. Only the terms of trade marks (#McDonald, #KFC) and #Canada are included in the listening report.

Next, in Fig. 2 the term #Bitcoin is also included in the social media listener.

As can be seen in previous figures, the number of mentions and 'likes' of KFC is much higher than McDonalds in Canada. When the term Bitcoin is included, McDonalds is hardly mentioned in the social media, while KFC has a peak of 180,000 mentions. In addition, it provides a positive opinion about the trade mark.

Fig. 2 Impact of social media and Bitcoin references. **a** McDonalds Canada; **b** KFC Canada. Data retrieved from Brand24 in 04/02/2019

5 Conclusions

This work describes a point of view of using virtual currencies for business based on its utility as marketing instrument. This is a preliminary work with the objective to present this idea. Results of an illustrative example from a marketing social network have been retrieved for supporting the assumptions.

Further future work should be done to rigorously demonstrate this theory. Thus, this work can be expanded in different ways and carry out some initiatives to provide new evidences. Some examples could be the following ideas: in first place, other social network can be analyzed in order to look for wider correlations between popularity and accepting virtual currencies; next a survey can be conducted among users and business to know the motivations of using and accepting virtual currencies; and finally, a correlation analysis between virtual currencies in circulation and number of business that accept them could be done to prove if this number boots the utilization of virtual currencies in real economy.

Finally, in connection with the development of emerging services for citizens [32], an interesting research line is the connection between smart city concept and virtual currencies in order to implement e-government initiatives.

References

1. Baur, A.W., Bühler, J., Bick, M., Bonorden, C.S.: Cryptocurrencies as a disruption? Empirical findings on user adoption and future potential of Bitcoin and Co. In: Lecture Notes in Computer Science (including subseries Lecture Notes in Artificial Intelligence and Lecture Notes in Bioinformatics) (2015)
2. Mendoza-Tello, J.C., Mora, H., Pujol-López, F.A., Lytras, M.D.: Social commerce as a driver to enhance trust and intention to use cryptocurrencies for electronic payments. IEEE Access **6**, 50737–50751 (2018)
3. Mendoza-Tello, J.C., Mora, H., Pujol-López, F.A., Lytras, M.D.: Disruptive innovation of cryptocurrencies in consumer acceptance and trust. In: Information Systems and e-Business Management (2019)
4. Adner, R.: When are technologies disruptive? A demand-based view of the emergence of competition. Strateg. Manage. J. (2002)
5. Sicilia, M.A., Visvizi, A.: Blockchain and OECD data repositories: opportunities and policy-making implications. Library Hi Tech **37**(1), 30–42 (2019)
6. Lytras, M.D., Visvizi, A., Daniela, L., Sarirete, A., Ordonez De Pablos, P.: Social networks research for sustainable smart education. Sustainability **10**, 2974 (2018)
7. European Central Bank: Virtual currency schemes—a further analysis (2015). https://www.ecb.europa.eu/pub/pdf/other/virtualcurrencyschemesen.pdf. Last accessed 13 Sept 2018
8. Bitcoin: A Peer-to-Peer Electronic Cash System. https://bitcoin.org/bitcoin.pdf. Last accessed 5 Nov 2018
9. Bitcoin for Businesses: https://bitcoin.org/en/bitcoin-for-businesses. Last accessed 6 Nov 2018
10. CoinMarketCap—Bitcoin: https://coinmarketcap.com/es/currencies/bitcoin/. Last accessed 21 Nov 2018
11. Who Accepts Bitcoins As Payment? List of Companies, Stores, Shops: https://99bitcoins.com/who-accepts-bitcoins-payment-companies-stores-take-bitcoins/. Last accessed 20 Oct 2018
12. Companies That Accept Bitcoin Payments: https://unblock.net/companies-that-accept-bitcoin/. Last accessed 25 Oct 2018
13. Coinmap: https://coinmap.org/. Last accessed 2 Oct 2018
14. SpendBitcoins: http://spendbitcoins.com. Last accessed 20 Oct 2018
15. Major Companies That Accept Cryptocurrency: https://www.nasdaq.com/article/7-major-companies-that-accept-cryptocurrency-cm913745. Last accessed 22 Oct 2018
16. Pay In Bitcoins: 17 Major Websites That Accept Bitcoins: https://coinsutra.com/who-accepts-bitcoins/. Last accessed 23 Oct 2018
17. How to use Bitcoin to add money to your Microsoft account: https://support.microsoft.com/en-us/help/13942/microsoft-account-how-to-use-bitcoin-to-add-money-to-your-account. Last accessed 23 Oct 2018
18. No, Big Companies Aren't Really Accepting Bitcoin: http://time.com/money/3658361/dell-microsoft-expedia-bitcoin/. Last accessed 22 Oct 2018
19. Bitcoin Terms & Conditions: https://www.expedia.com/Checkout/BitcoinTermsAndConditions. Last accessed 23 Oct 2018
20. Help Center—Bitcoin: https://help.overstock.com/help/s/article/Bitcoin. Last accessed 23 Oct 2018
21. PayBear.io: https://www.paybear.io. Last accessed 30 Oct 2018
22. CoinGate: https://coingate.com. Last accessed 30 Oct 2018
23. BitPay: https://bitpay.com. Last accessed 30 Oct 2018
24. Coinbase Commerce: https://commerce.coinbase.com. Last accessed 30 Oct 2018
25. Report: Coinbase made 43 percent of its 2017 revenue in December alone. https://www.theverge.com/2018/2/27/17058306/coinbase-bitcoin-cryptocurrency-revenue. Last accessed 24 Oct 2018
26. Coinbase had an amazing 2017: Can they sustain their growth? https://www.linkedin.com/pulse/how-do-you-buy-bitcoin-behavioral-study-coinbase-users-jonathan-meiri. Last accessed 25 Oct 2018

27. Datapo, Inc, How many companies are there in the world? https://datapo.com/en/news/how-many-companies-are-there-in-the-world/. Last accessed 25 Oct 2018
28. Coinmap REST API: https://coinmap.org/api/#api-Venue-GetVenue. Last accessed 2 Nov 2018
29. Andoni, M.: Blockchain technology in the energy sector: a systematic review of challenges and opportunities. Renew. Sustain. Energy Rev. **100**, 143–174 (2019)
30. Mora, H., Pujol-López, F.A., Mendoza-Tello, J.C., Morales-Morales, M.R.: An education-based approach for enabling the sustainable development gear. Computer in Human Behavior (2018)
31. Chen, J.: KFC Canada Now Accepting Bitcoin As Payment, Cryptocurrency News. https://cryptocurrencynews.com/daily-news/bitcoin-news/kfc-canada-now-accepting-bitcoin-as-payment/. Last accessed 04 Feb 2019
32. Lytras, M.D., Visvizi, A.: Who uses smart city services and what to make of it: toward interdisciplinary smart cities research. Sustainability **10**, 1998 (2018)

Conceptual Modeling as a Tool for Corporate Governance Support: State of the Art and Research Agenda

Yves Wautelet and Manuel Kolp

Abstract Early days of information systems engineering saw practitioners implementing a software solution without proper analysis of the business processes, organizational environment and partnerships' context. Due to a tremendous failure rate, more and more effort has, over the years, been devoted to accurate domain analysis with the aim of representing the as-is and to-be organizational and data settings through conceptual models. The ability of conceptual models to exhaustively and adequately represent organizational setting behavior has hugely been enhanced through the use of goal-oriented modeling; the latter allows to represent stakeholders as well as their intentions and goals with realization scenarios. Such a tool nevertheless offers more potential than software development only. Researches have used conceptual models to sustain organization theory and represent strategic relationships among organizations in competition, cooperation and coopetition. Also, goal modeling allows to represent the long-term strategy as of a set of business objectives allowing to support reasoning for corporate and IT governance decisions. The paper overviews and investigates the use of conceptual modeling for strategic management to better understand the relationships among organizations, evaluate alternatives and estimate the consequences of IT governance decisions in terms of business and IT alignment. A research agenda is also provided.

1 Introduction

Conceptual modeling has been used for information systems development for more than 50 years. The driving principle is to use abstractions of organizational settings, user requirements, software system behavior and structure, database schemas, etc.

Y. Wautelet (✉)
KU Leuven, Brussels, Belgium
e-mail: yves.wautelet@kuleuven.be

M. Kolp
UCLouvain, Louvian-la-Neuve, Belgium
e-mail: manuel.kolp@uclouvain.be

A. Visvizi and M. D. Lytras (eds.), *Research & Innovation Forum 2019*,
Springer Proceedings in Complexity,
https://doi.org/10.1007/978-3-030-30809-4_49

These so-called models are made of different concepts helping stakeholders to build, document and improve the modeled reality. Most often used into the perspective of representing and documenting an information and technology (IT) system to-be built, such an approach can nevertheless take roots on a more abstract and cognitive level. Goal-Oriented Requirements Engineering (GORE) suggests to represent an organizational setting in terms of actors pursuing individual and collective goals or intentions. As the name indicates, goals in GORE are mostly used to represent the user requirements and show how these can be satisfied/fulfilled within the IT system to-be built.

While conceptual GORE models are traditionally used for system development as the highest-level abstractions on the basis of which the software architecture and design are based, these can also be used as a tool for strategic analysis and decision making (i.e., governance). Two main fields of study have indeed emerged during the last two decades connecting goal-based conceptual modeling with strategic management, i.e., (i) studying strategic relationships and partnerships among organizations; and (ii) business and IT strategy representation to sustain the rationale of corporate and IT governance decisions. All of the contributions to strategic management using conceptual modeling studied here have been realized using the i* framework [35]; the latter is briefly described in Sect. 2.

Organization theory and strategic alliances – i.e., the commercial dependencies in which the company is involved (e.g., outsourcing, strategic alliances, joint ventures, etc.) – have been studied in management sciences for decades. Mostly studied theoretically, they fail to provide a tool for the representation of the stakeholders, their goals, intentions and dependencies which would allow to analyze the rationale of organizational and partnership choices into a specific (custom) context. This has nevertheless been (at least partially) provided by the use of goal-based modeling as will be seen in Sect. 3.

Following Henderson et al. [11], the business strategy defines the scope of the company; how its offer of products or services is positioned onto the market. Through the business strategy, a company is seeking competitive advantages with respect to its competitors (e.g. lower price, better product or quality of service). The business strategy is the domain of overall corporate governance (of which IT Governance is the part devoted to IT). Using conceptual modeling, the business and IT strategies can be represented as a set of long-term objectives to be pursued. All of the (governance-level) decisions on IT acquisitions, performance and compliance will have an impact on their realization; goal-based modeling tools also allow to evaluate the impact of evolutions in the organizational setting on the realization of these objectives. In other words, it can be used as a tool to evaluate the business and IT alignment (BITA) which is one of the crucial aspects of IT governance. This will be developed in Sect. 4.

After the overview of the state of the art of conceptual modeling for organization theory, strategic alliances and IT governance, Sect. 5 depicts the future research agenda in these fields and Sect. 6 concludes the paper.

2 Theoretical Background: The I* Framework

In *i** (which stands for "distributed intentionality"), stakeholders are represented as (social) actors who depend on each other for goals to be achieved, tasks to be performed, and resources to be furnished [35]. The framework and its applications are described in detail in [35]; it includes:

- The Strategic Dependency Model (SD) that describes a network of dependency relationships among various actors in an organizational context. Actors are usually identified within the context of the model. This model shows who an actor is and who depends on the work of an actor. An SD consists of a set of nodes and links connecting actors. Nodes represent actors and each link represents a dependency between two actors. The depending actor is called *Depender* and the actor who is depended upon is called the *Dependee*;
- The Strategic Rationale Model (SR) provides an intentional description of processes in terms of process elements and the rationale behind them. While the Strategic Dependency (SD) model maintains a level of abstraction by modeling only the external relationships among actors, the SR model forgoes that abstraction in order to allow a deeper understanding about strategic actors' reasoning about processes to be explicitly expressed. The SR model describes the intentional relationships that are"internal" to actors, such as means-ends relationships that relate process elements, providing explicit representation of "why" and "how" and alternatives. Rationales are at strategic level, so that the process alternatives being reasoned about are strategic relationships, i.e., SD configurations. Using knowledge represented in and organized by these modeling concepts, process alternatives can be systematically generated and explored to help actors to find new process designs that better address their interests, needs, and concerns. The SR model is a graph, with several types of nodes and links that work together to provide a representational structure for expressing the rationale behind processes.

Figure 1 distinguishes the i* elements and their graphical representation used in the diagrams of the paper. Even if we point to Tropos here, the contribution can nevertheless be generically used in any software engineering method for modeling the organizational setting producing an architectural model of the future software system.

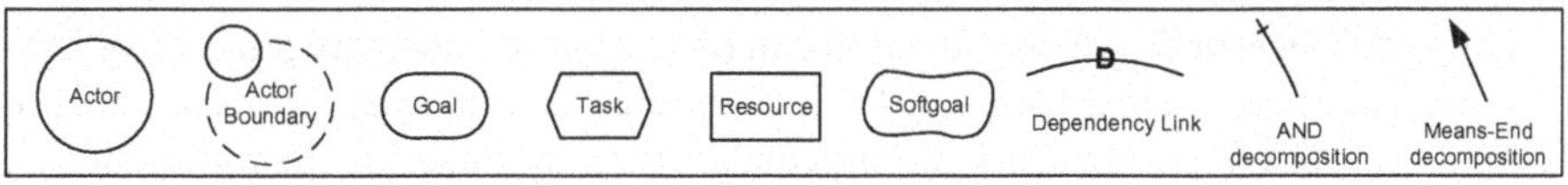

Fig. 1 i* elements and their graphical representation

3 Representing Strategic Relationships Among Organizations Using Conceptual Models

Organizational theories study organizational structures and alternatives among them. These consist of stakeholders – individuals, groups, physical or social systems – that coordinate and interact with each other to achieve common goals. Today, organizational structures are primarily studied by two disciplines: *Organization Theory* (e.g., [5, 18, 24, 34]), that describes the structure and design of an organization and *Strategic Alliances* (e.g., [8, 10, 19, 25, 26]), that model the strategic collaborations of independent organizational stakeholders who have agreed to pursue a set of business goals.

Both disciplines aim to identify and study organizational patterns. These are not just modeling abstractions or structures, rather they can be seen, felt, handled, and operated upon. They have a manifest form and lie in the objective domain of reality as part of the concrete world.

Many organizational styles are fully formed patterns with definite characteristics. Others are not well specified, implemented and evaluated. Michael Porter's generic strategies [23] are examples of such patterns.

In this paper, we are interested to identify organizational styles that have been already well-understood and precisely defined in organizational theories and represented using GORE conceptual models. Representing organizational settings with those styles allows organizations to better characterize their own organizational behavior as well as to look for and evaluate alternatives. We highlight a few interesting findings hereafter.

3.1 Organization Theory

"*An organization is a consciously coordinated social entity, with a relatively identifiable boundary, that functions on a relatively continuous basis to achieve a common goal or a set of goals*" [19]. Organization theory examines the development and implementation of these social entities.

A set of organizational styles are depicted and abstracted into an i* model in [16]. We here illustrate the structure-in-5 only and illustrate it with the case of hospital (see [31] for a complete description of the case study).

The Structure-in-5. An organization can be considered an aggregate of five substructures, as proposed by Minztberg [18]. At the base level sits the *Operational Core* which carries out the basic tasks and procedures directly linked to the production of products and services (acquisition of inputs, transformation of inputs into outputs, distribution of outputs). At the top lies the *Strategic Apex* which makes executive decisions ensuring that the organization fulfills its mission in an effective way and defines the overall strategy of the organization in its environment. The *Middle Line* establishes a hierarchy of authority between the Strategic Apex and the Operational

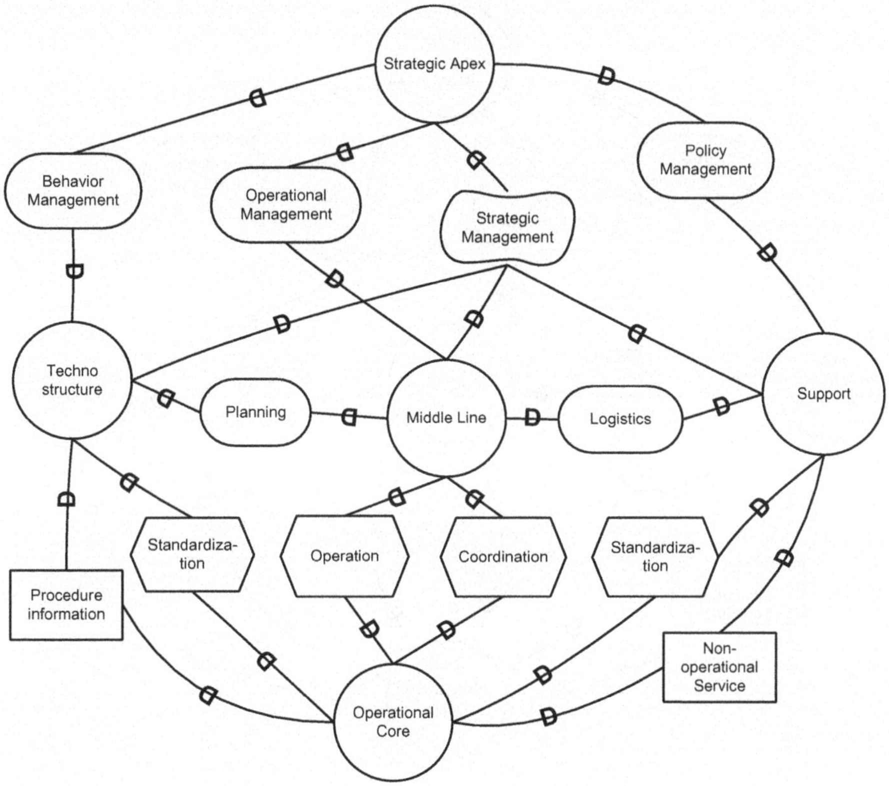

Fig. 2 The structure-in-5 style (from [15])

Core. It consists of managers responsible for supervising and coordinating the activities of the Operational Core. The *Technostructure* and the *Support* are separated from the main line of authority and influence the operating core only indirectly. The Technostructure serves the organization by making the work of others more effective, typically by standardizing work processes, outputs, and skills. It is also in charge of applying analytical procedures to adapt the organization to its operational environment. The Support provides specialized services, at various levels of the hierarchy, outside the basic operating work flow (e.g., legal counsel, R&D, payroll, logistics).

Figure 2 abstracts the Structure-in-5 style as composed of five generic actors. Case studies representing organizations modeled through a structure-in-5 have been performed in the steel industry at Carsid [14], in the automobile industry at Volvo [6] or in higher education organization [16].

We develop here, as a custom contribution of this paper, the application of the structure-in-5 onto the case of an hospital (see [31] for the case study context). The hospital internal functioning can indeed be represented as a socio-technical system following the mentioned pattern as shown in Fig. 3. The *Strategic Apex* is played

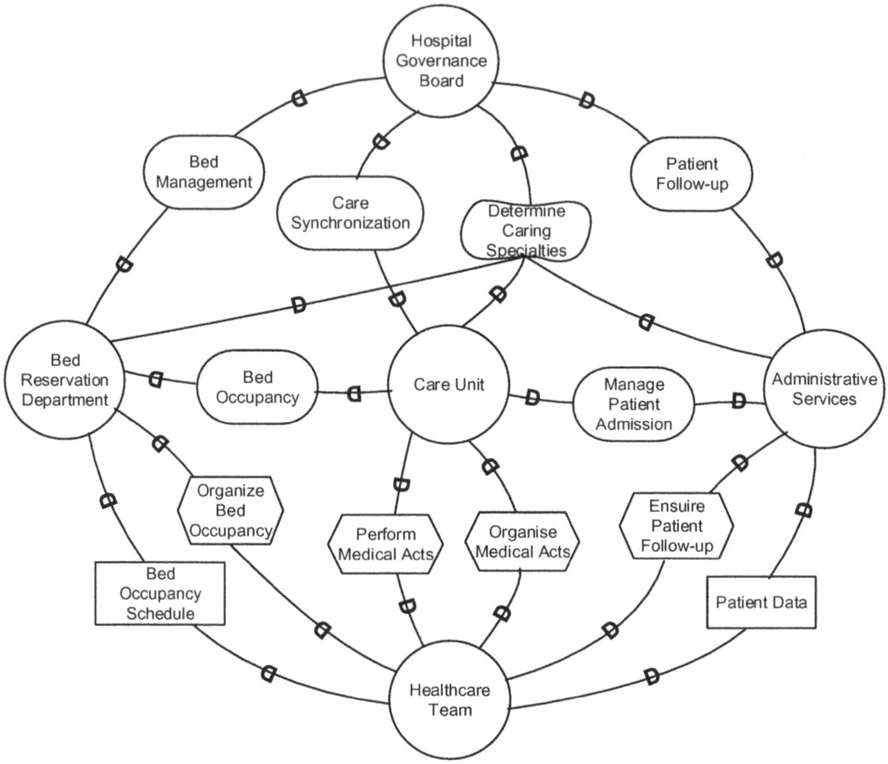

Fig. 3 Stakeholders as a structure in 5

here by the *Hospital Governance Board*, the *Operational Core* by the *Healthcare Team*, the *Technostructure* by the *Bed Reservation Department*, the *Middle Line* by the *Care Units* and the *Support* by the *Administrative Services*.

3.2 Strategic Alliances

Strategic alliance tie specific aspects of different organizations. At its core, this structure is a trading partnership that enhances the effectiveness of the competitive strategies of the participant organizations by providing for the mutually beneficial trade of technologies, skills, or products based upon them. Such an alliance can take several forms; these are described in [16] and can be represented with conceptual models. Outsided of the ones described in [16], we also identify in literature:

The coopetition style involves cooperation and competition between two or more organizations at the same time. Cooperation allows the involved organizations to increase their joint welfare and competition allows them to maximize the individual profit. The i* framework has been used to represent organizations in coopetition and their dependencies [20–22]. Essentially, the i* representations are used in order to align inter-organizational dependencies with a series of IT related design choices. Being in coopetition induces that information systems of the different organizations need to be aligned or at least that they can communicate in an efficient manner with a consequence on data models, integration technologies, organizational processes, etc. Conceptual modeling offers here a tool for expressing and evaluating the relationships between organizations [22] in order to make effective IT governance and management related decisions.

4 Conceptual Modeling for Corporate and IT Governance

Du Plessis et al. [7] indicate that corporate governance refers to the system by which corporations are directed and controlled. The governance structure specifies the distribution of rights and responsibilities among the different stakeholders; it also specifies the rules and procedures for making decisions in corporate affairs. Also, governance provides the structure through which corporations set and pursue their objectives while reflecting the context of the social, regulatory and market environment. IT governance is a subset discipline of corporate governance, focusing on IT, its performance and risk management. The interest in IT governance is due to the on-going need within organizations to focus value creation efforts on an organization's strategic objectives and to better manage the performance of those responsible for creating this value in the best interest of all stakeholders.

GORE models (and goal-driven organizational models) are here used to represent strategic objectives defined by an organization and that allows it to reach a competitive advantage in the long run [27]. Indeed, a strategic objective is defined as *a target that the organization aims to reach within a long-term time horizon (up to 5 years)* [27]. These objectives are part of the business or IT strategy pursued by the organization. Governance decisions are mostly taken in order to satisfy the long-term objectives set by the top-level management. Conceptual modeling can thus offer a tool to sustain corporate and IT governance decisions.

Bleistein et al. [2] introduced the *Strategy-Oriented Alignment in Requirements Engineering* (*SOARE*) approach for e-business systems. The purpose of their framework is to align the supporting system's requirements with the business strategies. With respect to conceptual modeling, SOARE uses GORE to depict the business strategy in the form of long-term objectives. SOARE focuses on building e-business systems. B-SCP [3, 4] is a further development of SOARE; the latter aims to ensure BITA using 3 distinguish views and themes, i.e., *Strategic*, *Context*, and *Process*. As far as conceptual models are concerned, B-SCP relies on i*-like representations to model the business strategy; as well as so-called *Role Activity Diagrams* (*RAD*) [33] for business process modeling and *Problem Frames* [12, 13] for context modeling of the strategy implementation. The main interest of B-SCP in terms of conceptual modeling is that it builds both the business strategy and the system to-be within the same model meaning that there is a unique representation; the different layers are linked using specific arrows.

Wautelet [27] proposes MoDrIGo, a model-driven corporate and IT governance framework allowing to evaluate the alignment of an organization with strategic (business and IT) objectives. MoDrIGo suggests to support governance by modeling the (long-term) business and IT objectives and studying their operational support by so-called business IT services. The business and IT objectives are determined by interviewing C-level executives and traced with operational execution representations of services to determine their match/mismatch. While B-SCP tries to represent the entire organization at once, MoDrIGo focuses on services (more specifically *business IT services*) as scope elements. In other words the governance and management layers are aligned through the services. Typically, a *business IT service* encapsulates a business process that needs to be sustained by IT. Using services as scope elements with an internal representation through i* has already been done in a software development management perspective with I-Tropos [28, 29, 32]. Business IT services are executed in a certain way leading to behavior that will sustain or hamper the realization of the business strategy. The use of i* in MoDrIGo permits to fully depict the behavior of business IT services at runtime so that alignment or misalignment (the ability of operational elements to support or hamper the realization of a business or IT objective) can be evaluated and the degree to which BITA is achieved can be determined. Using business IT services as scope elements allows to align with the service-oriented architecture paradigm and manage the development of the to-be system through independent/asynchronous projects. Also, organizational representations are kept limited which is required to deal with scalability issues. B-SCP prescribe to represent the behavior of the operational realization of a to-be IT system. MoDrIGo favors a middle-out approach where the tactical i* representations can be aligned with the long-term strategy but stakeholders' expectations represented at tactical level can also be used to define or refine business and IT objectives. Also, the concrete implementation of the modeled operational behavior is independent of its i* representation inducing that the BITA evaluation is completely independent of any technological paradigm. Figure 4 represents the alignment of a particular business IT service (Bed Management) onto the Business and IT strategies of an hospital. The internals of the same business IT service are represented through an i* diagram in Fig. 5; more explanations are given in [27].

Because of its service-oriented nature, MoDrIGo is aimed to be used through a service portfolio. Pipeline services are then by far the most important ones because it is precisely for them that improvements made with respect to BITA need to be evaluated for development decision. The behavior of catalog services can also be adjusted in order to provide higher BITA: the impact of modifications of infrastructure or processes can be studied to improve BITA.

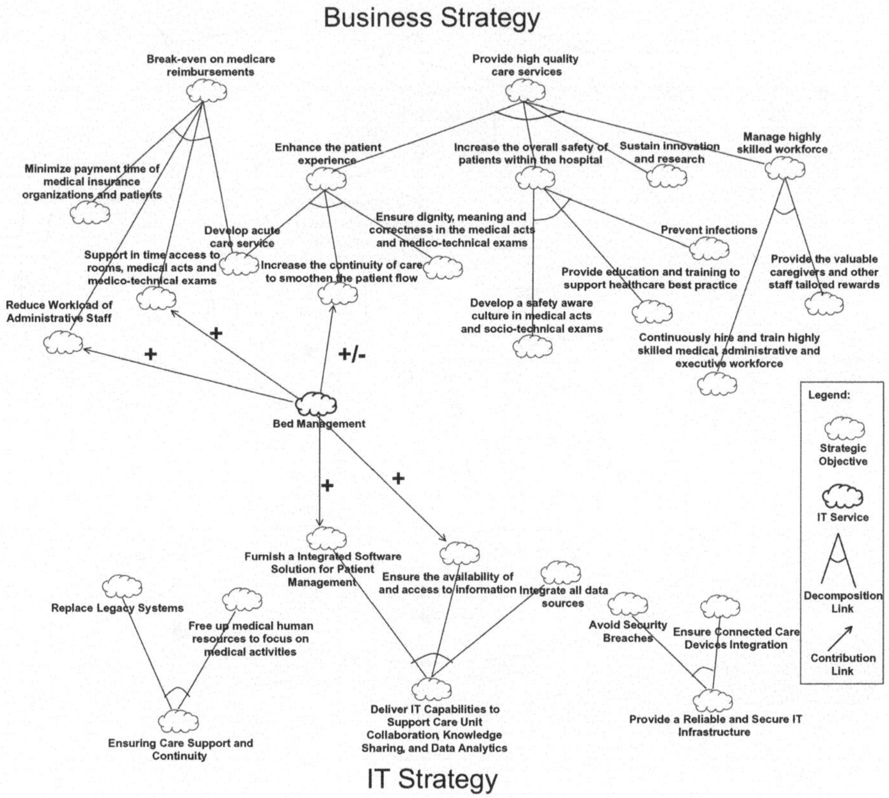

Fig. 4 IT and business strategies NFR decomposition model and bed management service contributions (from [27])

5 Research Agenda

5.1 Using Conceptual Models for Representing Organizational Styles and Strategic Alliances

Both for Organizational Styles and Strategic Alliances we will develop a tool allowing to guide organizations into the selection of a pattern at best aligned with their own situation. The idea is to define, through a systematic literature review, and quantitative studies what are the metrics or Key Performance Indicators (KPIs) that impact the adoption of an organizational style or a strategic alliance so that organizations are able to evaluate their own situation and adopt the patterns that best fit it. These quantitative indicators will be organized in a balanced scorecard such as those used for similar purposes in management accounting.

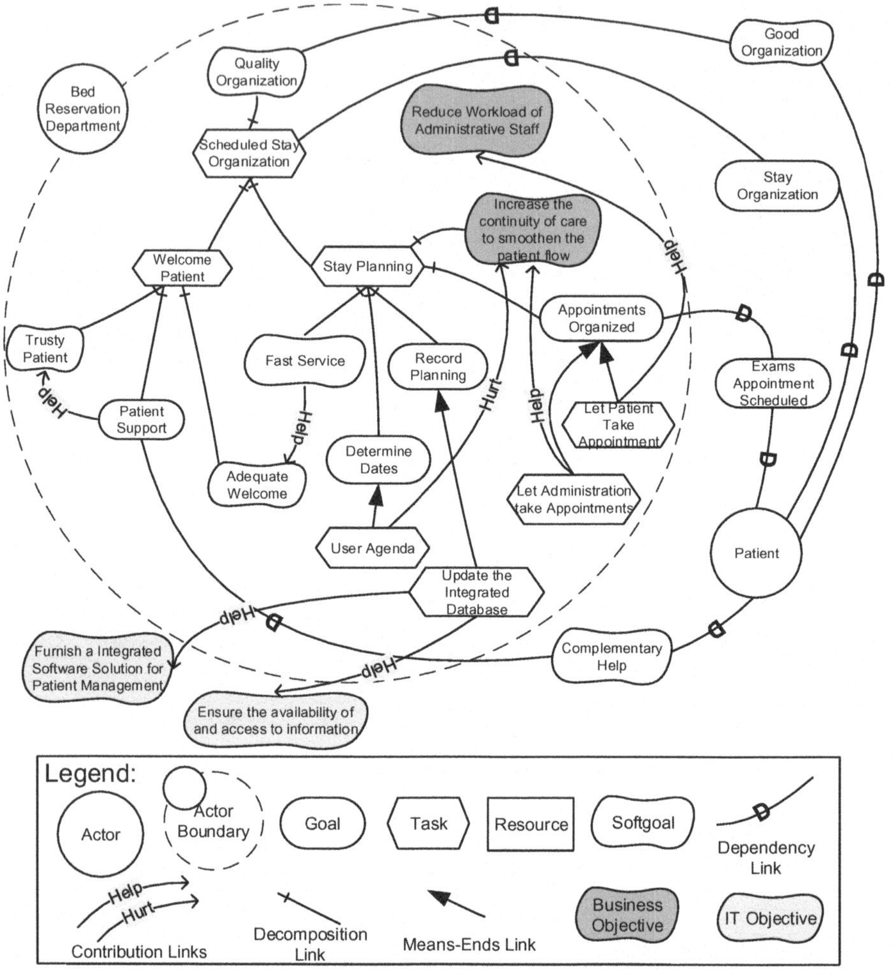

Fig. 5 The bed management service's impact on strategy: management level rationale (from [27])

5.2 Using Conceptual Models to Drive the Strategic Alignment Model

The Strategic Alignment Model (SAM) [1, 11] highlights that an organization is dealing with a business and an IT strategy. As seen, these strategies can be represented in the form of sets of business and IT objectives (both are strategic objectives) using conceptual modeling. Once represented, the impact of one strategy on the other can be highlighted in order to study possible strategic alignments. The SAM specifies that the primary alignment of the business and IT strategies is realized on the strategic level before the organizational and IT infrastructure levels are involved.

When we consider the business and IT strategies and take the SAM as guidance, four potential alignment modes are possible (*strategy execution*, *technology transformation*, *competitive potential* or *service level*). Future research will explore how these 4 alignment modes can be formally sustained with the use of the MoDrIGo framework. More specifically, with the use of the representations of the business and IT strategies on the one side and with the organizational and IT infrastructures on the other side, one can study if these align or not and how to obtain a better alignment in function of the SAM mode the organization finds itself into.

5.3 Towards a Stakeholder-Based Model of Corporate Governance

Stakeholder-based governance is founded on the key principle that governance decisions need to focus on maximizing the value for all of the stakeholders involved in the organization and not the shareholders only [9]. Stakeholders include employees, clients, suppliers, local communities, government, unions, etc. Fixing strategic objectives and supporting these through adequate decisions thus needs to be done to maximize shared value for all of them.

MoDrIGo supports such a type of governance. Indeed, even if the business and IT strategies are often built on the basis of the input of C-level executives, it can also be realized in a bottom-up fashion on the basis of the input of all the required stakeholders. The exercise can then be made to select the relevant elements for each type of stakeholder and include them as strategic objectives. Another approach consists in keeping the strategic objectives defined by top-management and considering/evaluating the alignment of these objectives with the individual goals pursued by all stakeholders and represented at tactical-level. Also, the impact of governance decisions can be considered not only between the strategic and tactical level but also the consequence of the implementation decision on the individual (tactical-level) goals of stakeholders to ensure shared added value. A more formal approach to this is left for future work.

5.4 Combining Strategic-Driven Governance with Agile-Based Software Development

Wautelet et al. [30] proposes a graphical approach for structuring user-story sets. User stories (US) are short, simple descriptions of a feature told from the perspective of the person who desires the new capability, usually a user or customer of the system. US are generally used in agile methods (like Scrum) to represent all of the user and stakeholders requirements. They are very important in the sense that they define what problem should be solved. The Rationale Tree proposed in [30] specifically links the operational elements described in user stories with stakeholder goals to provide a rationale to the functions described in user stories.

Future work is intended to introduce an Agile-MoDrIGo framework. The latter suggests to merge the approaches of MoDrIGo and the Rationale Tree. Concretely, business IT services are defined in a top-down fashion and a set of goals to be met in their development are cascaded from the IT and business objectives of the organization. The alignment of the functions to develop for a specific service (as they are depicted in the user story set) with the business and IT objectives defined at strategic level are then evaluated during each sprint in order to determine the functions that best align. The latter offer maximal long-term value to the business and receive highest development priority in release planning.

6 Conclusion

Conceptual modeling has evolved from analysis support to socio-technical aspects of IT development only to a tool for organizations' strategic decision making support. In literature, multiple usages of GORE conceptual models have been found. One of them is the use of patterns to describe organizational styles and strategic alliances, other ones are the representation of the business and IT strategies to evaluate BITA useful for governance decisions support.

More work can be done for an optimal use of conceptual modeling for strategic management consulting. We have notably pointed out the use of a reference model guiding organizations into the adoption of a particular style or strategic alliances as well as the adaptation of the MoDrIGo framework for support of the SAM and stakeholder-based governance.

Concretely the domain may lead to more fundamental but also applied research. The support and use of the MoDrIGo framework for stakeholder-based governance may also lead to take into account the interest of a larger range of actors for the benefit of the common good. We can take the example of smart cities [17] and e-government for where the number and profile of stakeholders (citizens, (public) administrations, politicians, private companies, etc.) is very high and taking consistent governance decisions is very challenging. Research projects are presently being set-up to support these application domains with the tools presented in the paper.

Finally, to easily put business IT services in production, more work will be done to allow their development in an agile fashion. To this end, conceptual modeling and GORE will be used as a bridge between the operational expectations of users expressed as user stories and the long-term objectives of the organization.

References

1. Avison, D.E., Jones, J., Powell, P., Wilson, D.: Using and validating the strategic alignment model. J. Strateg. Inf. Syst. **13**(3), 223–246 (2004). https://doi.org/10.1016/j.jsis.2004.08.002
2. Bleistein, S.J., Aurum, A., Cox, K., Ray, P.K.: Strategy-oriented alignment in requirements engineering: Linking business strategy to requirements of e-business systems using the SOARE approach. J. Res. Pract. Inf. Technol. **36**(4), 259 (2004)
3. Bleistein, S.J., Cox, K., Verner, J.M.: Validating strategic alignment of organizational IT requirements using goal modeling and problem diagrams. J. Syst. Softw. **79**(3), 362–378 (2006). https://doi.org/10.1016/j.jss.2005.04.033
4. Bleistein, S.J., Cox, K., Verner, J.M., Phalp, K.: B-SCP: a requirements analysis framework for validating strategic alignment of organizational IT based on strategy, context, and process. Inf. Softw. Technol. **48**(9), 846–868 (2006). https://doi.org/10.1016/j.infsof.2005.12.001
5. Daft, R.: Organization Theory and Design. Cengage Learning (2012)
6. Do, T.T., Faulkner, S., Kolp, M.: Organizational multi-agent architectures for information systems. In: ICEIS 2003, Proceedings of the 5th International Conference on Enterprise Information Systems, Angers, France, pp. 89–96 , 22–26 April 2003
7. Du Plessis, J.J., Hargovan, A., Harris, J.: Principles of contemporary corporate governance. Cambridge University Press, Cambridge (2018)
8. Dussauge, P., Garrette, B.: Cooperative Strategy: Competing Successfully Through Strategic Alliances. Wiley (1999)
9. Freeman, R.E.: Strategic management: a stakeholder approach. Cambridge University Press, Cambridge (2010)
10. Gomes-Casseres, B.: The alliance revolution: the new shape of business rivalry. Harvard University Press, Cambridge (1996)
11. Henderson, J.C., Venkatraman, H.: Strategic alignment: leveraging information technology for transforming organizations. IBM Syst. J. **32**(1), 472–484 (1993)
12. Jackson, M.: System behaviours and problem frames: concepts, concerns and the role of formalisms in the development of cyber-physical systems. In: Irlbeck, M., Peled, D.A., Pretschner, A. (eds.) Dependable software systems engineering, NATO science for peace and security series, D: Information and communication security, vol. 40, pp. 79–104. IOS Press (2015)
13. Jackson, M.A.: Problem frames—analysing and structuring software development problems. Pearson Education (2000). http://www.pearsoned.co.uk/Bookshop/detail.asp?item=100000000004768
14. Kolp, M., Wautelet, Y., Faulkner, S.: Sociocentric design of multi-agent architectures. In: Yu, E., Giorgini, P., Maiden, N., Mylopoulos, J. (eds.) Social modeling for requirements engineering. MIT Press, Cambridge (2011)
15. Kolp, M., Giorgini, P., Mylopoulos, J.: Multi-agent architectures as organizational structures. Auton. Agent. Multi-Agent Syst. **13**(1), 3–25 (2006). https://doi.org/10.1007/s10458-006-5717-6
16. Kolp, M., Wautelet, Y.: Human organizational patterns applied to collaborative learning software systems. Comput. Hym. Behav. **51**, 742–751 (2015). https://doi.org/10.1016/j.chb.2014.11.094
17. Lytras, M., Visvizi, A.: Who uses smart city services and what to make of it: toward interdisciplinary smart cities research. Sustainability **10**(6), 1998 (2018)
18. Mintzberg, H.: Structure in fives: designing effective organizations. Prentice-Hall, Englewood Cliffs (1992)
19. Morabito, J., Sack, I., Bhate, A.: Organization modeling: innovative architectures for the 21st century. Prentice Hall, Englewood Cliffs (1999)
20. Pant, V.: Strategic coopetition—a conceptual modeling framework for analysis and design. In: Ralyté, J., Wautelet, Y. (eds.) Proceedings of the doctoral consortium papers presented at the 11th IFIP WG 8.1 Working conference on the practice of enterprise modelling (PoEM 2018), Vienna, Austria, 31 October–2 November 2018. CEUR Workshop proceedings, vol. 2234, pp. 1–14. CEUR-WS.org (2018)

21. Pant, V., Yu, E.: Coopetition with frenemies: Towards modeling of simultaneous cooperation and competition among enterprises. In: The practice of enterprise modeling—9th IFIP WG 8.1. Working conference, PoEM 2016, Skövde, Sweden, 8–10 November 2016, Proceedings, pp. 164–178 (2016)
22. Pant, V., Yu, E.S.K.: Modeling simultaneous cooperation and competition among enterprises. Bus. Inf. Syst. Eng. **60**(1), 39–54 (2018). https://doi.org/10.1007/s12599-017-0514-0
23. Porter, M.: Competitive advantage: creating and sustaining superior performance. The Free Press, New York (1985)
24. Scott, W.: Organizations: rational, natural, and open systems. Prentice Hall, Englewood Cliffs (1998)
25. Segil, L.: Intelligent business alliances : how to profit using today's most important strategic tool. Times Business (1996)
26. Todeva, E., Knoke, D.: Strategic alliances and models of collaboration. Manage. Decis. **43**(1), 123–148 (2005)
27. Wautelet, Y.: A model-driven it governance process based on the strategic impact evaluation of services. J. Syst. Softw. **149**, 462–475 (2019)
28. Wautelet, Y., Achbany, Y., Kolp, M.: A service-oriented framework for MAS modeling. In: Cordeiro, J., Filipe, J. (eds.) ICEIS 2008—Proceedings of the tenth international conference on enterprise information systems, vol. ISAS-1, Barcelona, Spain, 12–16 June 2008, pp. 120–128 (2008)
29. Wautelet, Y., Kolp, M.: Business and model-driven development of BDI multi-agent systems. Neurocomputing **182**, 304–321 (2016)
30. Wautelet, Y., Heng, S., Kiv, S., Kolp, M.: User-story driven development of multi-agent systems: a process fragment for agile methods. Comput. Lang. Syst. Struct. **50**, 159–176 (2017)
31. Wautelet, Y., Kolp, M., Heng, S., Poelmans, S.: Developing a multi-agent platform supporting patient hospital stays following a socio-technical approach: management and governance benefits. Telematics Inform. **35**(4), 854–882 (2018). https://doi.org/10.1016/j.tele.2017.12.013
32. Wautelet, Y., Kolp, M., Penserini, L.: Service-driven iterative software project management with i-tropos. J. UCS **24**(7), 975–1011 (2018). https://www.jucs.org/jucs_24_7/service_driven_iterative_software
33. Wilson, J.M.: Business processes: modelling and analysis for re-engineering and improvement. J. Oper. Res. Soc. **47**(4), 595–596 (1996)
34. Yoshino, M., Rangan, U.S.: Strategic alliances : an entrepreneurial approach to globalization. Harvard Business School Press (1995)
35. Yu, E.: Modeling strategic relationships for process reengineering, chap. 1–2, pp. 1–153. MIT Press, Cambridge, USA (2011)

Price Determinants of Tourist Accommodation Rental: Airbnb in Barcelona and Madrid

Josep Lladós-Masllorens and Antoni Meseguer-Artola

Abstract The emergence of Airbnb in many of the most populated destinations for tourism and leisure has disrupted the market for rental accommodation. Despite pricing is one of the most critical factors in the evolution and success of the accommodation industry, only a few researchers have focused on the main factors determining prices in these digital platform for non-hotel accommodation. We have analyzed the activity in two principal Spanish cities, which are also object of an intense tourist activity: Barcelona and Madrid. The research has been carried out differently for non-commercial and multi-hosting (or commercial) hosts. Different dimensions of pricing had been analyzed: location, host's characteristics, property attributes and quality signaling factors. In addition, we also focus on the role of trust and reputation because traditional hotels have a clear competitive advantage in reducing risks through standardization, business reputation and safety regulations. The results show that site-specific characteristics, property qualities and online reviews explain most of pricing in Airbnb, with hosts also capitalizing their good reputation and professional status. Finally, the interaction of these accommodations with the evolution of rents in both cities has been also investigated. The results do not seem to confirm the complementary nature of the platform.

1 Introduction

1.1 Theoretical Background

Recent years have seen a proliferation of online peer-to-peer marketplaces originating the emergence of different business models. Some of them directly affect

J. Lladós-Masllorens · A. Meseguer-Artola (✉)
Open University of Catalonia (UniversitatOberta de Catalunya), Av. Tibidabo 39-43, 08035 Barcelona, Spain
e-mail: ameseguer@uoc.edu

J. Lladós-Masllorens
e-mail: jlladosm@uoc.edu

A. Visvizi and M. D. Lytras (eds.), *Research & Innovation Forum 2019*, Springer Proceedings in Complexity,
https://doi.org/10.1007/978-3-030-30809-4_50

the accommodation sector for tourism [1]. The economic consequences had made Airbnb a disruptive force for the traditional hotel industry and this online platform has become an increasing center of attention for academic research.

Many investigations had emphasized the multiple and diverse benefits of sharing economy-based accommodation rental, both from users and suppliers [2–4]. However, although pricing is one of the most critical factor defining success in the hospitality industry [5–7], only a few researches analyze price determinants of these rental services based on sharing economy models [8–11].

The examination of pricing configuration could provide significant insights about the business model and the economic consequences of these networked hospitality services. Airbnb is providing the most successful digital marketplace for tourism accommodation rental. The company offers accommodation in several of the most flourishing tourist destinations and its employs a self-assessment strategy to disclose information about the different qualities of the hospitality services traded in its marketplace.

In contrast to the hotel industry, trust and reputation become the weaker link in the value chain of tourist accommodation rental, because these digital platforms lack the competitive advantages based on standardization, ranking and brand [12]. The mutual review system of hosts and guests could be the foundation of trust in Airbnb transactions, creating value as reputational capital allowing for higher prices [13–16].

This network platform is also becoming a topic of increasing concern for policy-makers because these accommodation rental services for visitors could be accelerating rental prices and limiting the goal of affordable housing for low-income residents, as well as increasing rent gaps and aggravating the process of urban gentrification [17, 18]. In addition, the growing presence of commercial hosts, who own multiple listing of entire homes in central locations could suggest that Airbnb would actually be more like a rental marketplace rather than a spare-room sharing platform [19, 20].

The aims of this study would be the following:

- to identify the main factors determining the price of accommodation rentals using Airbnb as a digital marketplace,
- to check the contribution of this online marketplace for accommodations to the rise of rents in destinations supporting a heavy load of tourist activity,
- to find out how consumers perceive the role of trust and reputation in this digital community marketplace,
- to identify the main sources of value creation for them, and
- finally, to analyse the influence of professional agents in the digital marketplace for tourist rental accommodation.

Our integrated model to analyzing the determinants of price on Airbnb listings considers a variety of variables which measure different key factors. According to [10], we classify price determinants into these categories: location, host's characteristics, property attributes and quality signaling factors.

We have used variables to measure the location of each housing. In each city, we have chosen the districts with most favourable location to enjoy tourist amenities together with the neighbourhoods that have experienced the higher increase of rents.

These variables are dichotomic, containing information about the neighbourhood where the offer is placed.

Taking into account the fact that privacy perception is relevant when evaluating different housing alternatives, we included another dichotomic variable *Entire apartment/house* which has a value of 1 if the house is not shared with neither other users nor the host, and 0 otherwise.

Other important elements with a potential influence on price are the specific characteristics of the housing. Consequently, we have introduced the construct *Characteristics*, measured through the following three items: number of accommodates, number of beds and number of bedrooms.

With the objective of including information about the users' quality perception, we have considered information coming from the users' reviews. First, we have included the variable *Number of reviews* to measure the robustness of the issues evaluated by the users. Secondly, we have defined the construct *Perceived quality* from the following five review scores (items): accuracy, cleanliness, check-in, communication and location.

As it has been explained before, trust plays a central role in the purchase decision and this implies that it has to have a clear impact in the price determination. To include this concept in our model, we have built an additive index *Trust* from the following three hosts' attributes: host has been verified (or not) by Airbnb, host has a city identification (or not) and/or host has a jumio identification (or not). Consequently, this index ranges from 0 to 3.

2 Research Method

To test the hypothesized relationships between all independent variables and Price, and to check the internal reliability of the constructs used in the model, we have used SPSS 23.0 software and data gathered from Inside Airbnb (http://insideairbnb.com/). Raw data compiled on 10th of October, 2018 for Barcelona and Madrid was obtained from this web page. The data base was cleaned by considering only the registers having identification and containing non-missing data in the variables included in the model. This process resulted in a valid sample of 7006 registers in the case of Barcelona and of 7672 registers in the case of Madrid.

With the objective of analyzing the differences between private hosts and professional hosts, each sample was divided into two subsamples. One subsample contained information about those hosts having just 1 or 2 offers listed in Airbnb, and the other subsample information about those professional hosts having 3 or more housing offers in Airbnb. As a result of this split, we got that 37.70% of the listed offers came from professionals in Barcelona. In the case of Madrid, we had a very similar result: 37.73% of the registers are offered by professionals hosts.

3 Analysis of Results

Once the internal reliability of the constructs used in the analysis was checked for each subsample, we tested the proposed model for each 2 $\times$ 2 cases using multiple regression analysis. Results in Tables 1 and 2 show that in all four cases the model permits to explain around 50% of the price variability. Additionally, all the ANOVA analyses resulted with significant F-statistics. We can consider then that the models have an acceptable goodness of fit and are significant globally. Consequently, that are appropriate to explain the price of the Airbnb offers.

Regression results show that there are great similarities between Barcelona and Madrid. While the variables Entire apartment/house, Characteristics, Number of reviews, and Perceived quality are significantly different from zero (p-value <0.05) in both cases, the private case and the professional case, the variable Trust is not significant for the professional subsamples.

As expected, the number of people accommodated and the provision of real beds, bedrooms and bathrooms cause a more expensive lodging. The preference for privacy is also clearly confirmed, because the provision of a entire house or an apartment leads to higher prices.

Location is also a significant factor determining pricing. The further the accommodation from the tourist amenities, the lower its price. So, in relation to the neighborhoods, we can see that some of the central quarters are significant in both subsamples

Table 1 Regression results of the Airbnb model in Barcelona

Variables	Private	Professional
Intercept	45.939*	59.047*
Ciutat Vella	9.196*	17.332*
Modernism	18.187*	38.145*
SantMartí	3.670*	9.186*
SantAndreu	−8.072*	−21.399*
NouBarris	−9.226*	−17.584*
Poble sec	1.537*	−1.840*
Sants	−0.760*	2.270*
SagradaFamília	−0.044*	18.962*
Entire apartment/house	30.386*	30.764*
Characteristics	19.442*	35.798*
Number of reviews	−0.044*	−0.064*
Perceived quality	1.016*	2.978*
Trust	0.870*	1.070*
R^2	0.486*	0.513*
F	305.266*	189.587*

*p-value < 0.05

Table 2 Regression results of the Airbnb model in Madrid

Variables	Private	Professional
Intercept	42.138*	52.279*
Centro	9.660*	13.163*
Retiro	10.616*	9.794*
Chamberí	10.322*	9.797*
Salamanca	13.390*	49.704*
Tetuán	−1.265*	−7.262*
Carabanchel	−11.072*	−12.448*
PuenteVallecas	−13.557*	−16.447*
Villaverde	−12.902*	−6.034*
Barajas	2.037*	−0.024*
Moratalaz	−6.448*	−18.204*
Entire apartment/house	24.593*	38.425*
Trust	0.702*	−1.462*
Number of reviews	−0.087*	−0.169*
Characteristics	19.819*	26.859*
Perceived quality	2.942*	2.650*
R^2	0.522*	0.509*
F	311.647*	172.932*

*p-value < 0.05

(CiutatVella, Modernism and Sant Martí, in the Barcelona case, and Centro, Salamanca, Carabanchel and Puente Vallecas, in the Madrid case). On the contrary, there are some of them that are not significant neither for the private subsample nor for the professional subsample: SantAndreu, Poble sec and Sants, in Barcelona, and Tetuán, Barajas and Moratalaz, in Madrid. The other neighborhoods have a different behavior depending of the subsample.

Once we control the other factors, online reviews effectively play a significant role [21]. However, the number of reviews has a negative effect on price, because most tourists probably choose rent sharing economy-based accommodation to reduce costs. This data suggest that the most affordable sites probably suffer a higher rotation.

The results also confirm that the value of host identity verification is not only significant for trust empowerment, but also for price strategies. Hosts with verified identities usually benefit from premium prices because Airbnb users perceive this verification as a quality signal. The level of hosts' trustworthiness affects listings' prices and probably of being chosen, when other variables are controlled for [22]. In addition, as price is a critical factor that enhances consumers' perceived value [23], trust and reputation could be also critical for the repurchase intention of Airbnbn consumers.

This finding supports the conclusion previously in other studies: that hosts capitalize on a good reputation and professional status [8–10, 15]. However, we detect that this response essentially happens in the case of non-professional hosts. Since host verification is more common among commercial hosts, probably this variable is not significant for pricing.

Regarding to the contribution of Airbnb to the rise of rental prices in the urban core of both cities, this influence is clearly confirmed. Professionals hosts, mostly located in the hot spots, are accelerating the increase of prices and the social perception of shortage in rental accommodation.

Although the digital platform is not directly motivating the sharp increase of rents in the peripheral districts of these cities, probably is instigating a side effect on prices due to the displacement of residents from central places searching for an affordable rent.

4 Concluding Discussion

Airbnb, an online marketplace for accommodation rentals, has recently experienced a huge growth accompanied by a vivid academic and political discussion about its economic consequences and also giving place to diverse and heterogeneous regulations around the world. Extracting information from the website http://insideairbnb.com/, a large sample of tourist accommodation (more than 14.600 individuals) has been analyzed, related to the two most important Spanish cities, Barcelona and Madrid which are also object of an intense tourist activity. This research has been carried out.

The main objective of this research, conducted differently for non-commercial and multi-hosting (or commercial) hosts, is to investigate the price determinants in this marketplace. We confirm that, similarly to the hospitality industry, pricing in tourist accommodation rental is significantly influenced by hosts attributes, location, property characteristics and services, rating and the number of online customer reviews.

These variables mostly define the creation of value in the digital marketplace. However, the contribution of trust and reputation is particularly important for non-commercial hosts to perceive higher incomes through the building of reputational capital.

The decisive influence of commercials hosts on the evolution of rents is verified, essentially due to their disproportionately high participation in the supply of entire homes and accommodations located in the city center. They use to charge higher prices on accommodations located in central places, directly contributing to the rise of rents. In addition, this digital marketplace is inciting a displacement of residents and a side effect on rents in non-central locations is also confirmed. These results broaden the scope of previous research about the consequences on the urban landscape.

This study has an important limitation. The effect of amenities and rental rules on prices has not been tested. Probably, some services provided by hosts and the degree of flexibility in accommodation rules could have an impact on prices. In addition, although this is not the specific aim of this research, the study does not focus on the demand side of the marketplace. Some attributes of Airbnb consumers could also influence the price of transactions.

References

1. Sundarajan, A.: From Zipcar to the sharing economy. Harvard Bus. Rev. Retrieved from https://hbr.org/2013/01/from-zipcar-to-the-sharing-eco/ (2013)
2. Botsman, R., Rogers, R.: What's mine is yours: the rise of collaborative consumption. HarperCollins, New York, NY (2010)
3. Guttentag, D.: Airbnb: Disruptive innovation and the rise of an informal tourism accommodation sector. Curr. Issues Tourism **18**(12), 1192–1217 (2015)
4. Zervas, G., Proserpio, D., Byers, J.: The rise of the sharing economy: estimating the impact of Airbnb on the Hotel Industry. J. Mark. Res. **54**(5), 687–705 (2017)
5. Hung, W.T., Shang, J.K., Wang, F.C.: Pricing determinants in the hotel industry: quantile regression analysis. Int. J. Hospitality Manage. **29**(3), 378–384 (2010)
6. Zhang, H., Zhang, J., Lu, S., Cheng, S., Zhang, J.: Modeling hotel room price with geographically weighted regression. Int. J. Hospitality Manage. **30**(4), 1036–1043 (2011)
7. Yang, Y., Mueller, N.J., Croes, R.R.: Market accessibility and hotel prices in the Caribbean: the moderating effect of quality-signaling factors. Tour. Manag. **56**, 40–51 (2016)
8. Gutt, D., Herrmann, P.: Sharing means caring? Hosts' price reaction to rating visibility. In: ECIS 2015 Research-in-Progress Papers. Paper 54 (2015)
9. Li, J., Moreno, A., Zhang, D.J.: Agent behavior in the sharing economy: evidence from Airbnb. Ross School of Business Working Paper Series. Working Paper 1298 (2015)
10. Wang, D., Nicolau, J.L.: Price determinants of sharing economy based accommodation rental: a study of listings from 33 cities on Airbnb.com. Int. J. Hospitality Manage. **62**, 120–131 (2017)
11. Zhang, Z., Chen, R.J.C., Han, L.D., Yang, L.: Key factors affecting the price of Airbnb listings: a geographically weighted approach. Sustainability **9**, 1635 (2017)
12. Oskam, J., Boswijk, A.: Airbnb: the future of networked hospitality businesses. J. Tourism Futures **2**(1), 22–42 (2016)
13. Finley, K.: Trust in the Sharing Economy: An Exploratory Study. University of Warwick, Centre for Cultural Policy Studies (2013)
14. Wu, J., Ma, P., Xie, K.: In sharing economy we trust: the effects of host attributes on short-term rental purchases. Int. J. Contemp. Hospitality Manage. **29**(11), 2962–2976 (2017)
15. Ikkala, T., Lampinen, A.: Monetizing network hospitality: hospitality and sociability in the context of Airbnb. In: Proceedings of the 18th ACM Conference on Computer Supported Cooperative Work & Social Computing, ACM, Vancouver, pp. 1033–44 (2015)
16. Teubner, T., Hawlitschek, F., Dann, D.: Price determinants on Airbnb: how reputation pays off in the sharing economy. J. Self-Governance Manage. Econ. **5**(4), 53–80 (2017)
17. Lee, D.: How Airbnb short-term rentals exacerbate los angeles's affordable housing crisis: analysis and policy recommendations. Harvard Law Policy Rev. **10**, 229 (2016)
18. Wachsmuth, D., Weisler, A.: Airbnb and the rent gap: gentrification through the sharing economy. Environ. Planning A Econ. Space **50**(6), 1147–1170 (2018)
19. Gurran, N., Phibbs, P.: When tourists move in: how should urban planners respond to Airbnb? J. Am. Plan. Assoc. **83**(1), 80–92 (2017)
20. Ke, Q.: Sharing means renting?: an entire-marketplace analysis of Airbnb. In: Proceedings of the ACM on Web Science Conference, Cornell University, pp. 131–139 (2017)
21. Zervas, G., Proserpio, D., Byers, J.: A First Look at Online Reputation on Airbnb, Where Every Stay is Above Average. Available at SSRN: https://ssrn.com/abstract=2554500. or http://dx.doi.org/10.2139/ssrn.2554500 (2015)

22. Ert, E., Fleischer, A., Magen, N.: Trust and reputation in the sharing economy: the role of personal photos on Airbnb. Tourist Manage. **55**, 62–71 (2016)
23. Liang, L.L., Choi, H.S.C., Joppe, M.: Understanding repurchase intention of Airbnb consumers: perceived authenticity, electronic word-of-mouth, and price sensitivity. J. Travel Tourism Mark. **35**(1), 73–89 (2018)

Towards Personality Classification Through Arabic Handwriting Analysis

Mohamed A. Mostafa, Muhammad Al-Qurishi and Hassan I. Mathkour

Abstract Classification of personality types based on Arabic handwriting is a challenging task. It has multiple practical application since handwriting is a unique attribute for each person that provides as much differentiation as fingerprints. Accurate analysis of handwritten documents is useful in diverse areas including human resource management, criminal justice, forensics, security, archaeology, and countless other spheres of life. Writer identification is a big challenge mostly due to a limited human capability for observing and recognizing different styles of writing. Arabic is a language used by millions of people. Recognition of handwritten Arabic characters would, therefore, be of tremendous help in various sectors like mail sorting, verification of checks, etc. even in countries where Arabic is used only occasionally. For this purpose, we collected around 160 samples of Arabic handwriting from 83 different writers, where every writer wrote the same paragraph in both normal and fast writing. Carl Jung's and Isabel Briggs Myers' personality type theory with 16 different personality type was used for classifying the writers. The writers were asked to complete the personality test that contains 64 questions related to Jung-Myers' typology. Specific features of Arabic language handwriting were used to match each written document with a personality type of the writer based on a set of classification techniques. We used four different machine learning classification algorithms with 4-fold cross validation technique. We conducted two experiments and achieved a maximum accuracy of 68.67%.

M. A. Mostafa (✉) · H. I. Mathkour
Computer Science Department, King Saud University, Riyadh, Kingdom of Saudi Arabia
e-mail: momostafa@ksu.edu.sa

H. I. Mathkour
e-mail: mathkour@ksu.edu.sa

M. Al-Qurishi
Elm Company—Research & Innovation, Riyadh, Kingdom of Saudi Arabia
e-mail: malqurishi@elm.sa

A. Visvizi and M. D. Lytras (eds.), *Research & Innovation Forum 2019*, Springer Proceedings in Complexity,
https://doi.org/10.1007/978-3-030-30809-4_51

1 Introduction

Since its onset, graphology was primarily used to select the right person for a particular task or responsibility. It's not hard to envision the negative consequences in case the outcome of the predicted trait did not match the real behavior of the person sampled. With a more accurate model, recruiting agencies and numerous other specialist organizations would have a much easier job. The idea that handwriting features can be leveraged to predict a person's traits dates back to ancient times [1]. However, scores of these systems are plagued by inaccurate results as there is a significant drift from the preexisting Graphology principles.

Analysis of handwriting can be a reliable method for writer identification because handwriting is a unique attribute of each individual, equally unique and informative as fingerprints. Analysis of handwritten documents is commonly applied in human resource management, criminal justice, forensics, security & access control, archaeology and many other fields. Accurate recognition of any form of writing is quite crucial for the ability to acquire the intended information, which can be stored and shared across diverse platforms. In most languages around the world, accurate script recognition does not present enormous problems since the majority of characters are unique in both style and shape. However, this is not the case with the Arabic language, which is highlighted by scores of significant variances within the language and similarity of a number of characters. Despite significant efforts by recent researchers [2–4], varying writing styles among people and segmentation problems still represent major obstacles to accurate handwriting analysis.

This work is vital since Arabic is a language used by hundreds of millions of people globally, with the greatest concentration of speakers in the Middle East. Recognition of handwritten Arabic characters would, therefore, with results that truly reflect the gender of the sampled handwriting, provide massive benefits to the aforementioned sectors and empower development of advanced biometric systems. Very little research has been commissioned on the topic of handwritten Arabic characters' recognition despite the fact there are a number of languages that use Arabic characters such as Urdu, Pashto, Persian and many others. This work is essential given that Handwritten character recognition systems can be extremely valuable mail sorting, verification of checks, creation of electronic libraries, etc. In this regard, it is essential to create a reliable system for personality recognition based on Arabic handwriting.

The main challenges that we faced in this work were: a shortage of public datasets to experiment on, as well as the fact that one trait can have a multiplicity of variations with diverse users, arriving at a meaningful threshold that gives meaningful results. In the next section, we summarize the related work on handwriting analysis in general, and then specifically in the context of Arabic language. In Sect. 3, we will explain our research methodology and finally in Sect. 4, we will discuss the results.

2 Related Work

Although a lot of work has been done in the field of graphology, its scope is still narrow while the manual testing process is still quite slow and poorly suited for the contemporary digital world. To make matters worse, the skill level of the responsible analyst plays a crucial role in the accuracy of the outcome. The idea of demonstrating an autonomous model that combines both the handwriting and signature features to reliably predict the personality of the respective individual is still a considerable challenge.

In [5], Djamal et al., present a research on the capability of predicting a person's behavior and social background using both handwriting and signature features. Sen and Shah [6] also used graphology to implement an automated handwriting analysis system. This system extracts six features and then processes them in MATLAB, and a GUI is used to present the character traits of the writer. Gattal et al. [7] present a computerized model that can reliably predict the gender of an individual based on sampled handwriting. Hadler et al. [8] were interested in writer identification in the realm of Bangla, so they used visual handwriting-based features and a combination of multi-layer perceptron and simple logistic classifiers for the purpose of decision making in the algorithm.

When referring specifically to graphology for Arabic handwriting aimed at detecting writer personality, we found some research papers attempting to identify writers. In addition, insufficient research has been conducted on handwritten Arabic characters' recognition, limiting the progress of handwriting analysis. There are several different approaches to the problem of offline handwriting recognition and classification introduced in the literature.

In [2], Boufenar et al., investigated the feasibility of using what is known as Artificial Immune System (AIS) for interpreting Arabic characters. However, from the experimental results, it seems that this work did not outperform the state of the art method using the same dataset. The work by El-Sawy et al. [9] used a Convolution Neural Network (CNN) that draws inspiration from how information processing is performed by the human brain. Another work done by the same authors [10], proposed certain optimization techniques that can be included to improve the performance of CNN. Loay et al., presented a paper [11], which introduced a new unsupervised learning approach with stacked Autoencoders, basically representing an artificial neural network with a number of layers.

In [3], the authors proposed a statistical approach to Arabic text recognition based on the Hidden Markov Models (HMM) Toolkit (HTK). Even though the primary advantage of this method is the absence of prior segmentation, either the steps should be rearranged or a different set of parameters must be incorporated to obtain effective feature extraction. Ahmed et al. [4] presented a clustering approach to Arabic writer identification. In another work [12], De Sousa proposed a learning algorithm to improve speed and accuracy in Arabic handwritten character and digit recognition.

3 Proposed Methodology

The proposed methodology of our work is different from what we found in the literature review. In this research, we adopt a personality classification based on Jung-Myers' typology. Carl Jung (1875–1964) was a globally renowned scientist in the fields of psychiatry, psychology, psychotherapy, and analytical psychology [13]. In his book [14], 'Psychological Types', Jung classified people according to general attitude types into two main attitude groups: extraversion (E) vs introversion (I). In addition, people in each group were further divided into two primary types according to psychological functions: perceiving functions [sensation (S) and intuition (N)], and judging functions [thinking (T) and feeling (F)]. Briggs and Myers in their book [15] divided each of the attitude types into two classes—[Judging (J) and Perceiving (P)].

3.1 Collecting Dataset

In order to classifying the personality types of people according to Jung- Myers typology, we had to build a new real-world handwriting dataset from scratch, collecting the information from Arabic speakers with a questionnaire. This questionnaire was used to assign any person into one of the previously defined groups. i.e. the result of the questionnaire should determine whether the person is an extravert (E) or introvert (I), whether his/her perceptive style is based primarily sensation (S) or intuition (N), whether he/she is predominantly a thinking (T) or feeling (F) person, and finally whether judging (J) or perceiving (P) attitude is displayed by the person.

We designed a handwriting sample (see Fig. 1) consisting of one paragraph (at the top) that needs to be reproduced by the participant in the two rectangles underneath, one with normal writing speed and other written with a time limit.. The participant is also asked to fill the questionnaire related to the personality test at the same time. In total, there were 83 participants in this study, with 160 paragraphs written at normal speed collected along with 158 paragraphs written quickly. From the handwriting documents, we can extract the handwriting features while matching them with personality assessments from the questionnaire part.

3.2 Handwriting Features

There are certain parameters that needed to be taken into consideration when collecting data. We considered four parameters: gender, nationality (as we have multiple Arabic countries), age, and education level. We included people of 6 different nationalities, mostly from Saudi Arabia and Egypt. Also, we categorized the age parameter to be in a nominal form. The handwriting features have been extracted from our

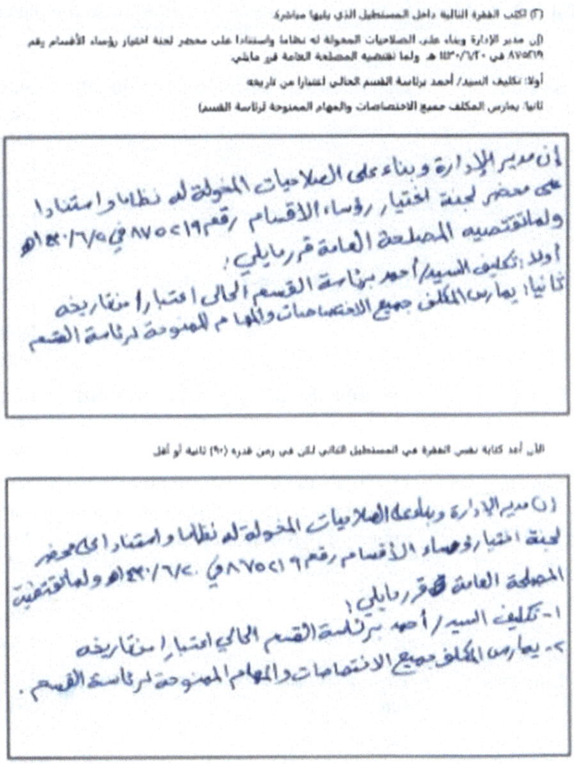

Fig. 1 Example of handwriting document collected from our study participant

experimental dataset. We applied some features from general attitude of handwriting according to paragraphs and words. In addition, we discovered an additional useful feature specific for Arabic scripts, which is writing style. Arabic scripts have two types of writing styles: Naskh (letters written in a complete manner) and Ruqah (letters written in abbreviated form). Table 1 gives a list of selected features that applied in our model.

4 Experiment and Results

As mentioned at the beginning of the previous section, we implemented our personality classification system by dividing the model into four classifiers. Classifier 1 is responsible for identification of the main attitude of a person, indicating whether he/she is an extravert (E) or introvert (I). Classifier 2 is responsible to describe the perceptive functions of the person, indicating whether sensation (S) or intuition (N) is dominant. Classifier 3 is intended for evaluating the judgment functions, with two categories: thinking (T) or feeling (F). Finally, Classifier 4 is tasked with recognizing

Table 1 Table of handwriting features extracted from the dataset

Feature	Description	Possible Values
Baseline	Describes the slope of the writing line	Straight Descending Ascending False Ascending Concave Erratic Mix
Slant	Describe the slope of letters	Vertical Leftward Rightward Variant
Letter size	Describe the size of letters and words	Large Medium Small
Margin	Describes the space around the written paragraph inside the rectangle appeared in the document	No margin Narrow All around Left Right Upper Lower
Line space	Describes the vertical space between lines	Narrow Wide
Word space	Describes the horizontal space between words	Narrow Wide
Writing style	Describes the Arabic writing style	Naskh Ruqah Mix
Writing speed	Describes the speed of writing whereas we have the normal and fast writing paragraphs	Fast Slow

the judging (J) and perceiving (P) attitudes. We therefore trained four machine learning algorithms and applied 4-fold cross-validation to train and test each classifier over the ground truth set. Given the dataset D, we want to select the best model m that provides us with the best representation of D. In order to accomplish that, we need to minimize the loss error. The used cross-validation technique is managed as follows:

- Permute the dataset randomly.
- Divide the dataset into N folds that are equally sized (in our experiment, $N = 4$).

In the validation phase, the number of rounds is as the same as the number of folds. One of the four folds is selected for validation in each round, in order to test the classifier, while the remaining three folds are used to train the classifier. For each classification algorithm, we selected the model that minimizes the loss error.

The results were obtained with classifiers based on several different learning algorithms, including Simple Logistic (SL), Decision Tree (J48), K-Nearest Neighbours Classifier (IBK), and Random Forest (RF).

In Fig. 2, the results are shown for two experiments. We conducted the experiment 1 for all classifiers by adapting the dataset with handwriting features beside the four parameters while the dataset in experiment 2 contained the handwriting features only. For experiment 1, the classifier 1 achieved high accuracies of 68.67%, and 66.26% with algorithms J48 and SL respectively. However, it accomplished far lower accuracies of 49.39%, and 56.62% using IBK, and RF algorithms respectively. Classifier 2 achieves peak accuracy with SL 56.62%, while for IBK it is measured at 55.42%, and for RF at 55.42% of the SL, IBK, and RF algorithms respectively, while 51.8% was achieved with J48. Classifier 3 achieved accuracies of 61.44%, 60.24%, 60.24%, and 59.03% with algorithms IBK, SL, RF, and J48 respectively. Finally, the results for classifier 4 highlight the highest accuracies in the entire set, reaching 71.08%, and 66.26% for J48, and SL respectively while IBK and RF algorithms achieve the same accuracy level at 60.24%. Results of the experiment 2 (see Fig. 2) indicate slight differences from experiment 1 among classifiers and algorithms used, except for two cases in classifier 1 for IBK algorithm and in classifier 2 for J48 algorithm.

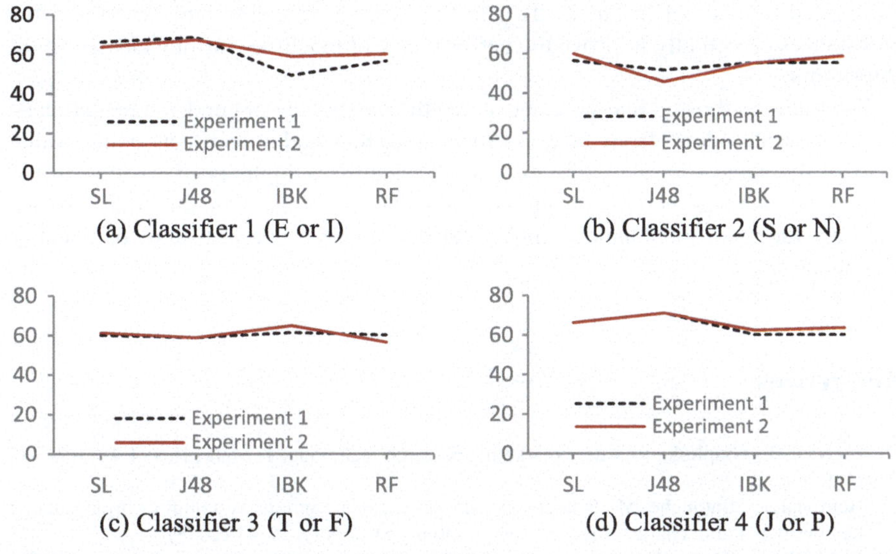

Fig. 2 Comparison between selected classifiers algorithms

5 Conclusion

In this study, Arabic handwriting features were used for personality classification according to the Jung-Myers typology. The number of participants in the study was 83, while total number of collected paragraphs was 160 normal, and 158 high-speed. We used four different machine learning classification algorithms: Simple Logistic (SL), Decision Tree (J48), K-Nearest Neighbours Classifier (IBK), and Random Forest (RF). To validate the model, 4-fold cross validation technique was used. We identified some relevant parameters (gender—age—nationality—education level) in addition to basic handwriting features (baseline—slant—letter size—margin—line space—word space—writing style—writing speed). We conducted two experiments: in the first experiment the dataset included personality parameters on top of the handwriting features, while the second one included the handwriting features only. Due to the nature of our objectives, the model contains four classifiers to obtain the four different personality dimensions.

Results of both experiments display a narrow range of values with only slight differences between methods. We achieved a maximum accuracy of 68.67% in the classifier 1, 56.62% in the classifier 2, 61.44% in the classifier 3, and 71.08% in the classifier 4. The authors believe that the main reason for so similar results is the limited number of instances used in the experiments, since increased number of instances can greatly improve the performance of machine learning classification algorithms.

The findings showed that we can exclude the parameters of gender, age, nationality, and education level from the experiment since they had little effect on the results. Also, at the same way we could exclude the feature of high speed writing. In the future work, we will increase the number of instances compared to this study, and we will introduce additional handwriting features to find out their impact on personality identification.

References

1. McNeal, J.: Graphology: a new marketing research technique. J. Mark. Res. **4**(4), 363–367 (1967)
2. Boufenar, C., Batouche, M., Schoenauer, M.: An artificial immune system for offline isolated handwritten arabic character recognition. Evolving Syst. **9**(1), 25–41 (2018)
3. El Moubtahij, H., Halli, A., Satori, K.: Using features of local densities, statistics and HMM toolkit (HTK) for offline Arabic handwriting text recognition. J. Electr. Syst. Inf. Technol. **4**(3), 387–396 (2017)
4. Ahmed, A., Al-Tamimi, M., Al-Sanjary, O., Sulong, G.,: Classification of Arabic writer based on clustering techniques. In: Saeed, F., Gazem, N., Patnaik, S., Balaid, A., Mohammed, F. (eds.) International Conference of Reliable Information and Communication Technology, IRICT, vol. 1, pp. 48–58. Springer, Cham (2017)
5. Djamal, E., Risna D., Sheldy N.: Application image processing to predict personality based on structure of handwriting and signature. In: International Conference on Computer, Control, Informatics and Its Applications, IC3INA, pp. 163–168. IEEE (2013)

6. Sen, A., Harsh S.: Automated handwriting analysis system using principles of graphology and image processing. In: Innovations in Information, Embedded and Communication Systems, ICIIECS, pp. 1–6. IEEE (2017)
7. Gattal, A., Djeddi, C., Siddiqi, I., Chibani, Y.: Gender classification from offline multi-script handwriting images using oriented Basic Image Features (oBIFs). Expert Syst. Appl. **99**, 155–167 (2018)
8. Halder, C., Obaidullah, S., Santosh, K., Roy, K.: Content independent writer identification on Bangla Script: a document level approach. Int. J. Pattern Recognit. Artif. Intell. **32**(09), 1856011 (2018)
9. El-Sawy, A., EL-Bakry H., Loey M.: CNN for handwritten Arabic digits recognition based on LeNet-5. In: Hassanien, A., Shaalan, K., Gaber, T., Azar, A., Tolba, M., (eds.) International Conference on Advanced Intelligent Systems and Informatics, (AISI), pp. 566–575. Springer, Cham (2016)
10. El-Sawy, A., Loey M., EL-Bakry H.: Arabic handwritten characters recognition using convolutional neural network. In: Transactions on Computer Research, WSEAS, vol. 5, pp. 11–19 (2017)
11. Loey M., El-Sawy, A., EL-Bakry H.: Deep Learning Autoencoder Approach for Handwritten Arabic Digits Recognition. arXiv preprint arXiv:1706.06720 (2017)
12. De Sousa, I.: Convolutional ensembles for Arabic handwritten character and digit recognition. PeerJ Comput. Sci. **4**, e167 (2018)
13. Carl Jung: https://en.wikipedia.org/wiki/Carl_Jung. Last accessed 18 Dec 2018
14. Jung, C.: Psychological Types, vol. 6. Rascher Verlag (1971)
15. Myers, I., Myers, P.: Gifts Differing: Understanding Personality Type. Davies-Black Publishing (1995)

Author Index

A
Abel, Marie-Hélène, 199
Abreu de, Wilson Correa, 411
Ahmed, Samara, 403
Alabyooki, Samar, 255
Al-Jayyousi, Odeh, 255
Alkatheiri, Mohammed Saeed, 315, 391
Al Omairi, Manar, 433
Alowibdi, Jalal S., 315, 391, 487
Al-Qurishi, Muhammad, 557
Ammetller, Gisela, 117
Attallah, Suha Abdullah Ali, 255
Azorín-López, Jorge, 515

B
Bellini, Emanuele, 171
Benaouda, Bensaid, 65
Brahimi, Tayeb, 65, 277
Buenaño-Fernandez, Diego, 125

C
Carbonaro, Antonella, 109, 223
Ceravolo, Paolo, 171
Chacón-Rivas, Mario, 441
Chagas, Daniel, 135
Ciclosi, Francesco, 171
Colomo-Palacios, Ricardo, 343
Comin-Nunes, Rafael, 419
Cordeiro, Luana Ibiapina, 411

D
Damiani, Ernesto, 171
Desouky, Tamer F., 479
Duarte, João Batista Furlan, 15

E
Espín-León, Aldrin, 515

F
Ferati, Drilon, 351
Ferrández, Antonio, 211, 441
Ferreras-Garcia, Raquel, 91
Filho, Egidio Gomes, 411
Filho, Marum Simão, 369, 419
Furtado, Elizabeth, 135

G
Garay, Ignacio Salamanca, 37
Gilart-Iglesias, Virgilio, 305
Gil, David, 125, 211
Gregor, Michal, 55
Grimaldi, Mara, 495

H
Hejres, Jehad, 433
He, Ling, 459
Hernández-Lara, Ana Beatriz, 91
Ho, Jung-tsung, 189
Hořejší, Petr, 55

I
Ibarra, Marco Moreno, 287
Ilyas, Muhammad U., 315, 391, 487

J
Jimeno-Morenilla, Antonio, 231, 515
Jussila, Jari, 153, 377

A. Visvizi and M. D. Lytras (eds.), *Research & Innovation Forum 2019*,
Springer Proceedings in Complexity,
https://doi.org/10.1007/978-3-030-30809-4

K
Kauffmann, Erick, 211
Khalifa, Sajid, 65
Khan, Muhammad Murtaza, 315, 391, 487
Kolp, Manuel, 535

L
Lima, Karla Krupskaia Ferreira, 369
Li, Siying, 199
Lladós-Masllorens, Josep, 451, 549
Llorens-Largo, Faraón, 75
Loia, Francesca, 495
Luján-Mora, Sergio, 125

M
Madhala, Prashanth, 153
Ma, Hung-wen, 189
Maione, Gennaro, 495
Mamlook, Rustom, 255
Martínez, Ana, 245
Mathkour, Hassan I., 557
Meijers, Stijn, 325
Meseguer-Artola, Antoni, 117, 181, 549
Mir, Nighat, 433
Molina-Carmona, Rafael, 75
Mollá Sirvent, Rafael, 47
Mora, Higinio, 47, 231, 245, 305, 525
Mostafa, Mohamed A., 557
Muirhead, Bill, 27

N
Naranjo-Zeledón, Luis, 441
Negre, Elsa, 199
Novo-Corti, Isabel, 269, 469
Nunes, Luciano Comin, 419

P
Partanen, Atte, 377
Peral, Jesús, 211, 441
Pereira, Maria Lúcia Duarte, 411
Pérez-delHoyo, Raquel, 305
Pérez, Magdalena Saldaña, 287
Pertegal-Felices, María L., 47, 75, 515
Picatoste, Xose, 269, 469
Pinheiro, Mirian Caliope Dantas, 411, 419
Pinheiro, Pedro Gabriel Caliope Dantas, 411, 419
Pinheiro, Plácido, 15, 369, 411, 419
Polcar, Jiří, 55
Pompeu, Monica, 419
Pontes, Herleson Paiva, 15
Pujol, Francisco A., 245
Pujol-Lopez, Francisco, 231
Pulkkinen, Jukka, 377

Q
Quintana, María Graciela Badilla, 37

R
Rajput, Adil E., 101, 277, 479
Real-Fernández, Alberto, 75
Reda, Roberto, 109, 223
Rimbau-Gilabert, Eva, 117
Robertson, Lorayne, 27
Rodríguez-Ardura, Inma, 117, 181
Rohlíková, Lucie, 55
Ruiz, Miguel Torres, 287

S
Saleem, Sajid, 315, 391, 487
Sanchez-Romero, Jose-Luis, 231
Sarirete, Akila, 277, 433, 479
Sciarrone, Filippo, 3
Sellers, Ricardo, 211
Serradell-López, Enric, 91
Shen, Chen-wen, 189
Signes-Pont, María Teresa, 47
Sirshar, Muneeba, 391
Sirvent, Rafael Mollá, 305, 525
Spruit, Marco, 325, 351

T
Temperini, Marco, 3
Țîrcă, Diana Mihaela, 269, 469
Troisi, Orlando, 495
Trotskii, Igor, 377

V
Vyšata, Jiří, 55

W
Wang, Wei, 459
Wautelet, Yves, 535
Wu, Yenchun Jim, 459

The manufacturer's authorised representative in the EU is Springer Nature Customer Service Centre GmbH, Europaplatz 3, 69115 Heidelberg, Germany. If you have any concerns regarding our products, please contact ProductSafety@springernature.com

Printed and bound by CPI Group (UK) Ltd, Croydon, CR0 4YY
05/10/2026
02228130-0019